Richard Mendelson Spirit in Metal

Denis + Anne

I hope you enjoy
this Book on my
art journey. It's
been incredibly
enriching.
All the best,

May 2023

Margin Release, 1996

Dedication

Two special men inspired me to pursue sculpture: Jack Chandler, gifted landscape architect and metal sculptor, and Rene Di Rosa, the father of viticulture in the Carneros region of the Napa Valley and a champion of Bay Area art.

I met Jack in 1992, and he quickly became my friend and later my mentor. Without Jack, I never would have tried my hand at metalworking. He started me on the journey, and he has stuck with me throughout. Like all good mentors, Jack regularly criticizes me and my work, but he also appreciates, as do I, that he planted a seed in me that has now fully sprouted. That would not have been possible without his inspiration, generosity and love.

Rene Di Rosa was the first person to buy my art when I decided to show it for the first time in 1996. He subsequently bought many of my other sculptures. Rene was born the same year as my father and was my father figure in the art world. Yale graduate, journalist by training and pioneering vineyardist, Rene loved art as much as he did grapes, wine and the written word. His interests mirrored my own. And his sense of adventure, his eccentricities and his passion were beacons for me and many other emerging artists. Rene passed away in 2010. His spirit lives on.

I dedicate this book to Jack and Rene, two legends of Bay Area art.

Val de Grâce Books, Inc.
Napa, California

Printed by Toppan Printing Company (SZ) Ltd., China

ISBN: 978-0-9817425-2-6

Library of Congress Control Number: 2011925892

Author: Richard Mendelson
        www.MendelsonArts.com

Editorial Director: Paul Chutkow
Creative Director: Robert M. Bruno
Graphic Design: Jin Son
Principal Photographers: Robert M. Bruno, Kurt-Inge Eklund

Publisher: Val de Grâce Books, Inc.
        www.ValDeGrace-Books.com

Richard Mendelson Spirit in Metal

Val de Grâce Books, Inc.
Napa, California

Awakening

When I was 23, I made my first trip to Japan. In the town of Hakkone, looking across to Mt. Fuji, I went to an outdoor sculpture garden beside Lake Ashi. I was amazed by the diversity of the works on display, from classical to contemporary, stabile to mobile, steel to stone to glass. I had always been interested in visual things, and I loved looking at sculpture in books and museums, but now, seeing these pieces outside, alive in that exquisite setting, made a deep impression on me.

Still, at that stage I never imagined making art with my own hands. I was immersed in my studies, which ultimately led me to the law and a law practice in Napa, California. I love the law. So much of being a lawyer is helping people with problems, advising them, strategizing, thinking through all the possible solutions. As a lawyer, you're a service provider, and I recognize this has enormous social value. For a full decade, I poured my energy into my law practice, specializing on a product for which I had a real passion: wine. I felt I had something special to offer in this realm: experience working for a wine producer in Burgundy, France; a decent palate; a fine educational background; and a good work ethic. By the start of the 1990s, I had built up a successful practice and was going full bore. I'd leave for lunch and come back with one hundred phone messages waiting for me. I had a lot of responsibility.

Although my cases and clients were interesting and challenging, I soon reached a point of diminishing returns; I mean *emotional* and *spiritual* returns. My practice was now less satisfying: the stress curve had overtaken the learning curve. I broadened my interests and involvement to help keep me motivated. I assumed the role of managing partner of our law firm, planted our own vineyard and later started to make wine. But the practice of law is demanding, and I kept pushing forward, head down, driven, marching down what I thought was my defined course in life.

Then my mother died.

She died very unexpectedly: ill on a Friday, dead by Sunday. I never really got to say goodbye. Her death, bearing with it the stark realization that life is so short, came as a profound shock to me. But then something else unexpected happened. We buried my mom at a cemetery in our family's hometown of Jacksonville, Florida, and when I returned one year later for the unveiling of her tombstone, I spotted, within a mile of my mom's gravesite, a scrap yard filled with heaps of metal, all of it broken and discarded.

Curious, and looking for some form of relief and comfort, I decided to have a look inside. I wandered around the scrap yard for awhile, looking through the mounds of metal, and then I spotted it: a section of an old, forgotten conveyor belt. It was junk, dead metal, nothing more than a series of mangled metal strips, fixed in parallel along a single rope of steel. I pulled the conveyer belt out of the pile and suddenly, in my hands, I saw that stretch of steel as a human spine, its discs out of line, all of the surrounding flesh brutally stripped away. I held it up and I could see this spine had character, its wear and tear gave it a distinctive feel and a real sense of individuality. I did nothing to the metal, other than cut out the section I wanted. I then found in nearby scrap heaps, as if predestined, the makings of a head and pelvic girdle. By the time I left the scrap yard, I knew I had designed a work of art, *Slipped Disc*, an *objet trouvé*, a found object, a silent expression of my pain and grief, metal brought to life by the breath of sudden inspiration. And in my bones I could feel it happening: a new journey was beginning, though I had no idea how significant this journey would soon become.

Slipped Disk, 1995

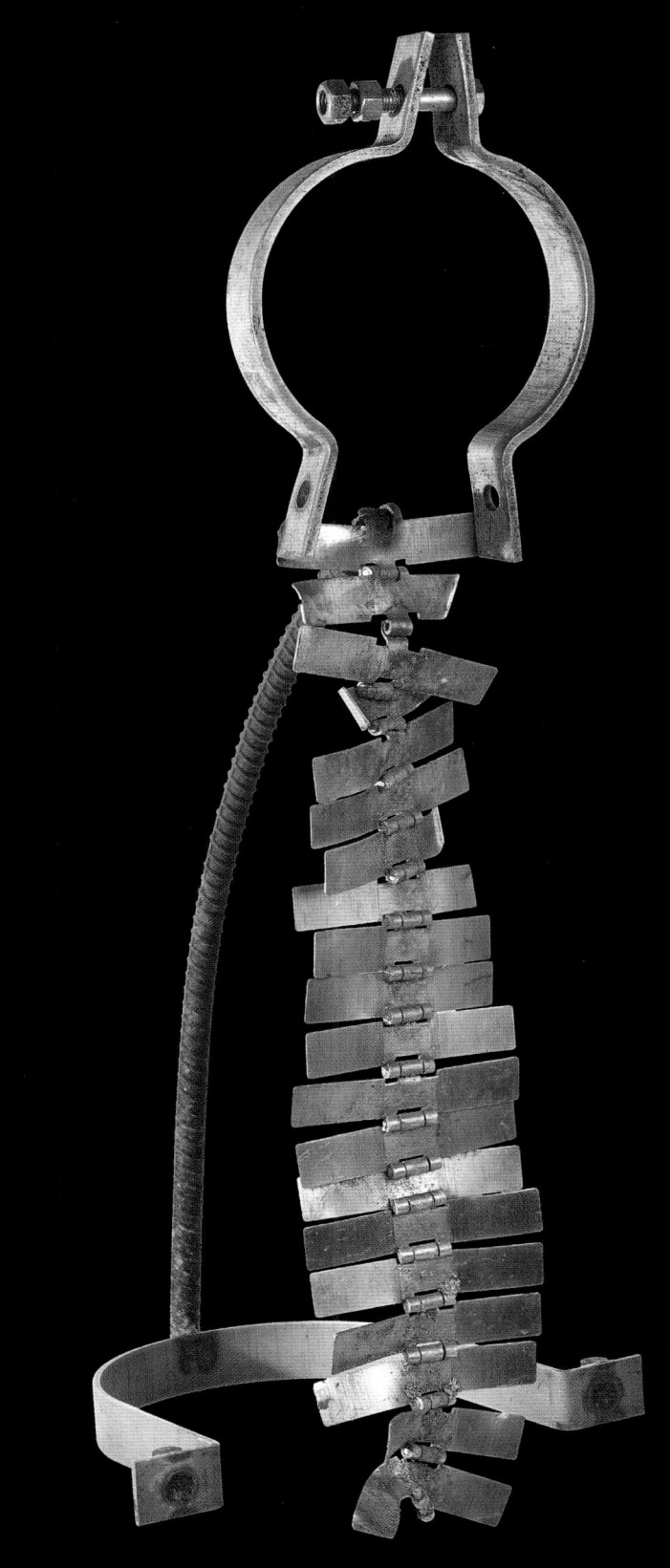

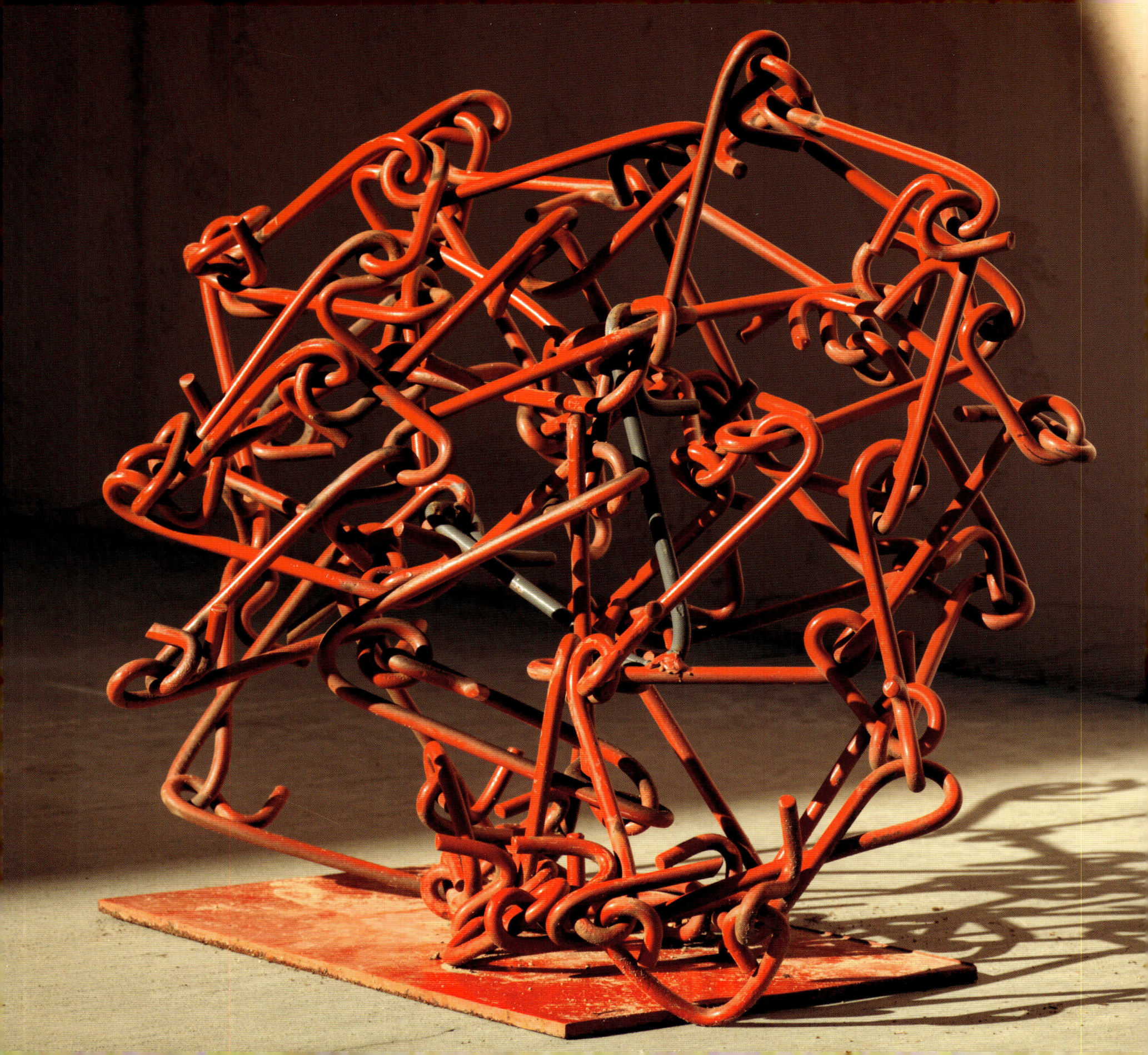

Spring Tooth, 1997

Blond Curl, 1998

Spiral, 1999

Father, Mother and Child, 1998

La Famiglia, 2001

Ball & Chain, 1996

Raptor, 1999 | **Jump for Joy**, 2001

Legs, 1996

The Hanging, 1997

Catching Fire

My mom's death provided the first spark; Jack Chandler provided the second.

I had known Jack from his work as a landscape designer in the Napa Valley, and I had always admired his creative eye. He has a fantastic sense of dimensionality and scale. He is equally adept at architecture, construction and gardening. Also, I had done some work for Jack as a lawyer and, in a very natural way, we had become good friends.

One day Jack and I were having lunch together at his house in Calistoga. Afterwards, we went out to his studio and his "bone yard," where he kept all of his raw steel and found objects. As part of his landscape work, Jack does a lot of work with metal, and that dovetails perfectly with his accompanying passion for crafting sculpture out of metal. In his studio, welding and art happily co-exist, and the place has a wonderful feel. Jack showed me around and introduced me to Miguel, who was busy working on one of Jack's projects. Jack saw that I was intrigued. "Are you interested in this stuff, Richard?" he asked. Instantly I replied yes; I didn't even have to think about it.

"I'll tell you what," Jack said. "Come up on Friday; you can work with my guys for the day. You don't know what welding or cutting is, but they'll help you. Just have at it and see what you think."

I went home and immediately set to work. I started by cutting out pieces of cardboard and trying to fit them together in interesting three-dimensional planes. I was not following any clearly defined path; I was just experimenting with shapes and "negative space," the empty space between the intersecting metal parts, the space that fills the void. Playing with my cardboard cutouts, I usually had a clear vision in my mind, and I would turn the pieces of cardboard into a maquette, a model for what I wanted to create.

All my life I have been fascinated by the human body and by movement – my devotion to yoga and tai chi grew out of that interest. So did my first experiments with cardboard cutouts. Now, working at home, I came up with a cardboard model I liked. The following Friday, I went to Jack's studio and Miguel introduced me to the crafts of welding and working with metal. That's how *Eleganza* was designed and fabricated.

The feeling this produced was something entirely new for me. As a lawyer, I love to help people and to develop a strong relationship with my clients. Together we might spend months hard at work creating a winery, but in the end the winery did not belong to me. I had nothing concrete to show for all that work other than the relationship I had forged and the money I was paid. Working with metal, by contrast, gave me immediate and very tangible satisfaction. That first day in Jack's studio I went home with something that I had created with my own hands. This was really very special to me. I loved it, and I couldn't wait for more.

There was something else. One day when I arrived at the studio, there was Jack hard at work, melting candles onto the top of a rusted metal bookcase. He was working with a "rosebud," a small tool used for heating metal or, in this case, wax. I was struck by the image. There was no functionality at all to what Jack was doing. He obviously had something in mind that he was tapping into, but he was totally free, following his wit and whim with no definite vision or end-result in mind. It was almost like a deep meditation; he was completely engaged and completely open to whatever might arise.

This was also entirely new to me. I loved the sense that in art you can go wherever your vision takes you, tap into whatever impulse seizes you at the moment. For me, that would not be wax on top of a steel bookcase; that was Jack's vision. But now, through art, I saw that you could apply all of your powers of instinct, intuition and concentration to bring your personal vision to life, whether it was serious or just pure whimsy. All of this brought a big smile to my face. Compared to practicing law, this was total freedom, and I felt something inside me let go and burst open.

From then on, I went to Jack's studio on a regular basis, and there I made a great discovery: the unbridled joy of working with metal. And I realized that metal was the perfect way for me to express my own instincts and intuition, in part because metal is so cold. You take this very cold, immutable object and you make it come to life. With tools, craft and fire, you evoke movement, feeling and personality, and when you learn how to master metal, you can even find the spirit inside.

Now I could feel something huge stirring inside me – and it was then that Jack gave me the boot. Good mentor that he was, he said, "Richard, you've got to get out of my studio. Go take a welding class. Go learn the craft." Jack, bless him, saw it long before I did: The fire was lit, I was ready to go.

Eleganza, 1996

Rolling Hands, 1998

Four I's, 1996

Blue Twist, 1998

Duet, 1997

Revelations

My next stop was Napa Valley College.

There, I learned how to weld, and I found a whole team to help me. At the college, I would hang out with the metalworking and machine shop teachers and students, and although they were not familiar with metal art, they were always willing to give me a hand and share their expertise. I learned from skilled craftsmen like Tom Smeltzer and Dennis Humphrey. We had fun, and they got a huge kick out of watching me change in and out of my law-office clothes, often with the black smudge of the shop still on my hands and, alas, sometimes on my suits.

One day, for instance, I raced out of the shop and back to my law office for an important meeting. I slipped in the back door and ran smack into a major client of ours from New York. He was dressed in a three-piece suit and looked at me somewhat strangely. I had no idea why – until I looked at my face in the mirror and saw that it was covered in black. He never said a word, but to this day he always asks about my sculpture, and I can almost hear him chuckling inside.

At first, in working with Jack Chandler and then at Napa Valley College, my approach to metal design and fabrication was intellectual, a function of the mind. Like writing a legal brief, I started with an idea in my head, then I'd sketch it on paper, and next I would cut out the design in cardboard, to see how it would look in three dimensions. If the cardboard pattern worked, I would replicate the design in metal, cutting the metal by hand or using a plasma pattern cutter. The cut steel came out flat. Then I would bend and roll and twist the pieces of steel applying intense heat and using several different pieces of equipment, including breaks, presses and rollers. Finally, I would weld the pieces together to assemble the sculpture. After that, the piece was sandblasted and either painted, powder-coated or acid-washed and left to rust.

Over time, I assembled a team of exceptional helpers, each with a special set of skills. Nothing could be done solo. Even the assembly of a maquette requires four hands: two to position the steel pieces and two to weld them in place. Some of these team members have been with me from the outset, like painter Eric Schneider and hauler Guillermo Alvarez. Everyone's contribution is respected. When I began to sell my work, the entire team was proud; we shared the work and we shared the triumphs too.

Then came *Dali*.

This was a sculpture that I did not plan or cut out of cardboard; it burst into life by sheer accident. I was designing a piece that would be a cloaked man, what I envisioned as a shrouded Giacometti-like figure. I had no name for the piece at that point; I almost never do when I draw or even when I fabricate. In the case of the cloaked man, my touch on the cutting torch was not very deft, and as I was cutting a thin steel maquette, first the cloak fell off, then both arms, and then I severed the leg. I was not happy, to say the least. But as I prepared to start over, from scratch, I looked at the damaged man and suddenly saw it in a whole new way. The fragility of this accidental *Dali*, the graceful tilt of its body, its jagged edges, captured the spirit of the man. I welded the leg back together, smoothed over the rough edges, removed some of the slag (the crust that forms on top of molten metal during welding) and voilà, *Dali* was born.

This proved to be a wonderful lesson: *Dali* taught me that art happens when you least expect it. Trusting your instincts, being spontaneous and open to whatever comes are important steps in the creative process. Conversely, doing things by formula and by rote can often get in the way and stifle the process. As I look back now, I can see that *Dali* was an important turning point for me, a "mistake" that liberated me and inspired me to play and to experiment in new ways.

Dali, 1996

After *Dali*, I started to bring interpersonal relationships into my work: two people united in a single piece of sculpture. Sometimes the figures, cut from different pieces of steel, would intertwine like *Lean on Me*; other times the two figures would emerge from the same plate of cut steel as in *Love in the Round*. In this phase, I also began doing multiple variations on a single theme. No two works in those series ever came out the same. Each had its own mood, color and eccentricities. The angles of their bodies, where they touched and crossed, the feelings they conveyed varied from one piece to the next. Still, I never made more than a few copies of any one piece. I had too many ideas in my mind, and I was always eager to move onto the next one, to deepen my exploration.

Now I fell into a new way of working, far removed from the way I had worked with Jack and Miguel. I would draw in my sketchbook for months at a time and never set foot in the shop. Many of my works were designed abroad, outside of my normal surroundings. Later I would enter the studio, often having filled several sketchbooks full of drawings. Typically I would cut all the metal pieces in advance or have them cut by water jet, which emits a powerful stream of water that cuts through steel with incredible precision and no slag. Where precise features such as hands or facial details are important, a water jet is the tool of choice. When I finally went to the studio, I would not bring my drawings. I would examine the parts and let my mind freely play from there. Sometimes a new idea or whole new piece would emerge from what I saw spread on the shop floor.

Once I had entered this phase of the work, I often would spend all or part of every day in the shop over a period of a month or six weeks, whatever it took to let the process unfold. As a lawyer, I was used to this kind of immersion. The focus helped to develop my skills and my eye and prompted me to dig deeper inside myself, mining territory, new and old, for fresh sources of inspiration.

Love in the Round, 1996

42

Lean on Me, 2000

Four's a Crowd, 2000

Lady, 1999 | **Classic**, 1997

Caraibes, 1998

Inescapable Grace, 1998

Dan Tien

Nothing happens in a welding studio without sparks – and so it is in life. After college, sparks flew at me from all directions, involving new places, new people and new endeavors that would totally change my life. These sparks from my past now became my next sources of artistic inspiration. Let me explain.

In 1975 I moved to England for a post-graduate scholarship at Oxford University, but the things that consumed me there were not at all academic. I became engrossed in yoga and gradually learned the postures (*asanas*), the breathing (*pranayama*) and the concept of balance, both physical and mental. The concentration and diligence that yoga requires came easily to me; surprisingly, I also had a good feel for the art of movement. I learned that all motion emanates from a central core, what the Chinese call the *dan tien*, located at the navel center (three fingers below the belly button and two fingers inside the body). The *dan tien* is both our center of gravity and the internal reservoir of our life force, our prana or chi.

Along with yoga, I discovered wine at Oxford, and it quickly became a passion. I was blessed to have access to one of the largest wine collections in all of Europe, housed in the cellars underneath my college; those wines were made available to students at unbelievable prices. The sensual side of wine, the fact that wine expresses so clearly its origin and is tied so directly to food, art and culture—all of this excited me.

After Oxford, I found a way to pursue yoga and wine simultaneously. First, I went to India to study with B.K.S. Iyengar, a legendary yoga teacher. From India, I moved to France to work at a winery in Burgundy. There, I spent my days learning about wine and my nights teaching yoga. Over time, I learned how to correct students' poses with the touch of a well-placed finger or hand. I discovered that, in body movement, inaction is as important as action. For example, in backbends, if you force the pose in even a small way, "muscling through it" as they say, you end up hurting yourself. But if you open the upper chest fully, the spine has room to unfurl. The balance between making something happen (*faire*) and allowing something to happen (*laissez faire*) was itself a revelation.

Shortly after moving to France, a third, more powerful spark—the spark of love—changed my life forever. And that spark came to me in a most unusual way. I worked the grape harvest at Domaine Dujac in the fall of 1977 with a New Zealander, Hugh Skyrme. After he left Burgundy, Hugh traveled to Marseille to visit his Australian friend, Marilyn Knight. When Marilyn mentioned her plan to visit Spain during an upcoming holiday, Hugh told her that, by chance, I was planning to drive there. She asked Hugh to call me to see if she could hitch a ride to Madrid. I happily consented.

That trip never happened. I had to cancel it when a medical problem forced me to return to the States unexpectedly. Before leaving, I left a message with Marilyn's housemate explaining the situation, but in my faltering French I didn't get it quite right and told her I had to return to the States immediately because of a drug problem!

I returned to France a couple of weeks later to find a letter from Marilyn. I guess she was curious about my strange admission and sudden disappearance. Marilyn, I learned, had won a year-long fellowship to teach English at a French high school. We started exchanging letters and quickly discovered that we shared many interests – yoga, travel, foreign languages, a love of culture, food and wine, and a passion for India. This was all before the Internet and email. The letter-writing phase of our relationship lasted almost three months. Our letters were frequent, long and intimate. I think our destiny was sealed even before we ever met.

We finally did meet at the Gare de Dijon. I had driven up from Beaune to meet Marilyn, who was taking the train from Marseille. The only subject we had failed to discuss was what we looked like; we had not exchanged any photos. So I hid behind a trash can, hoping to catch a glimpse of her beforehand. She was a bit of a hippy, with curly hair, blue eyes, tall grey boots and an old overcoat with fur trim, nothing like the girls from my home state of Florida. She looked lovely and intriguing to me.

Marilyn and I had a wonderful courtship in one of the most romantic places on earth, *la belle France*. We explored all the towns and countryside between Beaune and Marseille, and we drank fantastic wines from the centuries-old cellars of my employer, Bouchard Aîné et Fils. After we both had finished our jobs, we traveled to Spain for a month and then to Paris where

Detached, 1998

for several months we lived, loved and practiced yoga together, intensively and passionately—just as we did everything else.

I knew when we met that our relationship would endure. For me, it was love at first sight. I liked Marilyn's energy level, intellectual curiosity and sense of adventure. She was then, and remains now, trusting, open, warm and reliable. And there was a spiritual side to Marilyn that I didn't fully fathom at the time; perhaps neither did she. Whatever insights I did have about her, I couldn't have realized then how much Marilyn would enrich my life and that of our kids, Margot and Anthony.

Once I started making sculpture, all of these experiences in England, India and France provided a constant wellspring of inspiration. My yoga training was essential: it provided me with an insight into how bodies move—and don't move. That insight already had served me in unexpected ways outside of the world of art. Once, when I injured my right knee playing basketball, I was faced with the unhappy prospect of surgical repair. But my doctor suggested we wait, because he said I had good "proprioceptive" abilities. I had never heard that word before. It means "the unconscious perception of movement and spatial orientation," in short a keen awareness of where one's body is in space and time. Proprioception is important in yoga, providing an innate feeling about how movement or relaxation in one part of the body affects the rest of the body. It also is critical to navigating our way through life and across all the bumps in the road. In my case, it meant I stood a good chance of healing properly without surgery. The doctor proved right.

My interest in body movement naturally translated into my later work in metal. When Rene di Rosa said to me, "Make metal come to life," I knew exactly what I had to do: bring my understanding of the body and movement directly into play, make the metal bend and flex as naturally and gracefully as an accomplished dancer or yoga practitioner. This was exciting for me, and I poured the insights I had gained from yoga into my work in metal.

It's one thing to draw a pose and cut the linear form out of metal. It's another matter altogether to work the metal, to make it supple, and then to capture the essence of the pose in cold steel. That's a matter of touch and feel, of learning to work the metal until you find the

Sarvangasana, 2006

Snake Creeps Down, 1999

life inside but never to overwork it. I learned to trust my instincts and my proprioceptive skills. This way of working was a far cry from where I had begun my journey, with my drawings and my cardboard cut-outs. Also, following my instincts meant that each piece would be different, even if the poses I was rendering were identical. This seemed right to me: after all, no two people stand or do a yoga pose in the same way. They come to a yoga studio with their own propensities, their own inclinations, their own moods and feelings.

Ultimately, what I came to perceive as the static nature of yoga led me to tai chi and chi gong (sometimes referred to as *qi gong*), two forms of movement where the slow dance never stops. Here I found an unending sequence of moves with names like "snake creeps down," "reeling the silk," "building the jade palace," and "grasp tail of monkey." What an incredibly sophisticated and subtle world this is, the world of movement, a world based on balance, tranquility and expression. My teacher, Sifu Michelle Dwyer, taught me about the rivers (meridians) running through the body, each of which originates in the *dan tien*, and as I experienced this in my own body, I sought to express it in my work. Michelle agreed to be my model for one of my favorite tai chi poses, *Snake Creeps Down*, which I portrayed as two intersecting spirals, the yin and the yang, one spiral tied to the upper body and the other to the lower body. The *dan tien* is the intersection of the two spirals, the invisible source of all movement.

Later I took the source of movement out of the yoga and tai chi context. I no longer had to capture the pose itself in steel. I could capture its essence in other ways, using geometric patterns, abstraction and vibrant colors, as in the *Harlequin* series and *Roly-Poly*. That itself was liberating. I found other sources of inspiration in dance, classical and contemporary, from gymnasts and contortionists and even from the fashion runway.

And right then I began to learn a lesson that would expand and deepen throughout the course of my life: all the arts flow together. The art of yoga, the art of dance, the art of drawing, the art of sculpture, even the art of making fine wine—they all draw on the same inner resources at the core of our being; they are all rivers flowing from the *dan tien*. As I was now learning, one river, one spark, one art nourishes the others, and they all nourish the mind, body and spirit. What a marvelous journey this had become!

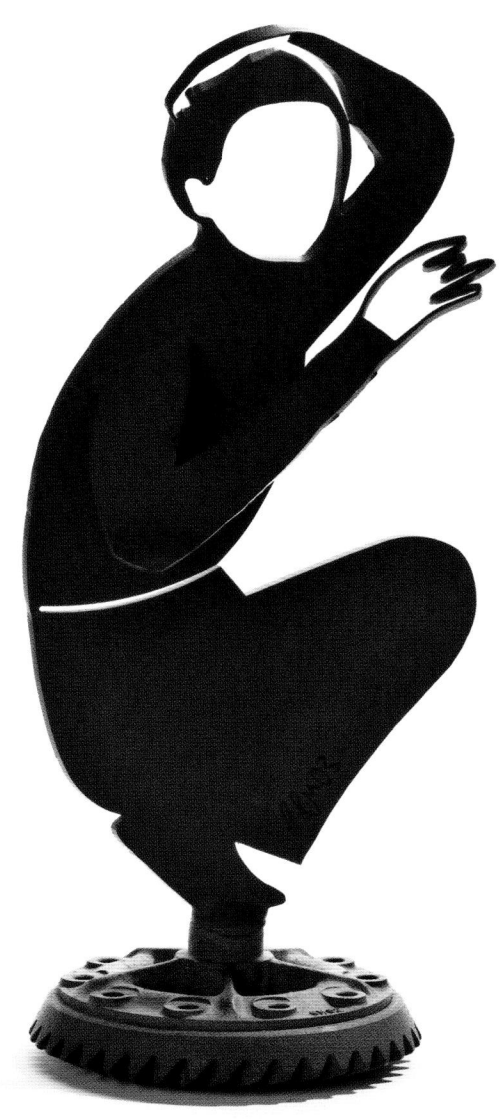

Crouch, 1999

Sedona, 2005

Seated Woman, 2001

Roly Poly, 1999

Cercle des Femmes, 1999

Trikonasana, 1999

Untitled, 2005 | **Harlequin II & III**, 1998 & 1999

Cutting Free

Nourishing my spirit gave me the confidence and the courage to explore whatever adventure or fancy might cross my path. I realized that I was free to play, and I went at it with delight. With that realization, I now found inspiration in the most unexpected places.

In a neighborhood scrap yard, I found two discarded hand drills that magically turned into *Breast Drills in Tandem*. I was given a discarded artillery shell casing that became the base of *Geishas*. I also found inspiration in the cacti of the Galapagos Islands and the vineyards of Napa Valley, in sacred art and totems, and even in tattoos.

Marilyn, bless her, was aware that there might be an artistic detour at any moment – to inspect junk alongside the road or to collect a random piece of discarded equipment to be disassembled later. She was usually understanding: she called me an *être absorbé*, a being totally consumed by my artistic explorations.

As my confidence grew, I dug deeper into the themes that had surfaced in my earlier works – spheres and curves, chains and shackles, and big "empty" heads formed by a steel perimeter with nothing inside. For example, a steel shackle anchors *Scales of Justice*; a broken chain on one side symbolizes justice gone awry. And *Colanzi*, a spherical female figure, consists of several empty ovals on different planes, representing her head, torso and limbs. Only her crossed arms connect the various ovals; without them, *Colanzi* would fly apart. A collector once asked me, rather puckishly, if all these empty heads of mine were an antidote to an overly active legal mind. Perhaps. But for me this interior void, this black hole, is also a sacred place, the mysterious realm where thoughts and emotions work their magic, where spirit dwells.

Breast Drills in Tandem, 2004

Endless Loop, 2005

This period of intense experimentation brought with it adventures of the sort that would make Rene di Rosa proud. Rene loved the unexpected, the eccentric and the absurd. He once told me that his favorite pastime was to board a bus to who knows where, get off at an unknown location and rummage around incognito in search of art by emerging artists. Rene also had a beautiful way of ensuring that people don't take themselves too seriously, that they don't become uppity or pretentious. Mid-sentence he would boisterously bellow "blah, blah, blah." He did that in a board room or a party, it mattered not. And Rene's bellowing disarmed everyone who heard it.

Rene's sense of adventure was infectious. One day Marilyn and I were in Monterey, walking around Point Lobos, and then we had an early dinner at a restaurant located right on the Monterey Pier. After dinner, in the fading light, I spotted something unusual perched at the end of the pier: an enormous sphere of rusted steel. Of course, I had to check it out. The sphere stood five feet tall and I couldn't begin to get my arms around it. I was able to move it ever so slightly, though—indicating it was hollow. But I still had no idea what it was or why it was there. The next morning, I returned to the pier and asked some fishermen about the giant steel ball. They explained that it was a buoy that had been pulled ashore after many years at sea. They pointed to other buoys bobbing in the ocean, a few hundred yards away.

Geishas, 2004

Scales of Justice, 2001

Intrigued, I went over and examined the buoy on the pier more closely in the morning light. I was amazed how the ocean salts had left a greenish patina on the rust, how the steel looked like the surface of the moon, craggy and cratered. I found the person in charge of the pier and asked if I could buy the buoy that had been brought ashore. "You bet," he responded quickly. "You can buy as many as you like." "Where are the others?" I asked. Pointing to the ocean, he nonchalantly informed me that his crew would simply cut the buoys free from their anchors and pluck them out of the ocean. I was amazed and immediately ordered three.

That was before I considered how I would get them home. They wouldn't exactly fit in our car. Soon we found a solution: we hired a trailer truck to haul the buoys to our home in Napa. It would leave later that same morning, arriving at our home a few hours after us. Marilyn was in disbelief, but I couldn't believe my luck. In their silent way, those buoys reached out and grabbed me.

By the time we got back to Napa, I had worked out where the buoys would be unloaded and how I would roll each one to the rear of our property where they would be displayed. That plan didn't work. After we had pushed the three buoys off the trailer and onto our driveway, I realized that they weren't truly round, so you couldn't control which direction they would move. As we found, too, those buoys had a strong mind of their own; they would only go where they wanted to go. Try as we might, those buoys refused to roll in a straight line; lest they roll over our vines, we left them where they were, blocking our driveway. What to do next? Finish the piece, name it, and call Rene di Rosa, of course.

I welded shut several gaping holes in the metal to ensure that rain wouldn't enter without a way to exit, welded back into place the chains that had fallen off two of the buoys during transport, and then I signed the pieces, *Buoy I, II and III*. The next day I called Rene and told him the story. He came over and bought my *Buoys* on the spot. These ocean symbols now reside in Rene's outdoor art preserve, next to his vines. I secretly think of them as "Blah, Blah, and Blah."

Colanzi, 1996

Buoys I, II and III, 2002

My Buoys lark was an important event for me. It confirmed that a creative and adventurous spirit is vital to art and that this spirit is accessible to each of us. This realization, this freedom, was like an elixir, unrivalled by even the thrill of courtroom victory. It made me want to celebrate the gloriousness of the human spirit and to capture that magnificence in steel. With hands uplifted, *Moi* does just that. *Moi* represents self-confidence, triumph and liberation. And so does *Grandeur*, a regal and stoic man cut out of inch thick steel, presiding like a prince or a priest over his people below.

Now when I was creating, I had a soaring feeling, and I often felt like I was standing, well, on top of the world.

Moi, 1999

La Vigne, 2000

Barrel Tasting, 2000

Racers, 2000 | **The Gondolier**, 1998

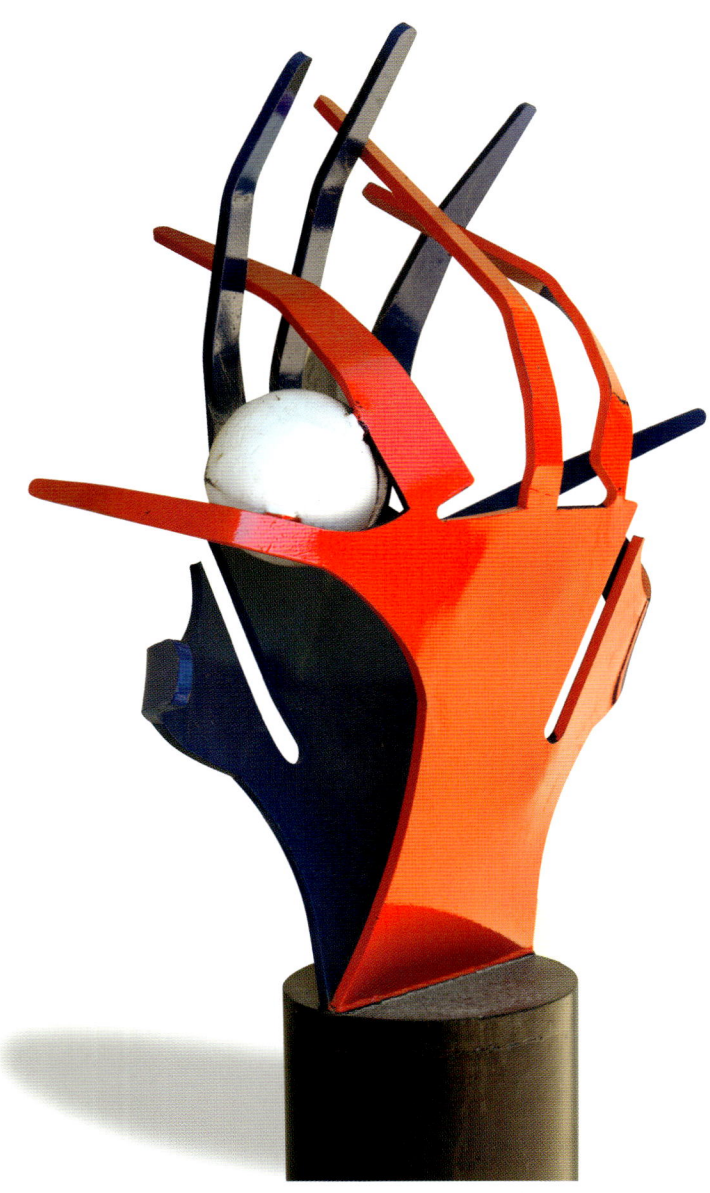

Red, White & Blue II, 2001 | **Juggling Hands**, 1997

84

Sink or Swim, 2000 | **In the Middle**, 1998

Top of the World

I was in Tokyo when the Twin Towers were attacked. For hours I just sat in my hotel room watching the TV images of the planes crashing into the skyscrapers, the rescue teams rushing in, the shocked faces of the bystanders, the sickening implosions that brought those majestic towers crashing down. I simply could not believe that terror on this scale could happen on our home soil. Nor could I fathom how those heavily reinforced, steel-girded buildings could crumble so quickly—so powerful one moment, destroyed by a single blow the next.

When I wandered out of my room to check the local reactions, everyone I saw was equally sad and anxious. The shock of the incident, all the senseless killing—I couldn't absorb it and no one else could either. The brutal reality of what happened was like a thick cloud that filled every space; it couldn't be comprehended but it also couldn't be avoided.

No one knew what other structures might be targeted. My hotel was within a block of Tokyo's World Trade Center, one very similar to our own. At 163 meters, this was the city's tallest building and it was also a bustling commercial hub. In Japan it stood as a towering symbol of the nation's prosperity, and the Japanese feared that it too might come under terrorist attack. Everyone was on edge, including me.

My flight home, scheduled for the following day, was cancelled. So I stayed in my room, glued to the TV news. Unlike in the States, some of the stations in Japan showed people leaping out of the flames to their deaths. It was horrifying. I had no idea how long I would be stranded in Japan. Marilyn and Anthony were at home in Napa; Margot was at Harvard. We kept in close contact.

In all of this emotion and turmoil, my previous two weeks in Japan, selling our wines, seemed like a distant memory. I had dined with extremely talented Japanese chefs who paired our dessert wines innovatively and elegantly with traditional Japanese food and Asian fusion cuisine. Now, though, the world of food and wine seemed entirely insignificant. Still, I couldn't help but think of the many times I had visited the Twin Towers, then ridden up to the 107th floor to the restaurant Windows on the World, where I would show our wines to the wine buyer there. I had been there countless times; I knew the faces of the managers and the waiters, and now I tried desperately to think what time the restaurant opened. How many of them might have been arriving for work when the deadly plane struck?

I had a sketch pad with me in Japan, as I did on all my trips. Typically, when I see a person, place or object that intrigues me, I sketch it. But after the Twin Towers fell, I put the sketch pad away. I could not fathom the tragedy of 9/11. As I wandered around the city, it didn't seem appropriate or possible to draw the grief and anxiety that I saw all around me. I didn't forget about art; I just couldn't fathom its relevance at that moment. Still, I knew that an intensive art period lay ahead for me. At the end of the month I would participate in the Napa Valley Open Studios, and the metal from all my recent drawings, which was being cut while I was away, would be awaiting me in Napa. I looked forward to a few intensive weeks in the studio. It would be a needed distraction.

After I did return to Napa, the media were full of images of the massive clean-up effort that followed 9/11. I fixated on the photographs of the metal remains of the Twin Towers, a smoldering morass of charred, crumpled and mangled steel. Those raw remains reminded me of the dirt, fire, sparks and rising steam that are common-place in a metal fabrication shop. But now those symbols of wondrous construction became symbols of monstrous destruction. Ironically, though, even in that scene of brutal devastation I found an eerie sense of beauty. I cut out photographs of the towers' remains and carried them with me as a reminder of the death and decay. I wondered if, from these very remains, I could craft a fitting tribute to those who had fallen.

Grandeur and his drop, 2000

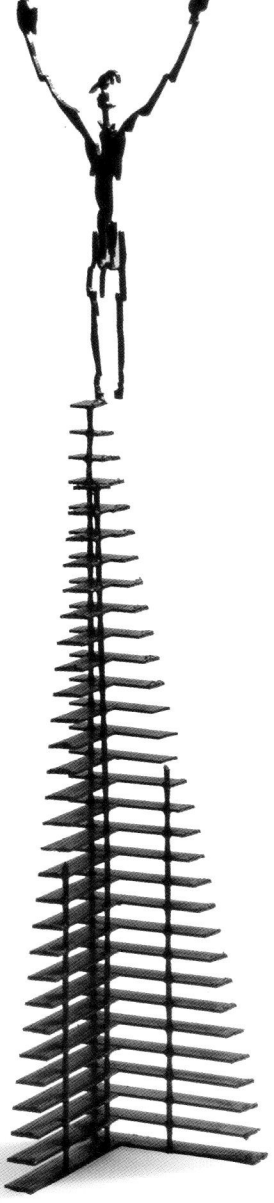

When I returned to the studio, I immediately set to work. The corner of the shop floor was covered with newly cut steel and the remnants of past projects. After *Dali*, I never threw out any steel. You never knew when the magic of coincidence would strike again. And now, in this time of pain and anguish, strike again it did.

I spotted a slender piece of steel with jagged edges in the heap of rubble. As I pulled it out to inspect more carefully, I recognized it as a "drop," as we call it, the pieces that drop out as a steel figure is cut.

This drop was from an earlier piece I had done, a tall figure who stands proudly with his arms uplifted in triumph. I called that piece *Grandeur*. The drop was *Grandeur*'s inner void, his central meridian, the sanctuary of his chi. I saw immediately, though, that the drop also was a man, a sinewy figure with a narrow, undulating body. He stands on slender legs with small, pointed toes.

The contrast between the two figures was stunning. They had been cut from the same piece of steel, one from inside the other, and each represented a different extreme of the human experience. *Grandeur*, arms aloft, is a symbol of man's solidity and self-confidence; he is triumphant and grandiose. *Grandeur*'s drop reveals the man inside: fragile, vulnerable, totally asleep to the terrible precariousness of his perch in life.

From there, in the flash of an eye, I conceived *Top of the World*. In it, *Grandeur*'s drop is perched atop the steel frame of a lofty tower. One foot stands on the tower's narrow crest, the other dangled over the abyss. But he never looks down. With arms raised like the *Grandeur* from whom he emerged, his inner self is completely self-absorbed, reveling in his proximity to the divine and his manifest destiny. Never had I imagined that *Grandeur*'s essence, his grandiosity, would become the source of his downfall—and of my lasting image of 9/11.

Top of the World, 2001

Finishing my vision was relatively easy. To recreate the crumpled, rusted remains of the Twin Towers, I built a pyramidal rather than rectangular base. As the structure rises to the sky, supported by steel columns, each steel floor grows narrower. At the apex, the figure stands with one leg on a square inch plate of steel. The floors look like steps, rising to heaven. I cut all the steps roughly out of raw steel, no straight lines, no finishes, just jagged edges, rust and slag. Here the skyscraper would be laid bare, its corrosion—indeed its destruction—happening day by day, evident to anyone who cares to look.

Top of the World is my 9/11 memorial. From the raw steel base to the building's crest, *Grandeur's* drop stands forthrightly at the top, unaware of his fragility or his predicament. It's a sad story. The process of conceiving and constructing *Top of the World* proved to be cathartic, as I had hoped. But even when I was finished, there was still a gaping hole inside me, a hole that I knew I had to inspect and explore. But how?

Innately, I knew where to go, but it was not comfortable terrain for me. I was at ease in the classroom, in court, in the welding shop, in a winery. But now I had to look elsewhere, and the voice inside me reverberated loudly: Go to the heart. Go to the heart.

Tears, 2004

Before Need

Even after making my memorial to 9/11, I felt empty. The feeling was exacerbated by the growing realization that sculpture, after all, is just a series of inanimate objects; it can't change the world or erase the savagery of man. At least the law is in service of people, I told myself, but I wasn't finding that totally satisfying either.

It was Marilyn who led the way out of this funk. And maybe this is how it should be. In a typical Hindu marriage ceremony, the bride and groom, adorned in garlands, walk around the sacred fire four times. The first three times the groom leads the bride, representing his leadership in the material world. But the last time the bride leads, signifying that the woman, not the man, leads the way in the spiritual realm.

And so it was with Marilyn. Long before 9/11, she had been attending yoga classes and practicing meditation at an ashram in Sonoma, a short drive from our home in Napa. I saw her meditation and spiritual practice grow, but I really wasn't interested. I was still totally wrapped up in my day-to-day world.

I'm not sure what veered me in the direction that Marilyn had charted. Perhaps it was our return to India in 2003. Although each of us had spent significant periods of time there before our marriage in 1979, we didn't return to India after that. Margot, Anthony and our jobs consumed all of our time. When we could find the time, we would return *en famille* to France. India remained a distant memory.

But that would soon change, and dramatically so.

Lotus, 2001

Aghoreshwar, 2007

In 2003, I was invited to speak at a UN conference in New Delhi on one of my favorite topics: recognizing and protecting unique products of place. What better place to deliver that talk than India, home to Darjeeling tea, Basmati rice and Ayurvedic medicine. These authentic products and this traditional knowledge are tied directly to Indian culture, to the land and to local farmers and craftsmen; they are art and native wisdom, entirely worthy of our attention and support. They also are an important means of economic development and empowerment. That trip reignited Marilyn's and my passion for India.

The following year, we planned another trip to India that would include a visit to Bal Ashram, the sister ashram to the one Marilyn was attending in Sonoma. Bal Ashram is an oasis amidst the crowds and chaos of the holy city of Varanasi. Set on the banks of the Ganges River, which is itself regarded as a goddess (*Ganga Ma*), the ashram is home to twenty-one orphan children, most of whom were abandoned by their parents. Both the Sonoma Ashram and Bal Ashram are the creation of Baba Harihar Ramji, the spiritual leader and teacher of Aghor, a branch of Hinduism.

Before the trip, Marilyn and I met with Baba. We would be staying at Bal Ashram free of charge and we wanted to help out. Baba assigned me the perfect task, working with his brother G.N. Pandey, an accountant, to figure out how much it costs to support an orphan child each year, all costs considered. We had regularly contributed money to Save the Children, but I never knew how that money was actually spent, how much of it got to the children rather than the program administrators. Now I would know.

When we arrived in Varanasi, we were met by an ashram resident and transported to Bal Ashram. I was looking forward to beginning my job and wandering around the ancient city. But first we met the orphan boys. They all lined up to greet us, kids from four years old to 16, each bowing to touch our feet as a sign of respect and each saying "Namaste." Then and there, something important happened inside me. I walked to the corner of the ashram grounds and cried uncontrollably. That's all it took. Heart open, I was on the path. And I've never looked back.

Aghor, as I was learning, is all about opening the heart, listening and talking from the heart, not the mind. To open up in this way requires stilling the mind. This is what mantra meditation and devotional practices are all about. Baba is our teacher, our guru. Of course, our practice doesn't allow us to comprehend a tragedy like 9/11, but at least we know that what we give will affect others and that it even will come back to us. Love awakens love, just as hatred awakens hatred.

Aghor is also about fearlessness and courage. Fearlessness in the face of a certain death and the courage born of conviction that we are part of the divine and that our spirit will endure after our body disintegrates. During this time, I began to wonder about the arc of life, about self-discovery, about divinity.

As my spiritual life emerged, I would often reflect on my mother's passing. Mothers are the embodiment of selfless giving and love, cardinal precepts of Aghor. I recalled how, as a college freshman, I would write my parents detailed letters about my experiences at Harvard, about my aspirations and my limitations. After my mom died in 1994, I found all those letters in the bottom drawer of her dresser. As I reread them, I was amazed by how little I had changed since 1972 when I wrote them. But I was also reassured. To me, those letters were proof positive that, try as we might, we carry with us throughout our lives a distinctive make-up, a genetic map perhaps, that defines our character. Call it what you will, our *être*, our *samskaras*, our code. But it remains constant, a remarkable reminder of our individuality, but also of our collective humanity.

With that knowledge, in 2004 I began working on a series of sculptures to capture in steel each individual's essence. I examined myself first, by investigating the *asanas* and tai chi poses which I felt best expressed who I am—my being and my baggage. The project never came to fruition, at least not in the way I expected.

Unbeknownst to me, I wrote my epitaph at the age of 18, in a letter to my mother. I rediscovered that letter after she died in 1994. To my surprise, I would have written the same words then... and now. Richard Mendelson 30 September 2001

Before Need, 2005

During my explorations, I realized that one of the letters I had written to my parents as a 19 year old came about as close to capturing my spirit as any pose could ever do. That letter was my epitaph, and suddenly the idea came to me: I would write my epitaph in steel—*before need*. I would also make this piece one that would disintegrate over time. To that end, I crafted a wood coffin and two images of my own body, one in solid sheet metal and a second outline figure rising above it. In the inner spaces of my body outline, I carved in steel the words of that letter to my parents.

The name of the sculpture leaped out at me. I remembered that a friend of mine had given me, several years before, a small metal plaque that was used in cemeteries to reserve a gravesite in advance. It bore two simple words in steel letters: BEFORE NEED. I found it in my shop and nailed it into the coffin. Those two words seemed to me a fitting title for the piece.

Over the years, I have watched *Before Need* erode with time. Starting as a pristine wood box and unrusted steel, the ensemble of coffin and body and epitaph have slowly decayed and collapsed. This is as it should be, and I am comforted by it. I know where my journey will end—where every journey ends—but that only gives me renewed energy to explore and create, to live every moment with passion and gratitude.

Wheel of Life

One day as I was driving my truck to the ashram in Sonoma, I happened to spot something unusual in someone's front yard: a massive steel spool. Right away I parked the truck and went over for a closer look. As it turned out, the owner of the house was having a garage sale, and he had put the spool out on his lawn on the off chance that some eccentric collector or forager might want to buy it.

The spool was made from what looked like two giant bicycle wheels, with a square metal brace in the middle. As the owner explained to me, the spool had been used to store and roll out heavy cable; he had found it at a nearby military base. As I walked around the piece and observed the symmetrical, rusted spokes, a distinct image came surging into my mind: *Vitruvian Man*, that iconic work of Leonardo da Vinci.

I had seen *Vitruvian Man*, also known as the *Canon of Proportions*, a few years before at the da Vinci Museum in Florence, where Leonardo grew up. Now it seemed to me that Leonardo's genius had made its way to Sonoma County. I had to have the piece.

It took a little negotiating; two other potential buyers were hot after the piece. Nonetheless, it was meant for me: a few minutes later the big steel spool was in the back of my truck, headed for some new incarnation. At that moment, I had no idea what I would do with it. I wasn't doing much sculpture at that time; instead I had taken a pause from making art and had entered a period of introspection, meditation and writing.

Riding home, I wondered whether Leonardo, the scientist and logician, had to him a whimsical side. It didn't appear so. *Vitruvian Man* is about symmetry and order, a perfectly proportioned man whose trunk and limbs are carefully aligned like the hands of a clock.

Then it hit me. I would introduce humor and playfulness to Leonardo's seriousness and stoicism. Two figures, not one, would be caught up in the spool. The figures would be long-limbed and supple, not upright and regal, and they would lounge and cavort in the middle of the wheels, undulating around and through the central spool. This playfulness brought an immediate smile to my face, and I wondered what Leonardo would have thought about my converting his rigid *Canon of Proportions* into a free-flowing and evanescent wheel of life.

The name of the piece was immediately obvious to me: *Leonardo's Wheelhouse*. I made the piece so that the wheels and the figures would roll freely, spinning round and round. This seemed a fitting and playful representation of the human journey.

Leonardo's Wheelhouse was another lesson in the spontaneity of creation and the importance of trusting one's instincts and vision. And this spurt of creativity brought me out of my artistic pause.

A few days later, I stopped by one of my favorite scrap yards in Petaluma. In a dumpster I spotted three pieces of jettisoned metal, rolled into unusual shapes. I saw life in those pieces of scrap, just as I had at the beginning of my journey when I visited the scrap yard in Jacksonville near my mother's gravesite. I played with an assemblage in the back of my truck to try to capture that vision. I then headed directly to the shop to weld the steel parts in place. But the pieces shifted during the trip, and I struggled to recreate the same shapes and angles that I had envisioned. I never got it right, so I let it go. In the resulting void, though, another, more powerful image appeared from the depths of my memory, inspired by another master, Henry Moore.

Leonardo's Wheelhouse, 2010

I had long admired Moore's series of reclining nudes. I had seen my first one in Harvard Yard. Moore's nudes are usually segmented into body parts and have an undulating quality. Now, with my three pieces of discarded metal, I saw that same undulation. In the blink of an eye, I reassembled and welded them into place. Then I quickly fabricated a head and *voilà*, there she was: my own *Reclining Nude*. To my surprise, the figure balanced perfectly on one of the sawhorses that we were using to fabricate it. I then fabricated a replica sawhorse out of metal, and it now serves as the base, and balance point, for my homage to Henry Moore.

Then came Henri Matisse and Edgar Degas. I have long been attracted to dance, all forms of it, and I love to dance myself. The flow of the dancer's body and the emotional expressiveness of the movements, so beautifully captured by Matisse and Degas, are captivating. I decided to cut ten dancers at one time in varying sizes and poses and from different materials— steel, aluminum and stainless steel. Then I would see how they would interact, without any preconceived notions. Would they resemble Matisse's *The Dance?*

I quickly concluded that Leonardo was right about his *Canon of Proportions* and the need for symmetry, so I placed the dancers of the same general size together in three maquettes. Then I made the full-sized *Les Trois Mains Ascendantes*, with three dancers in close quarters, hands joined and rising upward, each dancer looking and leaping out in a different direction. It's as if the dancers gain their power through intimacy and then direct it outward.

Working in the realm of the great masters gave me a fresh infusion of confidence and creative energy, and these came into my life at a most opportune time. In the years following *Before Need* and my first visit to the ashram in India, I had tried to make sense of a troublesome dichotomy: my growing attraction to the inward-looking world of spirituality and my continuing involvement in the workaday world of law, wine and business. These worlds seemed increasingly difficult to bridge. One moment I would immerse myself in action, the next I would retreat into silence. As a result, my life had become exhausting, almost schizophrenic.

Reclining Nude, 2010

Slowly, I came to realize from my spiritual practice and my own life experience that this dichotomy wasn't real, that we don't have to choose between the inner and the outer world or between involvement and detachment. Instead, we can hold and honor both perspectives simultaneously, carrying with us at all times an inner core of stillness and a peaceful heart. It's a matter of stepping back, carving out the space and time—even between breaths—to engage at a deeper level, literally and figuratively breathing in the world of beauty, mystery and divinity.

My sculptures inspired by the old masters brought this realization into sharp focus, in large part because fresh ideas like these seemed to come out of nowhere. Of course, that is not the case, but I would never have perceived or reacted to the sources of inspiration that were all around me without being as present and open as I had become.

This realization has been both liberating and reinvigorating. It now allows me to move forward in the world courageously, enthusiastically and without fear of losing my center of balance, physically or spiritually. It unites Rene di Rosa's sense of whimsy, Jack Chandler's sense of scale and Baba's commitment to the divine, all of which flow together from the same source of energy and inspiration.

Perhaps I knew all of this at the beginning of my journey in the world of art. Certainly, I trusted my instincts and intuitions when I first visited the scrap yard near my mother's grave and when I accepted Jack's invitation to try my hand at metal sculpture. But I didn't have any direct experience with this kind of creativity or the deep trust that's necessary to capture and express it.

After the works inspired by the masters, I began to explore sacred art through sacred geometry. My fascination with this subject began when I made *Lotus*, pictured on the front cover of this book. After my Buddha-like meditating man was completed, I learned that all representations of the Buddha must obey certain rules of geometry; otherwise, they are considered sacrilegious. For example, there is a prescribed distance between the bottom of Buddha's ear lobe and the top of his shoulder. As a believer in the spiritual power of numbers, I don't object to this need for symmetry.

Les Trois Mains Ascendantes, 2010

Wheel of Life, 2008

Whether my focus of the moment is sacred art, the old masters or whimsy, it's all part of the wheel of life. As the wheel turns, there appears to be a beginning and an end to the journey, but there's also a silent, timeless center. I see a certain symmetry here in my artistic journey. I was led into art by the death of my mother, and as I was completing this book, I received the sad news of the passing of Rene di Rosa at the end of a full and generous life. The wheel turns, in metal as in life, but the spirit lives on.

If past is prologue, my art will continue to move in tandem with my life and my spiritual evolution. I don't know what the next chapters hold in store. But a few things I do know. As my spiritual practice grows, my heart continues to open, and I value more than ever a life of peace, courage and service. Art is now a vital part of me. It has provided me with many valuable insights and lessons and is a powerful pathway for exploration and expression. It makes me laugh and it makes me cry.

I often wonder what my mother would think of my art journey. Her death provided the critical first spark, followed by the guidance and inspiration of Jack, Rene, Marilyn and many others. I doubt that they realize what a life-affirming influence they had on me, but I thank them all for steering me in this direction and then letting me go. I am forever grateful for their foresight, generosity and grace.

Richard Mendelson
January 2011

Fearless and Free, 2010

Soaring Spirits, 2009

Acknowledgements

Without the inspiration and commitment of several friends, this book never would have happened. I owe a tremendous debt of gratitude to creative director and photographer Robert M. Bruno, editorial director and publisher Paul Chutkow and photographer Kurt-Inge Eklund.

My art journey has brought me into contact with several people who have contributed their craftsmanship and comradeship along the way: Guillermo Alvarez, Lance Arbogast, his wife Linda and his sons Chad and Kevin, Toby Gewertz, Dennis Humphrey, Mark Morgenlaender, Robert Ramirez, Eric Schneider, Glenn Self and Tom Smeltzer. I have had the distinct pleasure of collaborating with woodworking artist Kim Taylor on *Moi* (page 79) and *Barrel Tasting* (page 81).

I also am indebted to my master teachers: in law, the late John Barton and Ellis Horvitz; in yoga, Kofi Busia, B.K.S. Iyengar and Noelle Perez-Christiaen; in tai chi and chi gong, Sifu Michelle Dwyer and Master Zhang Hao; and in Aghor, Baba Harihar Ramji.

Finally, I deeply thank my wife Marilyn, daughter Margot and son Anthony for their support of my foray, relatively late in life, into the world of art. Their love is deep and unwavering, and so is mine for them.

Photo Credits: ROBERT M. BRUNO

Lotus, book front cover

Margin Release, page 4

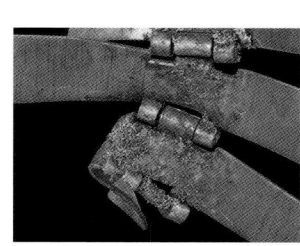

Slipped Disc (detail), page 10

Slipped Disc, 1995, page 13

Blond Curl, 1998, page 16

Spiral, 1999, page 17

Father, 1998, page 18

Mother, 1998, page 18

Child, 1998, page 18

Ball and Chain, 1996, page 21

Ball and Chain (detail), 1996, page 20

Jump for Joy, 2001, page 22

cutouts, page 29

Rolling Hands, 1998, page 34

Four I's, 1996, page 35

Blue Twist, 1998, page 36

Duet, 1997, page 37

Dali, 1996, page 40

Lean on Me, 2000, page 44

Caraibes, 1998, page 48

Detached, 1998, page 53

Snake Creeps Down, 1999, page 56

Sarvangasana, 2006, page 55

Crouch, 1999, page 58

Sedona, 2005, page 59

Rajakapotasana, 1997, page 60

Bakasana, 2002, page 61

Roly Poly, 1999, page 63

Untitled, 2005, page 66

Harlequins II & III, 1998 & 1999, page 67

Breast Drills in Tandem, 2004, page 70

Endless Loop, 2005, page 71

Geishas, 2004, page 72

Scales of Justice, 2001, page 73

Buoys I, Ii, & III, 2002, pages 76–77

Moi, 1999, page 78

Sink or Swim, 2000, page 86

In the Middle, 1998, page 87

Grandeur and his drop, 2000, page 92

Top of the World, 2001, page 93

Tears, 2004, page 97

Aghoreshwar, 2007, page 102

Before Need, 2005, page 107

Leonardo's Wheelhouse, 2010, page 112

Reclining Nude, 2010, page 115

Les Trois Mains Ascendantes, 2010, page 117

Wheel of Life, 2008, page 118

Fearless and Free, 2010, page 120

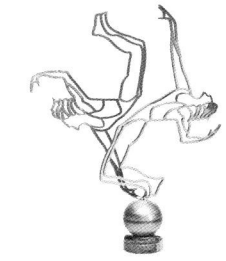

Soaring Spirits, 2009, page 121

KURT-INGE EKLUND

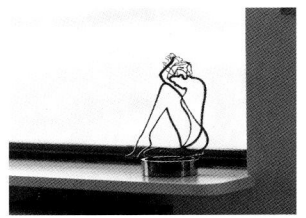

Sedona, book back cover

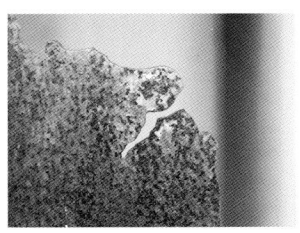

Dali (detail), pages 2–3

Spring Tooth, 1997, pages 14–15

La Famiglia, 2001, page 19

Legs, 1996, page 24

The Hanging, 1997, page 25

Working Shot, page 26

Eleganza, 1996, page 31

Working Shot, page 32

Working Shot, page 33

Dali (detail), page 38

Four's a Crowd, 2000, page 45

Lady, 1999, page 46

Classic, 1997, page 47

Caraibes, 1998, page 49

Inescapable Grace, 1998, page 50

Seated Woman, 2001, page 62

Buoys (detail), page 68

Colanzi, 1996, page 75

La Vigne, 2000, page 80

Barrel Tasting, 2000, page 81

Racer, 2000, page 82

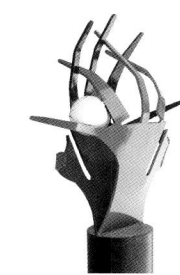

Red, White and Blue II, 2001,
page 84

Juggling Hands, 1997, page 85

Top of the World (detail),
pages 90–91

Top of the World, 2001, page 94

Top of the World, 2001, page 95

Top of the World, 2001, page 96

Before Need (detail), page 98

Lotus, 2001, pages 100–101

Before Need, 2005, pages 108–109

SUZANNE BECKER BRONK

Working Shot, page 110

FAITH ECHTERMEYER

Raptor, 1997, page 23   *The Gondolier*, 1998, page 83

LAURA HUNT

Before Need, 2005, page 105

STEVE MCCURRY, MAGNUM PHOTOS

World Trade Center Attack, 2001, Magnum Photos, pages 88–89

JEFF TANGEN

Love in the Round, 1996,   *Trikonasana*, 1999, page 65
pages 42–43

Instructor's Manual to accompany

LIFE: The Science of Biology

by Purves and Orians

Prepared by

Joan R. Callahan
University of Georgia

Russell Davis
University of Arizona

Arlene F. Foley
Wright State University

Elizabeth J. Mallon
State University of New York, Stony Brook

John H. Tullock
University of Tennessee

Gerald L. Vaughan
University of Tennessee

for

P.S. Associates, Inc.
Sterling, Massachusetts

SINAUER ASSOCIATES, INC. / WILLARD GRANT PRESS

Sinauer Associates, Inc.
Sunderland, Massachusetts 01375

Willard Grant Press
20 Providence Street
Boston, Massachusetts 02116

Printed in U.S.A.

The pair of red-winged blackbirds (<u>Agelaius
phoeniceus</u>) on the cover is a painting by
J.F. Lansdowne and appears as Plate 94 in his
book, <u>Birds of the Eastern Forest: 2</u> (©1968
by M.F. Fehely Artists Company Limited. All
rights reserved.)

ISBN 0-87150-770-6

87 86 85 84 83

9 8 7 6 5 4 3 2 1

CONTENTS

iv   Contents

This comprehensive Instructor's Manual has been designed to help you adapt LIFE: THE SCIENCE OF BIOLOGY, by William Purves and Gordon Orians, to your needs and make it a more effective tool for your course. Each chapter in the manual presents a variety of teaching aids for use with the corresponding chapter of LIFE.

Chapter Summaries and Objectives. A brief summary of each chapter section, followed by specific performance objectives your students should be able to complete after reading that section. The summaries and objectives are identical to those appearing in the corresponding chapter of the Purves and Orians Study Guide. If you are not assigning or recommending the guide, you may find it useful to reproduce the Chapter Summaries and Objectives given in the manual and distribute copies to your students. This material highlights the important concepts in each chapter.

Article Resources. A list of articles ranging from scientific to popular, including several that deal with the teaching of selected text topics. Each reference has been annotated to show how it relates to the material in the chapter. In addition, articles are organized by chapter section.

Essay Questions. There are 4 to 6 per chapter, with guidelines for grading each question included. These essays may be used for quizzes and examinations or to generate class discussion.

Multiple-Choice Test Sets. Two 15-item sets per chapter, ranging from simple to complex. Answers appear in the margin, along with the pages in LIFE on which full explanations of each question may be found. The presence of two similarly constructed test sets for each chapter allows you to use Test Set A (or items from it) with one class and Test Set B with another, or to employ the two test sets in alternate terms or years. In addition, this feature affords you the flexibility to use the tests in a self-paced, or individualized, course, wherein students who fare poorly on Test Set A are given the opportunity to restudy the chapter and then display their mastery of the chapter on Test Set B. Item answers and page references have been placed in the margin to the left of each question to allow you to use the test sets as they appear in the Instructor's Manual by masking the answers and reproducing the questions in the quantity needed for your class.

Supplementing the specific chapter-related summaries and resources described above are a bibliography of Resources and References for General Biology (beginning on page vii) and a section entitled Using Films in Biology (beginning on page xiv). The former includes a wide spectrum of background sources in biology, including dictionaries and encyclopedias, books on the teaching of biology, readers and reprint sources, suppliers of laboratory equipment and materials, and topical trade books (mostly popular) that you might assign to students or use yourself in preparing for lecture. All of these resources are designed to help you in setting up your course or rearranging it around the Purves and Orians text. The latter section offers tips on using films in your course, a selection of general suppliers of films in biology, a section-by-section annotated list of recent films in biology, and the names and addresses of distributors of these films.

> Transparency Masters for 120 text figures are included in the back of this manual. They have been selected and designed to give optimal schematic reinforcement to major concepts presented in each chapter. Each master includes the figure number, title, and text page on which the original may be found. A complete list of the figures included can be found on p. 423.

STUDY GUIDE

Designed to serve as an extension of LIFE, the Study Guide helps the student identify central concepts and organize the multitude of facts inherent in a one-year course in general biology. Each chapter of the Study Guide parallels a chapter in the text, and each of the Guide's five components serves a specific purpose.

> Chapter at a Glance gives a quick overview of the general organization and major topics in the chapter.

> The Key Terms section identifies the most important terms that students will need to know to thoroughly understand major concepts.

> Organizing and Testing Your Knowledge provides a review of each section of the chapter, followed by specific objectives for the student to accomplish and exercises ("Spot Checks") for the student to complete.

> Practice Examinations consist of a pair of similarly constructed quizzes of approximately 10-15 multiple-choice questions per quiz.

> Integrating and Reinforcing Your Knowledge requires students to test their knowledge and to demonstrate an ability to work with factual information drawn from throughout the chapter.

Answers to all chapter exercises are provided at the end of the guide. Text page references are also given next to the answers to the multiple-choice Practice Examinations. Hence students can review sections covering the questions they answered incorrectly.

If you have not as yet received a copy of the Study Guide, contact Willard Grant Press.

RESOURCES AND REFERENCES FOR GENERAL BIOLOGY*

TEACHING BIOLOGY

The following list includes sourcebooks on the teaching of biology, reference volumes, sources of book and audiovisual reviews, and other materials of interest to instructors of biology.

AAAS Science Book List. 3rd ed. AAAS, Sales Dept., Box RS, 1515 Massachusetts Avenue, N.W., Washington, D.C. 20005. Published in 1970, this volume includes thousands of reviews of trade and textbooks in the pure and applied sciences.

AAAS Science Book List Supplement. Extends the coverage of the AAAS Science Book List from 1969 through 1977.

AAAS Science Books and Films. Published quarterly. Reviews recent books and films on biology and other disciplines.

Anderson, O. R. Teaching Modern Ideas of Biology. New York: Teachers College Press, 1971.

Asimov, I. A Short History of Biology. Westport, Conn.: Greenwood, 1980.

Ayers, D. M. Bioscientific Terminology. Tucson, Ariz.: University of Arizona Press, 1972.

Bamuel, H. B., and J. J. Berger. Biology: Its People and Its Papers. National Science Teacher's Association, 1742 Connecticut Avenue, N.W., Washington, D.C. 20009.

Biological Science: Interaction of Experiments and Ideas. 2nd ed. Englewood Cliffs, N.J.: Prentice-Hall, 1971.

Bradley, L. D. Career Education and Biological Sciences. Boston: Houghton Mifflin, 1975.

Burns, John M. Biograffiti: A Natural Selection. New York: Norton, 1980. Nonsense poems for light lecture.

Catalog: Periodicals, Books, Tapes, and Reprints. Washington, D.C.: AAAS.

Cheremisinoff, P. N. Bioconversion Sourcebook. Ann Arbor, Mich.: Ann Arbor Science, 1982.

Current. Journal of the National Marine Education Association, c/o Virginia Institute of Marine Science Education Center, Gloucester, Va. 23062. Published four times a year, $15.00. Includes some reviews of college-level teaching materials, texts, and media in marine biology, but is directed mostly at K-12 level.

Dictionary of Life Sciences. New York: McGraw-Hill, 1976.

Eakin, R. Great Scientists Speak Again. Berkeley, Calif.: University of California Press, 1975. Includes impersonations of Darwin, Pasteur, Mendel, and Harvey as done in lecture by Professor Eakin.

Education in Science. Association for Science Education, College Lane, Hatfield, Hertsfordshire AL10 9AA, England. Published ten times a year. Includes excellent articles on teaching biology by the British A level (comparable to our first year of college) plus reviews of British films, texts, and teaching materials.

Encyclopedia of Biological Terms. Woodbury, N.Y.: Barron, 1982.

Gray, P. The Dictionary of the Biological Sciences. Melbourne, Fla.: Krieger, 1982.

_____. The Encyclopedia of the Biological Sciences. Melbourne, Fla.: Krieger, 1981.

Grobman, A. B. (ed.). Social Implications of Biological Education. Princeton, N.J.:
    Darwin Press, 1975.

Guidelines and Suggested Titles for Library Holdings in Undergraduate Biology. Publi-
    cation Number 32 (1971). CUEBS, 3900 Wisconsin Avenue, N.W., Washington, D.C.
    20016.

Haensch, G., and Haberkamp De Anton, G. Dictionary of Biology 2nd ed. New York:
    Elsevier, 1981.

Hickman, F. M., and J. B. Kahle. New Directions in Biology Teaching: Perspectives
    for the 1980s. Reston, Va.: National Association of Biology Teachers, 1982.

Jaeger, E. D. A Sourcebook of Biological Names and Terms. Springfield, Ill.: Charles
    C. Thomas, 1972.

Jones, E. S. How to Use Biological Abstracts. Cambridge, Mass.: Harvard University
    (Cabot Science Library, 1 Oxford Street, Cambridge, Mass. 02138). Includes
    slides and audio cassettes.

Journal of College Science Teaching. National Science Teacher's Association, 1742
    Connecticut Avenue, N.W., Washington, D.C. 20009. Published five times a year.
    Includes articles on teaching biology, plus reviews and synopses of recent arti-
    cles, books, films, and teaching materials.

Mayer, W. V. Biology Teacher's Handbook: BSCS 3rd ed. New York: Wiley, 1978.

Mayr, E. The Growth of Biological Thought: Diversity, Evolution, and Inheritance.
    Cambridge, Mass.: Harvard University Press, 1981.

Modules: The Use of Modules in College Biology Teaching. Publication Number 31.
    Washington, D.C.: CUEBS, 1971.

New Trends in Biology Teaching. Volume 4. New York: Unipub, 1978.

Preservice Preparation for College Biology Teachers. Publication Number 24. Washing-
    ton, D.C.: CUEBS, 1970.

Roe, K. E., and Frederick, R. G. Dictionary of Theoretical Concepts in Biology,
    Metuchen, N.J.: Scarecrow, 1981.

Selmes, C. (ed.). New Movements in the Study and Teaching of Biology. Levittown,
    N.Y.: Transatlantic Arts, 1974.

Tootill, E. (ed.). The Facts on File Dictionary of Biology. New York: Facts on File,
    1981.

JOURNALS AND MAGAZINES

The following publications provide up-to-date information on recent research in
biology.

Biology Digest. Data Courier, Inc., 620 Fifth Street, Louisville, KY 40202.

BIOSIS-Biological Abstracts. Marketing Bureau, Biosciences Information Service, 2100
    Arch Street, Philadelphia, PA 19103.

Proceedings of the Annual Convention of the American Institute of Biological Sciences.
    AIBS, 1401 Wilson Boulevard, Arlington, VA 22209. Published yearly.

The following are good sources for lecture ideas, information on current research,
overviews of important areas of research, and book reviews in biology. Many of the
articles from these periodicals are suitable for assignment in class.

American Scientist. Sigma Xi, 155 Whitney Avenue, New Haven, CT 06510.

Bioscience. AIBS, 1401 Wilson Boulevard, Arlington, VA 22209.

Mosaic. National Science Foundation, Superintendent of Documents, Government Printing
    Office, Washington, D.C. 20402.

Natural History. American Museum of Natural History, Central Park West and 79th
    Street, New York, NY 10024.

Nature. Macmillan (Journals) Ltd., Brunel Road, Basingstoke, Hampshire, England.

New Scientist. 126 Long Acre, London WC 2, England.

New Technical Books. New York Public Library, Fifth Avenue and 42nd Street, New York,
    NY 10018.

Quarterly Review of Biology. The Stony Brook Foundation, SUNY at Stony Brook, Stony
    Brook, NY 11790.

Science. AAAS, 1515 Massachusetts Avenue, N.W., Washington, D.C. 20005.

Science Books. AAAS, 1515 Massachusetts Avenue, N.W., Washington, D.C. 20005.

Science 83. AAAS, 1515 Massachusetts Avenue, N.W., Washington, D.C. 20005.

Science News. 1719 N Street, N.W., Washington, D.C. 20036.

<u>Scientific American</u>. 415 Madison Avenue, New York, NY   10017.

Several magazines and newspapers contain regular reports and general articles on biological topics. Included in this group are <u>Time</u>, <u>Fortune</u>, <u>Newsweek</u>, <u>Psychology Today</u>, <u>Omni</u>, <u>Harper's</u>, <u>Science Digest</u>, <u>National Geographic</u>, <u>Atlantic Monthly</u>, <u>The Wall Street Journal</u>, the <u>New York Times</u>, and the <u>New York Times Magazine</u>.

REPRINTS

Three valuable sources of reprints for use in class are:

<u>The Bobbs-Merrill Reprint Series in the Life Sciences</u>. The Bobbs-Merrill College Division, Indianapolis, IN   46268. Write for the catalog/index.
<u>Psychology Today Reprints</u>. Psychology Today Reader Service, P.O. Box 278, Pratt Station, Brooklyn, NY   11205. Write for the complete list. Includes many articles relevant to biology.
<u>Scientific American Offprints</u>. W. H. Freeman, 660 Market Street, San Francisco, CA 94104. Write for the complete list.

READERS

<u>Annual Editions, Readings in Biology '78-'79</u>. Guilford, Conn.: Dushkin, 1978.
Head, J. J. (ed.). <u>Carolina Biology Readers</u>. Carolina Biological Supply Co., 2700 York Road, Burlington, N.C.   27215.

As a group, the <u>Scientific American</u> readers are uniformly superb, although the level of some articles is probably too high for many introductory students. Included among the SA readers are the following:

<u>Animal Societies and Evolution</u>. Introduction by H. Topoff. 1981.
<u>Genetics</u>. Introduction by C. I. Davern. 1981.
<u>Immunology</u>. Edited by F. M. Burnet. 1976.
<u>The Insects</u>. Edited by T. Eisner and E. O. Wilson. 1977.
<u>Life in the Sea</u>. Introduction by A. T. Newberry. 1982.
<u>Life: Origin and Evolution</u>. Introductions by C. E. Folsome. 1979.
<u>Molecules to Living Cells</u>. Introduction by P. C. Hanawalt. 1980.
<u>Vertebrates: Adaptation</u> and <u>Vertebrates: Physiology</u>. Introductions by N. K. Wessells. 1980.

RECENT BOOKS OF GENERAL INTEREST IN BIOLOGY

Applewhite, P. B. <u>Molecular Gods: How Molecules Determine Our Behavior</u>. Englewood Cliffs, N.J.: Prentice-Hall, 1981.
Asimov, I. <u>In the Beginning...Science Faces God in the Book of Genesis</u>. New York: Crown, 1981.
Ayensu, E. S., and P. Whitfield (eds.). <u>The Rhythms of Life</u>. New York: Crown, 1982.
Billingham, J. (ed.). <u>Life in the Universe</u>. Cambridge, Mass.: MIT Press, 1981.
Borek, E. <u>The Atoms Within Us</u>, rev. ed. New York: Columbia University Press, 1980.
Bylinski, G. <u>Life in Darwin's Universe: Evolution and The Cosmos</u>. Garden City, N.Y.: Doubleday, 1981.
Campbell, P. N. <u>Biology in Profile: A Guide to the Many Branches of Biology</u>. Elmsford, N.Y.: Pergamon, 1981.
Colinveaux, P. <u>The Fates of Nations: A Biological Theory of History</u>. New York: Simon and Schuster, 1980.
Dawkins, R. <u>The Selfish Gene</u>. New York: Oxford University Press, 1977.
Ehrlich, P., and A. Ehrlich. <u>Extinction: The Causes and Consequences of the Disappearance of Species</u>. New York: Random House, 1981.
Eisely, L. <u>Darwin and the Mysterious Mister X: A New Look at the Evolutionists</u>. New York: Dutton, 1979.
_____. <u>The Star Thrower</u>. New York: Times Books, 1978.
Fox, M. W. <u>The Soul of the Wolf</u>. Boston: Little, Brown, 1980.

Galston, A. W. Green Wisdom: The Inside Story of Plant Life. New York: Basic Books, 1981.

Goldsmith, D., and T. Owen. The Search for Life in the Universe. Menlo Park, Calif.: Benjamin/Cummings, 1980.

Goodfield, J. An Imagined World: A Story of Scientific Discovery. New York: Harper & Row, 1981.

Gould, S. J. Ever Since Darwin: Reflections in Natural History. New York: Norton, 1977.

_____. The Panda's Thumb: More Reflections in Natural History. New York: Norton, 1980.

Harrison, G. Mosquitoes, Malaria and Man: A History of the Hostilities Since 1880. New York: Dutton, 1978.

Hoagland, M. B. Discovery: The Search for the DNA's Secrets. Boston: Houghton Mifflin, 1981.

Kitcher, P. Abusing Science: The Case Against Creationism. Cambridge, Mass.: MIT Press, 1982.

Leakey, R. E. The Making of Mankind. New York: Dutton, 1981.

Leakey, L., and R. Lewin. Origins. New York: Dutton, 1977.

_____. People of the Lake: Mankind and Its Beginnings. New York: Anchor Press/ Doubleday, 1978.

Line, L. and Reiger, G. The Audubon Society Book of Marine Wildlife. New York: Abrams, 1980.

Lionni, L. Parallel Botany. New York: Knopf, 1977.

Lopez, B. H. Of Wolves and Men. New York: Scribner's, 1978.

Margulis, L., and K. V. Schwartz. Five Kingdoms: An Illustrated Guide to the Phyla of Life on Earth. San Francisco: W. H. Freeman, 1982.

Mayr, E. The Growth of Biological Thought: Diversity, Evolution, and Inheritance. Cambridge, Mass.: Harvard University Press, 1982.

McKinnell, R. G. Cloning: A Biologist Reports. Minneapolis, Minn.: University of Minnesota Press, 1979.

Medawar, P. B., and J. S. Medawar. The Life Science: Current Ideas of Biology. New York: Harper & Row, 1977.

Milne, L. J., and M. Milne. Insect Worlds: A Guide for Man on Making the Most of the Environment. New York: Scribner's, 1980.

Montagu, A. Sociobiology Examined. New York: Oxford University Press, 1980.

Moore, D. M. (ed.). Green Planet: The Story of Plant Life on Earth. New York: Cambridge University Press, 1982.

Parker, S. P. (ed.). Synopsis and Classification of Living Organisms. Vols. I and II. New York: McGraw-Hill, 1982.

Reader, J. Missing Links and the Men Who Found Them. Boston: Little, Brown, 1981.

Sagan, C. Broca's Brain: Reflections on the Romance of Science. New York: Random House, 1979.

_____. The Dragons of Eden: Speculations on the Evolution of Human Intelligence. New York: Random House, 1977.

Silk, J. The Big Bang: The Creation and Evolution of the Universe. San Francisco: W. H. Freeman, 1980.

Stanley, S. M. The Evolutionary Timetable: Fossils, Genes, and the Origin of Species. New York: Basic Books, 1981.

Stebbins, G. L. Darwin to DNA, Molecules to Humanity. San Francisco: W. H. Freeman, 1982.

Stone, I. Origin. Garden City, N.Y.: Doubleday, 1981. (A superb biography of Darwin.)

Thomas, L. The Lives of a Cell. New York: Viking, 1977.

_____. The Medusa and the Snail. New York: Viking, 1979.

Verney, P. Animals in Peril. Provo, Utah: Brigham Young University Press, 1979.

Welty, J. C. The Life of Birds. New York: CBS College Publishing, 1982.

Wilson, E. O. On Human Nature. Cambridge, Mass.: Harvard University Press, 1978.

Wunderlich, K., and W. Gloede. Nature as Constructor. New York: Arco, 1981.

LAB MANUALS

There are dozens of general biology lab manuals, most of which are designed to accompany a specific text but nevertheless are suitable for independent use. Among the more recent are:

Abramoff, P., and Thomson, R. G. Biology and Man: Laboratory Manual. Englewood
    Cliffs, N.J.: Prentice-Hall, 1975.
_____. Laboratory Outline in Biology III. San Francisco: W. H. Freeman, 1982.
Chicson, J. A., et al. The Laboratory Experience: A Principles of Biology Manual.
    2nd ed. Minneapolis, Minn.: Burgess, 1976.
Eberhard, C. Biology Laboratory. 2nd ed. Philadelphia: Saunders College Press,
    1982.
Evert, R., et al. Laboratory Topics in Biology. New York: Worth, 1979.
Fiedler, A., and L. Fulton. Exploring Biology in the Laboratory. New York: Canfield/
    Harper & Row, 1974.
Freeman Laboratory Separates. San Francisco: W. H. Freeman, 1969-1983. Write for
    a complete list of available separates.
Gunstream, S. E., and Babel, J. S. Explorations in Basic Biology. 3rd ed. Minneapo-
    lis: Burgess, 1982.
Howard, L. D. Principles of Biology Laboratory Manual. Westport, Ct.: AVI, 1980.
Kaplan, E. H. Experiences in Life Science: A Laboratory Guide. 2nd ed. New York:
    Macmillan, 1976.
_____. Problem Solving in Biology: A Laboratory Workbook. New York: Macmillan,
    1976.
Keeton, W. T., and McFadden, C. M. Laboratory Guide for Biological Science. 2nd ed.
    New York: Norton, 1983.
Leonard, W. H. Laboratory Investigations in Biology. Minneapolis, Minn.: Burgess,
    1982.
Noland, G. B., and W. C. Beaver. Lab Manual in General Biology. St. Louis, Mo.:
    Mosby, 1975.
Research Problems in Biology: Investigations for Students. 2nd ed. New York: Oxford
    University Press, 1976.
Schiff, S. O., and D. A. Griffith. 21 Afternoons in Biology: A Lab Manual. 2nd ed.
    Boston: Allyn and Bacon, 1979.
Unbehaun, L., et al. Principles of Biology Laboratory Manual. Minneapolis, Minn.:
    Burgess, 1980.
Winchester, A. M. Biology Laboratory Manual. 6th ed. Dubuque, Iowa: William C.
    Brown, 1979.
Wodselak, J. E., et al. General Biology Laboratory Guide. 4th ed. Dubuque, Iowa:
    William C. Brown, 1980.

MODULES/LABORATORY KITS

In the past few years a number of publishers have introduced modular programs in
biology for use in either the laboratory or the classroom. Some contain written
material only; others combine booklets with slides, filmstrips, audio cassettes, and/or
laboratory materials. Included in this group are:

BioKit. Burlington, NC: Carolina Biological Supply, 1975. A hands-on approach to
    biology, developed primarily for senior high school use, but appropriate for
    college.
Biology Media. 918 Parker Street, Berkeley, CA 94710. Audiovisual instructional
    materials for biology teaching, including slide-tape modules, multi-image lec-
    tures, videocassettes, slide sets, and laboratory kits.
Bio-Modes. Distributed by Rand McNally Bio-Centre, P.O. Box 40, Somerset, WI 54025;
    1977. Series I: Five modules, consisting of duplicating master student work-
    sheets, instructor's guide, and audio cassette, for laboratory use.
BioStereo: Biology in Three Dimensions. Distributed by Hubbard Scientific, Box 105,
    Northbrook, IL 60062. 1975. Forty-four stereo cards on eight topics, with
    inquiry booklets, instructor's guide, and three-dimensional specs.
Biotech. Distributed by Prentice-Hall Media, 150 White Plains Road, Tarrytown, NY
    10591. Modules consist of color filmstrip and matching 35-mm slide, plus an
    audio cassette and laboratory study manual. Designed for pre-lab preparation,
    self-study, review, and lecture support.
Druger, M. M. Individualized Biology. Reading, Mass.: Addison-Wesley, 1971 and
    later. 13 units.
Ehrlich, P. R., et al. Biocore. New York: McGraw-Hill, 1974. 23 booklets.
Inquiries into Life. New York: Cambridge Book Co., 1973. 10 units.
Luria, S. E. 36 Lectures in Biology. Cambridge, Mass.: MIT Press, 1975.

Melecca, B. B., et al. <u>Bio-Learning Guide and Bio-Learning Notes</u>. 2nd ed. Minneapo-
    lis, Minn.: Burgess, 1975.
<u>Minicourses</u>. Philadelphia: Saunders, 1976. Developed by S. M. Postlethwait and
    BSCS. A complete audiotutorial laboratory course.
<u>Oxford Biology Readers</u>. Burlington, N.C.: Carolina Biological Supply, 1975. Printed
    modules only. Prentice-Hall Media, 150 White Plains Road, Tarrytown, NY 10591.
    (Instructional modules on various biological topics.)
Samuels, E. <u>Biotutorial: A Modular Program for Introductory Biology</u>. Boston:
    Houghton Mifflin, 1975. Thirteen audio-visual modules, student manual, and
    instructor's manual.
<u>Ward's Solo-Learn System</u>. Rochester, N.Y.: Ward's Natural Science Establishment,
    1976. Fifty semiprogrammed single-topic study units, with instructor's aide.
White, J. M., and R. D. Barnes. <u>Hands-on Biology</u>. New York: Hamilton, 1974. A
    series of scripts for an audiotutorial "learning system."

LAB MATERIALS

Bio Quip Products, Box 61, Santa Monica, CA 90406. Insects.
Carolina Biological Supply Co., Burlington, NC 27215. General.
Combined Scientific Supplies, Box 125, Rosemead, CA 91770. Insects.
Connecticut Valley Biological Supply Co., Inc., P.O. Box 326, Southampton, MA 01703.
    Excellent source of live materials.
Fifco Laboratories, Detroit, MI 48232. Reagents and culture media.
Ealing, 22 Pleasant Street, South Natick, MA 01760. Lab equipment.
Fisher Scientific Co., 4901 Le Moyne Street, Chicago, IL 60651. General.
Hubbard Scientific Co., Box 105, Northbrook, IL 60062. Audiotutorial program and
    general.
International Biologics, Inc., 1991 Sharondale Avenue, St. Paul, MN 55113.
Jewel Industries, Inc., 5005 West Armitage Avenue, St. Paul, MN 55113. Aquaria and
    cages.
Lab-Aids, Inc., 130 Wilbur Place, Bohemia, NY 11716. Models.
Macmillan Science Company, Inc., Dept. BI, 8200 South Hoyne Avenue, Chicago, IL
    60620. Distributor of <u>Turtox</u> teaching aids.
Nasco, Fort Atkinson, WI 53538. Animals.
Parco Scientific Co., 316 Youngstown-Kingsville Road, S.E., Vienna, OH 44473.
Rand McNally Bio-Centre, P.O. Box 40, Somerset, WI 54025. General.
S.R.M., Inc., Box 1422A, Janesville, WI 53545. Models.
Technilab Instruments, Inc., 6 Industrial Road, Pequannock, NJ 07440. Centrifuges
    and other equipment.
Thornton Associates, Inc., 87 Beaver Street, Waltham, MA 02154. Modular labs.
Ward's, Box 1712, Rochester, NY 14603. The Ward's catalog is a must.
Wheaton Scientific, 1000 North Tenth Street, Milville, NJ 08332. General.

GAMES

<u>The Cell Game</u>. Tombstone, Ariz.: Tecolote Press, 1974. Designed to familiarize
    students with the structures and functions of cell organelles and inclusions.
Newtown, D. <u>Biology Bingo</u>. Portland, Me.: J. Eston Welch, 1976. Four separate
    games: famous biologists; biological laws, theories, and principles; biological
    terms (two).
O'Neill, M. D. <u>Biological Puzzles and Puzzlers</u>. Portland, Me.: J. Weston Welch,
    1976.
U. S. Department of Agriculture. <u>Understanding the Game of the Environment</u>. Agri-
    cultural Information Bulletin No. 426, U. S. Forest Service, Washington, D.C.,
    1979. Don't be fooled by the source--it's highly creative and innovative.

SLIDES AND TRANSPARENCIES / FILM STRIPS AND LOOPS / AUDIO CASSETTES

<u>AAAS Cassettes</u>. c/o CEBAR Communications, Inc., 2735 Central Street, Evanston, IL
    60201. Tapes of recent AAAS annual meetings on subjects in the areas of agri-
    culture and ecology, biological science, energy, environment, medicine and health,

and science and public policy. $9 per cassette.

BFA Ealing Science Film Loops. Color Film Loops. BFA Educational Media, Box 1795, 2211 Michigan Avenue, Santa Monica, CA 90406. Areas include animal behavior, plant physiology, biochemistry, botany, classification, the animal kingdom, the plant kingdom, comparative anatomy, evolution, laboratory techniques, physiology, and reproduction and development.

Biology Enrichment Filmstrips/Cassette Program. Morristown, NJ: Silver Burdette, 1974.

Biology Media. 918 Parker Street, Berkeley, CA 94710. Audiovisual instructional materials for biology teaching, including slidetape modules, multi-image lectures, videocassettes, slide sets, and laboratory kits.

Biophoto Transparencies. Carolina Biological Supply Co., Burlington, NC 27215. High-quality reproductions of photographs selected for content and teaching value. Each transparency is accompanied by description sheets.

Carolina Filmstrips. Carolina Biological Supply Co., Burlington, NC 27215. Wide range of color strips, with accompanying teacher's manual, which contains a complete narrative and operator's code. More than twenty filmstrips in the series.

Oxford Slide Kits. Carman Educational Associates, Box 205, Youngstown, NY 14174.

Science Software Systems, Inc., 11899 West Pico Blvd., West Los Angeles, CA 90064. Write for catalog of AV instructional programs, including 33-mm color slides, sound filmstrips, and 16-mm motion pictures.

Scientific American Filmstrips. Scientific American, Inc., 415 Madison Avenue, New York, NY 10017.

35-mm Color Slide Programs. Science Software Systems, Inc., 11899 West Pico Boulevard, West Los Angeles, CA 90064. Combines live photography with art and graphics accompanied by a detailed study guide. Areas covered include animal behavior; animal sensory systems; biological history and evolution; cells, tissues, and their function; chemical mechanisms in geology; ecological science; reproduction and development; scientific methods and instrumentation; the animal world; and the human organisms.

USING FILMS IN BIOLOGY

For each section of the Purves and Orians text, several films are recommended below
for you to consider using in class. In addition to a brief annotation for each film
entry, the film distributor (abbreviated), year of release (if available), length of
film, and designation "color" or B&W" (black and white) are given in parentheses after
the film title. Addresses of the film distributors appear on page xxvii. Although
this film list is based in part on published reviews of recent biology films and on
the preparers' first-hand experience with films in biology, not all of the films have
been evaluated. It is therefore recommended that you preview those films you are
considering. In addition to helping you screen films for their appropriateness to
your course, the previewing process will allow you to prepare notes and follow-up
questions (to be answered in writing by your students, if you wish) that help focus
post-viewing discussion on important issues raised by the film. To counter the natural
tendency of students to respond passively to films, you might consider stopping films
at various points throughout their viewing and asking students to discuss what they
have just seen. Or you might use students' written answers to your follow-up questions as the basis for further class discussion.

ORDERING FILMS

It is essential that you plan your film schedule and place your orders well before the
dates of intended use, because the best films are likely to be in great demand. If you
are just starting to use films in your course, begin by familiarizing yourself with
the audiovisual department at your school. Most colleges not only offer equipment and
trained projectionists but run a film library as well. If this is the situation at
your school, all you may have to do is to tell the audiovisual department what you
need. If they don't own the film, they will go ahead and order it from the distributor. On the other hand, if your A-V department is more modest (or nonexistent), you
will have to take it upon yourself to order films, borrow a projector, and run the
projector yourself.

The film distributors listed below and on page xxvii fall into two general categories.
The majority are the actual producers of the film (such as CRM/McGraw-Hill or Encyclopaedia Britannica), whereas some simply purchase films from the producer and rent them
out (such as Indiana University). Since companies in the first group act as direct
distributors of their own films, they are usually happy to supply free copies of their
catalogs and to send films out, on loan, for previewing by prospective users. Should
you then decide to show one or more of their films, you (or your audiovisual department) may either rent or buy the film or films you have chosen. The companies in the
second group are in the rental business only, their margins are small, and most of
them charge for their catalogs and do not allow free previewing of films. However,
they also tend to charge less for their films than do the firms in the first group.

If you choose to make extensive use of films in your course and wish to supplement the films listed here, you may wish to order from the distributor or borrow from your library or audiovisual center publications from the following resources, in addition to those appearing on page xxvii.

Audio Film Center, 2138 East 75th Street, Chicago, IL 60649.
Blue Book of Audiovisual Materials. Educational Services and Audiovisual Guide, 434 Wabash Avenue, Chicago, IL 60605.
Educational Sound Film Directory, DuKane Corporation, Audio Visual Division, St. Charles, IL 60174.
Educator's International Guide to Free and Low-Cost Health Audio-Visual Teaching Aids, Pharmaceutical Communications, Inc., 15 Crescent Street, Long Island City, NY 11101.
Egan, R. S. (ed.), Topic-Aids: A Catalog of Instructional Media for College Biology. College Station, Tex.: Texas A&M, 1978.
Films in the Sciences (M. M. Newman and M. A. McRae, eds.), AAAS Sales Department, Dept. FL-A, 1515 Massachusetts Avenue, NW, Washington, DC 20005. (A comprehensive guide to approximately 1,000 science films produced between 1974 and 1980 for elementary, junior and senior high, college, and general audiences.)
Index to Biology: Multimedia, National Information Center for Educational Media, University of Southern California, University Park, CA 90007.
King Features, Inc., 235 East 45th Street, New York, NY 10017. Distributors of selected Nova episodes.
National Film Board of Canada, 1251 Avenue of the Americas, New York, NY 10020.
NICEM Index to 16mm Educational Films, University of Southern California, University Park, CA 90007.
Universal Education and Visual Arts, 425 North Michigan Avenue, Chicago, IL 61822.
University of Illinois, Audio Aids Center, Division of University Extension, Champaign, IL 61822.
University of Iowa, Audio-Visual Center, Iowa City, IA 52240.
University of Michigan, Audio-Visual Education Center, 416 4th Street, Ann Arbor, MI 48105.
University of Minnesota, Department of Audio-Visual Extension, General Educational Division, 2037 University Avenue, Minneapolis, MN 55455.
U. S. Government Films: A Catalog of Motion Pictures and Filmstrips for Sale by the National Audio-Visual Center, National Archives, GSA, Washington, DC 20409.

ANNOTATED FILM LIST*

Chapter 1   The Science of Biology

Life? (NMAC, 1976, 14½ min., color). Deals with common characteristics of all life, on both the macroscopic and the microscopic levels. Stimulates thought about life in an environment different from the earth's.
Biology: The Study of Life (IU, 18 min., color). General introductory film covering the scope of the science of biology.
A Question of Life (NAVC, 1977, 29 min., color). Emphasizes the Viking life-detection experiments and the significance of finding life elsewhere in the universe.
Living Machines (TL, 1980, 57 min., color). Examines the natural engineering found in a variety of animal and plant forms. For each of these natural designs (such as the effects of sun and wind on the shape of leaves), a researcher in the field describes the significance of the design and the research tools and methods used to study the phenomenon. Thus the film serves as an excellent introduction to scientific methodology applied to biological research and would be appropriate for viewing early in the course.
Notes of a Biology Watcher: A Film with Lewis Thomas (TL, 1981, 60 min., color). Part of the Nova series. "Every human being and virtually every living creature is, in a sense, owned and operated by legions of prehistoric organisms, hordes of them in each cell in the body." Thomas explains what he means.

Part I   The Cell

Charles Darwin (From the Great Scientists Speak Again series) (EMC, 1974, 24 min.,
    color). Professor Richard Eakin of the University of California at Berkeley,
    dressed in nineteenth-century costumes, impersonates Darwin and traces the
    scientist's discoveries, methods, theories, and personal philosophy.
Molecular Biology (COR, 1981, 17 min., color). Part of Coronet's Biological Sciences
    series, this beautifully produced and animated film deals with the sequence of
    events leading up to contemporary life and with aspects of cellular control.
    Current concepts of the origin of life are described.
The Living Cell: An Introduction (EBF, 1975, 20 min., color). The cell structure and
    biochemical processes are examined. The historical background of the discovery
    of the cell, the role of the nucleus, and protein synthesis are described and
    explained.
The Living Cell (H&R, 27 min., color). Cytological techniques are discussed, and an
    overview of cell structure and function is provided.
Cell Biology (COR, 1981, 17 min., color). From the Biological Sciences series. Traces
    the evolution of cells from early unicellular prokaryotes to modern multicellular
    eukaryotic forms. Highly creative visuals and accurate narrative combine to make
    this an excellent introductory film.
The Cell: A Functioning Structure. Parts I and II (CRM, 1972, 30 min. each, color).
    Organelle structure and function are covered well. Other topics included are
    microscopy, metabolism, reproduction, gene structure, and gene function.
Bags of Life (FI, 1981, 50 min., color). Focuses on investigations into the chemical
    and physical nature of the plasma membrane, with an emphasis on the experimental
    use of artificial membranes and liposomes to determine the nature of the membranes
    of living cells. Includes diagrams and excellent electromicrographs.
Photosynthesis Series: Chemistry of Food-Making and Biochemical Process (COR, 13½ min.
    and 16½ min., color). A general overview of the photosynthetic process is pro-
    vided in the first film, and the biochemical processes of photosynthesis, at a
    slightly higher level, are portrayed in the second film.
Riddle of Photosynthesis (AEC, 26 min., color). A thorough description of Melvin
    Calvin's work that led to insight into carbon dioxide function, complete with
    filming in Calvin's lab.
Plants Make Food, 2nd ed. (CF, 1981, 13 min., color). Uses animation and live-action
    photography to explain in simple terms the nature and significance of photosyn-
    thesis.
Photosynthesis, 3rd ed. (EBF, 1982, 20 min., color). This excellent film centers on
    the research of Melvin Calvin and Daniel Arnon into the nature of the carbon
    reduction cycle. Creative cinematography and time-lapse photography are combined
    with computer-generated animation to illustrate laboratory techniques used in
    studying the complex mechanisms of the light and dark reactions of photosynthesis.
Water in Biology (BFA, 21 min., color). The properties of water that make it essential
    to life are described.
A Is for Atom (AEC, 19 min., B&W). Overview of the structure of the atom, as well as
    identification of the practical use of atomic structures in our society.
Learning About Cells (EBF, 1976, 16 min., color). An introduction to the cell princi-
    ple, cell division, cell cooperation, and specialization, with a minimum of
    terminology and excellent photography of cultures in vivo. Notes the taxonomic
    approach to biology before the invention of the microscope and the history of cell
    biology.
Movements of Organelles in Living Nerve Fibers (NMAC, 1976, 12 min., B&W). Demonstrates
    the retrograde transport of small particles and mitochondria in nerve fibers.
    Suitable for medical, neurobiology, or college students and professors.
Cell Division: Mitosis and Meiosis (CRM, 1974, 20 min., color). Animation and cine-
    microscopy are combined to take the viewer inside living cells. Mitosis, meiosis,
    and fertilization are studied.
Mitosis (EBF, 1980, 14 min., color). Uses animation and time-lapse photography to
    trace the orderly process of cell division. Other topics covered include cloning
    and nuclear transplantation. An excellent introductory film.

Part II   Information and Heredity

Gregor Mendel (EMC, 1974, 24 min., color). Professor Richard M. Eakin of the University
    of California at Berkeley utilizes impersonation, including words, dress, and
    manner of the time, to convey Mendel's discoveries, methodology, theories, and

personal philosophy.

History of Genetics (General Genetics Series) (MF, 19 min., color). The history of genetics from Mendel to Müller is introduced skillfully. The processes of meiosis and mitosis are also described.

Chemistry of Heredity I: Identification of Genetic Material (COR, 1979, 14½ min., color).

Chemistry of Heredity II: Protein Synthesis (COR, 1979, 9 min., color). The first of this two-part series covers the experiments of Griffith, Avery, and MacLeod and McCarty on transformation; the isotope-labeling experiments of Hersey and Chase; the Watson-Crick model for DNA structure; and Meselson and Stahl's work on the nature of DNA replication. Part II follows the path of protein synthesis from nucleus to ribosome, including the nature of the genetic code, how it is established, and why it shows degeneracy.

Cracking the Code of Life (ACS, 24 min., color). Uses animation and examples to thoroughly explain the structure and function of DNA. The relationship of cancer to DNA coding is explained.

The Living Cell: DNA (EBF, 1976, 20 min., color). Using Syntha Vision, which stimulates microphotography, the film describes how DNA codes and directs the functioning of an organism.

DNA (McG, 11 min., color). Models, diagrams, and photomicrographs are used to introduce the structure and replication of DNA.

Protein Synthesis: An Epic on the Cellular Level (H&R, 10 min., color). Stanford students, Afro-jazz music, and narrator Paul Berg simulate protein synthesis.

Life Cycle of a Bacteriophage (EMC, 1961, 1 min., B&W). The complete life cycle of a bacteriophage from contact with the bacterial host to reproduction of several hundred virus particles.

The Chromosomes: General Considerations (Chromosomal Series) (MF, 14 min., color). The normal human karotype preparation, nomenclature, normal variations, and aberrations are demonstrated and/or described.

Enzyme Defects and DNA (General Genetics Series) (MF, 15 min., color). Both the one-gene/one-polypeptide concept and DNA as the genetic material are emphasized. DNA structure and replication, as well as clinical symptoms of genetic diseases, are surveyed.

Autosomal Recessive Inheritance (MF, 1976, 8¼ min., color). Albinism, a lack of body pigmentation in humans, serves as an illustration of inheritance patterns.

Chromosomal Errors (MF, 1976, 10 min., color). A genetic counseling situation provides the setting in which to introduce meiosis, mitosis, fertilization, and fetal development. Human chromosomal defects such as Down's syndrome, trisomy 13, and Klinefelter's syndrome are discussed.

Genetic Defects: The Broken Code (IU, 1976, 90 min., color). The genetic defects of cystic fibrosis, Huntington's chorea, hemophilia, combined immune deficiency, and sickle-cell anemia--common to 7 percent of the population of the United States--are discussed.

Genetics and You (MF, 1976, 11 min., color). The ethical issues of cloning, controlled mating, breeding, and test-tube babies are presented.

X-Linked Inheritance (MF, 1976, 11½ min., color). Color blindness is used to demonstrate the inheritance of a recessive trait.

The Sickle Cell Story (MF, 1976, 16 min., color). The historical background and physiology of sickle-cell disease are described. A distinction is made between sickle-cell anemia and sickle-cell trait.

Sickle Cell Fundamentals (NIH, 1978, 30 min., color). This two-part film describes both the genetics of and the medical problems of individuals with sickle-cell disease.

The Blueprint of Life (MF, 24 min., color). Using a human case study of a family with a Down's syndrome child, this film describes and explains human inheritance for such defects as Down's, Klinefelter's, harelips, and so on.

Autosomal Dominant Inheritance (MF, 1976, 12 min., color). Achondroplasia, a form of human dwarfism, is used to explain this inheritance pattern.

The Effect of Viruses on a Cell Line of Human Origin (PD, 28 min., color). The distinct effects of four viruses on a human cell line are demonstrated through use of photomicrographic techniques and tissue cultures.

Chromosomal Errors (MF, 1977, 10 min., color). Follows a middle-aged expectant mother, who already has one Down's syndrome child, through a genetic counseling situation. Examines widely used counseling procedures, such as amniocentesis and karytone analysis, and other aneuploid conditions, both autosomal and sex-linked.

The Gene Engineers (TL, 1977, 57 min., color). A Nova film tracing the controversy surrounding the development of gene-splicing technology from the original voluntary ban promulgated by Paul Berg to the Cambridge (Massachusetts) City Council

hearings in 1976. Includes interviews, shot on location in laboratories, with
  many of the world's great molecular biologists.
Genetics: Man the Creator (DA, 1977, 19 min., color). Reports on future possibilities
  in genetics, such as human sperm and egg banks, artificial wombs, the creation of
  human-animal hybrids, asexual reproduction, and cloning, with footage of lab work
  in each area. Also notes important advances in the prevention of birth defects.
Evolution by DNA: Changing the Blueprint of Life (DA, 1977, 23 min., color). Dr. David
  Suzuki of Columbia University and his assistants are shown in the lab drastically
  changing the life of Drosophilia by genetic engineering. Suzuki represents a new
  breed of scientists concerned with the ethical and political consequences of their
  work and attempting to minimize the misuse of genetic knowledge.
Cancer: The Wayward Cell (DA, 1976, 24 min., color). Designed to remove fears and
  myths about cancer. Current research on the diagnosis, causes, prevention, and
  cure of cancer are surveyed.
Body Defenses Against Disease (EBF, 1978, 14 min., color). Graphically depicts the
  several lines of defense the body uses against disease, including mucus and
  ciliated epithelial cells of the respiratory system, the skin, digestive juices,
  phagocytes, antibodies, killer lymphocytes, and antibiotics. The film begins
  with a look at what life would be like without natural immunity.
Immunization Now (AEF, 1976, 20 min., color). The history and need for immunization
  are explained.
The Immune Response (CRM, 1980, 20 min., color). Cellular and subcellular levels of
  the human defense system against disease are presented.
Visceral Organ Transplants (UP, 30 min., color). Deals with three major problems
  involved in the transplantation of visceral organs: surgical, immunological, and
  supply. Profuse animation is used to simplify the explanation of the rejection
  mechanism.

Part III   Multicellular Life

Liverwort: Alternation of Generations (COR, 15 min., color). The complete cycle from
  gametophyte to sporophyte and back to gametophyte is traced, and environmental
  factors are discussed.
Development in Volvox (NAVC, 1977, 15 min.). Focuses on the developmental stages of
  male, female, and asexual embryos in V. carteri f. nagaiensis.
Encyclopaedia Britannica Plant Life Series (EBF, 10-20 min., color). Included in this
  series are films on green plants, nongreen plants, and plant structures.
Flowers: Structure and Function (COR, 10½ min., color). The specialized parts of a
  flower are discussed in relation to their function of pollination and seed
  production.
The Many Worlds of Nature: Tree Blossoms (SS, 1980, 16 min., color). Discusses and
  illustrates monoecious, dioecious and perfect flowers, citing appropriate trees
  as examples.
Plant Tropisms and Other Movements (COR, 10½ min., color). This film studies three
  kinds of plant movements and nutational movements.
The Green Machine (TL, 1978, 49 min., color). Treatment of such topics as plant growth
  and development, phytochrome and photosynthesis, biological rhythms, and plant
  hormones is slick and well done, though the final section of the film (primitive
  brains) is a bit sensational.
Carnivorous Plants (SF, 1978, 11 min., color). Plants such as the Australian pitcher
  plant, sundew, Venus's flytrap, and aquatic bladderwort are featured to illustrate,
  through the use of time-lapse and close-up photography and photomacrography, the
  basic types of trapping mechanisms.
Man, the Incredible Machine (NGS, 1975, 16 min., color). An intriguing, well-made film
  that focuses on human systems: digestion, respiration, speech, motion, the
  skeleton, muscles, skin, hearing, and the heart and circulation.
Encyclopaedia Britannica Human Life Series (EBF, 10-20 min., color). The films in this
  series cover many body functions, including those of the circulatory and excretory
  systems.
Incredible Voyage (McG, 26 min., color). Describes the endoscope, a tiny instrument
  that makes it possible to film inside the human body, and then shows some fascinating "trips" through the circulatory system, respiratory system, heart, digestive
  system, skeletal system, and brain.
Regulating Body Temperature (EBF, 1972, 22 min., color). Demonstrates how physiological and behavioral mechanisms regulate body temperature in mammals.
The Animal and the Environment (McG, 28 min.). Investigates the role of homeostatic

mechanisms as they affect the functioning of various elements in the transport systems of higher animals.

Respiration in Man (EBF, 26 min., color). Animation, photomicrography, x-rays, and laboratory experiments are used to depict the structure, function, chemistry, and physiological controls of the respiratory system. Carbon dioxide and oxygen exchange in the lungs is examined. The effects of air pollution on respiration and the body are discussed.

The Respiratory System (CE, 1976, 9 min., color). More of an anatomical than a physiological treatment of the dissection of a rat's respiratory system. Includes diagrams and some demonstrations, such as inflation of the lungs, but does not cover the mechanics of breathing and the physiological control of the rate of respiration.

Smoking/Emphysema: A Fight for Breath (CRM, 1975, 12 min., color). Real-life scenes and animation show the entire respiratory system, the difference between diseased and healthy lungs, and the microscopic activity of respiratory structures.

I Am Joe's Heart (PYR, 1972, 25 min., color). Combines live color sequences with three-dimensional animation to create a vivid understanding of the heart.

Blood (EBF, 16 min., B&W). Live action, animation, and photomicrography illustrate the composition of blood, its circulation, and the work it does to support cell processes. Explains blood typing and the Rh test.

The Lymphatic System (IFB, 1979, 15 min., color). Accurate, straightforward presentation of the lymphatic system of mammals.

The Human Body: Digestive System (COR, 1980, 15½ min., color). Depicts the transformation of food as it traverses the mouth and gut to the cells wherein assimilated food molecules are metabolized. Includes some interesting x-ray cinematography of chewing, swallowing, and peristalsis.

The Digestive System (CE, 1976, 7 min., color). Follows the dissection of a laboratory rat, with some animated diagrams in addition, to describe the anatomy and physiology of the vertebrate digestive system. Accompanied by a booklet with vocabulary, text, discussion topics, and suggestions for further investigations.

Work of the Kidneys (EBF, 1972, 11 min., color). Uses animation and microcinematography to show the many functions of the kidney and urinary system. Explains how the kidneys maintain uniformity in the blood and tissues, how they keep water content, salt concentration, acidity, and osmotic pressure at optimum levels, and how they eliminate wastes.

Developmental Biology (COR, 1981, 17 min., color). Another in Coronet's excellent Biological Sciences series, this film blends molecular, cellular, and organismic biology in a visually exciting and scientifically accurate survey of recent research into differentiation and morphogenesis, including a discussion of cloning.

Nuclear Transplantation (IU, 1976, 12 min., color). Outlines the technique for transplanting nuclei, from collecting fertilized eggs to the removal and transplantation of their nuclei into unfertilized eggs activated by electric shock following ultraviolet destruction of their own nuclei. Must be approached with a background in developmental biology and the potentiality of nuclei from blastula stage cells.

How Life Begins (CRM, 1980, 45 min., color, 2 parts). An ABC News special. Illustrates reproduction in fish, birds, and mammals.

When Life Begins (CRM, 1980, 14 min., color). Live photography within the uterus during fertilization and various stages of development of the human fetus, ending with a natural childbirth.

The Miracle of Life (PYR, 1977, 12 min., color). Closeup views of mammalian gamete formation, fertilization, and early embryonic development through the use of cinephotomicrography with an endoscopic lens. Outstanding visual work.

The Fabric of Life (MGD, 1981, 24 min., color). Illustrates the genetic control of differentiation and morphogenesis, using the human embryo as a major subject but also including nuclear transplant experiments in frogs, embryogenesis of celery in tissue culture, and chromosome puffs in Drosophila for further illustration.

Development and Differentiation (CRM, 1974, 20 min., color). Traces the embryogenesis and the processes of development in the frog embryo.

Chick Embryology (NGS, 1974, 12 min.). A brief presentation of embryonic development from day one to hatching, including the development of characteristic chordate structures and their role in vertebrate evolution, nutritional and respiratory functions, and the dynamic aspects of the organ system.

The Reproductive System (CE, 1976, 7 min., color). Dissections of abdominal cavities of mature cycling female rat, pregnant female rat, and mature male rat. No detailed description of functions, hormonal controls, or physiology of the various parts.

Human Reproduction, 3rd ed. (CRM, 1981, 20 min., color). One of the best general
    introductions to the subject, including discussions of the human male and female
    reproductive systems, fertilization, development, and parturition. Employs a
    wide variety of techniques--animation, electron microscopy, drawings, and so on--
    to illustrate stages in development.
Beginning of Life (BF, 1975, 30 min., color). Traces prenatal development to birth
    with stunning photography.
The Miracle of Life (PYR, 1975, 12 min., color). Live-action film shot through micro-
    scope photography and showing the processes of fertilization, cell division, and
    growth to the first heartbeat.
The Body Human: The Miracle Months (TC, 50 min., color). Captures for the first time
    on film the earliest phases of human development, including ovulation, a living
    forty-day embryo, and the world of a twelve-week fetus as its hands and feet are
    taking shape. Glimpses of development through thermography and ultrasound
    pictures.
Birth Control: The Choices (CF, 1976, 25 min., color). Methods of birth control are
    presented along with their uses, limitations, and side effects. Also included are
    discussions of vasectomy, tubal ligation, and abortion.

Part IV   Integration and Behavior

Prostaglandins: Tomorrow's Physiology? (UP, 22 min., color). A status report that
    focuses on the origin, development, theoretical mode(s) of action, and potential
    clinical applications of these powerful and fascinating hormones.
The Life Cycle of Insects: Complete Metamorphosis (IFB, 1977, 20 min., color).
    Describes and illustrates reproductive and metamorphic processes in a selection
    of insects, primarily the butterfly.
The Life Cycle of Insects: Incomplete Metamorphosis (IFB, 1977, 17 min., color).
    Locusts, grasshoppers, stick insects, cockroaches, aphids, and dragonflies serve
    as examples of insects that develop by incomplete metamorphosis. The film focuses
    on the life cycle of the locust, with detailed narration of concepts depicted.
The Ionic Basis of Action Potential (NAVC, 1977, 11 min.). Computer-animated film
    describes the dynamic relationship of channel activity, ion movement, and charge
    separation at a nerve's axon membrane during an action potential.
The Nerve Impulse (EBF, 1970, 21 min., color). Straightforward physiological descrip-
    tion of neural activity, all-or-none activity, and synaptic function.
Fundamentals of Nervous System (EBF, 1961, 16 min., color). Photography, animation,
    and medical demonstrations show functions of the nervous system. Laboratory
    experiments help to clarify concepts.
The Nervous System in Man (IU, 1966, 18 min., color). Interactions of nervous and
    endocrine systems; types of neurons; functions of various parts of the brain;
    reflexes; divisions of the nervous system.
Automonic Nervous System (IFB, 1976, 17 min., color). New techniques vividly illustrate
    the functioning of such automatic structures as the stomach, bladder, and lungs.
The Peripheral Nervous System (IFB, 1977, 19 min.). Demonstrates an infant's reflex
    action, then explains the components of the spinal reflex arc and the importance
    of conscious bodily control. Covers the nature of the nerve impulse, examined in
    micrographs and animation, and introduces the transmission rate of myelinated
    nerves. Students in the lab are shown determining the velocity of transmission
    of nerve impulses in a frog.
Frog Skeletal Muscle Response (BFA, 8 min.). Shows normal muscle response as well as
    tetany and fatigue as registered on a kymograph.
The Human Body: Muscular System (COR, 1980, 22 min., color). The main functions of
    skeletal, smooth, and cardiac muscles are shown in animation in this brief, basic
    film.
Muscle: A Study of Integration (CRM, 1972, 25 min., color). Types of muscle, the
    properties of each, and the mechanics of muscle contraction are presented, as well
    as the accompanying role of the nervous system.
Drugs and the Nervous System (CF, 18 min., color). Explains how drugs affect the
    central nervous system and other organs. Discusses stimulants, depressants,
    hallucinogens, and narcotics.
Marvels of the Mind (NGS, 1980, 23 min., color). Brain structure and function are
    matched in this fairly brief National Geographic film. Communication between
    neurons, stimulation of the limbic system, split-brain surgery, acupuncture
    anesthesia, and the brain's own pain killers (the endorphins and enkephalins)
    are also covered.

Mysteries of the Mind (NGS, 1980, 59 min., color). A National Geographic special that uses superb photography and clear, accurate narrative to provide a general introduction to the structure and function of the human brain at a level appropriate to college students. Explores several topics of recent investigation, including narcolepsy, meditation, hypnosis, acupuncture, and the role of endorphins.

Mind of Man (IU, 1971, 119 min., 4 parts, color). Explores recent discoveries concerning the powers and workings of the mind. Considers control of "involuntary" functions, sleep and dreaming, drugs, infant mind development, and brain structure.

The Hidden Universe: The Brain (CRM, 1977, 45 min., 2 parts, color). An overview of the functions of the brain and some of the latest research results. Includes "split brain" studies, biofeedback, and some diagnostic and treatment methods of brain malfunctions.

The Keys of Paradise (TL, 1979, 57 min., color). A Time-Life film centering on the nature and function of endorphins and enkephalins, the brain's natural pain killers. A well-made presentation, though already slightly dated due to the rapid changes in this field.

Colors and How We See Them--Color Vision (SSS, 1976, 20 min., color). This film is based on the Young-Helmholtz three-component color vision theory and shows in detail how and why the rods and cones of the retina relay their specific messages to the brain.

The Skin as a Sense Organ (IFB, 1976, 12 min., color). The sensory receptors of the skin are microscopically examined.

The Sensory World (CRM, 1971, 33 min., color). Unusual animation sequences demonstrate clearly how the eyes, ears, sense of touch, and proprioception systems work.

The Human Eye (IFB, 1978, 14 min., color). The human optical system is compared with that of other animals and with the camera. Good, brief introduction to eye structure and physiology.

What Time Is Your Body? (TL, 1976, 23 min., color). A study of human circadian rhythms and their biological clock.

Biological Clocks (GC, 25 min., color). Fascinating study of how animals, including humans, regulate their lives according to natural rhythms.

Biological Rhythms: Studies in Chronobiology (EBF, 1977, 22 min., color). Focuses on circadian rhythms, using skillful photography and time-lapse sequences to study time zone changes on fiddler crab coloration, the effect of clock-shifting on pigeons' homing interest, moth-pupa transplantations, augiogenic responses in mice, and other chronobiological phenomena.

Bird Brain: The Mystery of Bird Navigation (TL, 1976, 27 min., color). A study of the mechanism of bird migrations.

Animal Communication (TL, 30 min., color). Looks at the variety of signals used by a dozen species, from insects to baboons.

The Mystery of Animal Behavior (NGS, 52 min., color). Portrays instinctive and learned behavior of animals, including courtship displays, reproductive sequences, and a bullfinch that has been taught a German folk song.

Animal Migration (BFA, 1977, 12 min., color). Introduction to animal migration using several species of birds, insects, fishes, and mammals. Includes theories of motivation, orientation, and navigation techniques.

Horses Without Man (FI, 1981, 50 min., color). This film uses three populations of free-ranging horses--the New Forest ponies, the white horses of France, and the ponies of North Carolina's Shackleford Islands--to explore the adaptive aspects of the social structure of horse herds. Included are sequences on the general mating behavior of horses, on male and female choice, and on the establishment of a new harem by a young stallion.

The Fruit Fly: A Look at Behavior Biology (CRM, 1973, 21 min., color). Examines the relationship of genes to behavior in the classical sense, using Drosophila as a test subject.

Strategy for Survival: Behavioral Ecology of the Monarch Butterfly (H&R, 1977, color). Basic life cycle from egg to adult monarch butterfly serves as an introduction to an analysis of the monarch's adaptations to its environment. Also demonstrates scientific methods and motivations for study. Covers the monarch's migrations, defenses, predators, habitats, biology.

The Guanaco of Patagonia: A Study of Behavior and Ecology (IFB, 1981, 28 min., color). This fine film for illustrating and discussing territoriality, social communication, and the adaptive significance of social systems uses the guanaco, cousin of the llama and alpaca, as its subject. The narration is accurate and highly informative (if slightly overstated), and the film is visually appealing.

Now You See Me, Now You Don't (AMP, 1977, 20 min., color). Covers the spectrum of
     protective strategems found among insects, from mimicry and camouflage eye spots
     to osmetria and noxious gas. Shots of feeding and predation, minimum of techni-
     cal terms.
Animal Camouflage (BFA, 9 min., color). Camouflage is seen to take three forms:
     protective coloration, protective resemblance, and mimicry. Film both provides
     specific animal examples of the three types and reveals how the animals' behavior
     is related to camouflage and its purpose.
Color in Nature (EBF, 1980, 12 min., color). Examines the role of color in sexual
     reproduction, feeding, defense, and other behaviors. A high-quality film.
Protective Coloration (SS, 12 min., color). The ways in which some animals use color
     to aid them in avoiding predators.
Jackson's Chameleon: Locomotion and Prey Capture (UCLA, 1977, 13 min., color).
     Describes chameleon biology and behavior, emphasizing the animal's various
     specializations for arboreal life. Slow-motion sequences, still photographs, and
     a diagram of tongue anatomy detail and clarify the facts presented.
The Predators (STP, 1977, 50 min., color, study guide). States the case for the
     predator from the standpoint of the predator, including the threat to the survival
     of many "unsuccessful" predators. Includes catch-and-kill sequences.
Come into My Parlour, Said the Spider... (AMP, 1977, 20 min., color). Illustrates
     some of the complex and highly specific predatory habits of spiders, covering a
     variety of spiders and exploiting exceptional photographic techniques.
Kingsnake Predation on Rattlesnakes (UCLA, 1978, 10 min., color). Demonstrates the
     major features of kingsnake predation on rattlesnakes.
Predatory Behavior of Snakes (UCLA, 1978, 16 min., color). Illustrates various feeding
     specializations found in snakes. Informative, well-organized, and visually
     superb.
Nesting Time (AVED, 1976, 26 min., color). Covers the diversity of habitats and
     feeding habits of a number of birds (ducks, grebes, shore birds, loons, hawks,
     owls) during their nesting periods.
The World's Largest Nest: The Sociable Weaverbird (UCLA, 1977, 10 min., color).
     Exposition on the weaverbird of the Kalahari Desert, presenting its life history,
     discussion of the construction and advantages of the next, and nest-building
     behavior as a unique adaptation that can be contrasted with the habits of the
     English sparrow, a close relative.
Why Do Birds Sing? (TL, 1976, 27 min., color). Introduces viewers to a number of
     scientists and their discoveries about the function and causes of bird song and
     to the methods ethologists use in both field and laboratory experiments. Sound
     spectrograms are presented in animation while the calls are heard.
Sea Turtles (TL, 1976, 13 min., color). Features the nesting habits of the Ridley
     turtle and the migration of the young turtles from the hatching grounds back to
     the sea. Depicts various hazards of the process and notes the relation of the
     nesting habits to the survival of the species.
Jane Goodall: Studies of the Chimpanzee (NGS, color).
     Feeding and Food Sharing (1976, 23 min.).
     Hierarchy and the Alpha Male (1977, 23 min.).
     Infant Development (1976, 23 min.). Following the development of specific chimps,
          shows dimensions of the matrifocal group and the influences of heredity and
          learning on maternal behavior, the relationship among siblings, and the
          long-term maintenance of the mother-son dyad.
     Introduction to Chimpanzee Behavior (1977, 23 min.).
     Tool Using (1976, 23 min.).
Care of the Infant--Animal and Human (PE, 1977, 22 min., color). Stresses the survival
     factor of parental care and protection and shows scenes of parent-child interac-
     tion in several animal species in the wild, notably elephants; brief treatment of
     humans.
Sociobiology (DA, 1977, 20 min., color). Harvard University biologists speak about
     their special fields: Robert Trivers discusses his theory of sex-determined
     behavior; Irven Devore presents the competitive drive for status among males of
     any species and the more probable survival of the strongest's genes; and Edward
     O. Wilson's zoological work is explained and illustrated with laboratory scenes.

Part V   The Diversity of Life

Diversity of Life (IU, 1981, 27 min., color). One of the finest general introductions
     to the variety of life forms on earth. Illustrates over 50 species within the

framework of the five-kingdom system of classification and covers both general
evolutionary advances and unusual specializations.

Dive to the Edge of Creation (NGS, 1980, 59 min., color). Presents the fascinating
voyage of the submersible vessel Alvin to deep-sea rifts off the Galapagos,
where the heat and nutrients produced by molten rock seeping from the spreading
ocean floor have produced an ecosystem unlike anything else in the world. Incredible, oversized tube worms and other types of marine life that have evolved in
this habitat are captured on film.

The Invisible World (NGS, 1980, 59 min., color). A National Geographic special that
uses slow-motion photography, time-lapse photography, stroboscopic lighting,
microphotography, astronomical photography, ultraviolet and infrared photography,
x-ray, scanning electron microscopy, and other methods to illustrate creatures
and phenomena that are either too big, too small, too fast, too slow, or otherwise outside our normal sensory capacity for us to detect them unaided. A good
film for showing early in the course.

Animal Classification (AVED, 11 min., color). Explains the necessity of the scientific
classification of plants and animals. Covers the work of Linnaeus and the Linnaean
hierarchy leading to modern classification. Uses the horse as an example.

Protozoa (One-Celled Animals) (EBF, 11 min., color). Presents a close-up exploration
of the world of one-celled animals. Provides examples of symbiosis, parasitism,
and colonial organizations.

Protists: Threshold of Life (NGS, 1974, 12 min., color). Study of the life in a single
drop of pond water.

Plankton (NGS, 1976, 12 min., color). Exploration of the microscopic life of the sea
and the role it plays in the ocean's food webs.

Microorganisms That Cause Disease (COR, 10½ min., color). Fungi, bacteria, viruses,
rickettsiae, and protozoa are studied using animation and photomicrographs.

Bacteria (EBF, 19 min., B&W). Unusual photomicrography reveals the anatomy of bacteria
and the processes of conjugation and cell division. Film shows different kinds of
bacteria, how they feed by enzyme activity, and how they are used in the study of
heredity.

Kingdom Protista Series (WNSE, 1975, color).

A New Look at Leeuwenhoek's "Wee Beasties" (12 min.). Shows the movement of one-
celled organisms (protozoa and algae) with microphotography, occasionally
quoting Leeuwenhoek's observations.

Protist Behavior (11 min.).

Protist Ecology (12 min.).

Protist Physiology (13 min.).

Protist Reproduction (10 min.).

Algae (IU, 17 min., color). Time-lapse cinematography shows major varieties, size
ranges, and habitats of algae, as well as actual examples of fission, spore
release, fragmentation, and daughter-colony formation.

Evolution of Vascular Plants--Ferns (EBF, 17 min., B&W). Illustrates how ferns can be
used to investigate the evolution of basic life processes. Photomicrography,
animation, and live photography are used to show the development of vascularity
in fern sporophytes and gametophytes, and to show how vascularity has developed
in the sporophyte generation.

Gymnosperms (EBF, 16 min., B&W). Animation and time-lapse photography trace the life
cycle of pine through growth of pollen and seed cones, pollination, fertilization,
and germination of seeds.

Angiosperms--Flowering Plants (EBF, 21 min., B&W). Animation shows the complete
fertilization process from pollination through development of the embryo.

Carnivorous Plants (SF, 1981, 11 min., color). Exceptional time-lapse photography
of various carnivorous plants: cobra lily, pitcher plant, sundew, cephalotus,
Venus's flytrap and bladderwort.

Fungi (EBF, 16 min., B&W). Time-lapse photography, photomicrography, and animation
show how fungi grow and how they get their food; their importance as converters
of dead matter; and their economic significance as sources of food, as agents in
food processing, and as causes of disease in plants and animals.

Ants and Aphids (AVED, 1976, 26 min., color). Discusses gall-forming insects, such as
wasps and aphids; describes the relationship of some ants and aphids, with special
details like the live birth of an aphid; focuses on ants and ecological aspects of
their underground communities.

The Life of the Honeybee, 2nd ed. (EBF, 1980, 24 min., color). Portrays the complete
life cycle of the honeybee, including such activities as nectar and gathering
pollen, constructing combs, raising larvae, defending the colony, swarming, and
communicating within the hive.

Kingdom Animalia Series (WNSE, 1976, color).
    Annelids (10 min.).
    Arthropods (13 min.). Integrates basic concepts of arthropod ecology with general
        classifications; presents a description of each class with a list of known
        species and pictures of common representatives.
    Coelenterates (10 min.).
    Echinoderms (10 min.).
    Flatworms (10 min.).
    Molluscs (10 min.). Depicts chief morphological features, feeding methods,
        locomotion, and breeding habits of certain molluscs, mostly common or soft-
        bodied.
Insects Helpful to Man (IFB, 1977, 17 min., color). Comprehensive survey of insects
    that are commercially, agriculturally, and ecologically useful to humans, high-
    lighting the honeybee and the silk moth and detailing both commercial processes
    and insect biology. Also considers insect diversity and major insect scavengers
    and predators.
Insects Harmful to Man (IFB, 1977, 16 min., color). Examines a wide range of insects
    having a negative health or economic impact on humans. Discusses insects that
    transmit diseases directly affecting public health (mosquitoes) and those that
    indirectly contaminate food (flies). Economically pernicious insects include
    animal parasites and wood, agricultural, and granary pests.
Sharks: Terror, Truth, Death (FI, 1976, 25 min., color). Narrated by Jaws author
    Peter Benchley, this film explores current attitudes of fear and aversion toward
    sharks, brings out popular misconceptions, and presents nonaggressive species and
    their habitats.
Kodiak Island (TL, 1976, 16 min., color). Includes in-stream shots of salmon, descrip-
    tion of the final stages of the salmon life cycle with a review of the physical
    changes occurring immediately before and after spawning, the effect of the fishery
    on the size and behavior of the Kodiak brown bear and on scavengers like the gull,
    magpie, fox, and eagle.
Amphibians: What, When, and Where (IFB, 1976, 14 min., color). This realistic descrip-
    tion of the natural history of amphibians focuses on reproductive habits, meta-
    morphosis, and care in captivity. Highlights ethics of conservation and encourages
    respect for all forms of animal life.
Komodo Dragons and Monitor Lizards (Mac, 1976, 22 min., color). Depicts characteris-
    tics and environment of Komodo dragons and other monitor lizards, showing reac-
    tions with others in their class and with their special environments. Includes
    reasons for their near-extinction and shows population maintenance work.
Deadly African Snakes (TL, 1976, 15 min., color). The essential role of poisonous
    and constrictive snakes in the forest and grassland of Africa is demonstrated by
    showing snakes in battle and hunting. Ingestion sequences detail the process of
    jaw disassociation; the role of venom in making serum is discussed; and the
    milking process is demonstrated.
Return of the Winged Giants (AVED, 1976, 15 min., color). Describes a subspecies of
    the Canada goose: relates its return from near-extinction, migratory habits, life
    cycle, mating and nesting behavior, defenses against predation and natural disas-
    ter, and early learning. Slow-motion sequences show geese in flight.
Osprey (BEAF, 1980, 34 min., color). William Wylie, who reviewed this film for
    Science Books and Films (March/April 1981, pp. 226-227), called it "undoubtedly
    the finest, most beautiful and best produced film that I have ever had the
    privilege of viewing." The film portrays the life cycle of ospreys from the Loch
    region of Scotland. The photography is unbelievable and the biology first-rate.
Spirits of the Wing (MDF, 1979, 32 min., color). The natural history of the raptors--
    birds of prey. Includes feeding and nesting habits and the role of raptors in
    the food web of the mountain country of Montana.
Raptors: Birds of Prey (BFA, 1976, 14 min., color). Describes various species of rap-
    tors, their habitats, sizes of offspring, food-gathering techniques, problems of
    population control, pesticide accumulation, and loss of living spaces.
The Pocket Gopher: Adaptations for Living Underground (UCLA, 1977, 13 min., color).
    Presents the adaptations of the pocket gopher for fossorial living, using live
    animals in laboratory viewing chambers, prepared skeletons, and hand-held animals
    to demonstrate the digging and tunneling habits and anatomical specializations of
    Thomomys botanae.
Saga of the Sea Otter (PICT, 1976, 26 min., color). Presents otter biology, behavior,
    and tool usage and focuses on the competitive interrelationships among sea urchins,
    kelp, abalone, otters, and abalone fishermen.

The Great Whales (NGS, 1978, 59 min., color). This National Geographic special, which
first appeared on public television, covers many aspects of current cetacean
research and attempts to place our changing conceptions and attitudes of whales
in a historical framework. Disturbing, informative film.

Cry Wolf! (TL, 1976, 15 min., color). Covers the timber wolf's lifestyle from vocali-
zation through feeding habits, social relationships, reproduction, and growth to
final scenes depicting the actual nonaggressive relationship between humans and
wolves. Birth scenes are filmed in an actual den, and the wolves' impact on deer
and caribou populations is noted.

The Lion (Mac, 1976, 22 min., color). General-awareness film depicting lions' behavior,
physical characteristics, adaptation to the environment, and disruption by human
intervention. Also discusses the efforts being made toward maintenance of the
species.

Encyclopaedia Britannica Animal Life Series--Vertebrates (EBF, 10-20 min., color).
Included are films on vertebrate evolution, fish, amphibians, reptiles, birds,
and mammals.

Etosha: Place of Dry Water (NGS, 1981, 59 min., color). This National Geographic
special focuses on life in the Etosha National Park of Namibia, where highly
variable rainfall results in a cyclically appearing and disappearing lake that
may reach 50 miles across during the height of the rainy season. The adaptations
of various animal forms to these changing conditions are the subject of this
visually stunning film.

The Living Sands of Namib (NGS, 1978, 59 min., color). Life in one of the driest
areas on the earth is examined in this National Geographic special. The adaptations of plants and animals to these conditions are dealt with beautifully.

The Indiana Dunes (IFB, 1981, 14 min., color). The history of the Dunes, from the
formation of Lake Michigan to the present, with an emphasis on modern flora,
ecological relationships, and the dangers posed by the human presence.

Island of the Moon (Mac, 1981, 28 min., color). Offers a view of the unique flora and
fauna of the island of Madagascar, with emphasis on the evolution of the lemur.

Memories from Eden (TL, 1978, 57 min., color). Both the length and theme of this
beautiful film may preclude its use in an introductory biology course, but its
penetrating examination of the function of zoos in our society is guaranteed to
generate class discussion. Its message: zoos solely for the entertainment of
people are obsolete, and more must be done to preserve what little of nature is
still left.

Evolution and the Origin of Life (CRM, 1972, 36 min., color). This film does an
excellent job of tracing the chemical evolution of life on earth and Darwin's
theory of evolution.

The Origin of Life (EBF, 11 min., color). Oparin's and Miller's work are included in
this presentation on the chemical evolution of life on earth.

Archeological Dating: Retracing Time (EBF, 1976, 18 min., color). The relative and
absolute dating techniques used by archeologists are described. These include
tree rings, shifts in the magnetic north pole, and radioactivity levels.

Dating Game: How Old Are We? (EMC, 29 min.). Documents the work of Jeff Bada, whose
development of an archeological dating technique based on the reaction of amino
acids in an organism after its death offers a tool that can penetrate prehistory
beyond the 40,000-year limit of the radio-carbon dating method.

The Apes (FF, 1976, 12 min., color). General overview of apes and monkeys, filmed
principally in zoos with simulated natural backgrounds; no discussion of apes'
place in the scheme of primate physiology.

The Coming of Man (McG, 13 min., color). The evolution from primate to modern humans
is presented.

Ascent of Man Series: Lower than the Angels (TL, 1974, 52 min., color). Jacob
Bronowski demonstrates how a number of anatomical and cultural adaptations inter-
acted to produce modern humans.

Evolution of Man (COR, 13½ min., color). Human evolution is traced from Darwin's
theory through fossil evidence that supports the theory of human evolution.

Part VI   The Strategy of Evolution

Genetic Polymorphisms and Evolution (MF, 1976, 16 min., color). The British moth
Biston betularia and sickle-cell hemoglobin in humans demonstrate the role of
natural selection in producing these polymorphisms.

Natural Selection: Evolution at Work (MGD, 1981, 24 min., color). Focuses on two
examples of evolution in progress: the development of resistance to the poison

warfarin by rats and the development of tolerance of meadow grass to copper mine
runoff. In each case the issue is clearly stated and the use of scientific
methodology in arriving at an explanation is demonstrated.

Evolutionary Biology (COR, 1981, 17 min., color). Gives a general and fairly elementary treatment of natural selection, adaptation, sexual reproduction, and other
aspects of evolution, with some excellent photography and unusually clear narrative. Part of Coronet's Biological Sciences series.

Skeletal Adaptations: Variations on a Theme (MGD, 1981, 24 min., color). Illustrates
the concepts of convergence and divergence through analysis of the skeletal
adaptations of mammals, birds, and fish.

Galapagos: The Enchanted Islands (JF, 1981, 25 min., color). Part of Journal Films'
The Natural Environment series, this film celebrates the one-hundredth anniversary
of Charles Darwin's death by capturing the beauty and unique ecology of the
Galapagos. Demonstrates a variety of evolutionary and ecological principles
through analyses of such examples as the shell shape of tortoises and sexual
dimorphism in frigate birds (but excludes a discussion of the famous finches).

Every Care But No Responsibility (AMP, 1981, 20 min., color). How do insects lay
eggs and promote their survival? Focusing on a range of Australian insects, this
film tells how, and in stellar fashion.

Encyclopaedia Britannica Ecology and Environmental Science Series (EBF, 11 min., color).
A variety of films covering ecological concepts.

Ecological Biology (COR, 1981, 17 min., color). Part of Coronet's Biological Sciences
series, this film surveys a broad range of ecological concepts, focusing on some
extreme habitats (hot springs, alpine) and on intertidal zonation.

The Nature of Things: Puffins, Predators and Pirates (FL, 1979, 28 min., color).
Documents the manner in which the increasing population of herring gulls on an
island off the coast of Newfoundland is adversely affecting the island's breeding
population of Atlantic puffins. Highly recommended.

Man and His Environment (EMC, 1976, 29 min., color). This film displays nature at work
as it investigates food webs, the flow of energy, predator-prey relationships, and
the checks and balances that work to control population numbers, including those
of humans.

Nature's Ever-Changing Communities (JF, 1975, 14 min., color). Uses live scenes,
animation, and graphics to explore the concepts of renewal, change, and inter-
dependence in the natural world.

Communities of Living Things (AB, 1975, 15 min., color). Concentrates on the six major
land biomes.

Life and Death of a Tree (NGS, 1979, 20 min., color). Portrays the life and death of
an oak tree and the ways in which it interacts with other biotic components of a
forest ecosystem.

Ecosystems: Desert Environment (AIMS, 1972, 8 min., color). Portrays the delicate
balance of plants, animals, insects, and reptiles that make up the desert biome.

Life in a Tropical Forest (TL, 30 min., color). An examination of the tropical rain
forest biome.

Marine Biology: Life in the Tropical Sea (BFA, 1976, 11 min., color). Demonstrates
how members of a tropical reef community interact with one another.

Funk Island (Mac, 1981, 28 min., color). Deals with the bird populations of the most
northeasterly island of North America, including the breeding and nesting of
razor bills, murres, puffins, kittiwakes, and herring gulls. Ecological relationships among the different species are emphasized.

Following the Tundra Wolf (FI, 1977, 44 min., color). With narration by Robert
Redford, this film attempts to dispel long-held myths about the evil nature of
wolves and to explain the wolf's beneficial role in the arctic ecosystem.

Gulf Islands: Beaches, Bays, Sounds and Bayous (NAVC, 1982, 28 min., color). Portrays
the varied habitats associated with the barrier islands and inshore waters along
the Gulf Coast. Emphasis is given to important plant and animal groups found in
each habitat and to the food webs comprised by these groups. The final section of
the film focuses on the precarious nature of these coastal ecosystems in the face
of human contact.

Tree of Thorns (FI, 1981, 50 min., color). The story of the acacia, the dominant tree
species of the Serengeti, is told in this fine BBC-produced film. The ecological
role of the acacia, as well as the threat posed by the encroachment of civiliza-
tion, is explored in detail.

The End of the Game (PHX, 1976, 28 min., color). Demonstrates predation and other
behavioral characteristics of several organisms on an African plain in their daily
struggle for existence; concludes the web of interacting life forms by introducing
the omnipotent human threat.

Pond-Life Food Web (NGS, 1976, 10 min., color). A study of the depths of a fresh-water
    pond, through which basic ecological principles and relationships are demonstrated.
    Traces several interrelated food webs from the microscopic phytoplanktonic pro-
    ducers through the zooplankton and small herbivores to the larger predatory
    carnivores.
Boreal Forest: Fall and Winter (LCA, 1980, 15 min., color). A view of the relatively
    streamlined boreal (northern) forest and the many adaptations of the flora and
    fauna of this zone to the rigors of cold and snow.
Mountain Life Zone Communities (IFB, 1976, 21 min., color). Presents a four-zone
    system for the ecological relationships within the Rocky Mountain habitats, with
    details of fauna family size, gestation period, camouflage, eating habits, plant
    and animal adaptation to altitude, and associated meteorological and climatic
    effects on communities.
Jungles: The Green Oceans (EBF, 1980, 23 min., color). Covers the terrestrial
    community structure of the jungle, including the extraordinary diversity, range
    of adaptations, and mutual interdependence of species found in jungles and the
    distinct vertical stratification in their community organization. Argues that
    the conservation and preservation of these natural ecosystems should be of global
    concern.
Ecological Realities: Natural Laws at Work (EMC, 1976, 13 min., color). A general
    introduction to the principles of ecology, using three diverse geographical
    settings (ocean, marsh, and plain) to describe concepts such as the food web,
    food and energy pyramids, competition, and natural cycles. Effective footage
    of microphotography of plankton and predator-prey sequences.
Life in Lost Creek: Fresh Water Ecology (COR, 1978, 15 min., color). From a small
    mountain stream to a lake bottom, the film explores the diversity of fresh-water
    ecosystems in a straightforward manner.
Man and the Animal Series (NGS)
    An Ecosystem: A Struggle for Survival (1975, 22 min.).
    The Right Whale: An Endangered Species (1976, 23 min.).
    Tracking the North American Mountain Lion (1976, 23 min.).
The Choice is Ours (EMC, 1976, 23 min., color). Uses the theme of a family reunion
    to examine a variety of crucial current topics from family size to energy sources,
    food production, and forms of our institutions. Illustrates the consequences of
    "nature's solutions" (increased death rate), biological control of pests in
    commercial agriculture, the complex energy pollution web, the meat vs. grain
    diet, and so forth.
Common Ground (CFND, 1978, 29 min., color). Deals with the various groups that are
    directly interested in our national forests, focusing on the conflicts that arise
    regarding the management and use of our natural resources.
Noah's Park (PHX, 1976, 27 min., color). Story of an Israeli conservation effort to
    repopulate the Negev Desert with species of animals that inhabited it in biblical
    times. Sequences depict maternal, mating, and dominance behaviors of the chosen
    grazing animals, and narration includes the history and anthropology of the area
    and biblical quotes.

ADDRESSES OF FILM DISTRIBUTORS

AB   Barr Films
        Box 5667
        3490 East Foothill Boulevard
        Pasadena, CA   91107

ACS   American Cancer Society
        (Contact the division nearest
        to you for film information.)

AEC   Atomic Energy Commission
        c/o Films
        Knoxville, TN   37920

AEF   American Educational Films
        132 Lasky Drive
        Beverly Hills, CA   90212

AIMS   Instructional Media Services, Inc.
        626 Justin Avenue
        Glendale, CA   91201

AMP   Arthur Mokin Productions
        17 West 60th Street
        New York, NY   10023

An asterisk (*) denotes major producers/distributors of biology films.

AVED   AV-ED Films
        7934 Santa Monica Boulevard
        Hollywood, CA   90046

BEAF   Beacon Films
        P.O. Box 575
        1250 Washington Street
        Norwood, MA   02062

BF   Benchmark Films, Inc.
        145 Scarborough Road
        Briarcliff Manor, NY   10510

*BFA   BFA Educational Media
        Box 1795
        2211 Michigan Avenue
        Santa Monica, CA   90404

*CE   Centron Educational Films
        Box 687
        1621 West 9th Street
        Lawrence, KA   66044

*CF   Churchill Films
        662 North Robertson Boulevard
        Los Angeles, CA   90069

CFND   The Conservation Foundation
        1717 Massachusetts Avenue, N.W.
        Washington, D.C.   20036

*COR   Coronet Instructional Media
        65 East South Water Street
        Chicago, IL   60601

*CRM   CRM/McGraw-Hill Films
        Box 641
        110 15th Street
        Del Mar, CA   92014

DA   Document Associates
        573 Church Street
        Toronto 285 Ontario, Canada

*EBF   Encyclopaedia Britannica Films
        425 North Michigan Avenue
        Chicago, IL   60611

*EMC   University of California
          Extension Media Center
        2223 Fulton Street
        Berkeley, CA   94720

FF   FilmFair Communications
        Box 1728
        10900 Ventura Boulevard
        Studio City, CA   91604

*FI   Films, Inc.
        733 Green Bay Road
        Wilmette, IL   60091

*FL   Filmmaker's Library, Inc.
        133 East 58th Street
        New York, NY   10022

GC   Graphic Curriculum, Inc.
        University of California Extension
          Media Center
        2223 Fulton Street
        Berkeley, CA   94720

H&R   Harper & Row Media
        Harper & Row Publishers
        10 East 53rd Street
        New York, NY   10022

*IFB   International Film Bureau
        332 South Michigan Avenue
        Chicago, IL   60604

*IU   Indiana University Audio-Visual
          Center
        Division of University Extension
        Bloomington, IN   47401

*JF   Journal Films
        930 Pitner Avenue
        Evanston, IL   60202

*LCA   Learning Corporation of America
        1350 Avenue of the Americas
        New York, NY   10019

*Mac   Macmillan Films
        34 MacQuesten Parkway South
        Mount Vernon, NY   10550

McG   (See CRM.)

MDF   Montana Department of Fish, Wildlife
          & Parks
        Film Production Unit
        MO Building
        Helena, MT   59601

MF   Milner-Fenwick, Inc.
        3800 Liberty Heights Avenue
        Baltimore, MD   21215

*MGD   The Media Guild
        11526 Sorrento Valley Road
        Suite J
        San Diego, CA   92121

*NAVC   The National Audiovisual Center
        National Archives and Records Service
        GSA, Information Services
        Washington, D.C.   20409

*NGS   National Geographic Society
        Educational Services
        17th and M Streets, N.W.
        Washington, D.C.   20036
        (Regional film libraries, Modern
        Film Rentals)

NIH   National Institute of Health
        7500 Wisconsin Avenue
        Bethesda, MD   20014

NMAC   National Medical Audiovisual
           Center
         Chamblee, GA   30005

PD   Parke-Davis and Co.
         Motion Picture Library
         Joseph Campau Avenue at the River
         Detroit, MI   48232

PE   Perennial Education
         P.O. Box 885
         Highland Park, IL   60035

PHX   Phoenix Films, Inc.
         468 Park Avenue South
         New York, NY   10016

PICT   Pictura Films
         111 Eighth Avenue
         New York, NY   10011

PYR   Pyramid Films
         2801 Colorado Avenue
         Santa Monica, CA   90404

SF   Stanton Films
         2417 Artesin Boulevard
         Redondo Beach, CA   90278

SS   Screen Scope
         Suite 2000, 1022 Wilson Boulevard
         Arlington, VA   22209

*SSS   Science Software Systems, Inc.
         11899 West Pico Boulevard
         West Los Angeles, CA   90064
         (also filmstrips)

STP   Stouffer Productions
         P.O. Box 15057
         Aspen, CO   81611

TC   Trainex Corporation
         12601 Industry Street
         Garden Grove, CA   92641

*TL   Time-Life Video Distributors
         100 Eisenhower Drive
         Paramus, NJ   07652

UCLA   UCLA Media Center
         405 Hilgard Avenue
         Los Angeles, CA   90024

UP   Upjohn Professional Film Library
         7000 Portage Road
         Kalamazoo, MI   49001

WNSE   Wards Natural Science Establishment,
           Inc.
         Modern Learning Aids to Vision
         Box 1712
         Rochester, NY   14603

CHAPTER ONE
THE SCIENCE OF BIOLOGY

CHAPTER SUMMARIES AND OBJECTIVES

1-1 <u>A Brief History of Earth</u> The earth is about 4.5 billion years old. The earliest life forms developed about 4.0 billion years ago. Early life was confined to areas protected from the ultraviolet radiation of the sun. Development of an ozone layer in the upper atmosphere shielded the surface from much of this radiation and life was able to spread. There have been many changes in global geography and climate since life on earth began. <u>Homo sapiens</u>, the human species, evolved during the most recent period of earth's history.

   1. List ten major events in the history of earth, from the cooling of the earth's crust to the evolution of the human species. The list should be in chronological order.

1-2 <u>Science as a Form of Human Behavior / Facts and Laws</u> Science proceeds via application of the scientific method, which is an orderly way of observing the universe. One way of making scientific investigations is by experiment, another is by direct observation. Science can study only those things that we can experience directly with our senses or through instruments that extend the power of our senses through space and time. The scientist observes some feature of the universe and formulates a hypothesis to explain the observation. The scientist then tests the hypothesis, attempting to refute it. A hypothesis that survives much such testing is tentatively accepted as a correct explanation for the observation.

Scientific facts are summaries of past observations used to make predictions about future events. Different scientists may interpret the same facts differently. Scientific laws are facts that experience has shown always to be true.

   2. List the steps involved in carrying out a scientific investigation.

   3. Define the term experiment. Why is it important that the scientist vary only <u>one</u> factor in any one experiment?

1-3 <u>Biology: The Science of Life / The Characteristics of Life / Major Organizing Concepts</u> Biology is concerned with living organisms, their organs and cells, and their activities. Organisms may be composed of one or many cells. Cells are of two basic types: <u>prokaryotic</u> cells lack a nucleus and the other internal structures that are found in <u>eukaryotic</u> cells. The different species, or kinds, of organisms are organized into five kingdoms: Monera, Protista, Fungi, Plantae, and Animalia.

Living organisms share several characteristics: (1) energy is taken up from the environment, (2) growth and reproduction take place, and (3) the structures that make up organisms are adapted to their functions. Adaptation is a unique attribute of organisms.

Our discussion of biology will be guided by several important concepts: (1) life can be explained in physical and chemical terms, (2) organisms are systems that take up energy from the environment and process it for their needs, (3) living things are highly organized and much information is needed to express this degree of organization, (4) all organisms are composed of cells and all cells come from preexisting cells, and (5) organisms become adapted to their environments through the process of evolution.

4. List the characteristics of each of the kingdoms of organisms.

5. Differentiate between autotrophs and heterotrophs.

6. Answer: What is adaptation and how do organisms become adapted?

1-4 Early Speculation about Evolution / Charles Darwin and Evolution / Objections to Darwin's Theory / The Importance of the World View As early as the eighteenth century, scientists had begun to speculate about the nature of adaptation. Jean Baptiste Lamarck was the first person to write extensively about evolution. He believed that organisms changed over generations through the inheritance of characters acquired via continued use by their ancestors. By 1859, the climate of scientific opinion was receptive to the idea of evolution as proposed by Charles Darwin and Alfred Russel Wallace.

Darwin's concept of evolution by natural selection is based on two observations: (1) There is a great amount of inherent variability within a species. (2) At least some variations are inherited. The key point in the theory is natural selection, the notion that slight variations affect the chances that an individual will survive. One major problem Darwin had was the lack of knowledge of the mechanism of inheritance during his time. Much study since Darwin's time has confirmed that the theory is correct, although some of the details of the process of evolution remain unknown.

There are three major scientific objections to Darwin's theory.

When Darwin's theory came to be accepted, biology experienced a major change in paradigms. Indeed, the theory of evolution is the major organizational concept for the science of biology.

7. Explain the general outline of Darwin's theory of evolution.

8. List the major scientific objections to Darwin's theory.

9. Answer: Why are objections to the theory on religious grounds not considered in the text?

10. Answer: What is a paradigm? How does the post-Darwinian paradigm of biology compare with the pre-Darwinian paradigm?

1-5 Levels of Organization in Biology Biologists may study life at several levels of organization: molecules, cells, organisms, populations, communities, ecosystems. Some problems require a reductionist approach, whereas others require a synthetic approach.

11. Answer: What are emergent properties?

12. Answer: Which level of organization is affected most profoundly by natural selection?

13. Differentiate between reductionism and the synthetic approach to a problem.

1-6 Questions Beyond Science Some questions about living organisms cannot be considered scientifically because they are not testable. This does not mean that they are unacceptable inquiries, but only that they are not scientific questions.

ARTICLE RESOURCES

A Brief History of Earth

Scientific American, 1978, 239(3). This entire issue is devoted to the topic of evolution. Nine articles discuss the history of life on earth. Worth reading by both instructor and student. Good graphic of geologic time in articles by Mayr. Dickerson gives a chart showing the distribution of the major chemical elements and then discusses the classic Miller and Avery experiments. Schopf discusses microscopic fossils and presents a time line of major events in Precambrian evolution. This issue can also be used for Chapters 30, 31, 35, 36, and 37.

Science as a Form of Human Behavior

Kramm, Kenneth R. Going to the dogs to test hypotheses. The Science Teacher, 1981, 48(8), 40-41. Clever demonstration to get students to generate and test hypotheses.

Facts and Laws

Morowitz, Harold. Facts and artifacts. Science 82, 1982, 3(3), 20-21. Illustrates the different ways in which the sciences and the humanities approach the same words, such as "culture" and "artifact." The connotations and emotional responses are completely different in the two scholarly disciplines.

Horner, Jack K., and Peter A. Rubba. The laws-are-mature-theories fable. The Science Teacher, 1979, 46(2), 31. Indicates that the true relationship between laws and theories is explanatory rather than maturational. A good short article clarifying the differences between law and theory.

Biology: The Science of Life

The Characteristics of Life

Major Organizing Concepts

Early Speculation about Evolution

Charles Darwin and Evolution

Bingham, Roger. On the life of Mr. Darwin. Science 82, 1982, 3(3), 34-39. An imaginative interview based on Darwin's writings and on historical facts. Gives an insight into Darwin the man, as well as Darwin the scientist.

Rensberger, Bozce. Evolution since Darwin. Science 82, 1982, 3(3), 40-45. Discusses the evolution of Darwin's theories up to today. Exposes concisely the components of Darwin's theories. Discusses succinctly the effects of Mendel's research, Thomas Henry Huxley's modern synthesis, and Stephen Jay Gould and Niles Eldridge's punctuated evolution. An excellent historical overview.

Bowser, Hal. Micromasterpieces. Science Digest, 1981, 89(11), 46-55. Beautifully illustrated article on Ernst Haeckel, one of Darwin's earliest defenders in Germany. He is the scientist who insisted that "ontogeny recapitulates phylogeny." His drawings of microscopic organisms assure him a place in the history of science.

Objections to Darwin's Theory

The Importance of the World View

Levels of Organization in Biology

Questions Beyond Science

There are three articles that instructors may wish to read before starting to teach this course. All are concerned with ways to reach students, develop their higher cognitive skills, and help them use scientific literature.

The first, by Peter Kovacic and Martin B. Jones, refers to a course in chemistry, but substitute "biology" for "chemistry" and it provides a helpful start for any biology instructor. The second, by S. H. Gehlbach, J. A. Bobula, and J. C. Dickerson, is directed at medical students but can also be used with beginning biology students with equal effectiveness. It is an excellent article that offers hints on the procedures to use in reading literature that has a specialized vocabulary and a concise manner of expression. The third, by Margaret A. Waterman and Jane F. Rissler, shows how students' cognitive skills can be improved by judicious selection and assignment of periodicals.

Kovacic, Peter, and Martin B. Jones. The opening chemistry lecture: Reaching students without demonstrations. Journal of College Science Teaching, 1982, xl(6), 365-366.

Gehlbach, S. H., J. A. Bobula, and J. C. Dickerson. Teaching residents to read the medical literature. Journal of Medical Education, 1982, 55 (April), 362-365.

Waterman, Margaret A., and Jane F. Rissler. Use of scientific reports to develop higher-level cognitive skills. Journal of College Science Teaching, 1982, xl(6), 336-340.

ESSAY QUESTIONS

1. Why do you think the processes that preceded the appearance of the first living cells are referred to as chemical evolution?

    Organic molecules produced by the action of energy on atmospheric gases and dissolved in the ocean became increasingly complex, evolving into the components of living systems and finally into living cells.

2. Why is it important to design the tests of a hypothesis (experiments) in such a way that at least one test is capable of refuting the hypothesis?

    If none of the experiments is capable of refuting the hypothesis, we have no way of knowing whether the hypothesis should be rejected in some case where we would otherwise expect it to hold true.

3. List the characteristics of each of the kingdoms of organisms.

    Monera: unicellular prokaryotic organisms
    Protista: unicellular eukaryotic organisms
    Fungi: multicellular eukaryotic organisms that absorb foods from their
        surroundings and digest them intracellularly
    Plantae: multicellular eukaryotic organisms that are autotrophic by means of
        photosynthesis
    Animalia: multicellular eukaryotic organisms that digest foods extracellularly
        and absorb the resulting compounds

4. Explain the general outline of Darwin's theory of evolution.

    In any species there is a range of variability in characteristics, at least some of which are inheritable. An individual with traits that favor its survival and its reproductive success in a given environment stands a better chance of leaving offspring than does an individual with less favorable traits. This is the process of natural selection. Natural selection results in the adaptation of a species to the environment, making it better fitted to thrive and reproduce there. (Hence the concept of survival of the "fittest.")

5. What is a paradigm? How does the post-Darwinian paradigm of biology compare with the pre-Darwinian paradigm?

    A paradigm is the general world view held by an individual, which determines the predictions one makes and one's interpretations of information. The pre-Darwinian paradigm saw the world as relatively young and viewed organisms as having been specially created to inhabit it in their observed forms. The post-Darwinian paradigm views the earth as about 4.5 billion years old, during which time the planet and its organisms have been evolving from forms quite unlike the present ones.

6. In the Great African Rift Valley, there is a small lake called Lake Nabugabo. Geological studies indicate that this lake was formed about 6,000 years ago by the extension of a sand bar across an arm of Lake Victoria. Of the several species of cichlid fishes within Lake Nabugabo, there are forms quite similar in appearance to certain cichlids in Lake Victoria. These fishes are sufficiently different from those in Lake Victoria, however, to be classified as different species. The Lake Nabugabo cichlids show various adaptations, especially involving the types of foods preferred and the structure of the jaws and teeth. Interpret this information in light of the modern theory of evolution.

   The Lake Nabugabo cichlids are clearly descended from (perhaps three or four) ancestral Lake Victoria species that became isolated within Lake Nabugabo. Natural selection has favored those forms that were able to exploit the foods and shelter available to them in their new home. It is likely that the food supply is more restricted in the smaller lake, which would have increased competition and resulted in the partitioning of the available food sources among several species of specialized feeders. Note: This is probably an example of sympatric speciation (see Chapter 37), but some questions still remain unanswered about these cichlids.

TEST SET 1-A

Name _____   Section _____

Choose the best answer to each of the following questions, and write the appropriate letter in the space provided.

Ans: a
p. 2

_____   1. The original atmosphere of the earth is thought to have con-sisted largely of: (a) hydrogen; (b) nitrogen; (c) oxygen; (d) carbon dioxide.

Ans: c
p. 3

_____   2. The UV radiation that initially prevented living organisms from using dry land was finally blocked by the production of: (a) heavy cloud cover; (b) carbon dioxide; (c) ozone; (d) atmos-pheric water vapor.

Ans: c
p. 5

_____   3. Scientific laws are: (a) principles that are known to be cate-gorically true; (b) facts that usually prove to be true; (c) relationships that have held true in the past and are thus expected to hold true for future events; (d) never proved untrue.

Ans: d
p. 5

_____   4. In general, a species may be considered to be: (a) a group of organisms that look alike; (b) a group of organisms that serve the same functions in nature; (c) a group of organisms that have identical anatomies; (d) a group of organisms that can interbreed only with each other.

Ans: a
p. 7

_____   5. The metabolism of an organism is: (a) the sum of all the chemi-cal reactions it performs; (b) the mechanism whereby nutrients are broken down; (c) its rate of growth; (d) those mechanisms it uses in its growth.

Ans: c
p. 10

_____   6. A simple definition for the term energy is: (a) a measure of the change of state of a system; (b) a measure of the tendency of a system to remain at equilibrium; (c) the capacity to do work; (d) the potential inherent in a system.

Ans: a
p. 10

_____   7. The German physician responsible for first succinctly stating the cell theory was: (a) Rudolf Virchow; (b) Theodor Schwann; (c) Louis Pasteur; (d) Gigante Verde.

Ans: b
p. 12

_____   8. Lamarck was an evolutionist who: (a) believed that humans did not evolve from apes; (b) believed that a characteristic was likely to be inherited if used or lost if not used; (c) ini-tially proposed natural selection; (d) believed that mutations were only theoretical.

Ans: d
p. 12

_____   9. Darwin's theory of evolution was based on two facts: (a) fossil finds and similarities between primates; (b) the inheritance of acquired characteristics and similarities among the primates; (c) the inheritance of color in birds and the variety of life forms in the world; (d) the existence of variability and the similarities between parents and offspring.

Ans: b
p. 14

_____   10. Scientific objections to the theory proposed by Darwin include: (a) the argument that evolution is a theory and can't be proved; (b) large variations in the apparent rate of evolutionary change; (c) the fact that the process he proposed no longer seems to be in effect; (d) the statement that mutation doesn't provide enough variability.

Ans: c
p. 18

_____ 11. From the ecological standpoint, communities are: (a) the local-ized environment where organisms live; (b) the sum of the factors that control populations; (c) the sum of the different popula-tions interacting in a given area; (d) the group of organisms that function similarly.

Ans: b
p. 10

_____ 12. Robert Hooke was responsible for first using, as a description of the cavities he observed in cork under the microscope, the term: (a) pore; (b) cell; (c) lacuna; (d) cisterna.

Ans: d
p. 8

_____ 13. A good example of an autotrophic organism is a(an): (a) amoeba; (b) shark; (c) mushroom; (d) rose.

Ans: a
p. 7

_____ 14. The kingdom Monera includes: (a) blue green algae; (b) oak trees; (c) mammalia; (d) earthworms.

Ans: b
p. 3

_____ 15. The continents are: (a) in the same positions they have occu-pied for billions of years; (b) constantly drifting on the earth's surface; (c) locked in place on the earth's surface after an initial period of continental drift; (d) entering a period of drift after having been stationary for billions of years.

TEST SET 1-B

Name _____   Section _____

Choose the <u>best</u> answer to each of the following questions, and write the appropriate letter in the space provided.

Ans: d
p. 3

_____   1. The earliest forms of life on earth were confined below the surface of the oceans because: (a) the original hydrogen atmosphere was lost; (b) the only nutrients to be found were there; (c) the atmosphere was too warm; (d) radiation levels were too high above water.

Ans: b
p. 2

_____   2. The major layers that make up the earth are the: (a) surface, intermediate layer, and inner layer; (b) crust, mantle, and core; (c) surface, mantle, and core; (d) crust, magma, and core.

Ans: b
p. 7

_____   3. A cell that lacks a nucleus is likely to be classified as: (a) eukaryotic; (b) prokaryotic; (c) heterotrophic; (d) protista.

Ans: c
p. 7

_____   4. An organism living its entire life below ground in a cave would have to be: (a) prokaryotic; (b) eukaryotic; (c) heterotrophic; (d) autotrophic.

Ans: c
p. 10

_____   5. A complex molecule like an enzyme can be characterized as having: (a) low information content; (b) high information capacity; (c) low information content and low information capacity; (d) high information content.

Ans: a
p. 11

_____   6. The cell theory was formally proposed by Virchow, but the accepted proof of the assertion was provided by: (a) Pasteur; (b) Hooke; (c) Schwann; (d) Buffon.

Ans: d
p. 12

_____   7. The proposition that evolution occurred in organisms as structures developed characteristics through increased use or lost them through disuse is attributed to: (a) Darwin; (b) Wallace; (c) Buffon; (d) Lamarck.

Ans: c
p. 12

_____   8. Part of Darwin's theory for evolution's mechanism involved differential survival as a function of slight differences between organisms. This mechanism is commonly given the name: (a) mutation; (b) the origin of species; (c) natural selection; (d) comparative survival.

Ans: a
p. 16

_____   9. A paradigm is loosely defined as a(an): (a) world view or framework; (b) experimental method; (c) scientific fact; (d) measure of information capacity.

Ans: a
p. 18

_____   10. The process of pressing scientific investigation to ever more fundamental levels of organization (for example, from cells to organelles to membranes to molecules) in attempts to understand the matter being studied is called: (a) reductionism; (b) experimentation; (c) the synthetic approach; (d) rationalism.

Ans: a
p. 15

_____   11. Though generally supporting Darwin's theory of evolution, the fossil record can be used to raise certain objections, such as: (a) fossil evidence that rates of evolution vary greatly, contrary to Darwin's contention; (b) the lack of a complete record; (c) the absence of evolutionary activity in recent ages; (d) inaccuracies arising from inability to date fossil forms correctly.

Ans: b
p. 14

_____ 12. The heritable changes in character that allow evolution to proceed are thought to stem from: (a) interspecific breeding; (b) mutation events; (c) altered environmental pressure; (d) natural selection.

Ans: c
p. 10

_____ 13. The basic functional unit of living organisms is the: (a) chromosome; (b) organ; (c) cell; (d) molecule.

Ans: a
p. 7

_____ 14. The kingdom of organisms from which all the others are thought to derive is the kingdom: (a) Monera; (b) Protista; (c) Procaria; (d) Eucaria.

Ans: d
p. 1

_____ 15. Most planets, including the earth, are thought to have arisen from: (a) large chunks of the primordial ball of matter that were scattered in the "big bang" event; (b) the aggregation of asteroids; (c) the aggregation, through mutual gravitational attraction, of material in comets; (d) the aggregation, through mutual gravitational attraction, of cold dust particles.

CHAPTER TWO
SMALL MOLECULES

CHAPTER SUMMARIES AND OBJECTIVES

2-1 Atoms, Elements, Particles, and Isotopes / Radioisotopes in Biological Research
All matter is composed of atoms. Each atom has a nucleus that is made up of protons and neutrons and surrounded by a cloud of electrons. Protons have a charge
of +1 and a mass of 1. Electrons have a charge of -1 and essentially 0 mass.
Neutrons have the same mass as protons but no charge. A chemical element is a
substance composed of one type of atom.

The number of protons equals the number of electrons in a given atom, making atoms
electrically neutral. The number of protons or electrons is the atomic number of
the atom. The mass number of the atom is the combined mass of the protons and
neutrons. Some elements can exist in forms that have the same atomic number but
different mass numbers. Such forms are called isotopes. Some isotopes are radioactive, spontaneously giving off energy or subatomic particles. Radioisotopes are
important tools in biological research, because the fact that their radiation can
be detected makes them useful as tracers.

1. List the basic particles of matter, giving the charge, mass, and position in
    the atom of each.

2. Give several examples of the use of radioisotopes in biological research.

2-2 Energy and Stability / The Behavior of Electrons / Chemical Bonds Any system tends
to change in such a way as to minimize its energy content. This principle governs
much of the behavior of electrons. Chemical reactions are exchanges of electrons,
or changes in the pattern of sharing them, between atoms. Electrons are found in
orbitals, regions of space where the electron is most likely to be found. Orbitals
are grouped into shells, which surround the nucleus. The number of electrons in
the outermost shell determines the chemical behavior of the atom. An atom with a
full outer shell is very stable.

A chemical bond is formed when two atoms come together in such a way that an
attractive force is formed between them. When electrons are shared between two
atoms, a covalent bond exists. Sharing of electrons permits each atom to have a
full outer shell.

3. Sketch the arrangement of electrons for each of the following atoms, showing
    orbitals and shells: hydrogen, beryllium, carbon, oxygen, fluorine, aluminum,
    sulfur, sodium.

4. Explain why there is a characteristic distance (bond length) that separates
    atoms in a chemical bond, using the example of H_2, as outlined in your text.
    Would you expect each type of chemical bond to have a different bond length?
    Why?

11

2-3 <u>Molecules and Reactions</u> A compound is formed of two or more different atoms connected by a chemical bond. The number of covalent bonds a particular atom can form is determined by the number of electrons in its outer shell. A chemical reaction occurs when atoms change bonding partners. When covalent bonds are involved, group transfer occurs: Atoms are exchanged, but the total number of bonds remains the same. Energy may be given off or taken up when a chemical reaction occurs.

1. Answer: Why are hydrogen, oxygen, nitrogen and carbon such important atoms in living systems?

2-4 <u>Some Quantities</u>   The mass of an average atom of an element is the atomic weight of the element. The sum of the atomic weights of the atoms in a molecule is the molecular weight of the compound. In any chemical reaction, a specific number of molecules of one substance reacts with a specific number of molecules of the other substances in the reaction. To avoid dealing with individual molecules, we deal in moles, an amount in grams numerically equal to the molecular weight. One mole of any substance contains 6.02×10^{23} (Avogadro's number) particles of that substance. The basic unit of energy used in our studies is the calorie, the amount of heat required to raise the temperature of 1 gram of pure water from $14.5°C$ to $15.5°C$.

5. Calculate the molecular weight of each of the following substances: H_2O, $CaCO_3$, H_3PO_4, $Mg_3(PO_4)_2$, CH_3CH_2COOH.

6. Answer: How many grams are there in 1.0 mole of carbon dioxide? In 0.5 mole of methane? In 3.6 moles of aluminum hydroxide, $Al(OH)_3$?

2-5 <u>Water / Ions, Ionization, and Ionic Bonds</u> Water is one of the most important biological compounds because it has a number of unique properties. Water will dissolve more kinds of substances than any other solvent. It can exist as a solid, liquid, or gas at temperatures commonly found on earth. Ice is less dense than liquid water. On evaporation, water takes up a very large amount of heat. Liquid water has a high surface tension. Water molecules are readily attracted to other kinds of molecules, resulting in capillary action.

Some types of molecules break apart when dissolved in water, becoming ionized. An ion is an atom or group of atoms that has gained or lost some of its electrons. Ions thus carry a charge, forming crystals held together by ionic bonds. Ionic bonds are actually covalent bonds in which one partner has the electrons all the time.

7. List three of water's unique properties and give an example of how each property benefits biological systems.

8. Sketch a hydrogen atom, a chlorine atom, chloride and hydrogen ions, and a hydrochloric acid molecule, showing the different arrangements of electrons in each.

2-6 <u>Acidity and pH</u> Water ionizes, producing hydrogen ions and hydroxide ions. In pure water, the numbers of each of these are equal and the "solution" is said to be neutral. Adding other substances to the water may result in an excess of hydrogen or hydroxide ions. Acids release H^+ ions. Compounds that can accept H^+ ions are called bases. In pure water the concentration of H^+ ions, in moles per liter (molar concentration), is 10^{-7}. For convenience, pH is defined as the negative logarithm of the hydrogen ion concentration. A buffer is a solution of a weak acid and its conjugate base. Buffers are resistant to changes in pH, even when acid or base is added.

9. Define and give two examples of each of the following: acid, base, conjugate pair.

10. Be able to work written problems involving pH, molarity, and acids and bases.

2-7 <u>Polarity and the Hydrogen Bond</u> When electrical charges are not distributed evenly within a molecule, the compound is said to be polar. Water is such a compound. The polarity of water results from the strong attraction of the oxygen atom, as opposed to the hydrogen atoms, for the shared electrons. The attraction between the negative region of one water molecule and the positive region of another molecule is a hydrogen bond. Many of the important properties of water derive from its ability to form hydrogen bonds. Hydrogen bonds can form between many types of molecules, and they are important in biological systems. Molecules in which the electrical charge is evenly distributed are said to be nonpolar. Attractions between nonpolar molecules include van der Waals and hydrophobic interactions. These, too, are important in living systems.

11. Answer: With reference to the solubility of polar and nonpolar substances, a rule of thumb is "like dissolves like." Based on the information in this section, what is the rationale behind that statement?

2-8 <u>Some Simple Organic Compounds</u> Organic compounds contain carbon. These include hydrocarbons, which may be saturated or unsaturated; alcohols, with a hydroxyl group; amines, with an amino group; and amino acids, with both an amino group and a carboxyl group.

12. Define: asymmetric carbon, optical isomers.

13. Explain why ethanol, CH_3CH_2OH, is soluble in water, whereas octyl alcohol, $CH_3CH_2CH_2CH_2CH_2CH_2CH_2CH_2OH$, is not.

ARTICLE RESOURCES

<u>Atoms, Elements, Particles, and Isotopes</u>

Perl, Martin L. Electron, muon, and ton heavy lipton--Are these the true elementary particles? <u>The Science Teacher</u>, 1980, 47(9), 16-19. Concise discussion of elementary particles with definitions of elementary particles and their characteristics. Clarifies the forces acting between particles. Liptons and hadrons are explained.

Benfey, Otto Theodore, Geometric forms in chemistry. <u>The Science Teacher</u>, 1981, 48(6), 26-31. Discusses isomerism and the topography of organic molecules. Uses simple diagrams and photographs to illustrate the various properties of regular solids as they apply to organic molecules.

Gizara, Jeanne M. Bridging the stoichiometry gap. <u>The Science Teacher</u>, 1981, 48(4), 36-37. Step-by-step method of solving stoichiometry problems leading to greater student confidence in their ability to use dimensional analysis (factor label) methods.

<u>Radioisotopes in Biological Research</u>

Radioactive blood scouts, <u>Science Digest</u>, 1982, 90(5), 97. By using radioactive indium-111 on white blood cells, scientists are able to locate sites of infection in the human body. Fluorine-18 is used to determine whether portions of the brain are blood-deficient.

New carbon calendar. <u>Science Digest</u>, 1982, 90(6), 17. Brief (one page) report based on accelerator dating, which is a new and more sensitive method of determining the age of organisms. Only fragmentary particles are needed to determine the age of any sample with this method.

<u>Energy and Stability</u>

Webb, Michael J. Physical and chemical changes: What's the difference? <u>The Science Teacher</u>, 1982, 49(3), 39-40. Excellent clarification of these two concepts. Although the article is directed to junior high school teachers, undergraduate students will find it helpful in distinguishing between these two kinds of change.

<u>The Behavior of Electrons</u>

<u>Chemical Bonds</u>

<u>Molecules and Reactions</u>

<u>Some Quantities</u>

Water

Smail, James R. What's in the ocean? The American Biology Teacher, 1981, 43(6), 312-316. The most important component of the ocean is the water itself. Ninety-seven percent of the earth's total water supply is to be found in the ocean. The properties of water, including its ability to act as a solvent, are clearly explained. Dittmar's law of relative proportions is discussed. The ocean's role as a buffer and this feature as a stabilizing influence are considered.

Mansfield, Donald H., and Jay E. Anderson. Measuring plant water status: A simple method for investigative laboratories. The American Biology Teacher, 1980, 42(9), 541-544. Water is an absolute necessity for plant growth. It is a key factor in plant distribution, as well as in the quality and quantity of plant growth. A simple method of determining the relative water content (RWC) is delineated. Even without doing this investigation, the student should be able to grasp the concept.

Water's wondrous talents. Science Digest, 1982, 90(5), 103. Brief exposition of the qualities of water. This article is only a few paragraphs in length, but it is packed with information in a readable style.

Ions, Ionization, and Ionic Bonds

Acidity and pH

Polarity and the Hydrogen Bond

Some Simple Organic Compounds

Benfey, Otto Theodore. Geometric forms in chemistry. The Science Teacher, 1981, 48(6), 26-31. Discusses isomerism, which has incorporated topology into organic structural theory. This is a clearly written article.

ESSAY QUESTIONS

1. Explain how the various subatomic particles are arranged in atoms, isotopes, and ions.

    Protons and neutrons are found in the nucleus, with electrons in a surrounding cloud. The number of protons equals the number of electrons in an atom. Isotopes differ in the number of neutrons in their nuclei. Ions differ from atoms in that they have gained or lost an electron or electrons.

2. Sketch each of the compounds in the following list, showing the arrangement of outermost electrons and indicating the types of bonds present. For example, the answer for water, H_2O, would be

$$H:\ddot{O}:H$$

water

COMPOUNDS: methane, CH_4; ethylene, $H_2C=CH_2$; acetylene, $HC\equiv CH$; ammonia, NH_3; and glycine, $H_2N-\overset{H}{\underset{H}{C}}-\overset{}{C}=O$.
                                                OH

methane   ethylene   acetylene   ammonia   glycine

3. Why are hydrogen, oxygen, nitrogen, and carbon such important atoms in living systems?

    They are the smallest atoms that can form one, two, three, and four covalent bonds, respectively.

4. What is a mole? Why is it helpful to use moles when doing calculations involving

chemical reactions?   What is a molar solution?

> A mole is an amount in grams numerically equal to the molecular weight of a given substance. Because a mole of any substance contains the same number of atoms, substances react in whole-number molar ratios, and we do not have to deal with individual atoms. A molar solution is one mole of solute in one liter of solvent.

5. List three of water's unique properties and give an example of how each property benefits biological systems.

> High specific heat--responsible for the constancy of temperature of large bodies of water and for the effectiveness of evaporative cooling.

> Solid less dense than vapor--as long as a body of water does not freeze solid, there will always be water at 0°C underneath the ice, in which organisms can survive.

> Cohesiveness of water molecules--creates surface tension, aids in the transport of water through the conducting tissues of plants.

> Adhesiveness of water molecules--produces capillary action, aids in transport in plants.

> Excellent solvent--many substances are absorbed by cells from aqueous solution, such as oxygen and various ions.

6. Define and give two examples of each of the following: acid, base, conjugate pair.

> Acid--proton donor, HCl, CH_3CH_2COOH; base--proton acceptor, $NaOH$, HCO_3^- ion; conjugate pair--an acid and its corresponding base, acetic acid and acetate ion, carbonic acid and bicarbonate ion.

7. What is a buffer and why are buffers important in living systems?

> A buffer is a solution of a weak acid and its conjugate base that is resistant to changes in pH, even with the addition of acid or base. Buffers are important in maintaining pH within precise limits in living systems.

8. Explain why ethanol, CH_3CH_2OH, is soluble in water, whereas octyl alcohol, $(CH_3CH_2CH_2CH_2CH_2CH_2CH_2CH_2OH$, is not.

> The hydroxyl group of ethanol is sufficiently polar to make the molecule soluble. The long hydrocarbon chain of octyl alcohol makes the molecule hydrophobic and thus insoluble.

9. List three types of weak interactions between molecules and explain why these are important in biological systems.

> Hydrogen bonds, van der Waals interactions, hydrophobic interactions; all are important in determining the shapes assumed by macromolecules.

10. Calcium carbonate, $CaCO_3$, reacts with sulfuric acid, H_2SO_4, to yield calcium sulfate, $CaSO_4$, and carbonic acid, H_2CO_3. If 150 grams of calcium carbonate are allowed to react completely with sulfuric acid, how much calcium sulfate will be produced?

> The molecular weight of calcium carbonate is 100, so we are starting with 1.5 moles of $CaCO_3$. Because the one mole of sulfuric acid reacts with each mole of calcium carbonate to produce one mole of calcium sulfate, 1.5 moles of calcium carbonate will yield 1.5 moles (204 grams) of calcium sulfate.

$$CaCO_3 + H_2SO_4 \rightarrow CaSO_4 + H_2CO_3$$

Molecular weight =   100   98   136   62

TEST SET 2-A

Name _____   Section _____

Choose the <u>best</u> answer to each of the following questions, and write the appropriate letter in the space provided.

Ans: b
p. 23

_____ 1. The most abundant molecular species in living organisms is: (a) carbon; (b) water; (c) hydrogen; (d) oxygen.

Ans: a
p. 24

_____ 2. Elements are composed of atoms all having the same: (a) atomic number; (b) atomic weight and number; (c) number of neutrons; (d) size nucleus.

Ans: a
p. 26

_____ 3. Which of the following are radioisotopes of hydrogen? (a) tritium; (b) gallium and deuterium; (c) deuterium and tritium; (d) deuterium.

Ans: d
p. 29

_____ 4. A system is <u>most</u> stable when it: (a) approaches maximum energy content; (b) is losing energy; (c) is gaining energy; (d) approaches minimum energy content.

Ans: a
p. 31

_____ 5. The K shell: (a) can contain as many as two electrons; (b) contains electrons of the highest energy level; (c) can contain up to four electrons; (d) is one of the last to fill with electrons.

Ans: b
p. 32

_____ 6. A triple covalent bond (as in the gas nitrogen) involves the sharing of: (a) six pairs of electrons; (b) six electrons; (c) three electrons; (d) three electron triplets.

Ans: d
p. 34

_____ 7. Avogadro's number is equal to the number of: (a) atoms in a molecule; (b) grams in a mole of a substance; (c) moles in a liter; (d) molecules in a mole.

Ans: a
p. 35

_____ 8. A kilocalorie (the nutritionist's Calorie) is equivalent to: (a) the number of calories required to raise one kilogram of water from a temperature of 14.5°C to a temperature of 15.5°C; (b) 4.19 joules; (c) 419 joules; (d) the amount of heat energy required to raise one gram of water from a temperature of 14.5°C to a temperature of 15.5°C.

Ans: c
p. 35

_____ 9. Capillary action, the rising of water and solutions in narrow tubes, is associated <u>most</u> closely with: (a) water's high surface tension; (b) the response of the water molecule to changes in atmospheric pressure; (c) the attraction of water molecules to other kinds of molecules; (d) covalent bonds with water molecules.

Ans: c
p. 36

_____ 10. Ions are formed whenever: (a) a material goes into aqueous solution; (b) covalent bonds are broken; (c) an atom gains or loses electrons; (d) an atom has an unfilled orbital.

Ans: a
p. 38

_____ 11. The pH of a 1.0 molar solution of HCl is: (a) 0; (b) 2; (c) 4; (d) 1.5.

Ans: b
p. 38

_____ 12. Solutions that resist pH changes with the addition of acids and bases are called: (a) neutral; (b) buffers; (c) alkaline; (d) balanced.

Ans: d
p. 40

_____ 13. The solubility in water of compounds such as sugars depends on: (a) the nonpolarity of the compound; (b) the formation of ionic bonds; (c) van der Waals interactions; (d) the formation of hydrogen bonds.

Ans: b   _____   14. A hydrocarbon becomes an alcohol when: (a) a hydroxyl group is
p. 41   replaced by an alcohol group; (b) a hydroxyl group replaces a
                               hydrogen; (c) a hydroxyl group is replaced by a hydrogen; (d) a
                               hydrogen is changed into an amide.

Ans: c   _____   15. When a group of atoms are joined to form compounds whose only
p. 41   difference is that they are mirror images of each other, these
                               compounds are called: (a) stereotypic; (b) asymmetric com-
                               pounds; (c) optical isomers; (d) mirror compounds.

TEST SET 2-B

Name _____   Section _____

Choose the <u>best</u> answer to each of the following questions, and write the appropriate letter in the space provided.

Ans: b
p. 23

_____ 1. Of the common chemical elements, four account for all but about 1 percent of the composition of living material. These four are: (a) potassium, carbon, oxygen, and sodium; (b) carbon, oxygen, nitrogen, and hydrogen; (c) carbon, oxygen, sodium, and hydrogen; (d) hydrogen, carbon, potassium, and nitrogen.

Ans: c
p. 24

_____ 2. An atomic nucleus is: (a) electrically neutral; (b) composed of protons; (c) positively charged; (d) composed of neutrons.

Ans: a
p. 26

_____ 3. Neutrons are: (a) equal in mass to protons, but not charged; (b) equal in mass to protons, but opposite in charge; (c) smaller than protons, but not charged; (d) similar to electrons except for charge.

Ans: d
p. 26

_____ 4. Deuterium has the same atomic number as hydrogen, but it: (a) is not a stable element; (b) has an extra proton; (c) has smaller atomic mass number; (d) has a larger atomic mass number.

Ans: c
p. 30

_____ 5. An orbital: (a) defines the absolute position of an electron around the nucleus; (b) is composed of subdivisions called shells; (c) defines the spatial location of an electron with 90 percent confidence; (d) can define the location of two pairs of electrons.

Ans: b
p. 39

_____ 6. The chemical bond with the greatest typical strength is: (a) van der Waals; (b) covalent; (c) ionic; (d) hydrogen.

Ans: d
p. 33

_____ 7. Which of the following elements <u>cannot</u> form double bonds with oxygen, as is common in living materials? (a) carbon; (b) nitrogen; (c) phosphorus; (d) silicon.

Ans: b
p. 34

_____ 8. The molecular weight of a compound is: (a) Avogadro's number of atoms; (b) the sum of the atomic weights of its atoms; (c) the weight of one mole of the substance in grams; (d) the sum of the atomic numbers of its component atoms.

Ans: d
p. 35

_____ 9. One kilocalorie equals: (a) 4.19 joules; (b) 0.24 joules; (c) 240 joules; (d) 4,190 joules.

Ans: a
p. 36

_____ 10. Ions differ from atoms in that: (a) atoms in elemental form are neutral; (b) ions are more reactive because of positive charge; (c) ions have extra electrons; (d) atoms in elemental form are more easily involved in chemical reactions.

Ans: c
p. 37

_____ 11. An acidic solution: (a) can release hydrogen ions; (b) has a high pH; (c) has more hydrogen ions than hydroxyl ions; (d) has more hydroxyl ions than hydrogen ions.

Ans: d
p. 38

_____ 12. The negative logarithm of hydrogen concentration defines: (a) acidity; (b) alkalinity; (c) buffering; (d) pH.

Ans: b
p. 38

_____ 13. Some compounds are said to be polar because: (a) electrical charge is symmetrical from one molecular pole to the other; (b) electrical charge is not evenly distributed along the molecule; (c) their electron cloud is evenly distributed; (d) they are not water-soluble.

Ans: a
p. 40

_____ 14. Van der Waals interactions: (a) are weak, short-distance inter-actions; (b) are stronger than hydrogen bonds; (c) are roughly equivalent to ionic bonds; (d) are about half as strong as covalent bonds.

Ans: b
p. 42

_____ 15. D-amino acids: (a) cannot form peptide bonds to make protein; (b) are not commonly found in living materials; (c) are larger than L-amino acids; (d) are the stereoisomers commonly found in natural protein.

CHAPTER THREE
LARGE MOLECULES

CHAPTER SUMMARIES AND OBJECTIVES

3-1 <u>Carbohydrates: Monosaccharides and Disaccharides / Carbohydrates: Polysaccharides</u>
The most important molecules in biological systems are macromolecules, polymers of 100,000 molecular weight or more, which are made up of chemically similar subunits called monomers. The categories of macromolecules are carbohydrates, lipids, proteins, and nucleic acids.

There are three types of carbohydrates: monosaccharides, oligosaccharides, and polysaccharides. The most abundant monosaccharide is glucose, which is produced during photosynthesis and is metabolized in the process of cellular respiration. Important oligosaccharides are sucrose and lactose.

Polysaccharides are energy-storage compounds. Starch, a polymer of glucose, is found in plants. In animals another glucose polymer, glycogen, is the major energy-storage compound. The usefulness of these compounds lies in the fact that they can be digested to yield glucose.

Cellulose is a structural polysaccharide. It is perhaps the most abundant biological material on earth. Derivative carbohydrates are important in cellular metabolism and as components of structural compounds such as chitin.

1. Explain the difference between an α-linkage and a β-linkage between two glucose molecules. Of what significance is this difference to a living system?

2. Give an example of each of the following: monosaccharide, disaccharide, polysaccharide.

3. Answer: Why are starch and glycogen called storage carbohydrates?

3-2 <u>Lipids</u> Lipids constitute a diverse group of compounds that are insoluble in water but soluble in organic solvents. Large amounts of energy are released when lipids are metabolized. Each of these properties is important in living systems. Lipids are the primary compounds of membranes. The regulation of materials entering and leaving the cell depends largely on the relative solubility of various compounds in the membrane lipids. Lipids are excellent energy-storage compounds because they are the most highly reduced molecules in the cell.

There are several types of lipids. Triglycerides, composed of glycerol and fatty acids, include fats and oils. Phospholipids are important as components of membranes. Carotenoids are light-absorbing pigments. Steroids are important constituents of hormones.

4. Answer: How does the use of lipids (as opposed to polysaccharides) as energy-

storage compounds represent an adaptation in animals?

5. Differentiate between hydrophobic molecules and hydrophilic molecules.

6. Answer: What is a hormone?

7. Differentiate among saturated, unsaturated, and polyunsaturated fats.

8. Answer: Why are lipids said to be "highly reduced" compounds? Of what importance is this fact in the functioning of lipids as energy-storage compounds?

3-3 <u>Amino Acids and Proteins / Levels of Protein Structure</u> Proteins are large polymers of amino acids connected by peptide linkages. They are exceedingly diverse and are the most important molecules in biological systems. One major class of proteins includes the enzymes, catalysts required for all biochemical reactions in the cell. Other types of proteins may function as structural elements, as immunoglobulins, as channels for the movement of ions across membranes, or as mediators of muscle contraction.

Four levels of protein structure are recognized. The linear sequence of amino acids constitutes the primary structure of the protein molecule. The primary structure of each protein is specified by the information coded in the DNA molecule. The primary structure determines the shape the protein molecule will assume. Amino acids with hydrophobic side chains tend to associate in the interior of the molecule; those with hydrophilic side chains, on the exterior. Some amino acids with polar side chains may be found either inside or outside the molecule and can form hydrogen bonds or disulfide bridges with other amino acids. Hydrogen bonds and disulfide bridges are especially important in determining how the protein molecule folds.

The secondary structure of a protein results from the formation of hydrogen bonds and disulfide bridges between amino acid side chains. Three patterns of secondary structure are α-helix, β-pleated sheets, and γ-helix. These patterns are important in structural proteins.

The tertiary structure of a protein consists of the localization in space of all the atoms in the molecule. It determines the specific function of the protein. Proteins that are composed of one or more subunits are said to have quaternary structure. Tertiary and quaternary structure are important in the functioning of globular proteins, most of which are either enzymes or carriers.

9. Answer: What determines the amino acid sequence of a protein? How do the R groups of various amino acids ultimately determine the function of a particular protein?

10. Name and define the four levels of protein structure.

11. Answer: Carrier proteins are usually highly specific in the types of molecules they can carry. Based on the information contained in this section, can you suggest a reason for this phenomenon?

3-4 <u>Nucleic Acids</u> Nucleic acids are polymers of nucleotides. Deoxyribonucleic acid (DNA) carry the instructions for making protein molecules. Ribonucleic acids (RNA) interpret and execute the DNA instructions. Nucleotides are composed of a sugar, a phosphate group, and one of four nitrogenous bases, with the sugar and phosphate molecules alternating to form the "backbone" of the polymer. RNA is normally single-stranded; DNA, double-stranded.

Deoxyribose is the sugar found in DNA, whereas ribose is found in RNA. Four nitrogenous bases are found in DNA: adenine, guanine, cytosine, and thymine. RNA has all of these except thymine, in place of which it has uracil. Complementary base pairing is important in the structure and functioning of nucleic acids. Adenine always pairs with thymine or uracil, and guanine always pairs with cytosine. Base pairing makes possible the accurate copying of the DNA molecule, because one strand can act as a template for the synthesis of the complementary strand.

In DNA, the two complementary polynucleotide strands form a double helix. As with other macromolecules, the structures of DNA and RNA are intimately linked to their functions in the cell.

12. Explain how the structure of DNA, on the one hand, and that of protein molecules, on the other, are appropriate to the functions these molecules perform in the cell.

13. Give two reasons why guanine always pairs with cytosine, and adenine with thymine or uracil.

14. Answer: In what form is information about proteins stored in the DNA molecule?

ARTICLE RESOURCES

Carbohydrates: Monosaccharides and Disaccharides

Jenkins, David J. A., et al. Glycemic index of foods: A physiological basis for carbohydrate exchange. American Journal of Clinical Nutrition, 1981, 34(3), 362-366. Experiments to determine the effects of 62 commonly eaten foods on blood glucose. Blood glucose was measured over 2-hour periods.

Sharon, Nathan. Carbohydrates. Scientific American, 1980, 243(5), 90-116. History of the discovery of glucose, amino sugars, and sugar nucleotides. Good atomic models of monosaccharides and polysaccharides. Discusses the role of carbohydrates in such genetic diseases as galactosemia, Hurler's syndrome, Tay-Sachs, Hunter's syndrome, and diabetes. Various other roles of carbohydrates are shown, such as that of cell-surface saccharides.

Carbohydrates: Polysaccharides

Lipids

Rakshit, A. K. G., et al. Monolayer properties of fatty acids: Surface vapor pressure and the face energy of compression. Journal of Colloidal Interface Science, 1981, 80(2), 466-473. Recommended for the instructor. Saturated, unsaturated, and hydroxy fatty acids were measured in monolayers spread at the air-water interface. Acetylenic and hydroxy acids had significant chain-water reactions.

Mankin, Eric. The body's elusive master chemical. Science Digest, 1982, 90(2), 42-43, 94. Thorough discussion of prostaglandins. Gives a clear explanation of what they are (fatty acids) and what they do.

Amino Acids and Proteins

Levels of Protein Structure

Cellular protein factories. Science Digest, 1982, 90(3), 103. Brief article on protein synthesis. A fairly clear diagram is presented, showing the role of mRNA and tRNA.

Oldendorf, William H., and William Zabielski. Liquid lightning in your nerves. Science Digest, 1982, 90(5), 82-83, 116. Discussion of the workings of neurons in the brain, especially the chemical and electrical nature of these cells. Peptides within the neuron may be the key to the response mechanism by responding to the chemical messengers on the surface of the neurons.

Nucleic Acids

Gilbert, Walter. DNA sequencing and gene structure. Science, 1981, 214(4527), 1305-1312. Nobel lecture outlining the steps in determining the chemical sequence in DNA by labeling DNA with radioactive phosphorus, isolating it by electrophoresis and cutting it with a restrictive enzyme. Between 200 and 400 base sequences can now be read from the point of labeling. DNA is a linear information store, so it can be read easily. Proteins, however, are strings of amino acids brought into a wide variety of chemistries by nature. Protein fragments exhibit different properties and there is no way of keeping track of the total content of amino acids; contiguous coding sequences are the rule for prokaryotic genes. A complex exon-intron structure characterizes higher eukaryotes.

Chedd, Graham, Genetic gibberish in the code. Science 81, 1981, 2(9), 50-55. The history of the genetic code and how it was discovered. Good diagram of how DNA is programmed to transcribe base sequences to RNA. Includes photographs of several leading geneticists who were instrumental in decoding DNA.

ESSAY QUESTIONS

1. Explain the difference between α-linkage and β-linkage between two glucose molecules. Of what significance is this difference to a living system?

    These differ in the placement of H- and -OH groups attached to the C-1 carbon of the molecule. Enzymes that can attack an alpha linkage are unable to attack a beta linkage, and vice versa.

2. Why are starch and glycogen called storage carbohydrates?

    Each is made up of glucose monomers and can be digested to yield glucose, which can then be processed to release energy.

3. How does the use of lipids (as opposed to polysaccharides) as energy-storage compounds represent an adaptation in animals?

    Lipids are capable of storing more energy per gram than the more bulky polysaccharides and are thus more efficient for mobile organisms. Plants do not move, so excess bulk is of little consequence.

4. Some lipids cannot be synthesized by animals and must be obtained in the diet. What does that fact have to do with peanut butter?

    Untreated peanut butter is rich in unsaturated fatty acids, but it is runny and objectionable to some people. Processed peanut butter is hydrogenated to increase the saturated fat content and make it less runny. This also reduces the value of the peanut butter as a source of unsaturated fatty acids.

5. Name and define the four levels of protein structure.

    Primary structure: the linear sequence of amino acids in a polypeptide.

    Secondary structure: formation of an α-helix, β-pleated sheet, or γ-helix, as a result of hydrogen bonding between R groups.

    Tertiary structure: location in space of every atom in the protein molecule; maintained by interactions between charged groups, van der Waals interactions and hydrophobic interactions.

    Quaternary structure: a property of proteins composed of more than one polypeptide chain.

6. Carrier proteins are usually highly specific in the types of molecules they can carry. Based on the information contained in this section, can you suggest a reason for this phenomenon?

    The tertiary structure of carrier proteins determines which molecules will fit into the "gate" provided by the carrier (analogous to enzyme-substrate interactions).

7. Explain how the structure of DNA, on the one hand, and that of protein molecules, on the other, represent adaptations at the macromolecular level.

    "DNAs are similar and uniform and are read by a single type of machinery; proteins are diverse and interact with a diversity of other compounds." DNA is an informational molecule, while proteins are highly specific. (See Chapter 1 for a discussion of information capacity and information content.)

8. An important technique in the study of proteins is the process of electrophoresis. A sample of mixed proteins is placed on an appropriate carrier and subjected to an electric field. Some of the proteins will move toward the positively charged electrode, others will move toward the negative electrode, and still others will not move at all. From your knowledge of the structure of proteins and the nature of their component amino acids, explain why electrophoresis is an effective technique for separating even quite similar proteins.

Proteins are linear polymers of amino acids. Because each amino acid carries on amino group and a carboxyl group, as well as a side chain, individual amino acids have either a positive charge, a negative charge, or no charge. The algebraic sum of the charges on the amino acid is the net charge on the protein molecule. All else being equal, two proteins with sufficiently different primary structures will carry different net charges. Thus they are attracted to one or the other of the electrodes to a greater or a lesser degree and may be separated by electrophoresis.

TEST SET 3-A

Name _____   Section _____

Choose the best answer to each of the following questions, and write the appropriate letter in the space provided.

Ans: a
p. 46

_____   1.   Oligosaccharides are composed of: (a) a few simple sugars linked together; (b) a group of polysaccharides; (c) a dimer of glucose; (d) polymers of maltose.

Ans: c
p. 47

_____   2.   Glucose and fructose are: (a) oligomeric riboses; (b) dimeric pentoses; (c) isomeric hexoses; (d) monomeric pentoses.

Ans: d
p. 48

_____   3.   Maltose differs from cellobiose in that: (a) cellobiose is a dimer of fructose; (b) maltose is a dimer of fructose; (c) one is made of D-glucose and the other of L-glucose; (d) one is alpha-linked glucose and the other is beta-linked glucose.

Ans: a
p. 48

_____   4.   Amylose, a polysaccharide consisting of alpha-linked glucose monomers, is also called: (a) starch; (b) cellulose; (c) acetyl-chitobiose; (d) sucrose.

Ans: b
p. 51

_____   5.   Glycogen, sometimes called animal starch: (a) serves as the main energy reserve in animals; (b) provides a rapid-access energy store; (c) is slowly metabolized to provide energy; (d) is too stable to be digested easily.

Ans: c
p. 52

_____   6.   Because it is so highly reduced, the most efficient energy-storage substance for cells is: (a) glucose; (b) starch; (c) lipid; (d) protein.

Ans: b
p. 52

_____   7.   Triglycerides are composed of: (a) three glycerol molecules bound to a fatty acid; (b) three fatty acids bound to a glycerol molecule; (c) three glycerol molecules loosely bound to a carbohydrate; (d) a mixture of three isomers of glycerol.

Ans: a
p. 56

_____   8.   Cholesterol is a(an): (a) steroid; (b) hormone; (c) bile salt; (d) adrenal corticoid.

Ans: d
p. 55

_____   9.   Carotenoids are important in the human diet as precursors of: (a) vitamin C; (b) phototropic proteins; (c) vitamin D; (d) rhodopsin.

Ans: c
p. 57

_____   10.   In the polymerization of amino acids to protein, the bond between the amino acids is called a(an): (a) polypeptide bond; (b) amino linkage; (c) peptide linkage; (d) amide bond.

Ans: a
p. 57

_____   11.   The sequence of amino acid residues in a polypeptide chain defines protein structure at the: (a) primary level; (b) secondary level; (c) tertiary level; (d) quaternary level.

Ans: b
p. 58

_____   12.   The R groups of amino acids such as valine, isoleucine, and phenylalanine are usually found on the interior of a protein molecule because they are: (a) sensitive to perturbation; (b) hydrophobic in nature; (c) smaller than other R groups; (d) hydrophilic.

Ans: b
p. 60

_____   13.   At what level of protein structure do polypeptides form alpha-helices or beta-pleated sheets? (a) primary; (b) secondary; (c) tertiary; (d) quaternary.

Ans: d   _____   14.   A strong, rigid, unstretchable protein is found in cartilage,
p. 61   skin, cornea, and tendons. This material, which has a triple
                              helix for a secondary structure, is called: (a) myoglobin;
                              (b) keratin; (c) dermatin; (d) collagen.

Ans: c   _____   15.   Through hydrogen bonding, the two polynucleotide strands of DNA
p. 67   pair and twist to form a characteristic: (a) alpha helix;
                              (b) beta helix; (c) double helix; (d) triple helix.

TEST SET 3-B

Name _____   Section _____

Choose the best answer to each of the following questions, and write the appropriate letter in the space provided.

Ans: d
p. 51

_____ 1. Which of the following is not composed of polysaccharide? (a) glycogen; (b) cotton; (c) starch; (d) wool.

Ans: a
p. 47

_____ 2. Which of the following is an isomer of glucose? (a) fructose; (b) hexose; (c) sucrose; (d) lactose.

Ans: b
p. 48

_____ 3. The disaccharide maltose is composed of two molecules of: (a) sucrose; (b) glucose; (c) cellobiose; (d) fructose.

Ans: d
p. 48

_____ 4. Amylose is another name for: (a) cellulose; (b) cellobiose; (c) amylase; (d) starch.

Ans: b
p. 52

_____ 5. The main energy reserves in animals are maintained as: (a) glucose; (b) fat; (c) starch; (d) glycogen.

Ans: b
p. 52

_____ 6. Fatty acids are: (a) simple lipids; (b) carboxylic acids with a long hydrocarbon chain; (c) glycerol with a long-chain acid attached; (d) triglycerides.

Ans: c
p. 53

_____ 7. Peanut butter is usually hydrogenated to: (a) improve the taste of the lipid; (b) preserve the lipid; (c) saturate the oil; (d) make the product more nutritious.

Ans: a
p. 55

_____ 8. Which of the following is not a member of the class of steroid lipid materials? (a) beta carotine; (b) testosterone; (c) estrogen; (d) cholesterol.

Ans: d
p. 57

_____ 9. A polypeptide is a string of at least 100 amino acids joined by: (a) ionic bonds; (b) an alpha helix; (c) beta linkages; (d) peptide linkages.

Ans: b
p. 60

_____ 10. A disulfide bridge in a polypeptide can be formed between the side chains of two: (a) leucines; (b) cysteines; (c) phenyl-alanines; (d) glycines.

Ans: a
p. 60

_____ 11. Which of the following is the type of secondary structure characteristic of the keratin found in hair and animal claws? (a) alpha-helix; (b) beta-helix; (c) gamma-helix; (d) beta-pleated sheet.

Ans: d
p. 63

_____ 12. The tertiary structure of myoglobin is a good example of a: (a) planar protein; (b) long-chain protein; (c) random chain; (d) globular protein.

Ans: c
p. 63

_____ 13. Protein quaternary structure consists of: (a) refolding the tertiary structure; (b) the formation of helix structures; (c) the association of subunits; (d) the determination of posi-tion of disulfide bridges.

Ans: b
p. 66

_____ 14. The adenine and thymine residues in one strand of a DNA molecule always pair with: (a) cytosine and adenine, respectively; (b) thymine and adenine, respectively; (c) thymine and cytosine, respectively; (d) cytosine and guanine, respectively.

Ans: c
p. 66

_____ 15. Ribonucleotides differ from the bases in DNA in that they in-clude: (a) thymine and use ribose sugars; (b) ribose sugars instead of pentose; (c) uracil and use ribose sugar; (d) uracil and use hexoses instead of pentoses.

CHAPTER FOUR
ORGANIZATION OF THE CELL

CHAPTER SUMMARIES AND OBJECTIVES

4-1 <u>A Comment on Size</u> The cell theory states that all organisms are composed of cells and that all cells come from preexisting cells. Each cell is endowed with all the genetic information necessary to produce an entire organism. All living things are composed of one of two types of cells, either prokaryotic or eukaryotic.

Cells must be large enough to contain all the necessary biochemical machinery. The major factor that places an upper limit on cell size is surface-to-volume ratio.

1. Explain what is meant by surface-to-volume ratio, and show how this relationship limits cell size.

4-2 <u>The Prokaryotic Cell</u> All members of the kingdom Monera are prokaryotic cells. The prokaryotic cell is surrounded by a plasma membrane. Inside the plasma membrane are the nucleoid, composed of DNA, and the cytoplasm. The cytoplasm contains numerous ribosomes, which are the sites of protein synthesis. Some prokaryotes also have a cell wall outside the plasma membrane and there may be a capsule or sheath outside the cell wall. Photosynthetic prokaryotes have molecules of chlorophyll embedded in specialized areas of the membrane. Mesosomes may be present in some prokaryotes. Some prokaryotes are able to swim by means of a whiplike flagellum, which spins like a propeller.

2. Sketch a typical prokaryotic cell and clearly label all structures.

4-3 <u>Probing the Subcellular World: Microscopy</u> The resolving power of a microscope is the smallest distance between two objects at which they can be distinguished as separate. The light microscope has a resolving power of about 200 nm. The electron microscope has a resolving power of about 1.5 nm. There are two types of electron microscopes. In transmission electron microscopy, a beam of electrons passes through the object under observation; in scanning electron microscopy, the beam is reflected from the object, revealing surface features.

3. Answer: What are the drawbacks of each type of just described microscopy?

4. Distinguish between magnification and resolution.

4-4 <u>The Eukaryotic Cell</u> / <u>Membranes in Eukaryotic Cells</u> All organisms other than members of the kingdom Monera are composed of eukaryotic cells. These differ from prokaryotic cells in that they contain numerous membrane-bound organelles, which are specialized for various functions.

Membranes regulate the flow of materials across them, and they may perform a variety of other functions in the cell.

5. Give at least three examples of the specialized functions of membranes in eukaryotic cells.

4-5 Nucleus and Cytoplasm / Structures of the Nucleus The largest organelle in any cell is usually the nucleus. The material outside the nucleus is the cytoplasm of the cell. The nucleus controls the activities that are carried on in the cytoplasm.

The nucleus is surrounded by a nuclear envelope of two membranes. The outer membrane is continuous with the endoplasmic reticulum. Eukaryotic DNA is combined with proteins to form chromatin. At certain times the chromatin becomes organized into discrete strands called chromosomes. Also present within the nuclear envelope are spherical nucleoli, which are sites for ribosome assembly.

6. Differentiate among cytoplasm, nucleoplasm, and cytosol.

7. Answer: What is the function of chromosomes?

4-6 The Mitochondria / Plastids / Origin of the Eukaryotes The mitochondrion functions in the release of energy from food molecules. These are initially processed in the cytosol. The resulting products are taken up by the mitochondrion and used to manufacture ATP. Mitochondria are surrounded by two membranes. The inner one is infolded to form cristae, thus increasing the surface area available for the attachment of proteins involved in cellular respiration. The region enclosed by the inner membrane is the matrix.

Plastids are found only in plants and certain protists. The most common type is the chloroplast, which carries on photosynthesis. The chloroplast is surrounded by two membranes. The inner one is continuous with the grana, which are stacks of flat sacs called thylakoids. The photosynthetic pigments are located on the thylakoids. Chromoplasts contain the pigments that color certain flowers and fruits. Leucoplasts are sites for storage of starch and fats. All plastids develop from proplastids.

Most scientists agree that eukaryotes evolved from prokaryotes. One theory that has been offered to explain how this might have taken place is the endosymbiotic theory. According to this theory, mitochondria, chloroplasts and perhaps other eukaryotic organelles are derived from prokaryotes that originally lived within other primitive cells in a symbiotic relationship. There is much evidence to support this theory, as well as some objections to it.

8. Describe the similarities in structure of mitochondria and chloroplasts. How are these similarities related to their reciprocal functions of photosynthesis and respiration?

9. Define and give an example of symbiosis.

10. Discuss the endosymbiotic theory, citing evidence to support your statements. What is one of the primary objections to the theory?

4-7 Endoplasmic Reticulum / The Golgi Apparatus The endoplasmic reticulum (ER) is a network of membranous tubes running through the cytoplasm. It is continuous with the nuclear membrane. Ribosomes become attached to areas called rough ER. Areas lacking ribosomes are smooth ER, which functions to modify proteins. Many important enzymes are associated with ER. Only those proteins that are to be exported from the cell are synthesized on rough ER. Those that are to remain in the cell are synthesized on ribosomes floating free in the cytoplasm. The signal hypothesis has been proposed to explain how the cell determines which proteins are to be made where. According to this hypothesis, the first few amino acids of an export protein act as a signal to cause the ribosomes to become attached to rough ER.

The Golgi apparatus is a series of flattened, membrane-bound sacs (dictyosomes). It is a site for storage, chemical modification, and packaging of proteins. Some proteins processed by the Golgi apparatus are retained in the cell, while others are released to the environment.

11. List three functions of the endoplasmic reticulum.

12. Explain the process whereby export proteins move from their site of synthesis to the outside of the cell.

13. Briefly explain the signal hypothesis.

4-8 Endocytosis and Exocytosis / Lysosomes / Microbodies / Vacuoles   Materials to be exported from the cell are packaged in vesicles, which merge with the plasma membrane, expelling their contents to the environment. This process is called exocytosis. By the reciprocal process, endocytosis, materials from the environment may be taken up by the cell. The uptake of liquids is called pinocytosis; the uptake of solid matter is called phagocytosis. In either case, the ingested material becomes surrounded by a vesicle as it passes through the plasma membrane.

One of the functions of primary lysosomes is the digestion of materials taken up via endocytosis. Primary lysosomes begin as vesicles released by the Golgi apparatus. When a primary lysosome merges with a food-containing vesicle, a secondary lysosome is formed. The ingested material is digested by enzymes contained within the lysosome.

Microbodies are small, membrane-bound compartments derived from rough ER. They function to segregate biochemical processes from the rest of the cell.

Vacuoles are droplets of solution surrounded by a single membrane. Vacuoles function as storage sites for food, pigments, or wastes. A specialized type of vacuole, the contractile vacuole, functions to rid the cell of excess water.

14. Answer: Of what benefit to the cell is it to have certain enzymes contained within lysosomes? Within microbodies?

15. List three ways in which vacuoles may be involved in the "social" relationship between plants and animals.

16. Give two ways in which storage vacuoles are useful to nonwoody plants.

4-9 Microtubules and Microfilaments, Cilia and Flagella / Fine Structure of the Cytosol   Microtubules are long tubes of protein subunits. Flagella and cilia are arrays of microtubules arranged in a characteristic "9 + 2" pattern. Sliding of the microtubules past each other produces the motion of the flagella and cilia. Energy for this process comes from ATP. Microtubules also form the centriole, which may be associated with locomotor structures (and is then called a basal body), or may give rise to the spindle, which is involved in cell division. Microfilaments are fibrous proteins involved in movement at the cellular level, as in cell division.

The microtrabecular lattice forms a network within the cytosol to which other organelles are attached. It seems to be involved in maintenance of cell shape, in intracellular movement, and in cell movement.

17. Describe the structure of eukaryotic flagella. How do they differ from prokaryotic flagella?

18. Explain the role of the spindle in cell division.

19. Answer: What evidence suggested the existence of the microtrabecular apparatus prior to its actual discovery?

4-10 The Cell Wall   Plants, fungi, and some protists possess a cell wall. In plants it is composed chiefly of cellulose, often with waterproofing agents. In fungi, the cell wall contains chitin. The cell wall lies outside the plasma membrane and provides strength and rigidity to the cell. Plasmodesmata are channels through the cell wall that provide connections between the cytoplasm of adjacent cells.

20. List three functions of the cell wall.

21. Answer: What is the purpose of plasmodesmata?

4-11 <u>Eukaryotes, Prokaryotes, and Viruses</u> Eukaryotes differ from prokaryotes in several ways. (1) Prokaryotes have a nucleoid, whereas the eukaryote nucleus is
surrounded by a membrane. (2) Prokaryotes lack membrane-bound organelles found
in eukaryotes. (3) The cell walls of prokaryotes differ from those of eukaryotes
in composition and structure. (4) The DNA of prokaryotes is naked, whereas that
of eukaryotes is combined with protein.

Viruses cannot be considered cells, because they consist only of a small piece of
nucleic acid surrounded by a protein coat.

22. Cite the major differences between prokaryotic and eukaryotic cells.

23. Answer: What is the function of the protein coat in viruses?

4-12 <u>Fractionating the Eukaryotic Cell: Isolating Organelles</u> In order to study the
biochemistry and functioning of cellular components, it is necessary to isolate
and purify organelles. To do this, cells are first carefully broken open. Organelles are separated by centrifugation, which separates particles on the basis
of their rates of sedimentation under centrifugal force. Equilibrium centrifugation can be used to separate organelles that differ in density, whereas differential centrifugation can be used to separate organelles that differ in either density or radius.

24. Explain how either of the two centrifugation techniques discussed in this
section enables one to separate different types of organelles.

ARTICLE RESOURCES

<u>A Comment on Size</u>

<u>The Prokaryotic Cell</u>

Foltz, Charles V., and George C. Hartmann. Demonstration of cellular organelles
using a teflon slide culture chamber. <u>Journal of College Science Teaching</u>, 1982,
9(5), 301-302. Using phase microscopy, students are able to observe nuclear
structure, mitochondria, vacuoles, and other cell inclusions in in a mycelial frag-
ment from <u>Basidiobolus ranarum</u>. This technique can be used for studying the
growth of many filamentous structures as well. It can also be used to study the
effects of gibberellic acid on <u>Penicillium</u>.

<u>Probing the Subcellular World: Microscopy</u>

Ford, Brian J. Found: The lost treasure of Anton von Leeuwenhoeck. <u>Science Di-</u>
<u>gest</u>, 1982, 90(3), 88-92, 110. Specimens actually used by Leeuwenhoeck have been
found in the vaults of the Royal Society of London. Step-by-step directions for
making a Leeuwenhoeck microscope are included.

<u>The Eukaryotic Cell</u>

<u>Membranes in Eukaryotic Cells</u>

Lodish, H. G., and J. E. Rothman. The assembly of cell membranes. <u>Scientific</u>
<u>American</u>, 1979, 240(1), 48-63. Expands on textbook discussion of membranes.
Useful for instructor as well as student. Discusses the role of lipids and proteins in the formation of cell membranes. Cites the work of Milstein and Brownlee at Cambridge and of Ghosh and Toneguzzo at McMaster, which revealed the
existence of a signal sequence-- a group of amino acids-- essential to the distinction between membrane proteins and secreted proteins.

Staehelin, Andrew L., and Barbara E. Hull. Junctions between living cells.
<u>Scientific American</u>, 1978, 238(5), 140-152. Electron microscopy of animal tis-
sues reveals intercellular junctions within the plasma membrane. These latter
junctures (called "tight" junctures) are excellent examples of biological engineering at the cellular level. Well illustrated. Good, clear discussion of
how cell samples are prepared for electron microscopy.

<u>Nucleus and Cytoplasm</u>

<u>The Mitochondria</u>

<u>Plastids</u>

<u>Origin of the Eukaryotes</u>

<u>Endoplasmic Reticulum</u>

<u>The Golgi Apparatus</u>

<u>Endocytosis and Exocytosis</u>

<u>Lysosomes</u>

<u>Microbodies</u>

<u>Vacuoles</u>

<u>Microtubules and Microfilaments, Cilia and Flagella</u>

Dustein, Pierre. Microtubules. <u>Scientific American</u>, 1980, 242(2), 66-76. Beautifully illustrated article on these structural organelles. Gives a brief history of their discovery and investigation. Microtubules provide a scaffolding for the cell. The shape of many complex structures, especially in the protozoans, depends on the presence of microtubules. Discusses cilia and flagella as well as spindle fibers as microtubules.

<u>Fine Structure of the Cytosol</u>

<u>The Cell Wall</u>

Sperm's key to the egg. <u>Science Digest</u>, 1982, 90(6), 90. Reports by William Linarz, Daniel Rossignol and Glenn Decker, of Johns Hopkins University School of Medicine, on research on sea urchin egg fertilization. Efforts to determine exactly how a sperm is able to penetrate the egg membrane have led to a large, sperm-binding receptor molecule on the egg's membrane, and a protein, binder, forming the tip of the sperm.

Sandler, Naomi and Alex Keyhan. Cell wall synthesis and initiation of deoxyribonucleic acid replication in <u>Bacillus subtilis</u>. <u>Journal of Bacteriology</u>, 1981, 148(2), 443-449. A physical connection between the outer membrane of <u>E. coli</u> and the initiation of chromosome replication has been found. It was further substantiated in <u>B. subtilis</u> when these cells were treated with Vancomycin and penicillin. Both are inhibitors of cell wall formation. Penicillin binds to the cell wall. Vancomycin is found within the cell wall and plasma membrane. This affects the membrane binding of the DNA-initiation complex at the internal surface.

<u>Eukaryotes, Prokaryotes, and Viruses</u>

<u>Fractionating the Eukaryotic Cell: Isolating Organelles</u>

ESSAY QUESTIONS

1. Explain what is meant by the term "totipotent," and give one piece of evidence to support the conclusion that cells are totipotent.

   "Totipotent" refers to the fact that every cell contains all the necessary genetic instructions to regenerate an entire organism. Any form of cloning (as with carrots) demonstrates the totipotency of the cloned cells.

2. Distinguish between magnification and resolution.

   Magnification is the apparent increase in size of an object viewed through a lens. Resolution is the smallest distance between two objects such that they can be distinguished as separate.

3. What is the function of chromosomes?

Chromosomes carry the genetic information of the cell. Duplication of chromosomes and their subsequent separation during cell division ensures that each daughter cell will get the correct complement of genetic information from the parent.

4. Discuss the experiments on Acetabularia described in the text, and indicate what these experiments demonstrated.

Reciprocal grafting of the nucleated and non-nucleated portions of the cells of two species of Acetabularia demonstrated that the nucleus controls the activity of the cytoplasm. Regardless of the manner in which the graft is made, the regenerated cap is like that of the species which supplied the nucleus.

5. Define and give an example of symbiosis.

Symbiosis is the living together of two dissimilar organisms in a relationship stabilized by the mutual benefits conferred on the symbiotic partners. Examples include lichens and Myxotricha paradoxa.

6. Discuss the endosymbiotic theory, citing evidence to support your statements. What is one of the primary objections to the theory?

The endosymbiotic theory holds that mitochondria, plastids, and perhaps some other organelles are derived from originally endosymbiotic prokaryotes. Evidence includes: (1) Mitochondria and plastids are surrounded by a double membrane; one of these could have been the plasma membrane of the endosymbiont, while the other could have been acquired when the symbiont was ingested by the host cell. (2) Both chloroplasts and mitochondria contain ribosomes and DNA. These are more like the ribosomes and DNA of prokaryotes than they are like the ribosomes and DNA of the cells in which mitochondria and chloroplasts are found. (3) The primary structures of certain bacterial and mitochondrial enzymes are similar. (4) Some extant organisms contain endosymbionts that suggest evolutionary stages by which endosymbioses can develop. The major objections to the theory are that it does not satisfactorily account for the development of all the intracellular structures of eukaryotes and that most of the DNA required for synthesis of most of the enzymes of mitochondria and chloroplasts is located within the nucleus, not in the organelle itself.

7. Give three reasons why viruses cannot be considered cells.

(1) Viruses consist only of a piece of nucleic acid surrounded by a protein coat.
(2) Viruses cannot carry on metabolism in the absence of a host cell.
(3) Viruses are not surrounded by a plasma membrane.

8. Cite five major differences between prokaryotic and eukaryotic cells.

(1) Prokaryotes have no organelles; eukaryotes do.
(2) The eukaryotic nucleus is surrounded by a nuclear envelope; the prokaryotic nucleus is not.
(3) The cell walls of prokaryotes differ structurally and chemically from those of eukaryotes.
(4) The DNA of prokaryotes is naked; that of eukaryotes is complexed with proteins.
(5) The ribosomes of eukaryotes are larger and heavier than those of eukaryotes.
(6) The flagella of prokaryotes are composed of flagellin and spin like a propeller, whereas the flagella of eukaryotes are composed of tubulin and move by the sliding of the tubulin filaments across each other.

9. Explain the process whereby export proteins move from their site of synthesis to the outside of the cell. How is it determined whether a protein is for export?

Ribosomes synthesizing proteins for export attach themselves to the endoplasmic reticulum. When the protein is complete, it moves via the ER to the forming face of the Golgi apparatus. It then makes its way to the maturing face of that organelle, during which time chemical alterations in the molecule may occur. Vesicles bud off from the maturing face and merge with the plasma membrane, dumping their contents to the outside of the cell. It is believed that the first few amino acids of the export protein act as a signal to cause attachment of the ribosome to the ER.

10. List three ways in which vacuoles may be involved in the "social" relationship between plants and animals.

(1) Vacuoles may store toxic products that make the plants unpalatable to herbivores.
(2) Certain pigments contained in vacuoles may color flowers, thus attracting pollinators.
(3) Pigments within vacuoles may color fruits, making them attractive to animals, that then become agents of seed dispersal.

11. A prokaryotic alga called _Prochloron_ has recently been discovered that contains both chlorophyll _a_ and chlorophyll _b_. This discovery is interesting in view of the fact that no other photosynthetic prokaryotes contain both these types of chlorophyll, whereas _all_ photosynthetic eukaryotes contain both. Explain, then, why proponents of the endosymbiotic theory are especially happy to hear of the discovery of _Prochloron_.

Prochloron is precisely the organism one would want to have as the prokaryotic precursor of eukaryotic chloroplasts, according to the theory of the origin of eukaryotic cells via endosymbiosis.

TEST SET 4-A

Name _____   Section _____

Choose the best answer to each of the following questions, and write the
appropriate letter in the space provided.

Ans: a   _____   1. The most effective single factor limiting the size of cells and
p. 72   organisms is: (a) surface: volume ratio; (b) weight;
                           (c) metabolic rate; (d) acquisition of food for growth.

Ans: b   _____   2. Peptidoglycan, polymerized to form a single molecule for each
p. 72   cell; (a) serves as a cytoskeleton; (b) forms the cell wall
                           in bacteria; (c) forms the capsule for bacteria; (d) supports
                           the internal organization of the cell.

Ans: c   _____   3. In microscopy, the minimum distance at which two objects can be
p. 77   seen as separate structures is called: (a) magnification;
                           (b) separation ability; (c) resolving power; (d) distinguishing
                           distance.

Ans: c   _____   4. The nucleoli of eukaryotic cells function in: (a) control of
p. 87   heredity; (b) control of cell division; (c) ribosome assembly;
                           (d) cellular recognition.

Ans: a   _____   5. Those inner mitochondrial membranes where the structures in-
p. 88   volved with respiratory activity are located are known as:
                           (a) cristae; (b) matrix; (c) lamellae; (d) stroma.

Ans: d   _____   6. One piece of evidence supporting the endosymbiotic origin of
p. 92   some eukaryotic organelles is that: (a) the organelles contain
                           the DNA to control synthesis of the organelles, protein com-
                           plement; (b) they can live independently of the eukaryote host;
                           (c) they are identical to certain free-living forms; (d) they
                           are bound by a double membrane.

Ans: b   _____   7. One role for smooth endoplasmic reticulum is: (a) protein syn-
p. 94   thesis; (b) detoxification of various drugs; (c) formation of
                           secondary lysosomes; (d) oxidative metabolism.

Ans: d   ____   8. The Golgi apparatus is formed by: (a) fusion of vesicles from
p. 96   the plasma membrane; (b) fusion of vesicles from the nuclear mem-
                           brane; (c) fusion of vesicles from the rough endoplasmic reticu-
                           lum and the plasma membrane; (d) fusion of vesicles from the
                           smooth endoplasmic reticulum and/or the nuclear membrane.

Ans: c   _____   9. The fusion of an endocytotic vesicle with a primary lysosome
p. 97   results in: (a) formation of a vacuole; (b) loss of lysosomal
                           enzymes; (c) formation of a secondary lysosome; (d) synthesis
                           of new lysosomal enzymes.

Ans: a   _____   10. The flagellae of eukaryotic and prokaryotic cells are composed
pp. 100,101   of: (a) tubulin and flagellin, respectively; (b) flagellin and
                           tubulin, respectively; (c) actin and myosin, respectively;
                           (d) microtubulin and flagellin, respectively.

Ans: b   _____   11. The materials lignin and suberin are found in: (a) protist
p. 104   motile organelles; (b) certain plant cell walls; (c) animal
                           connective tissue; (d) bacterial cell walls.

Ans: c   _____   12. The cytosol of most cells seems to have a fine structure, or
p. 103   network, called the: (a) microtubular cytoskeleton; (b) micro-
                           filamentary lattice; (c) microtrabecular lattice; (d) cytoskele-
                           tal lattice.

Ans: b
p. 106

_____ 13. Equilibrium centrifugation separates organelles on the basis of: (a) mass; (b) density; (c) mass and density; (d) radius and density.

Ans: a
p. 98

_____ 14. The peroxisome is a microbody in the cell in which: (a) reactions involving toxic peroxides are isolated; (b) the peroxides necessary for metabolism are synthesized; (c) drugs are detoxified; (d) lysosomal enzymes are peroxidized and destroyed.

Ans: a
p. 97

_____ 15. Pinocytosis is one form of: (a) endocytosis; (b) exocytosis; (c) phagocytosis; (d) pericytosis.

TEST SET 4-B

Name _____   Section _____

Choose the best answer to each of the following questions, and write the appropriate letter in the space provided.

Ans: b
p. 71
_____   1. The fact that a single carrot somatic cell can contain all the information to produce a complete carrot under proper conditions means that the cell is: (a) genetically competent; (b) totipotent; (c) a germ cell; (d) a sporophyte.

Ans: a
p. 72
_____   2. The ratio of surface area to volume is probably the single factor placing the greatest limit on: (a) cell size; (b) metabolic rate; (c) rates of movement of materials through membranes; (d) the rate of cell growth.

Ans: d
p. 72
_____   3. The cell wall of a moneran would probably be composed of: (a) a capsule or sheath of "slime"; (b) cellulose; (c) lipoprotein; (d) peptidoglycan.

Ans: b
p. 76
_____   4. Escherichia coli is categorized with the bacteria: (a) cocci; (b) bacilli; (c) spirilla; (d) spirochetes.

Ans: d
p. 73
_____   5. A cell posessing a nucleoid and no nuclear envelope is: (a) protistan; (b) eukaryotic; (c) from a fungi; (d) moneran.

Ans: a
p. 85
_____   6. The shape of the cap of a cell of Acetabularia is controlled by the nucleus which is found in the: (a) rhizoid; (b) stem; (c) cap; (d) mycelium.

Ans: a
p. 87
_____   7. The basic assembly of ribosomes from protein and RNA is performed by: (a) nucleoli; (b) the ER; (c) the nucleus; (d) the Golgi.

Ans: c
p. 90
_____   8. Grana, DNA, and ribosomes in chloroplasts are suspended in: (a) matrix material; (b) a phospholipid membrane; (c) stroma; (d) vesicles.

Ans: c
p. 96
_____   9. The membranes of the forming face of a cell's Golgi apparatus are probably derived from: (a) mitochondrial membranes; (b) lysosomal membranes; (c) smooth ER or nuclear membranes; (d) phagocytic vesicles.

Ans: b
p. 97
_____   10. Export of material from the cell with concomitant transformation of the plasma membrane involves the process of: (a) phagocytosis; (b) exocytosis; (c) pinocytosis; (d) endocytosis.

Ans: d
p. 97
_____   11. The process of intracellular digestion generally involves the enzymatic activity inside: (a) lysosomes; (b) cytoplasm; (c) phagocytic vesicles; (d) secondary lysosomes.

Ans: b
p. 100
_____   12. In some protists an organelle is responsible for accommodating the different concentrations of salt between cytoplasm and the freshwater environment. This organelle is called a(n): (a) kidney; (b) contractile vacuole; (c) sodium pump; (d) osmotic vacuole.

Ans: a
p. 101
_____   13. The flagellae and cilia of eukaryotes always show: (a) a "9 + 2" arrangement of microtubules; (b) a "9 + 2" arrangement of microfilaments; (c) the same "9 + 2" arrangement of fibers as do the procaryotes; (d) a "9 + 2" arrangement of the protein called flagellin.

Ans: d   _____ 14. Revealed in the mid-1970s, a fine structure characteristic of
p. 103   cellular cytosol was named the: (a) filamentary cytoskeleton;
                        (b) cytoskeletal lattice; (c) cellular microskeleton;
                        (d) microtrabecular lattice.

Ans: c   _____ 15. A method for collecting large quantities of relatively pure cell
p. 106   components or organelles involves differential sedimentation of
                        fractured cells in: (a) a chromatography tank; (b) purified
                        organic solvents; (c) a centrifuge; (d) hyperosmotic solutions.

CHAPTER FIVE
MEMBRANES

CHAPTER SUMMARIES AND OBJECTIVES

5-1 <u>Membrane Structure and Composition / Membrane Lipids / Membrane Proteins / Membrane Carbohydrates</u> Membranes are thin layers of lipid molecules interspersed with proteins. The biological activity of a membrane is directly related to its physical structure. Three types of compounds are found in membranes: lipids, proteins, and carbohydrates.

Most membrane lipids are phospholipids, which have both hydrophobic and hydrophilic regions. The molecules orient themselves such that their polar heads face the aqueous interior and exterior of the cell. The hydrophobic portions form a layer sandwiched between.

There are two basic types of membrane proteins: intrinsic and extrinsic proteins. The former are embedded in the lipid bilayer and may protrude all the way through it, whereas the latter are outside the bilayer, attached to other membrane components by weak bonds. The nature of the protein, whether intrinsic or extrinsic, is determined by its tertiary structure.

Membrane carbohydrates occur as short chains of monosaccharides that are attached either to a lipid or to a protein molecule. The carbohydrates are extremely varied in structure, largely because they can form branching patterns. They are important in recognition reactions at the cell surface.

1. Diagram a typical membrane seen in cross section.

2. List three functions carried out by membrane proteins, two principal functions of membrane lipids, and three characteristics of membrane carbohydrates.

3. Explain how the chemical properties of phospholipid molecules result in the formation of the membrane bilayer.

4. Membranes are able to fuse with one another and to seal themselves spontaneously. Explain these phenomena in terms of the structure of phospholipid molecules.

5. Differentiate between intrinsic and extrinsic proteins.

6. Answer: How does the tertiary structure of a protein determine whether it will be an intrinsic or an extrinsic protein? Why do intrinsic proteins resist removal from the lipid bilayer?

7. Give two types of evidence to support the conclusion that some membrane pro-

teins are free to migrate within the lipid bilayer.

8. Distinguish between glycolipids and glycoproteins.

9. Give two reasons why membrane carbohydrates can exhibit such diversity although they are composed of only nine different monosaccharides.

5-2 Microscopic Views of Biological Membranes Membranes are best seen with the electron microscope. Freeze-fracturing, in which the membrane is frozen solid and then broken, is used to visualize the internal arrangement of the membrane. The break often splits the lipid bilayer, revealing the intrinsic proteins. In freeze-etching, the preparation is exposed to vacuum to bring out the texture of the membrane. With either of these techniques, the contours of the exposed surface may be brought out by shadowing with platinum.

10. Describe how one of the techniques mentioned in this section enhances the ability of biologists to study membranes with the electron microscope.

5-3 Movement of Materials in the Microscopic World / Crossing the Membrane Barrier / An Explanation of Simple Diffusion The random movements of ions or molecules in solutions results in the particles becoming evenly distributed throughout the solution. This process is called diffusion. All solutes diffuse at a rate determined by their physical properties, whereas the direction of diffusion depends on the concentration gradient of the solute. The presence of a membrane, however, affects the behavior of the solute molecules.

There are three types of movement across biological membranes: simple diffusion, facilitated diffusion, and active transport. In simple diffusion, molecules of a given substance pass across the membrane until the concentration is equal on both sides.

Hydrophobic molecules tend to pass through the lipid bilayer easily, while hydrophilic substances are prevented from doing so. Water and other small molecules (although not hydrophobic) freely pass, owing to the presence of small pores in the membrane.

11. List four factors on which the rate of diffusion of a particular solute molecule depends.

12. Define diffusion and explain how the random movement of molecules can result in the even distribution of a solute throughout a solution.

13. Define simple diffusion, facilitated diffusion, and active transport.

14. Answer: What is a concentration gradient?

15. Explain what is meant by "saturation" with respect to the transportation of molecules across membranes.

16. Answer: How are hydrophilic molecules such as water able to traverse the lipid bilayer?

17. Answer: With the exception of molecules like water and oxygen, what determines the rate at which a particular molecule diffuses across the lipid bilayer?

5-4 Facilitated Diffusion and Active Transport In facilitated diffusion, solute molecules combine with intrinsic proteins called carriers, which act as "gates" to permit the solute to cross the membrane. Because there is a limited number of carrier molecules, the system can become saturated when all of the carriers are "busy" moving solute molecules. Carrier proteins are very specific. Solutes may be moved in either direction, depending on the concentration gradient.

Active transport is also made possible by the activities of carrier molecules. The important difference is that active transport consumes energy, thus enabling cells to accumulate certain substances against the concentration gradient. Active transport systems work in one direction only, pumping materials either into or out of the cell.

18. Answer: Why does facilitated diffusion always result in the movement of molecules from a region of higher concentration?

19. Facilitated diffusion may be compared to the opening of a channel and active transport may be compared to the operation of a pump, even though both processes can become saturated. Justify the use of these analogies.

5-5 Osmosis The movement of water across the membrane is called osmosis. The direction of water movement is determined by the relative concentrations of solute molecules on either side of the membrane. The greater the solute concentration, the more negative the osmotic potential. Water tends to move in the direction of the more negative osmotic potential. Two solutions with the same osmotic potential are said to be isotonic. When the osmotic potentials differ, the solution with the more negative osmotic potential is hypotonic, while the one with the less negative osmotic potential is hypertonic. In cells with sturdy cell walls, osmosis is also regulated by pressure potential. As water is absorbed, pressure builds up and prevents further water from entering. The overall tendency for a cell to absorb water is called the water potential; it is equal to the algebraic sum of the osmotic potential and the pressure potential. Water always moves in the direction of the more negative water potential.

20. Distinguish between the members of each of the following pairs of terms: osmotic potential / osmotic pressure, pressure potential / water potential, hypertonic / hypotonic.

21. Explain why it is incorrect to say that a solution of 0.5 molar sodium chloride is "isotonic."

22. State the difference between simple diffusion and osmosis.

5-6 Membranes and Energy Transformations / Receptors on the Membrane Surface / Other Activities of Membranes Some membranes process energy. They are able to accomplish this because of their capacity to organize proteins and to provide for the separation of electrical charges.

Many substances can be recognized and bound to the membrane surface by membrane proteins and carbohydrates. In many cases, contact of a substance with its receptor protein causes the protein to become altered in tertiary structure, triggering changes in the interior of the cell. Many other biological activities are also associated with membranes.

23. Answer: Why does the organization of a series of proteins on the surface of a membrane facilitate the biochemical process in which those proteins are involved?

24. Answer: On what two factors does the ability of a membrane to separate electrical charges depend?

25. List four substances that are recognized by protein and carbohydrate receptors at the membrane surface.

26. Define and give an example of contact inhibition.

27. Answer: In what way is the structure of a membrane receptor protein or carbohydrate analogous to the structure of an enzyme?

28. List at least five functions of membranes that have been discussed in this chapter.

5-7 Membrane Formation and Continuity Membrane phospholipids are synthesized in the endoplasmic reticulum. Proteins are synthesized at ribosomes and transferred to the plasma membrane via the Golgi apparatus, where carbohydrates may be attached to them. Membrane fragments are constantly moving about within the cell. One source of turnover is the process of exocytosis and its reciprocal process, endocytosis.

29. Membrane proteins are synthesized in several steps. Trace the process of synthesis, giving the location in the cell at which each step occurs.

30. Answer: How is the Golgi apparatus involved in membrane formation and turn-over?

ARTICLE RESOURCES

Membrane Structure and Composition

Membrane Lipids

Membrane Proteins

Membrane Carbohydrates

   Costerton, J. W., G. C. Gusey, and K. J. Cheng. How bacteria stick. Scientific American, 1978, 238(1), 86-95. Profusely illustrated article on the mechanisms whereby polysaccharides on the surface of bacterial cells form a "glycocalyx" that provides adhesion of the bacteria to each other and to other cell surfaces in the vicinity. This glycocalyx is a major determinant in the initiation and progression of bacterial disease.

Microscopic Views of Biological Membranes

   Lodish, H. G., and J. E. Rothman. The assembly of cell membranes. Scientific American, 1979, 240(1), 48-63. Expands on the textbook discussion of membranes. Useful for instructor as well as student. Discusses the role of lipids and proteins in the formation of cell membranes. Cites work of Milstein and Brownlee at Cambridge and Ghosh and Toneguzzo at McMaster, who found the existence of a signal sequence--a group of amino acids--essential to the distinction between membrane proteins and secreted proteins.

Movement of Materials in the Microscopic World

Crossing the Membrane Barrier

   Staehelin, L. Andrew, and Barbara E. Hull. Junctions between living cells. Scientific American, 1978, 238(5), 140-152. Electron microscopy of animal tis-sues reveals intercellular junctions within the plasma membrane. These permit the passage of nutrients in intestinal linings and prohibit the passage of mole-cules across an epithelium. These latter junctures (called "tight" junctures) are excellent examples of biological engineering at the cellular level. Well illustrated. Good, clear discussion of how cell samples are prepared for electron microscopy.

An Explanation of Simple Diffusion

Facilitated Diffusion and Active Transport

   Keynes, Richard D. Ion channels in the nerve-cell membrane. Scientific American, 1979, 240(3), 126-134. Axons of animal nervous systems show an unequal distri-bution of sodium and potassium ions. By means of a voltage-clamp technique, it was possible to show that the potassium and sodium systems worked in totally different ways when the membrane was polarized from when it was depolarized. Includes an excellent diagram of ion channels in axon membranes.

Osmosis

   Dillner, Harry J. Osmosis in plants. The Science Teacher, 1978, 45(2), 33. Clever visual model to illustrate differences between cells in tap water and in 5% salt solution. Uses plastic bags, glass beakers and covers, and colored water.

Membranes and Energy Transformations

Receptors on the Membrane Surface

Other Activities of Membranes

Membrane Formation and Continuity

   Lodish, Harvey F., and James E. Rothman. The assembly of cell membranes.
   Scientific American, 1979, 240(1), 48-63. Excellent article to use to supplement "Membrane Formation and Continuity" in the text. The cell membrane has two
   distinct sides, the inner and outer. This helps to preserve the concentration
   gradients of ions, nutrients, hormones, and other chemical signals. These sides
   show a structural symmetry. Proteins have a fixed asymmetric orientation. Each
   protein of each type is aligned in the same direction. The lipids are responsible for the structural integrity of the membrane. They consist of hydrophobic
   and hydrophilic portions, which spontaneously form a bilayer of molecules.

ESSAY QUESTIONS

1. Choose two of the three primary components of membranes and discuss how their
    chemical and physical properties are related to the roles they play in membrane
    functioning.

       The important ideas are: (1) Phospholipids have both hydrophobic and
       hydrophilic regions, causing them to associate spontaneously in the bilayer
       pattern. The nature of the bilayer not only determines the permeability of
       the membrane but also dictates the nature of the proteins that can be asso-
       ciated with the membrane. (2) The tertiary structure of membrane proteins
       determines whether they will be intrinsic or extrinsic and the nature of
       their functions. (3) The ability of carbohydrates to form branched oligomers
       confers on them the great degree of specificity needed for molecules in-
       volved in recognition reactions.

2. Discuss the adaptations that are evident in the membranes of chloroplasts and
    mitochondria.

       The essential facts here are that both exhibit infoldings which serve to
       increase surface area (cristae and grana) and that the membrane provides a
       matrix in which a variety of enzymes can be organized to carry out compli-
       cated processes much more efficiently than would be the case if the enzymes
       were floating in the cytosol.

3. How does the random movement of molecules in a solution bring about an even
    distribution of molecules throughout the system?

       The movement of each ion or molecule is totally random. A molecule that
       happens to move against the concentration gradient will have a greater
       likelihood of colliding with another particle. Such collisions have the
       effect of causing the molecule to reverse direction. Molecules that move
       into regions of low concentration are less likely to encounter other
       molecules and thus are not pushed back. In this manner, the solute becomes
       evenly distributed over time.

4. Compare and contrast simple diffusion, facilitated diffusion, and active trans-
    port. Include in your answer the terms concentration gradient and carrier
    proteins, and name the source of energy that drives each process.

       In simple diffusion, substances move in the direction of the concentration
       gradient, and no outside energy source is required. The process is driven
       by the energy of molecular movement. In facilitated diffusion, specific
       types of molecules are transported by carrier proteins. Energy for this
       process is the same as for simple diffusion, with the major differences
       being specificity and saturability. Transport can be in either direction,
       depending on the concentration gradient. Active transport also involves

carrier proteins and is thus both specific and saturable. The major difference between this process and the other two is the use of ATP to drive transport against a concentration gradient. Transport is thus in one direction only.

5. Differentiate between the members of each pair of terms: osmotic potential / osmotic pressure, pressure potential / water potential, hypertonic / hypotonic.

Osmotic potential is the potential of a membrane-surrounded solution to take up water; it results from the presence of dissloved solutes. Osmotic pressure is the same phenomenon, but it is defined as a positive, rather than a negative, value. Pressure potential is the real pressure exerted on the walls of plant cells (or other walled cells) as a result of their water content, whereas water potential is the overall tendency of a cell or solution to take up water from pure water. Hypotonic and hypertonic are comparative terms. If solution A has a more negative osmotic potential than solution B, A is said to be hypertonic and B is said to be hypotonic.

6. Of what value is it to a cell to have the enzymes that are involved in a multistep biochemical pathway arranged in an orderly fashion on a membrane?

The organization of enzymes on membranes enables them to be brought into the right orientations for the products of one reaction to be immediately available to the adjacent enzyme, which makes for greater efficiency than would be possible if the enzymes were not held in close proximity. Such an arrangement also makes possible the coupling of reactions, again because the enzymes are lying in close proximity. In this way the cell can use the energy generated by the breaking of a chemical bond to synthesize ATP, for example.

7. How can the bonding of specific molecules to receptor sites on the <u>outside</u> of the membrane trigger a whole series of reactions <u>inside</u> the cell?

Contact of the receptor protein with its target molecule usually results in a change in the tertiary structure of the protein. This, in turn, causes a structural change in the intrinsic protein(s) to which the receptor is attached. The structural change may result in the activation of a previously inactive enzyme or the opening of a channel for the influx or outflux of ions or molecules, for example. In any case, the structural change in the intrinsic protein is the first step in the chain of events induced inside the cell. The intrinsic proteins involved in such reactions must obviously protrude through both sides of the bilayer.

8. As was stated in Chapter 4, the mitochondrion is the site of ATP synthesis in the cell. Studies of this process have led to the following conclusions about the mechanism by which ATP is produced.

    a. As reduced electron carrier molecules are passed from one member of a series of proteins to the next, protons are pumped into the space between the inner and outer mitochondrial membranes.

    b. The proteins involved in this process are all located on the mitochondrial membranes.

    c. The overall result of the process just described is that the mitochondrial matrix becomes electrically negative with respect to the space between the inner and outer membranes.

    d. Once the electrical potential difference reaches a certain point, protons rush from the space between the inner and outer membranes into the matrix.

    e. A specific protein, ATP synthetase, acts as a channel for the inflow of protons and requires very acidic conditions in which to function. This enzyme catalyzes the formation of ATP from ADP and inorganic phosphate. As the protons move back into the matrix, flowing through the channel formed by the ATP synthetase molecule, conditions around the active site of this enzyme become favorable for ATP formation. In this way, the energy from the electron carriers is used to manufacture ATP.

9. Cite the various properties of membranes, as discussed in this chapter, that make the process of ATP synthesis possible.

The positioning of protein molecules on the mitochondrial membranes illustrates the "pegboard" effect. Pumping of protons from the matrix to the space between the membranes illustrates active transport, driven not by ATP in this case, but by some of the energy released as electrons are transferred from one protein to another. The pumping of protons results in the separation of electrical charges, maintained in part by the insulating effect of the membrane, and in a difference in pH across the inner membrane, also maintained by the presence of the membrane itself, since protons do not readily diffuse through the nonpolar membrane layer. ATP synthetase must be an intrinsic protein, held in precise orientation by the membrane (another example of the pegboard effect). ATP synthetase is also participating in the facilitated diffusion of protons from the intermembrane space to the matrix. The process of ATP synthesis described here is called the "chemiosmotic theory" and is discussed in Chapter 7 of the text. The role of membranes in this process is absolutely critical to the functioning of the mitochondrion.

TEST SET 5-A

Name _____   Section _____

Choose the best answer to each of the following questions, and write the appropriate letter in the space provided.

Ans: a
p. 121

_____   1. Water is able to enter cells freely by means of: (a) small pores; (b) active transport; (c) simple diffusion; (d) facilitated diffusion.

Ans: d
p. 115

_____   2. Which of the following is not likely to be a good intrinsic membrane protein? (a) a carrier protein; (b) a hydrophobic protein; (c) an enzyme; (d) a hydrophilic protein.

Ans: c
p. 113

_____   3. That many proteins can migrate within the plane of the phospholipid bilayer can be demonstrated via the technique of: (a) freeze-fracture; (b) freeze-etch; (c) cell fusion; (d) vital staining.

Ans: c
p. 112

_____   4. The membrane component that accounts for much of the physical integrity of the membrane is: (a) intrinsic protein; (b) extrinsic protein; (c) lipid; (d) carbohydrate.

Ans: b
p. 111

_____   5. Amphipathic compounds contain: (a) oligomeric carbohydrate; (b) hydrophobic and hydrophilic regions; (c) lipid-soluble protein; (d) multiple charged groups.

Ans: d
p. 112

_____   6. Because of the strong hydrophobic interactions between the lipids they contain, biological membranes: (a) are difficult to repair when damaged; (b) can be caused to fuse only under extreme conditions; (c) are seldom subject to being punctured; (d) spontaneously reseal rips or punctures.

Ans: c
p. 126

_____   7. The two cell membrane components that are most likely to be implicated in cell surface recognition signals are: (a) protein and lipid; (b) phospholipid and protein; (c) carbohydrate and protein; (d) carbohydrate and lipid.

Ans: d
p. 116

_____   8. Protein is so diverse in structure because of the variety of ways in which twenty different amino acids can be assembled to form linear polymers. Diversity in carbohydrates such as those that are linked to membranes stems mainly from: (a) the vast number of different monosaccharides available; (b) the length of polysaccharide chains; (c) the association of monomers with other molecular types; (d) the ability to form branched structures.

Ans: a
p. 121

_____   9. Although a number of things are involved, the major factor that determines whether a material passes a biological membrane by diffusion is: (a) relative lipid solubility; (b) concentration gradient; (c) molecular size; (d) temperature.

Ans: d
p. 119

_____   10. Simple diffusion and facilitated diffusion: (a) are different in that simple diffusion requires a greater concentration gradient to be effective; (b) are similar in that both are "saturable" phenomena; (c) are different in that facilitated diffusion is unidirectional; (d) are similar in that both tend toward equilibrium distribution for the diffusing material.

Ans: b
p. 120

_____   11. Stored energy in the form of ATP is used to transport materials against the gradient in: (a) exchange diffusion; (b) active transport; (c) simple diffusion; (d) facilitated diffusion.

Ans: a
p. 124

12. A sample of cells is placed in a salt solution. The cells swell and burst. The solution is probably: (a) hypotonic; (b) isosmotic; (c) hypertonic; (d) hyperosmotic.

Ans: c
p. 122

13. Osmosis is considered a special case of diffusion because: (a) it works in only one direction; (b) less energy is required; (c) it involves water and membranes; (d) it does not involve molecular movement.

Ans: c
p. 125

14. Energy processing is a membrane-associated process that depends on: (a) the ready availability of ATP; (b) the diffusion of molecules within the plane of the membrane; (c) the separation of charges across a membrane; (d) the osmotic potential within the cell.

Ans: d
p. 128

15. The specificity of receptor protein and carbohydrate is determined by: (a) location on the cell surface; (b) the ratio of protein to carbohydrate; (c) the hydrophobic nature of the protein; (d) shape at the level of tertiary structure.

TEST SET 5-B

Name _____   Section _____

Choose the <u>best</u> answer to each of the following questions, and write the appropriate letter in the space provided.

Ans: c
p. 111

_____   1. The phosphorous-containing region of membrane phospholipid is situated to face the aqueous medium on both faces of the membrane because it is: (a) hydrophobic; (b) too bulky for the bilayer; (c) electrically charged; (d) electrically neutral.

Ans: b
p. 112

_____   2. Phospholipid provides the continuous phase for biological membrane and stabilizes membrane structure through: (a) covalent bonding; (b) hydrophobic bonding; (c) hydrophilic bonding; (d) ionic bonding.

Ans: a
p. 115

_____   3. Extrinsic proteins on biological membranes differ from intrinsic proteins in that they are: (a) hydrophilic and associated with the membrane surface; (b) hydrophobic and associated with the membrane interior; (c) hydrophilic and associated with the membrane interior; (d) hydrophobic and associated with the membrane surface.

Ans: d
p. 116

_____   4. The basic composition of biological membranes includes: (a) cholesterol, glycolipid, and phosphoprotein; (b) phospholipids, polypropylene, and glycoprotein; (c) cholesterol, phospholipids, and proteins; (d) lipid, carbohydrate, and protein.

Ans: d
p. 116

_____   5. The ability to form linkages from several points on the molecule and the ability to form branched oligomers foster the great diversity characteristic of membrane: (a) lipids; (b) proteins; (c) lecithins; (d) carbohydrates.

Ans: a
p. 117

_____   6. When the freeze-fracture process is employed, the cleavage plane sometimes splits the middle of the lipid bilayer. The exposed surfaces in such cases are "bumpy." The bumps are thought to be: (a) intrinsic proteins; (b) extrinsic proteins; (c) phospholipids; (d) cholesterol.

Ans: c
p. 121

_____   7. Biological membranes are thought to possess small pores or channels because: (a) cells continually leak small ions; (b) pores are usually observed in electron micrographs of membranes; (c) water and other small hydrophilic materials pass through membranes readily; (d) of studies using the freeze-etch process.

Ans: d
p. 124

_____   8. Plant cells tolerate a high pressure potential better than animal cells because: (a) of the presence of membrane pores; (b) of lower osmotic activity; (c) they are not exposed to hyposmotic solution; (d) they possess a cell wall.

Ans: b
p. 125

_____   9. A red blood cell will crenate in: (a) hypotonic solution; (b) hypertonic solution; (c) isosmotic solution; (d) hyposmotic solution.

Ans: b
p. 120

_____   10. A process observed in biological membranes that permits the movement of hydrophilic materials down a concentration gradient is: (a) simple diffusion; (b) facilitated diffusion; (c) active transport; (d) passive diffusion.

Ans: a
p. 119

_____ 11. Which of the following would not exhibit saturbility?
(a) simple diffusion; (b) facilitated diffusion; (c) active transport; (d) an intrinsic membrane enzyme.

Ans: d
p. 126

_____ 12. Which of the following is most likely to be hydrophilic?
(a) a membrane carrier molecule; (b) an intrinsic glycoprotein; (c) a receptor protein molecule; (d) an extrinsic glycoprotein.

Ans: b
p. 126

_____ 13. The discharge of electrical and pH gradients across membranes is frequently coupled to: (a) active transport of glucose; (b) synthesis of ATP; (c) the process of crenation; (d) cell fusion.

Ans: a
p. 126

_____ 14. When cells in culture become crowded, cell division is inhibited. This process, which is mediated by cell surface receptors, is called: (a) contact inhibition; (b) growth inhibition; (c) division control; (d) mitotic regulation.

Ans: c
p. 128

_____ 15. The specificity in signal-receptor interaction is inherent in the: (a) tertiary structure of receptor protein and lipid; (b) shape and character of receptor carbohydrate and lipid; (c) tertiary structure of receptor carbohydrate and protein; (d) shape and character of receptor protein and lipid.

CHAPTER SIX
ENERGY, ENZYMES, AND CATALYSIS

CHAPTER SUMMARIES AND OBJECTIVES

6-1 <u>Energy and the Laws of Thermodynamics</u> Life involves the conversion of raw materials from the environment into living matter. This activity constitutes metabolism. Many metabolic reactions require energy (the ability to do work), which cells obtain from their environment, either directly or as food. Energy can be neither created nor destroyed. This is the first law of thermodynamics. The various forms of energy are interconvertible. Entropy is a quantitative measure of the amount of disorder in a system. In the universe, or in any closed system, entropy constantly increases. This is the second law of thermodynamics. In all energy-conversion processes, some of the energy is lost as the unusable heat that is associated with an increase in entropy.

1. Give several examples of energy conversions carried out by living cells.

2. Explain the purpose of an automobile radiator in terms of the second law of thermodynamics.

3. Answer: Since energy cannot be created, and since every energy-conversion process involves some loss of energy as waste heat, how can living organisms continue to exist on earth?

6-2 <u>Chemical Equilibria / The Equilibrium Constant</u> All chemical reactions can run both backward and forward. A spontaneous reaction (one that can occur, given enough time, without an input of energy) always proceeds toward a state of equilibrium. At equilibrium, the forward reaction and the reverse reaction are taking place at the same rate. Thus no observable change in the concentrations of reactants and products occurs, although chemical bonds are being formed and broken.

Each reaction has a specific equilibrium point, which is defined in terms of the concentrations of reactants and products at equilibrium. The equilibrium constant (K_{eq}) is defined as the ratio of the concentration of the products to the concentration of reactants. A high value of K_{eq} means that the forward reaction will go very far toward completion. A low value of K_{eq} means the reaction will proceed only slightly--that is, the reverse reaction occurs more readily.

4. Be able to solve problems dealing with the equilibrium constant and the concentrations of reactants and products for various types of chemical reactions.

5. "Living cells prevent the attainment of equilibrium as long as they live." Explain why this statement is true.

6-3 <u>Free Energy and Equilibria</u> / <u>Free Energy, Heat, and Entropy</u> The position of the equilibrium is determined by the amount of free energy liberated by the reaction. Reactions that produce a large amount of free energy are exergonic. Those that require an input of energy are endergonic. The change in free energy resulting from a reaction is ΔG. This value is negative for exergonic reactions and positive for endergonic ones. A reaction that goes far to completion (has a large K_{eq}) will also have a large negative ΔG. In living cells, endergonic reactions occur only when they are able to gain energy from a sufficiently exergonic reaction that is occuring simultaneously.

Free energy, heat, and the energy associated with entropy are related to each other by the equation $\Delta G = \Delta H - T\Delta S$, where ΔG is the change in free energy, ΔH is the change in heat, ΔS is the change in entropy, and T is the absolute temperature. Both ΔH and ΔS contribute to the overall ΔG. ΔG determines the K_{eq} for the reaction.

6. State the second law of thermodynamics in terms of free energy and entropy.

7. Explain the differences between endergonic and exergonic reactions in terms of free energy, spontaneity, and equilibrium constants.

8. Answer: How can endergonic reactions, such as the synthesis of proteins from amino acids, occur in the cell?

9. Answer: For a given reaction, why is the effect due to entropy change greater at high temperatures than at lower temperatures?

6-4 <u>Reaction Rates</u> / <u>Rate Constants</u> Even though a reaction may be spontaneous, it may occur at a very slow rate due to the presence of an energy barrier between reactants and products. For such reactions, a small amount of energy (activation energy) is necessary to get the reaction going. The activation energy barrier represents the energy needed to bring the reacting molecules together, break the old bonds, and form new bonds. The higher the activation energy, the slower the reaction. The overall change in free energy for the reaction is unaffected by the amount of activation energy necessary.

The rate of a chemical reaction is proportional to the concentration of reactant, and for every reaction there is a specific rate constant (k). The rate constant may be increased by an increase in temperature or by catalysis.

10. Answer: Why does every reaction require a certain amount of activation energy?

11. Answer: How are the activation energy and the rate constant of a reaction related?

12. Explain why the application of heat increases the rate of a reaction.

6-5 <u>Enzymes and Selective Catalysis</u> / <u>Molecular Structure of Enzymes</u> / <u>Structures and Actions of Proteolytic Enzymes</u> Enzymes are biological catalysts. The molecules acted on by an enzyme are substrates of the enzyme. The substrate binds to a specific location called the active site, forming an enzyme-substrate complex. Once this complex is formed, it undergoes a change in shape. The effect of this change in shape is the stretching of the bonds of the substrate and thus a lowering of the activation energy for the reaction. The exact structure of the active site determines the types of molecules that can become substrates for the enzyme. Only those molecules that fit into the active site can be acted on.

Enzymes are protein molecules whose tertiary structure determines the nature of the active site. There may be one or several polypeptide chains comprising the active enzyme molecule. One extensively studied enzyme is carboxypeptidase, which hydrolyzes peptide bonds. Carboxypeptidase removes amino acids only from the carboxyl terminal end of a protein molecule.

13. Answer: How do enzymes speed up the rates of reactions in living cells?

14. The relationship between an enzyme and its substrate has been compared to that between a lock and key. Justify the use of this analogy.

15. Explain the specificity of the carboxypeptidase molecule in terms of the structure of its active site.

6-6   Prosthetic Groups   Some enzymes possess a prosthetic group: a metal ion, a metal ion held in a single organic molecule, or a coenzyme. Prosthetic groups aid in the functioning of the enzyme in several ways.

16. Answer: What are prosthetic groups and how do they assist enzymes?

6-7   Substrate Concentration and Reaction Rate   For most enzymes, the rate of the reaction being catalyzed increases with increased substrate concentration. However, the rate eventually reaches a plateau where increases in the substrate concentration produce no further changes in reaction rate. At this point, all of the enzyme molecules are bound to substrate molecules.

17. Explain how the rate of an enzyme-catalyzed reaction varies with the substrate concentration.

6-8   Enzyme Inhibition   Enzyme activity may be inhibited in three ways: competitive inhibition, noncompetitive inhibition, and irreversible inhibition. A competitive inhibitor resembles the substrate so closely that it can occupy the active site and thus prevent binding of the substrate. Noncompetitive inhibitors bind to some area of the enzyme molecule other than the active site. In so doing, they cause the enzyme molecule to change shape, distorting the active site and preventing the binding of the substrate. Irreversible inhibitors cause permanent chemical changes in the R groups of amino acids at the active site.

18. Define competitive inhibition, noncompetitive inhibition, and irreversible inhibition, and give an example of each.

6-9   Coupling of Reactions   Cells are able to carry out endergonic reactions only if energy is available from a sufficiently exergonic reaction occuring simultaneously. Thus exergonic reactions are often coupled to endergonic ones. An enzyme provides for a favorable orientation of the various molecules participating in the coupled reactions, ensuring that the energy liberated by the exergonic reaction will be available to drive the endergonic one.

19. Explain how the enzyme succinate dehydrogenase allows for the capture of energy that is liberated in the conversion of succinate to fumarate.

6-10 Subunits, Allosteric Effects, and Control / Control of Metabolism Through Allosteric Effects   An allosteric effector binds to an effector site on one of the subunits of a regulatory, or allosteric, enzyme. Allosteric enzymes exist in both active and inactive forms, which are in equilibrium with each other. The active form is able to bind substrate; the inactive form cannot. Both can bind to effector molecules. An allosteric inhibitor binds to the effector site of the inactive form of the enzyme and decreases the reaction rate. An allosteric promoter binds to the effector site of the active form of the enzyme and increases the reaction rate.

Metabolic pathways are sequences of reactions in which the products of one reaction become the reactants for the next step. Regulatory enzymes are those that catalyze reactions at points where two or more metabolic pathways diverge. Regulatory enzymes are often controlled by negative feedback. Such control of enzyme activity by small molecules allows adjustment to changes in metabolism or in the environment.

20. Explain how allosteric inhibitors and promoters work to regulate the activity of enzymes.

21. Using threonine synthesis in E. coli as an example, explain what is meant by the statement "An organism is only as perfect as its environment demands."

6-11 Sensitivity of Enzymes to the Environment   Enzymes are sensitive to the temperature and pH of their environment. Denaturation of enzymes occurs at extremes of temperature or pH. Each enzyme has an optimal temperature at which it is maximally active. Organisms adapt to changes in their environment by producing isozymes (groups of enzymes with the same catalytic properties but with different temperature optima).

22. Answer: What is denaturation?

23. Explain why the rate of an enzyme-catalyzed reaction depends on the pH at which the reaction is occurring.

24. Explain how isozymes enable organisms to adapt to changing environmental conditions.

ARTICLE RESOURCES

Energy and the Laws of Thermodynamics

Simonis, Doris A. Energy: The analogy approach. The Science Teacher, 1982, 49(2), 41-44. Analogies illustrating the laws of thermodynamics are given.

Chemical Equilibria

The Equilibrium Constant

Free Energy and Equilibria

Free Energy, Heat, and Entropy

Reaction Rates

Rate Constants

Beall, Paula. The water of life. The Sciences, 1981, 21(1), 6-9, 29. Water is the basis on which chemical reactions are founded. It can make hydrogen bonds. It can exist as a liquid at temperatures at which life processes take place. Water is one of only a few inorganic compounds to do this.

Enzymes and Selective Catalysis

Cheung, Wai Yiu. Calmodulin. Scientific American, 1982, 246(6), 62-70. Calmodulin mediates the regulatory functions of calcium ions. It is a single protein chain of 148 amino acid groups. This article traces the history of research into this molecule. Excellent charts and photographs.

Molecular Structure of Enzymes

Stent, Gunther S. To the Stockholm Station--Makers of the molecular-biological revolution. Encounter, 1980, 54(3), 79-85. History of the development of molecular biology.

Structures and Actions of Proteolytic Enzymes

Moog, Florence. The lining of the small intestine. Scientific American, 1981, 245(5), 154-176. The actual site of admission of nutrients into the body is the membrane covering the microvillus. This membrane has an array of enzymes incorporated into it. The enzymes maltase, sucrase, and lactase are present, as well as aminopeptidases. The pathways of the glucose molecule are shown, as are those of amino acids.

Prosthetic Groups

Substrate Concentration and Reaction Rate

Enzyme Inhibition

Coupling of Reactions

Subunits, Allosteric Effects, and Control

Control of Metabolism Through Allosteric Effects

Sensitivity of Enzymes to the Environment

Notkins, Abner Louis. The courses of diabetes. Scientific American, 1979, 241(5), 62-73. The theory that diabetes arises from a complex combination of genetic and environmental factors is explored. Maturity-onset diabetes can be linked to obesity. A decreased number of insulin receptors was found. When patients were placed on a weight-reducing diet, these receptors returned to normal.

ESSAY QUESTIONS

1. Explain the purpose of an automobile radiator in terms of the second law of thermodynamics.

    In the conversion of the chemical energy of gasoline to the energy of motion of the car, unusable heat associated with disorder is generated. This is dissipated by the radiator to prevent damage to the engine.

2. "Living cells prevent the attainment of equilibrium as long as they live." Explain why this statement is true.

    Living cells always prevent the attainment of equilibrium by changing the concentrations of reactants and products such that a given reaction proceeds in one direction only.

3. How can endergonic reactions, such as the synthesis of proteins from amino acids, occur in the cell?

    Endergonic reactions can occur in the cell only if they are coupled to a sufficiently exergonic reaction.

4. For a given reaction, why is the effect due to entropy change greater at high temperatures than at lower ones?

    This relationship is apparent from the equation $\Delta G = \Delta H - T\Delta S$. Multiplying ΔS by a larger value for T will obviously increase the contribution of the $T\Delta S$ term. At higher temperatures, molecular movement becomes more vigorous and hence more disorderly.

5. Why does every reaction require a certain amount of activation energy?

    Energy is required to bring the reacting molecules together, to stretch and break the bonds of the reactants, and to form the new bonds of the products. The sum of these processes is called activation energy.

6. How do enzymes speed up the rates of reactions in living cells?

    Because of the tertiary structure of the active site, the enzyme brings the reacting molecules together in a favorable orientation for the reaction to occur, thus lowering the activation energy. Often the induced fit between enzyme and substrate results in the stretching of bonds in the substrate, further lowering the activation energy.

7. The relationship between an enzyme and its substrate has been compared to that between a lock and key. Justify the use of this analogy.

    The tertiary structure of the active site of the enzyme is "molded" to the shape of the substrate so that the two fit precisely together, as a lock and key. Just as my house key will not fit my car ignition, a given enzyme will "accept" one substrate (or at most a few).

8. How does the enzyme succinate dehydrogenase allow for the capture of the energy that is liberated in the conversion of succinate to fumarate.

The enzyme couples the exergonic reaction (succinate ⟶ fumarate) with the endergonic reaction (FAD ⟶ $FADH_2$) by ensuring that the hydrogen atoms liberated by succinate are used to make $FADH_2$. One site on the enzyme binds succinate and one binds FAD. Succinate transfers two protons to amino acid side chains on the enzyme. These are oriented to donate the protons to FAD.

9. How do allosteric inhibitors and promoters work to regulate the activity of enzymes?

All allosteric enzymes exist in two forms, active and inactive, which are constantly being interconverted in the cell (that is, the two forms are in equilibrium with each other). In the presence of substrate, active enzyme binds to substrate molecules, preventing its conversion to inactive enzyme. Because inactive enzyme is still being converted to active enzyme, the concentration of active enzyme increases. In the presence of inhibitor, however, the inhibitor binds to the inactive form of the enzyme, preventing its conversion to the active form. Meanwhile, active enzyme is still being converted to inactive enzyme. Thus the concentration of inactive enzyme increases in the presence of the inhibitor, and the reaction slows down. An allosteric promoter binds to the active form of its enzyme and prevents conversion to the inactive form. Thus the concentration of active enzyme increases, and the rate of the reaction is increased.

10. How do isozymes enable organisms to adapt to changing environmental conditions?

Isozymes are sets of proteins with the same catalytic activity and different temperature optima. When environmental conditions change, the organism is able to continue to operate at optimum efficiency by producing the appropriate isozyme. In rainbow trout, for example, a different form of acetylcholinesterase is produced at 10°C than at 2°C. This enzyme is important in the proper transmission of nerve impulses, so producing the appropriate isozyme allows the trout to function normally when transferred from water of one temperature into water of another temperature.

11. The effects of three compounds, X, Y, and Z, on the following biochemical reaction are being investigated.

$$\text{Substrate A} \xrightarrow{\text{Enzyme Q}} \text{Product B}$$

The following results were obtained:

a. Addition of X to the reaction mixture caused a decrease in the rate of formation of Product B. This effect was reversed by the addition of a large amount of Substrate A.

b. Addition of Y to the reaction mixture caused a decrease in the rate of product formation. This effect persisted when additional substrate was added.

c. Addition of Z to the reaction mixture caused a rapid decrease in the rate of product formation. Analysis indicated that no enzyme activity remained, even when excess Z was removed from the mixture by chemical techniques.

d. When a control experiment was run, adding only distilled water instead of X, Y, or Z, it was found that the reaction proceeded normally until the concentration of B reached 10^{-2} molar, at which time the reaction rate decreased markedly.

Interpret each of these results in light of your knowledge of enzyme chemistry.

a. X is a competitive inhibitor.

b. Y is a noncompetitive inhibitor.

c. Z is an irreversible inhibitor.

d. B regulates the activity of enzyme Q through feedback inhibition and is thus also a noncompetitive inhibitor of the enzyme.

TEST SET 6-A

Name _____   Section _____

Choose the <u>best</u> answer to each of the following questions, and write the appropriate letter in the space provided.

Ans: b
p. 133

_____ 1. The second law of thermodynamics provides that: (a) the amount of energy in the universe is unchanging; (b) the amount of entropy in the universe is always increasing; (c) energy can be neither created nor destroyed; (d) entropy is zero at a temperature of zero degrees absolute.

Ans: c
p. 135

_____ 2. A reaction with a K_{eq} value of 20.0 is obviously: (a) very rapid; (b) at equilibrium; (c) spontaneous; (d) lacking energy to proceed.

Ans: b
p. 134

_____ 3. At equilibrium, a reaction wherein compound A converts to compound B with K_{eq} = 0.25 has: (a) the concentration of B at 75% that of A; (b) A at 4 times the concentration of B; (c) B at 4 times the concentration of A; (d) the concentration of A at 25% that of B.

Ans: c
p. 135

_____ 4. Which of the following reactions is endergonic? (a) a reaction with K_{eq} = 3.80; (b) a reaction where ΔG = -5000 kcal; (c) a reaction where ΔG = +2 kcal; (d) a reaction with a positive entropy change.

Ans: a
p. 137

_____ 5. Knowing the value for free-energy change in a particular reaction does <u>not</u> provide information about: (a) reaction rate; (b) the probability that the reaction will proceed; (c) equilibria; (d) whether the reaction is spontaneous.

Ans: d
p. 140

_____ 6. A material that lowers the apparent energy of activation for a reaction can be called a(an): (a) enzyme; (b) inhibitor; (c) competitive inhibitor; (d) catalyst.

Ans: a
p. 140

_____ 7. The change in shape to produce an induced fit between an enzyme and a substrate stretches substrate bonds and: (a) lowers the energy of activation to break them; (b) results in stabilization of the enzyme-substrate complex; (c) activates an allosteric site; (d) raises the energy of activation to break them.

Ans: b
p. 144

_____ 8. The hydrolysis of a peptide bond without an enzyme such as carboxypeptidase is exergonic, but it is very slow because: (a) the hydrophobic nature of most proteins excludes the water necessary for the reaction; (b) no convenient mechanism exists for a water molecule in neutral solution to attack a peptide bond; (c) enzymatic repair mechanisms replace the bond as fast as it is hydrolyzed; (d) most biological materials contain carboxypeptidase inhibitors.

Ans: c
p. 145

_____ 9. Coenzymes are: (a) a group of enzymes that are active only when acting in concert with others; (b) activating metal ions; (c) organic prosthetic groups; (d) vitamins.

Ans: d
p. 147

_____ 10. An enzyme inhibitor that does not bind the active site and cannot be overcome by excess substrate levels is probably a(an): (a) competitive inhibitor; (b) uncompetitive inhibitor; (c) reversible inhibitor; (d) noncompetitive inhibitor.

Ans: c
p. 148

_____ 11. The hydrogenation of FAD is endergonic and requires a large input of energy to proceed--an amount of energy that is supplied by the reaction of succinate to fumarate. In this case these reactions are said to be: (a) linked; (b) associated; (c) coupled; (d) reciprocal.

Ans: a
p. 150

_____ 12. An enzyme that exists in two or more distinct forms with different catalytic activities is said to be: (a) allosteric; (b) variable; (c) a regulatory enzyme; (d) coupled.

Ans: b
p. 150

_____ 13. A material that binds to an enzyme, locking it into an active form rather than allowing a change to an inactive form, is called a(an): (a) regulator; (b) promoter; (c) accelerator; (d) activator.

Ans: a
p. 152

_____ 14. The regulation of metabolic pathways by control of enzyme activity at a committed step near an important branch point rather than later in the sequence: (a) prevents the wasteful accumulation of unneeded intermediates; (b) is accomplished by positive feedback; (c) requires the presence of competitive inhibitors; (d) requires the presence of noncompetitive inhibitors.

Ans: c
p. 153

_____ 15. Dissimilar enzymes that catalyze the same reaction under different optimum conditions are called: (a) allozymes; (b) allosteric enzymes; (c) isozymes; (d) coenzymes.

TEST SET 6-B

Name _____   Section _____

Choose the <u>best</u> answer to each of the following questions, and write the appropriate letter in the space provided.

Ans: b
p. 133

_____   1. In any change of energy from one form to another, some portion
             is "lost" to an increase in entropy. This observation is the
             basis for: (a) the first law of thermodynamics; (b) the second
             law of thermodynamics; (c) the third law of thermodynamics;
             (d) the fourth law of thermodynamics.

Ans: c
p. 133

_____   2. When a reaction is at equilibrium: (a) no further chemical
             changes are possible; (b) it is very spontaneous; (c) the rate
             of forward reaction equals the rate of back reaction;
             (d) entropy has reached a minimum for the reaction.

Ans: d
p. 134

_____   3. If a reversible reaction at equilibrium has formed 0.02 mole
             per liter of product, leaving the substrate at a concentration
             of 0.04 mole per liter, the reaction has a K_{eq} value of:

             (a) 2.0; (b) 0.2; (c) 0.06; (d) 0.5.

Ans: c
p. 135

_____   4. The reverse reaction for any particular exergonic reaction
             should always be: (a) spontaneous; (b) capable of producing
             energy; (c) possible if enough energy is provided to run it;
             (d) exergonic as well.

Ans: a
p. 136

_____   5. In the complete combustion of 1 mole of glucose, 673 kcal of
             heat are given off and there is an entropy change of 43.3
             cal/degree. If the temperature is assumed to be 298 degrees
             absolute, the total free-energy change for the reaction is
             approximately: (a) -686 kcal; (b) -716.3 kcal; (c) -630.3
             kcal; (d) -626 kcal.

Ans: b
p. 138

_____   6. The energy of activation that must be provided to surpass the
             energy barrier in many spontaneous reactions is: (a) lost to an
             increase in entropy; (b) recovered as the reaction proceeds;
             (c) greater when a catalyst is used; (d) not a factor when
             enzymes are used.

Ans: d
p. 139

_____   7. In a reaction where A converts to B, the rate of formation of
             B is given by: (a) the ratio of product to substrate; (b) the
             value of K_{eq}; (c) A = k [B]; (d) B = k [A].

Ans: a
p. 140

_____   8. The characteristic of an enzyme that dictates which substrates
             it will utilize is called: (a) its specificity; (b) the
             enzyme-substrate complex; (c) its catalytic range; (d) its
             substrate capability.

Ans: c
p. 143

_____   9. The crystallization of the enzyme lysozyme and the determination
             of its three-dimensional structure by X-ray crystallography
             provided much of the first direct support for: (a) the discovery
             of enzymes as catalysts; (b) the induction model; (c) the
             lock-and-key model o f enzyme-substrate interaction; (d) the
             exact role of the reactions catalyzed by lysozyme in the cell.

Ans: c
p. 144

_____   10. The hydrolysis of a peptide bond by an enzyme requires: (a) a
             hydrophobic environment; (b) a hydrophilic environment; (c) the
             addition of water across the bond; (d) a solution of neutral pH.

Ans: b
p. 146

_____ 11. Addition of extra substrate to an enzyme reaction wherein the catalyst is saturated: (a) increases the rate of reaction; (b) leaves the reaction rate unchanged; (c) changes the value of K_{eq}; (d) reduces the rate of reaction.

Ans: b
p. 149

_____ 12. An allosteric enzyme is one that: (a) has an unusually high saturation level; (b) has different states with variable activities; (c) has two or more active sites; (d) seldom has more than one subunit in its structure.

Ans: c
p. 150

_____ 13. The addition of an allosteric inhibitor slows a reaction by: (a) binding the substrate to prevent interaction with the enzyme's active site; (b) increasing the rate of the back reaction; (c) locking the enzyme in an inactive form; (d) preventing the release of the substrate-product from the active site.

Ans: a
p. 153

_____ 14. Some enzymes, such as lactic acid dehydrogenase, are composed of subunits with different characteristics, such that rearrangements of those subunits produces enzymes that catalyze the same reaction under different optimal conditions. Enzymes of this type are called: (a) isozymes; (b) induced-fit enzymes; (c) alloenzymes; (d) regulatory enzymes.

Ans: d
p. 153

_____ 15. An enzyme that has been heated and has lost its normal tertiary structure, and thus its activity, has been: (a) hydrolyzed; (b) deformed; (c) depolymerized; (d) denatured.

CHAPTER SEVEN
CELLULAR RESPIRATION AND
FERMENTATION

CHAPTER SUMMARIES AND OBJECTIVES

7-1 <u>Anaerobic and Aerobic Energy Metabolism</u> Living systems require a continuous input of matter and energy. Matter is obtained from the environment. Energy is obtained through fermentation, cellular respiration, and photosynthesis. Fermentation involves the breakdown of glucose to pyruvic acid, liberating some of the energy from the sugar. Pyruvic acid is converted to other compounds. The first steps of the fermentation process involve the glycolysis pathway. All organisms carry out glycolysis. It is the sole energy source for anaerobes, organisms that can function only in the absence of oxygen. In some organisms, fermentation may be supplemented by cellular respiration when oxygen is available. These organisms are facultative anaerobes. In cellular respiration glucose is broken down completely, into carbon dioxide and water. Organisms that carry out cellular respiration and cannot survive without oxygen are called aerobes. All these processes take place in a series of steps. Energy released from glucose during some of these steps is used to synthesize the short-term energy-storage compound, adenosine triphosphate (ATP).

1. Distinguish among anaerobes, facultative anaerobes, and aerobes.

2. Distinguish among glycolysis, fermentation, and cellular respiration.

7-2 <u>Adenosine Triphosphate, the Cell's Energy Currency / The Flow of Energy Through the Living World</u> All organisms use ATP for short-term energy storage. Many enzymes catalyze the breakdown of ATP to form ADP and P_i (inorganic phosphate).

This reaction yields 10 kcal/mole of energy, which can be used to drive various endergonic reactions in the cell. In turn, ADP and P_i can be combined to form ATP if sufficient energy is available. This energy can be obtained from the exergonic reactions associated with the breakdown of glucose. Part of the ATP used by the cell is diverted to the synthesis of long-term storage compounds such as starch, glycogen, and fats.

Photosynthetic organisms manufacture their own food, using the energy from the sun. It is the constant influx of energy from the sun that enables living systems to maintain their high degree of organization. Nonphotosynthetic organisms utilize the sun's energy indirectly, either by consuming photosynthetic organisms or by feeding on large organic molecules from the dead bodies of other organisms.

3. Discuss the role of ATP in cellular energy metabolism.

4. Explain why living systems of all types do not violate the second law of theromodynamics.

7-3 An Overview of the Release of Energy from Glucose / The Transfer of Hydrogen Atoms and Electrons / ATP Revisited Energy is released from glucose in a series of processes that are physically separated within the cell and that evolved separately or at different times. Glycolysis, which occurs in all organisms, is carried out in the cytosol and involves the metabolism of glucose to pyruvic acid, reducing NAD and yielding a net of two molecules of ATP. In aerobic organisms, the pyruvate is oxidized to carbon dioxide and a number of electron-carrier molecules are reduced via the citric acid cycle. Much of the energy originally in the glucose molecule is now present in the covalent bonds of reduced electron-carrier molecules. This energy is released to form ATP by means of the respiratory chain. The respiratory chain is a series of redox reactions in which electrons are ultimately passed to oxygen to produce water. The energy released at certain points along the chain is used to form ATP. Another function of the respiratory chain is to reoxidize certain electron-carrier molecules so that they may be reused in glycolysis or the citric acid cycle.

The transfer of hydrogen atoms or electrons is a redox reaction. Oxidative half-reactions are exergonic and reductive half-reactions are endergonic. An oxidizing agent is any molecule that can accept electrons or hydrogen atoms or can donate an oxygen atom. A reducing agent is a molecule that can donate electrons or hydrogen atoms or accept an oxygen atom. The cell uses a number of oxidizing and reducing agents for the exchange of electrons. The most important of these is nicotinamide adenine dinucleotide (NAD). Others include flavin mononucleotide (FMN) and flavin adenine dinucleotide (FAD).

The energy change associated with the hydrolysis of ATP to ADP and inorganic phosphate (P_i) is about 10 kcal/mole. Even more energy is liberated when ADP is hydrolyzed to AMP (adenosine monophosphate). For this reason, the first and second bonds of ATP are often called high-energy bonds.

5. Summarize the events involved in the oxidation of glucose to carbon dioxide and water in a typical aerobe. How does the metabolism of glucose differ in anaerobes?

6. Explain the role of electron-carrier molecules such as NAD in cellular energy metabolism.

7-4 Glycolysis / Overview of the Citric Acid Cycle / Reactions of the Citric Acid Cycle Glycolysis is a series of ten reactions in which a molecule of glucose is converted to two molecules of the three-carbon compound known as pyruvic acid. The first five reactions actually use up energy, requiring two molecules of ATP to convert glucose to two molecules of a three-carbon sugar phosphate. Each of the remaining reactions occurs twice for each molecule of glucose. First, the three-carbon sugar phosphate is converted to 1,3-phosphoglyceric acid, with the incorporation of an inorganic phosphate ion. This is an exergonic oxidation reaction that releases about 100 kcal/mole. Part of this energy is used to reduce NAD^+ to $NADH + H^+$. Finally, the two phosphate groups are transferred to ADP to form ATP, leaving pyruvic acid.

The citric acid cycle receives pyruvate from the glycolysis pathway and oxidizes it to carbon dioxide, using the hydrogen atoms released to reduce carrier molecules. The free energy from pyruvate thus is transferred to these carriers. In the first step of the citric acid cycle, pyruvate is oxidized to form acetyl coenzyme A. Acetyl coenzyme A reacts with oxaloacetate to produce citrate. A series of reactions follows in which the citrate is broken down, releasing two of the carbon atoms as CO_2 and regenerating oxaloacetate. At several of the steps in this series, hydrogen atoms are used to reduce NAD and FAD, and one step is sufficiently exergonic to allow the formation of a molecule of ATP as well. The inputs to the citric acid cycle are pyruvate and oxidized carrier molecules, and the outputs are carbon dioxide, reduced carrier molecules, and ATP. Reoxidation of the carriers and use of the energy thus released to for ATP are the function of the respiratory chain.

7. Summarize the reactions of the glycolysis pathway and the citric acid cycle. Account for the energy and carbon atoms from a molecule of glucose that is

oxidized completely to carbon dioxide via these two processes.

7-5 The Respiratory Chain / Oxidative Phosphorylation and Mitochondrial Structure
The oxidative reactions of glycolysis and the citric acid cycle could not occur without the oxidizing agent NAD^+. In addition, the free energy stored in the $NADH + H^+$ molecule must be transferred to ATP. In the respiratory chain, $NADH + H^+$ is oxidized by FMN. The FMN is, in turn, oxidized by another carrier, which is oxidized by a third compound, and so forth. The final carrier in the chain is reoxidized by molecular oxygen from the atmosphere, producing water. Carrying out the overall reaction, $NADH + H^+ + \frac{1}{2}O_2 \rightarrow NAD^+ + H_2O$, in a series of steps allows the free energy to be liberated in small "bursts" some of which are sufficiently large to form ATP from ADP and P_i. Three molecules of ATP are formed as electrons from one molecule of $NADH + H^+$ pass through the chain. When a molecule of $FADH_2$ transfers its electrons to the chain, two molecules of ATP are subsequently produced. The process by which the operation of the respiratory chain causes the production of ATP is called oxidative phosphorylation.

The most recent model for the mechanism by which oxidative phosphorylation operates is called the chemiosmotic hypothesis. The enzymes involved with glycolysis are found in the cytosol of the cell. The carriers and enzymes of the respiratory chain are embedded within the inner mitochondrial membrane. Enzymes of the citric acid cycle are found in the mitochondrial matrix, with the exceptions of those catalyzing steps 1, 5, and 7, which are bound to the inner membrane. This arrangement of enzymes and carriers facilitates the pumping of protons from the matrix to the space between the inner and outer membranes, creating a difference in electrical charge across the inner membrane. The chemiosmotic hypothesis proposes that there exists in the inner membrane an ATP synthetase which requires unusually acidic conditions to produce ATP, and which also can act as a proton channel, allowing a discharge of the proton gradient across the membrane. Thus the operation of the respiratory chain results in the establishment of a proton gradient across the membrane, which, when discharged, provides the acidic conditions required by the ATP synthetase.

8. Summarize the operation of the respiratory electron transport chain.

9. Explain how the process of oxidative phosphorylation occurs, according to the chemiosmotic hypothesis.

7-6 Fermentation Fermentation involves the reduction of pyruvate to lactic acid or ethanol, using $NADH + H^+$ as the reducing agent, regenerating NAD^+. Fermentation produces just enough NAD^+ to carry another molecule of glucose through glycolysis. Continued production of a small amount of ATP is thus maintained. Different types of cells carry out different types of fermentation. Muscle cells reduce pyruvate to lactate, whereas many yeasts and plants produce ethanol.

10. Explain the process of fermentation. In your answer, indicate what value the process has to a cell deprived of oxygen?

7-7 Comparative Energy Yields / Connections with Other Pathways / Feedback Regulation
When fermentation occurs, the cell obtains a net of 2 molecules of ATP per molecule of glucose metabolized. Complete anaerobic oxidation gains the cell, in addition to the 2 molecules of ATP from glycolysis, 30 molecules from regeneration of 10 molecules of NAD^+ from $NADH + H^+$ (2 from glycolysis and 8 from the citric acid cycle). Additional ATP is obtained by the reoxidation of $FADH_2$ (4 ATP for 2 $FADH_2$ generated by the citric acid cycle). Two more ATP molecules are produced from utilization of 2 GTP produced during the citric acid cycle. This gives a total of 38 molecules of ATP per molecule of glucose completely oxidized. However, 1 molecule of ATP is required to transport each of 2 pyruvate molecules across the mitochondrial membranes. Therefore the net yield of ATP from the complete aerobic oxidation of a molecule of glucose is 36 molecules.

There is a constant interchange of molecules between the respiratory pathways and the other biochemical processes of the cell. Starch and glycogen are digested into their component monomers before entering the glycolytic pathway. Fats are digested to glycerol and fatty acids. Glycerol is converted to glyceraldehyde-3-phosphate and enters the glycolytic pathway, whereas fatty acids are converted to acetyl coenzyme A and enter the citric acid cycle. The reverse of each of these processes also occurs. Similarly, intermediates from the citric acid cycle are used as the starting point for the synthesis of important cellular constituents, such as chlorophyll and amino acids.

Glycolysis, the citric acid cycle, and the respiratory chain are regulated by allosteric control of their enzymes. There are a number of positive and negative feedback control systems in these pathways. For example, the enzyme phosphofructokinase is inhibited by ATP and activated by ADP or AMP. Other important control points include those reactions that involve pyruvate and acetyl coenzyme A. All of these controls have evolved because they increase efficiency and contribute to the success of the species that has them.

11. Compare the energy yield from complete aerobic oxidation of glucose to that obtained through glycolysis alone.

12. Give at least two examples of the operation of allosteric controls in the regulation of energy metabolism.

ARTICLE RESOURCES

Anaerobic and Aerobic Energy Metabolism

Adenosine Triphosphate and the Cell's Energy Currency

The Flow of Energy Through the Living World

An Overview of the Release of Energy from Glucose

The Transfer of Hydrogen Atoms and Electrons

Dickerson, Richard E. Cytochrome c and the evolution of energy metabolism. Scientific American, 1980, 242(3) 136-153. Photosynthesis and respiration are compared. The formation of ATP by means of a respiratory electron chain in mitochondria is shown.

ATP Revisited

Novak, Marietta. Fireflies light the way to better health. Science 82, 1982, 3(6), 96-97. The intensity of the glow of luciferase, when combined with luciferin and ATP from any living organism, will give a sensitive measurement of the amount of ATP present. It is being used at present to screen for urinary infections and to determine the ATP levels in malignant and normal cells.

Hinckle, Peter C., and Richard E. McCarty. How cells make ATP. Scientific American, 1978, 238(3), 104-123. This is a classic that should be read by student and teacher alike. The mechanisms of ATP formation are clearly explained.

Glycolysis

Overview of the Citric Acid Cycle

Reactions of the Citric Acid Cycle

The Respiratory Chain

Oxidative Phosphorylation and Mitochondrial Structure

Fermentation

Demain, Arnold L., and Nadine A. Solomon. Industrial microbiology. Scientific American, 1981, 245(3), 67-75. History of fermentation. Fermentation is dependent on the ability of microorganisms to reproduce rapidly and to do a wide variety of metabolic reactions.

Phaff, Herman, J. Industrial microorganisms. Scientific American, 1981, 245(3), 77-90. Pathways in fermentation are delineated and a number of types of fermentation--acetone butanol, heteropermentative, and homofermentative--are explored. Discusses a wide variety of microbial organisms with industrial products.

Rose, Anthony H. The microbiological production of food and drink. Scientific American, 1981, 245(3), 126-138. Explores the manufacture of such food items as yogurt, pickles, bread, cheese, beer, sauerkraut, spirits, and wine. A chart showing various types of cheeses, their country of origin, and the microorganism used to manufacture them is given. Similar diagrams show the production of beer, the fermentation of wine, and the distillation of spirits.

Comparative Energy Yields

Connection with Other Pathways

Feedback Regulation

ESSAY QUESTIONS

1. Briefly summarize the complete oxidation of 1 molecule of glucose by an aerobic eukaryotic cell.

    Glycolysis, which occurs in the cytosol, begins with a series of reactions in which ATP is used and 2 molecules of a three-carbon sugar phosphate are produced. In the remaining reactions of glycolysis, the three-carbon sugar phosphate is oxidized to 1,3-phosphoglyceric acid, reducing NAD, and the phosphates are transferred to ADP, forming 2 ATP molecules, and leaving pyruvate. Pyruvate is oxidized to carbon dioxide via the citric acid cycle, a series of reactions that captures the free energy of the pyruvate in NADH + H^+, $FADH_2$, and ATP. The free energy stored in NADH + H^+ and $FADH_2$ is used to form ATP through the process of oxidative phosphorylation, which is associated with the respiratory electron transport chain. The final acceptor of electrons is oxygen, which is reduced to water. The net gain in stored energy as a result of the complete oxidation of glucose is 36 molecules of ATP.

2. What is meant by allosteric control? Give an example of allosteric regulation of the glycolytic pathway.

    Allosteric control is exerted when the product of a later reaction in a biochemical pathway acts as an allosteric inhibitor or allosteric activator of the enzyme catalyzing one of the earlier steps in the pathway. For example, ATP is an allosteric inhibitor of phosphofructokinase, whereas ADP and AMP are allosteric activators of this enzyme.

3. Outline the chemiosmotic hypothesis of British biochemist Peter Mitchell.

    Mitchell proposed that the operation of the respiratory chain results in the pumping of protons into the space between the inner and outer mitochondrial membranes. This results from the positioning of the enzymes and pigments of the respiratory chain within the inner membrane. As the respiratory chain operates, a proton gradient is established across the inner membrane. Also located in the membrane is an ATP synthetase that requires unusually acidic conditions in which to function and can act as a proton channel through which the gradient across the membrane can be discharged. As protons flow through the channel to discharge the gradient, ATP is made from ADP and inorganic phosphate.

4. Give at least two illustrations of ways in which the energy metabolism of the cell is integrated with other cellular activities.

    Energy-storage compounds are digested into their component monomers, which then enter the glycolytic or citric acid cycle pathways. Conversely, intermediates in these pathways may be converted into energy-storage compounds for later use. Many intermediates of the citric acid cycle are used as the starting point for the synthesis of important cellular component such as chlorophyll and amino acids.

5. Explain why aerobic eukaryotes rapidly die when deprived of oxygen.

    There are two reasons. First, oxygen is the terminal electron acceptor for the respiratory chain. Cutting off the oxygen supply means that oxidative phosphorylation is stopped, resulting in a depletion of ATP. Without the energy stored in ATP molecules, critical endergonic reactions cannot occur. Second, the respiratory chain also functions to regenerate the oxidized electron carriers that are required for the reactions of the citric acid cycle and glycolysis. Once all of the electron carriers in the cell are reduced, neither of these pathways can operate, and the supply of ATP is interrupted completely. In short, life depends on a continuous supply of ATP for the cell's many energy-requiring processes.

TEST SET 7-A

Name _____   Section _____

Choose the best answer to each of the following questions, and write the appropriate letter in the space provided.

Ans: b
p. 180

_____ 1. A high level of citrate is associated with: (a) the activation of phosphofructokinase; (b) the increased synthesis of fatty acids; (c) a low level of ATP; (d) a low level of NADH + H$^+$.

Ans: b
p. 167

_____ 2. As glucose is converted to 2 molecules of glyceraldehyde-3-phosphate: (a) 2 molecules of inorganic phosphate are formed; (b) 2 ATP molecules are converted to 2 ADP molecules; (c) 2 ADP molecules are converted to 2 ATP molecules; (d) 2 NAD$^+$ molecules are formed.

Ans: b
p. 178

_____ 3. Which of the following reactions describes the Pasteur effect?
(a) Glycolysis increases as the respiratory chain operates.
(b) Glycolysis decreases as the respiratory chain operates.
(c) Glycolysis increases as the amount of glucose increases.
(d) Glycolysis increases as the amount of glucose decreases.

Ans: d
p. 163

_____ 4. Which of the following molecules is a reducing agent? (a) ATP; (b) NAD$^+$; (c) FAD; (d) NADH + H$^+$.

Ans: a
p. 177

_____ 5. Fermentation results in a net yield of ____ ATP molecules. (a) 2; (b) 4; (c) 5; (d) 36.

Ans: c
p. 168

_____ 6. One of the principal outputs of the citric acid cycle is:
(a) oxidized electron carriers; (b) oxidative phosphorylation; (c) CO_2; (d) H_2O.

Ans: a
p. 165

_____ 7. After hydrolysis, which of the following molecules would release more than 20 kcal/mole of free energy? (a) ATP; (b) ADP; (c) AMP; (d) cAMP.

Ans: d
p. 168

_____ 8. In glycolysis and cellular respiration, ____ molecules of reduced NAD are formed per molecule of glucose. (a) 4; (b) 5; (c) 8; (d) 10.

Ans: c
p. 179

_____ 9. The processes of fermentation and cellular respiration begin with the molecule: (a) ATP; (b) ADP; (c) pyruvic acid; (d) lactic acid.

Ans: b
p. 177

_____ 10. For each pair of hydrogen atoms passed from reduced NAD to the end of the respiratory chain, ____ molecules of ATP are produced. (a) 2; (b) 3; (c) 4; (d) 6.

Ans: d
p. 173

_____ 11. The cytochrome molecules transfer electrons by redox reactions involving: (a) ATP molecules; (b) FAD molecules; (c) NAD molecules; (d) Fe atoms.

Ans: c
p. 176

_____ 12. The steps in fermentation are important to a cell because:
(a) ATP is formed; (b) NAD$^+$ is reduced; (c) NADH + H$^+$ is oxidized; (d) alcohol is formed.

Ans: c
p. 178

_____ 13. The net yield of ATP molecules from the aerobic breakdown of 1 molecule of glucose is not 38 (as calculated) but 36, because: (a) the $FADH_2$ molecules lose 1 ATP in transporting electrons;

(b) the 2 GTP molecules are not converted to ATP molecules; (c) the 2 NADH + H^+ molecules produced by glycolysis yield a net of only 4 ATP molecules; (d) the 2 NADH + H^+ molecules produced in the formation of acetyl coenzyme A yield a net of only 4 ATP molecules.

Ans: c
p. 179

_____ 14. Phosphofructokinase is activated by: (a) fructose; (b) ATP; (c) ADP; (d) inorganic phosphate.

Ans: c
p. 172

_____ 15. One of the functions of the respiratory chain is to: (a) form oxygen; (b) break bonds in H_2O; (c) reoxidize NADH + H^+ and $FADH_2$; (d) reduce NAD^+ and FAD.

TEST SET 7-B

Name _____   Section _____

Choose the best answer to each of the following questions, and write the appropriate letter in the space provided.

Ans: d
p. 166

_____ 1. The first five reactions of glycolysis: (a) produce 2 ATP molecules; (b) produce 2 NAD^+ molecules; (c) are exergonic; (d) are endergonic.

Ans: a
p. 178

_____ 2. A key intermediate molecule to many metabolic pathways is: (a) acetyl-CoA; (b) lactic acid; (c) 1,3-diphosphoglycerate; (d) cytochrome $a3$.

Ans: b
p. 163

_____ 3. Which of the following molecules is an oxidizing agent? (a) ATP; (b) NAD^+; (c) NADH + H^+; (d) $FADH_2$.

Ans: d
p. 177

_____ 4. At the end of glycolysis and the citric acid cycle, how many molecules of ATP per molecule of glucose have been formed? (a) 2; (b) 3; (c) 4; (d) 6.

Ans: b
p. 168

_____ 5. In glycolysis, at which step are 2 NAD molecules reduced? (a) fructose 1,6-diphosphate to glyceraldehyde 3-phosphate; (b) glyceraldehyde 3-phosphate to 1,3-diphosphoglyceric acid; (c) pyruvate to lactic acid; (d) dihydroxyacetone phosphate to glyceraldehyde 3-phosphate.

Ans: b
p. 165

_____ 6. The two terminal phosphate bonds in ATP are referred to as "high-energy" bonds because: (a) 10 kcal/mole of energy are stored in each bond; (b) 10 kcal/mole of energy are released from each bond upon hydrolysis; (c) the last two phosphate groups are more positively charged than the first phosphate group; (d) the last two phosphate groups contain high levels of energy.

Ans: d
p. 169

_____ 7. In the breakdown of 1 molecule of glucose, _____ molecules of CO_2 are produced by cellular respiration. (a) 2; (b) 3; (c) 4; (d) 6.

Ans: d
p. 176

_____ 8. The production of lactic acid by muscle cells is an example of: (a) cellular respiration; (b) glycolysis; (c) aerobic respiration; (d) fermentation.

Ans: a
p. 173

_____ 9. For each pair of hydrogen atoms that move from reduced FAD to the end of the respiratory chain, _____ molecules of ATP are produced. (a) 2; (b) 3; (c) 4; (d) 6.

Ans: a
p. 170

_____ 10. Which of the following reactions occurs as pyruvate is converted to acetyl coenzyme A? (a) CO_2 is released. (b) H_2O is produced. (c) NAD^+ is oxidized. (d) ATP is produced.

Ans: d
p. 173

_____ 11. Enzymes involved in the conversion of glucose to pyruvic acid are found in the: (a) matrix of the mitochondrion; (b) inner membrane of the mitochondrion; (c) outer membrane of the mitochondrion; (d) cytoplasm of the cell.

Ans: b
p. 172

_____ 12. The final electron acceptor in the respiratory chain is: (a) H_2O; (b) oxygen; (c) cytochrome c; (d) FMN.

Ans: b
p. 162

_____ 13. The process of oxidative phosphorylation is important because: (a) ADP is produced; (b) ATP is produced; (c) NAD is oxidized; (d) FAD is oxidized.

Ans: a
p. 179

_____ 14. A high level of acetyl coenzyme A would result in increased synthesis of: (a) oxaloacetate; (b) pyruvate; (c) glucose; (d) fructose.

Ans: b
p. 175

_____ 15. The location of the ATP synthetase molecules suggests that it is possible experimentally to show that: (a) ATP is stored exclusively between the inner and outer mitochondrial membranes; (b) ATP production can be blocked without immediately affecting other reactions of the respiratory chain; (c) ATP cannot be formed unless the pH is extremely basic; (d) blocking ATP production involves activation of the synthetase molecules.

CHAPTER EIGHT
PHOTOSYNTHESIS

CHAPTER SUMMARIES AND OBJECTIVES

8-1 <u>Early Studies of Photosynthesis</u> Many organisms (including animals, fungi, many protists, and most monerans) are heterotrophs. They require partially reduced carbon compounds as a source of energy and carbon atoms. Organisms that do not need preformed organic substances are autotrophs. The principal autotrophs are photosynthetic, utilizing visible light as their source of energy. Early investigations of photosynthesis showed that water, carbon dioxide, and light were required for the process and that food (glucose) and oxygen were produced by plants. Studies with radioisotopes have more recently shown that water is both a reactant and a product of photosynthesis. The oxygen produced comes from water molecules, not from CO_2.

1. Write the overall equation for photosynthesis and explain why water appears on both sides of the equation.

8-2 <u>Basic Physics of Light / Pigments / Absorption Spectra and Action Spectra / The Photosynthetic Pigments</u> Light is the energy source for photosynthesis. It is a form of radiant energy and thus comes in discrete units called quanta, or photons. Light also has a wave nature. The velocity of light in a vacuum is one of the physical constants of nature. We may speak of the frequency of light, the number of wave peaks passing a given point in a given time, and of the wavelength--the distance between peaks. Frequency, wavelength, and the velocity of light are related in the following way:

$$c = \lambda \nu$$

where λ = wavelength (cm), ν = frequency (Hz), and c = velocity (cm/sec). The energy contained in a single photon is related to the wavelength by the equation $E = h\nu$, where h = Planck's constant, and E = the energy of the photon. Note that shorter wavelengths have greater energy.

Molecules that absorb radiation in the visible wavelengths are called pigments. When a photon of energy is absorbed by a pigment, the molecule is raised from a lower-energy (ground) state to a higher-energy (excited) state by the boosting of an electron to a higher orbital. This results in the electron's being held less tightly by its nucleus.

A graph of the wavelengths of light that a particular pigment may absorb is called an absorption spectrum. A graph of the biological effectiveness of light, as a function of its wavelength, is called an action spectrum. Action spectra have been useful in ascertaining which pigments are involved in the photosynthetic process.

The most important photosynthetic pigments are the chlorophylls, with chlorophyll <u>a</u> and chlorophyll <u>b</u> predominating. These absorb light at the blue and red ends of the spectrum. Accessory pigments increase the number of wavelengths that can be used in photosynthesis by absorbing photons intermediate in energy between the blue and red wavelengths and passing the absorbed energy to chlorophyll. The "antenna" formed by chlorophylls and accessory pigments absorbs energy over a large part of the visible spectrum.

2. Write the equation that relates the frequency and wavelength of light to its velocity, and that which relates the wavelength of a photon to its energy.

3. Define the term pigment, give at least two examples of pigments, and explain what happens when a pigment molecule absorbs a photon.

4. Explain the difference between an absorption spectrum and an action spectrum.

8-3 An Electron Carrier for Photosynthesis / The Subpathways of Photosynthesis Many redox reactions occur in photosynthesis. The electron carrier for photosynthesis is nicotinamide adenine dinucleotide phosphate, NADP, which exists in two forms: $NADP^+$ (oxidizing agent) and NADPH + H^+ (reducing agent).

The overall process of photosynthesis does not occur in a single step but in two subpathways. The "light reactions," or photophosphorylation, use the energy of sunlight to make ATP, NADPH, and O_2. The ATP and NADPH are utilized in the second subpathway, the "dark reactions" or Calvin-Benson cycle, to trap CO_2 and reduce it to sugar. Both reactions occur in the chloroplast and are tied together by the exchange of ADP/ATP and NADP/NADPH.

5. Name the primary electron carrier for photosynthesis.

6. Explain how the subpathways of photosynthesis interlock to utilize the sun's energy to manufacture sugar.

8-4 Working Out the Calvin-Benson Cycle Our ability to work out the steps involved in the conversion of atmospheric CO_2 to complex carbohydrates depended in large part on the development of improved laboratory techniques. The availability of radioactive carbon-14 made it possible to keep track of carbon atoms as they participated in biochemical processes. The development of paper partition chromatography and two-dimensional chromatography made possible the separation of complex mixtures of organic compounds. Small amounts of material separated by chromatographic techniques could be located by means of a third technique, autoradiography. Working with the alga Chlorella, Calvin and Benson and their co-workers at UC, Berkeley, were able to work out the photosynthetic carbon reduction cycle.

7. Explain how the techniques of paper partition chromatography, two-dimensional chromatography, and autoradiography are carried out.

8. Describe a typical experiment conducted by the Berkeley group in working out the photosynthetic carbon reduction cycle.

8-5 The First Stable Product of Carbon Dioxide Fixation / What Is the Carbon Dioxide Acceptor? / The Photosynthetic Carbon Reduction Cycle Much experimental work by the Berkeley group headed by Calvin and Benson has led to elucidation of all the intermediate steps in the photosynthetic carbon reduction cycle. This process may be summarized as follows: A molecule of ribulose biphosphate (RuBP) combines with a molecule of carbon dioxide to yield two molecules of 3-phosphoglycerate (3PG). The 3PG is reduced to the sugar phosphate glyceraldehyde-3-phosphate (G3P) in a reaction that requires ATP and NADPH + H^+. Some of the G3P is then used to form glucose and other products, while the remainder is used to regenerate RuBP in a complex series of reactions, one of which requires ATP. The cycle must "turn" six times, incorporating six molecules of CO_2, in order to form one molecule of glucose.

8. Summarize the photosynthetic carbon reduction cycle as presented in Figure 8-23 of the text. Pay particular attention to the steps that require energy and/or reducing power.

8-6 Alternate Modes of Carbon Dioxide Fixation / Photorespiration Some plants produce the four-carbon compound oxaloacetate, rather than 3PG, as the first product of carbon dioxide fixation. These C_4 plants carry out the Calvin-Benson cycle but have an extra step to increase the rate of photosynthesis at low levels of CO_2. C_4 plants contain the enzyme PEP carboxylase, which catalyzes the formation of oxaloacetate from CO_2 and PEP (phosphoenolpyruvate). Because PEP carboxylase has a higher affinity for CO_2 than does RuBP carboxylase, such plants are able to trap CO_2 more efficiently than "ordinary" plants. C_4 plants are generally found in relatively dry environments, and they keep their stomata closed much of the time to prevent water loss. This results in depletion of CO_2 within the leaf. Near the surface of the leaf are mesophyll cells, which trap CO_2 and form oxaloacetate. The oxaloacetate diffuses to the bundle-sheath beneath the mesophyll layer. In the bundle sheath cells the oxaloacetate is decarboxylated to yield CO_2, which is taken up by RuBP and enters the Calvin-Benson cycle.

Photorespiration is a light-dependent process that interferes with photosynthesis. Its function in the plant is unknown. C_4 plants carry out little photorespiration compared with C_3 plants, which may lose a large part of the CO_2 fixed in photosynthesis to photorespiration.

9. Describe the means by which C_4 plants are able to carry out carbon dioxide fixation even at very low levels of CO_2.

10. Explain what photorespiration is and why it is of interest to agricultural biologists.

8-7 The Activation of Chlorophyll: A "Light Reaction" / Cyclic Photophosphorylation / Noncyclic Photophosphorylation When a molecule of chlorophyll absorbs an electron to become excited chlorophyll (Chl*), it becomes a good reducing agent. This is because the electron is less tightly held by the chlorophyll molecule. Thus Chl* can interact with an oxidizing agent: $Chl* + A \longrightarrow Chl^+ + A^-$. This redox reaction would not occur in the dark--that is, in the absence of the energy input from light.

In some organisms, the formation of A^- (shown in the foregoing equation) is followed by a series of redox reactions, one of which is sufficiently exergonic to drive the formation of ATP. The last product in the series, compound D^- (see Figure 8-29 in the text) is used to reduce the Chl^+, converting it back to ground-state chlorophyll. This chain of reactions is called cyclic photophosphorylation.

Cyclic photophosphorylation is satisfactory for the formation of ATP, but is not sufficient for the needs of the Calvin-Benson cycle because no reducing agent is produced. The evolution of noncyclic photophosphorylation provided a means whereby the energy of light could be used to form both ATP and NADPH + H^+. In noncyclic photophosphorylation, electrons from water are elevated to a higher energy state by the action of light on chlorophyll. These electrons are transferred to $NADP^+$, reducing it to NADPH + H^+, through a series of electron carriers. One of the intermediate steps is sufficiently exergonic to allow the formation of ATP. Noncyclic photophosphorylation requires the participation of two molecules of chlorophyll (Photosystems I and II). The chlorophyll of Photosystem II becomes excited by light energy and passes an electron to an oxidized receptor molecule (J), reducing it. The electron lost by the Photosystem II chlorophyll is replaced from water. The electron acquired by J is passed through a series of redox reactions, forming ATP in the process. Photosystem I chlorophyll also absorbs light, and reduces a different oxidized receptor molecule (Q). The electron lost by the Photosystem II chlorophyll is replaced by the one that passed through the redox chain from Photosystem II. The electron transferred to compound Q is used, together with protons produced from the ionization of water in Photosystem II, to reduce $NADP^+$ to NADPH + H^+. Noncyclic photophosphorylation must operate twice (that is, two photons must be absorbed by each photosystem) to reduce one molecule of $NADP^+$. (See Figure 8-28 in the text.)

Noncyclic photophosphorylation is intimately associated with the Calvin-Benson cycle in two ways: (1) The reduction of carbon requires the input of energy in the form of ATP and NADPH + H^+. (2) The operation of noncyclic photophosphorylation requires the regeneration of ADP and $NADP^+$ by the Calvin-Benson cycle.

11. Explain how chlorophyll makes possible the reduction of oxidized receptor compounds, using energy derived from light.

12. Outline the process of noncyclic photophosphorylation. Explain why this process cannot participate in the manufacture of glucose.

13.   Outline the process of noncyclic photophosphorylation. Include in your explanation the role of water, Photosystems I and II, ADP, and $NADP^+$.

14.   Explain why noncyclic photophosphorylation and the Calvin-Benson cycle are interdependent.

8-8   ATP Formation in the Chloroplast / Limiting Factors and Compensation Points / Photosynthesis and Cellular Respiration   The formation of ATP in the chloroplast is believed to occur by a chemiosmotic process analogous to that which occurs in the mitochondrion (see Chapter 7). Electrons are pumped to the interior of the thylakoids and are followed by protons, making the inside acidic. Proton channels in the membrane, which function as ATP synthetases, allow the protons to move out again and are activated by the lowering of the pH as the protons move out of the thylakoid. Experimental evidence tends to confirm this hypothesis.

Photosynthesis ceases in the dark, but cellular respiration continues, and both processes occur in the light. The light intensity at which photosynthetic CO_2 uptake is balanced by respiratory CO_2 production is called the compensation point. Plants must have light of greater intensity than the compensation point in order to produce food for growth. As the intensity increases above this point, photosynthesis increases until the plant is light-saturated. At this point, the limiting factor in determining the rate of photosynthesis becomes CO_2 availability. Temperature exerts different effects on photosynthesis and cellular respiration. Thus the compensation point is dependent on temperature, and this effect helps determine plant distributions.

There are important similarities between the processes of photosynthesis and respiration, and these two processes serve to interconnect virtually all living organisms.

15.   Define compensation point, light-saturated, and limiting factor.

16.   Explain the chemiosmotic model of photosynthetic ATP formation. Cite supportive experimental evidence.

17.   Give three similarities between photosynthesis and cellular respiration.

ARTICLE RESOURCES

Early Studies of Photosynthesis

    Levine, R. P. The mechanisms of photosynthesis. Scientific American, 1969, 221 (6), 15. Excellent discussion of early research into mechanisms whereby photosynthesis occurs. This is a classic and will greatly enhance understanding of this chapter.

Basic Physics of Light Pigments

Absorption Spectra and Action Spectra

    Govindjee and Govindjee. The absorption of light in photosynthesis. Scientific American, 1974, 231 (6), 68-82. Basic but clear discussion of photosynthesis with emphasis on absorption spectra and action spectra. Includes the role of plant pigments in photosynthesis and of the physical aspects of photosynthesis.

The Photosynthetic Pigments

An Electron Carrier for Photosynthesis

The Subpathways of Photosynthesis

Working Out the Calvin-Benson Cycle

The First Stable Product of Carbon Dioxide Fixation

What Is the Carbon Dioxide Acceptor?

The Photosynthetic Carbon Reduction Cycle

Alternate Modes of Carbon Dioxide Fixation

Photorespiration

The Activation of Chlorophyll: A "Light Reaction"

Miller, Kenneth R. The photosynthetic membrane. Scientific American, 1979, 241 (4), 102-114. Membranous sacs called thylakoids in which chloroplasts are piled on top of each other form the grana. The light reaction of photosynthesis is carried out in these membranes. The dark reaction occurs in the nonmembranous, soluble components of the chloroplasts. A detailed discussion follows on the asymmetry of the thylakoid membrane, which makes the light reaction possible.

Arntzen, Charles J. Dynamic structural features of chloroplast lamellae. Current Topics in Bioenergetics, 1978, 8, 111-160. Detailed discussion of the chloroplast structures involved in photosynthesis. A bit difficult to read but very informative.

Cyclic Photophosphorylation

Hinckle, Peter C., and Richard E. McCarty. How cells make ATP. Scientific American, 1978, 238 (3), 104-122. Clarifies Mitchell's chemiosmotic theory of ATP synthesis. Discusses oxidative and photosynthetic phosphorylation, as well as how plants utilize proton gradients.

Boyer, Paul D., Britton Chance, Lars Ernster, et al. Oxidative phosphorylation and photophosphorylation. Annual Review of Biochemistry, 1977, (46), 955-1026. This detailed analysis is difficult reading but worth the effort.

Noncyclic Photophosphorylation

ATP Formation in the Chloroplast

Limiting Factors and Compensation Points

Photosynthesis and Cellular Respiration

Dickerson, Richard E. Cyctochrome and the evolution of energy metabolism. Scientific American, 1980, 242 (3), 136-153. Molecular evidence for the evolution of photosynthesis based on three-dimensional folding and the amino acid sequences of protein molecules. A phylogenetic tree tracing the evolution of bacterial photosynthesis and respiration is included.

ESSAY QUESTIONS

1. Explain how the chloroplasts of photosynthetic autotrophs and the mitochondria of aerobic heterotrophs "cooperate" to create cycles of carbon, oxygen, and energy through the biosphere.

Through their chloroplasts, the autotrophs utilize the energy of sunlight to reduce inorganic carbon (CO_2) to sugar, releasing molecular oxygen. Through their mitochondria, heterotrophs oxidize the sugar, utilizing molecular oxygen as the ultimate electron acceptor, to release carbon dioxide and energy. Thus there is a continuous cycle of reduction and oxidation of carbon dioxide, driven by the energy of the sun.

2. What are the three possible outcomes when a photon interacts with a pigment molecule? Which of these outcomes is significant in terms of photosynthesis? Why?

The photon may be reflected off the molecule, it may be transmitted through the molecule, or it may be absorbed by the molecule. Absorption of the photon is significant in terms of photosynthesis, because this results in the molecule's being raised from a lower to a higher energy state. In the case of chlorophyll, such excitation makes the molecule a very good reducing agent, enabling it to initiate the chain of events in photophosphorylation.

3. In a brief paragraph, summarize the events of photosynthesis, showing how the subpathways interlock to bring about the reduction of carbon dioxide.

There are two principal subpathways in photosynthesis: photophosphorylation and the carbon reduction cycle. In the former, the energy of the sun is used, through the action of light on chlorophyll, to elevate electrons from water to a higher energy state. These electrons are transferred to $NADP^+$, reducing it to $NADPH + H^+$, through a series of electron carriers. One of the intermediate steps is sufficiently exergonic to drive the formation of ATP. In the carbon reduction cycle, a molecule of carbon dioxide combines with a molecule of RuBP to yield two molecules of 3-phosphoglycerate. The 3PG is reduced to glyceraldehyde-3-phosphate in a reaction requiring NADPH and ATP. Some of the G3P is used to form glucose and other products, while the remainder is used to regenerate RuBP in a reaction chain, one link of which requires ATP. Thus the high-energy compounds generated by photophosphorylation are used in the carbon reduction cycle to reduce carbon dioxide to glucose.

4. Describe the experimental procedures involved in the determination, by Calvin and Benson and their co-workers, of the nature of the first stable product of carbon dioxide fixation.

Cultured algae were exposed to radioactive carbon dioxide. After two seconds, a sample was withdrawn from the culture and killed in boiling ethanol to extract the various organic compounds present. The mixture was then separated by paper partition chromatography, and the radioactive compounds thus separated were located by autoradiography. A number of the compounds were found to be radioactive, but the one produced most rapidly and in the largest amount was 3-phosphoglycerate, the first stable product of carbon dioxide fixation.

5. Explain the biochemical process used by C_4 plants to trap carbon dioxide. Why is this process advantageous to the plants in their environment?

C_4 plants contain the enzyme PEP carboxylase, which catalyzes the formation of oxaloacetate from carbon dioxide and phosphoenolpyruvate. Because PEP carboxylase has a higher affinity for CO_2 than does RuBP carboxylase, such plants are able to trap CO_2 more efficiently than plants lacking PEP carboxylase. C_4 plants are usually found in relatively dry environments, and they keep their stomata closed much of the time to conserve water. This results in a depletion of CO_2 within the leaf. By trapping CO_2 efficiently, C_4 plants are able to continue to carry out carbon reduction while keeping their stomata open a minimal amount of time. Thus they are able to live in environments that are unavailable to C_3 plants.

TEST SET 8-A

Name _____   Section _____

Choose the best answer to each of the following questions, and write the appropriate letter in the space provided.

Ans: b
p. 185

_____ 1. Photosynthetic organisms utilize which of the following waste products of aerobic respiration?   (a) O_2; (b) CO_2; (c) glucose; (d) NADH + H$^+$.

Ans: d
p. 184

_____ 2. Which of the following is an example of an autotroph?   (a) zebra; (b) amoeba; (c) yeast; (d) Chlorella.

Ans: a
p. 189

_____ 3. A graph of the effect of light on a biological activity as a function of wavelength is called a(an):   (a) action spectrum; (b) absorption spectrum; (c) molecular spectrum; (d) atomic spectrum.

Ans: c
p. 187

_____ 4. The energy of a wavelength is: (a) directly proportional to the increasing size of the wavelength; (b) directly proportional to the color of the wavelength; (c) inversely proportional to the increasing size of the wavelength; (d) not related to the size of the wavelength.

Ans: d
p. 191

_____ 5. Carotenoids absorb blue and blue/green wavelengths of light and are _____ in color.   (a) blue; (b) green; (c) red; (d) yellow.

Ans: c
p. 192

_____ 6. The pathway of photosynthesis that requires light and water is called: (a) the Calvin-Benson cycle; (b) the C_4 cycle; (c) photophosphorylation; (d) oxidative phosphorylation.

Ans: b
p. 194

_____ 7. In one-dimensional chromatography, each spot represents: (a) one kind of molecule; (b) a possible mix of several molecules; (c) the distance traveled by the solvent; (d) half the distance traveled by the solvent.

Ans: c
p. 199

_____ 8. How many CO_2 molecules are needed to produce 1 glucose molecule? (a) 1; (b) 2; (c) 6; (d) 12.

Ans: b
p. 201

_____ 9. In C_4 plants, oxaloacetate is decarboxylated in the: (a) mesophyll cells; (b) bundle sheath cells; (c) epidermal cells; (d) guard cells.

Ans: d
p. 203

_____ 10. Which of the following statements about the process of photorespiration is true?   (a) Photorespiration involves the production of ATP for the dark reactions.   (b) Photorespiration occurs more often in C_4 plants than in C_3 plants.   (c) Additional CO_2 is fixed in the process of photorespiration.   (d) Some CO_2 is released in the process of photorespiration.

Ans: c
p. 205

_____ 11. Chlorophyll can be changed from a ground state to an excited state by: (a) oxidizing another molecule; (b) reducing another molecule; (c) absorption of a photon of light; (d) fluorescence of a photon of light.

Ans: c
p. 205

_____ 12. In cyclic photophosphorylation, the series of reactions that occur from Chl* to compound D are: (a) enzymatically driven; (b) controlled by allosteric feedback; (c) exergonic; (d) endergonic.

Ans: a
p. 206

_____ 13. In Photosystem I, excited chlorophyll molecules are reduced by: (a) electrons from Photosystem II; (b) electrons from the splitting of H_2O molecules; (c) absorbing a photon of light; (d) the oxidizing agent, compound Q.

Ans: c   _____   14.   How many photons of light are required to produce 1 reduced NADP
p. 208   molecule? (a) 1; (b) 2; (c) 4; (d) 8.

Ans: b   _____   15.   When the photosynthetic rate exceeds the respiratory rate in
p. 210   plants, which of the following may occur? (a) The plant fails
                            to grow because the balance is upset. (b) The plant stores
                            glucose in the form of starch. (c) The plant converts starch
                            to glucose to increase the metabolic rate. (d) The respiratory
                            rate is inhibited by the products of photosynthesis.

TEST SET 8-B

Name _____   Section _____

Choose the best answer to each of the following questions, and write the appropriate letter in the space provided.

Ans: c
p. 186

_____ 1. The principal waste product of photosynthesis is: (a) H_2; (b) CO_2; (c) O_2; (d) light.

Ans: b
p. 188

_____ 2. Chlorophyll appears green in color because the pigment: (a) absorbs wavelengths of green light; (b) absorbs blue and red wavelengths of light; (c) is excited to the green color state; (d) reflects the color of surrounding pigments.

Ans: c
p. 187

_____ 3. The intensity of light with a wavelength of 420 nm can be expressed as: (a) blue; (b) low in energy; (c) photons; (d) a vacuum.

Ans: a
p. 191

_____ 4. Blue-green algae appear blue due to which of the following pigments? (a) phycobilins; (b) carotenoids; (c) chlorophyll a; (d) chlorophyll b.

Ans: a
p. 193

_____ 5. Which of the following processes does(do) not occur in higher plant cells during the day? (a) fermentation; (b) oxidative phosphorylation; (c) photophosphorylation; (d) the "dark reactions."

Ans: c
p. 199

_____ 6. The CO_2 acceptor in the Calvin-Benson cycle is: (a) 3-phosphoglycerate; (b) acetyl-CoA; (c) ribulose bisphosphate; (d) H_2O.

Ans: d
p. 200

_____ 7. The production of 1 molecule of glucose requires how many reduced NADPH molecules? (a) 1; (b) 2; (c) 6; (d) 12.

Ans: b
p. 199

_____ 8. Most of the enzymes of the Calvin-Benson Cycle are found in the: (a) thylakoid membranes of chloroplasts; (b) stroma of chloroplasts; (c) inner membrane of the chloroplast; (d) cytoplasm of the cell.

Ans: c
p. 200

_____ 9. The function of PEP carboxylase is to: (a) decarboxylate PEP; (b) decarboxylate oxaloacetate; (c) synthesize oxaloacetete from PEP and CO_2; (d) add CO_2 to oxaloacetate.

Ans: d
p. 203

_____ 10. The process of photorespiration occurs predominantly in: (a) mesophyll cells; (b) guard cells; (c) C_4 plants; (d) C_3 plants.

Ans: a
p. 207

_____ 11. The ultimate receptor for the electrons from the splitting of the water molecule is: (a) NADP; (b) ATP; (c) CO_2; (d) O_2.

Ans: b
p. 205

_____ 12. In cyclic photophosphorylation, excited chlorophyll molecules are reduced by: (a) the first compound in the electron transport chain; (b) the last compound in the electron transport chain; (c) ATP; (d) ADP.

Ans: b
p. 207

_____ 13. In Photosystem II, excited chlorophyll molecules are reduced by: (a) electrons from Photosystem I; (b) electrons from the splitting of H_2O molecules; (c) absorbing a photon of light; (d) the oxidizing agent, compound J.

Ans: c
p. 208

_____ 14. In higher plants, ATP is formed by protons moving across the: (a) inner membrane of the chloroplast; (b) outer membrane of the chloroplast; (c) thylakoid membranes; (d) microbody membranes.

Ans: a
p. 210

_____ 15. In a plant that is light-saturated, the rate of photosynthesis is increased by an increase in the: (a) amount of CO_2; (b) amount of O_2; (c) amount of NADPH + H^+; (d) number of photons of light available.

CHAPTER NINE
CHROMOSOMES AND CELL DIVISION

CHAPTER SUMMARIES AND OBJECTIVES

9-1 <u>Mitosis</u> Mitosis (karyokinesis) is a nuclear division resulting in the formation of two daughter nuclei identical to each other and to the original dividing nucleus with respect to chromosomal makeup and genetic constitution. It involves a progressive sequence of events wherein the nuclear material is noticeably active. Prior to mitosis, nuclear and cell materials are duplicated. Once prophase of mitosis begins, chromosomes may be seen to coil and shorten; they line up during metaphase attached to the spindle; the daughter chromosomes (formerly chromatids) separate from each other in anaphase; and the two new nuclei undergo what appears to be a "reverse prophase" during telophase.

1. Explain how chromosomal arrangement on the spindle at the mataphase plate is significant in the production of two identical cells.

2. Sketch a diploid cell that typically has four chromosomes and is in anaphase of mitosis. Show clearly the number and general shape of the chromosomes.

3. Describe centriole activity during mitosis.

9-2 <u>Chromosomal Structure</u> Chromatin changes dramatically in its organization from interphase through mitosis. During most of mitosis the chromatin is visible as short, thick, supercoiled structures; during interphase it is unwound and strung out thinly. Presumably each shape helps accomplish what must be done. DNA must be available to direct cellular activities during interphase (unwound), and chromosomes carrying each gene site must be delivered in packages (short and thick) to each daughter cell.

4. Describe the relationship between DNA and histones in chromosome structure.

9-3 <u>The Mechanism of Chromosomal Movement</u> Chromosomal movement before metaphase and during anaphase is brought about by shortening and movement of the spindle fibers. Microtubules comprising the spindle may shorten by disassembling at the ends nearest the poles, or pole-to-chromosome microtubules may slide on pole-to-pole microtubules--possibly involving the making and breaking of cross bridges between actin and myosin.

5. Compare the theoretical mechanics of chromosome movement to what is visible before metaphase and during anaphase.

9-4 <u>Cytokinesis</u> Division of cytoplasmic components usually follows mitosis; however, no mechanism exists for the precise distribution of all the organelles. So long

as each daughter cell has some of each component, additional ones will be manufactured. As would be expected, rigid plant cells and flexible animal cells complete the mechanics of cytokinesis differently.

6. Compare the ways in which plant and animal cells carry out the mechanics of
    cytokinesis.

9-5 The Cell Cycle The "cell cycle" of a living cell refers to the way it spends its
lifetime. Once differentiated, some cells--such as neurons--never divide again;
they metabolically maintain themselves until cell death without going through cell
division again. Many cells' life cycles are spent in various time periods of interphase or mitosis. Mitosis has already been discussed; a consideration of interphase completes the picture. Most of the time, cells carry out metabolic maintenance in a part of interphase known as Gap 1 (G1); dividing cells follow G1 with the
S or Synthesis stage, wherein DNA is manufactured. Once the S stage commits the
cell to division, it ordinarily undergoes the Gap 2 (G2) stage, wherein cell structures and molecules especially needed for cell division are manufactured.

7. Compare interphase of plant cells and of animal cells by describing their
    similarities and differences.

8. The S stage is considered a "committal" to mitosis. Explain the circumstances
    supporting that designation.

9. Describe the relationship between interphase and mitosis.

9-6 Asexual and Sexual Reproduction / Ploidy and the Karyotype Asexual reproduction is
accomplished without meiosis or fertilization; it leads to an organism genetically
identical to its parent (except for any mutations that may have occurred). A
population of genetically identical progeny is known as a clone. Sexual reproduction requires both meiosis and fertilization at some stage of the life cycle.
Haploid gametes fuse to produce diploid zygotes. Sexual reproduction fosters
diversity.

Ploidy refers to the number of sets of chromosomes present in a cell. A gamete has
one full set (haploid); a zygote formed at fertilization has two full sets (diploid). The collective physical characterization of all the chromosomes in a cell
is known as the karyotype of the cell.

10. Assume that an asexually produced organism originated from a cell that carried
    a new mutation making it unable to produce a particular amino acid. Describe
    how the organism had survived up to that point. Then make some astute comment
    about its progeny.

9-7 Meiosis / Mitosis, Meiosis, and Ploidy Reviewed Meiosis is the process whereby
diploid nuclei divide, producing haploid nuclei. Each haploid nucleus contains
one full set of chromosomes that have been separated from like forms, their
homologues. Two nuclear divisions are necessary in order to achieve these haploid
gametes. Interphase I prior to meiosis I is the stage in which virtually all DNA
for the two divisions is synthesized. Meiosis I is the division that reduces the
chromosome number from diploid to haploid. Centromeres continue to hold daughter
chromosomes together; homologues separate. Meiosis II results in the daughter
chromosomes being separated--the resulting cells are again haploid.

11. Compare meiosis with mitosis.

12. Assume that, for some reason, a plant produces a few diploid eggs and sperm
    (from pollen). If fertilization occurred between two of these diploid gametes
    and the plant produced was fertile, describe the plant as to ploidy. Then
    determine the ploidy of that plant's progeny.

9-8 Cell Division in Prokaryotes Prokaryotic cells are not so complex in structure as
eukaryotic cells--they have no nuclear membranes or complex protein permanently
associated with their chromatin. They do, however, reproduce faithfully, forming
clones during asexual reproduction.

13. Explain how prokaryotes get all their genes into their progeny without spindle

formation.

ARTICLE RESOURCES

Mitosis

Sundberg, Marshall, D. Making the most of onion root tip mitosis. The American Biology Teacher, 1981, 43(7), 386-388. Laboratory exercise to demonstrate the duration of successive mitotic stages. Calculations of class mitotic indices and average cell lengths are made.

Chromosomal Structure

Bauer, William R., F. H. C. Crick, and James H. White. Supercoiled DNA. Scientific American, 1980, 243(1), 118-133. Simple mathematical model for describing and analyzing supercoiling, which occurs in many forms of DNA. Topology and differential geometry are utilized. A discussion of how supercoiling occurs and a clear explanation of the physical structure of the helix are included.

Clark, Sherry A. A hands-on model of DNA. The American Biology Teacher, 1982, 44(2), 100-110. Ingenious model to illustrate DNA, codons, anticodons, and peptide bonds. A list of options for using the model is given. Clear instructions for constructing and using the model.

The Mechanism of Chromosomal Movement

Albrecht-Buehler, Guenter. The tracks of moving cells. Scientific American, 1978, 238(4), 68-76. Indications that the centriole plays an important role in guiding the direction of migrating cells in animals, particularly during cell division. Evidence is presented to show that daughter cells formed during mitotic division are mirror images.

Cytokenesis

The Cell Cycle

Asexual and Sexual Reproduction

Ploidy and the Karyotype

Meiosis

Mitosis, Meiosis, and Ploidy Reviewed

Kenkel, Leonard A. Teaching mitosis with playing cards. The Science Teacher, 1980, 47(6), 40-42. Seven decks of playing cards using four identical pairs from each deck are used to provide a complement of chromosomes for four cells. Diploid cells with eight chromosomes are formed. Stages of mitotic and meiotic division are illustrated.

Cell Division in Prokaryotes

Nash, Peter, and G. J. Epp. The bacteriophage: A functional model for demonstrating a viral life cycle. The American Biology Teacher, 1981, 43(5), 269-271 and 283. A clever way of depicting the life cycle of a virus using readily available materials. Can be used either for demonstration in the lecture hall or for students to use in the laboratory setting.

ESSAY QUESTIONS

1. Compare and contrast mitosis with meiosis I and meiosis II in diploid nuclei.

Mitosis	Meiosis
Occurs in somatic cells	Occurs in reproductive cells
Results in diploid products; there is no separation of homologues	Meiosis I is a reduction division resulting in haploid products; homologues pair and then separate
Begins and ends with diploid products	Meiosis II begins and ends with haploid nuclei

Both mitosis and meiosis have similar configurations at prophase, metaphase, anaphase, and telophase.

2. Describe interphase between mitosis and prior to meiosis I.

Gap 1--maintenance metabolism; Synthesis--synthesis of DNA and proteins for chromosomes; Gap 2--synthesis of molecules needed specifically for cell division.

3. Support the concept that sexual reproduction promotes diversity.

The only source of variation in asexual reproduction is mutation. Sexual reproduction results in the combining of genetic information from two different individuals; it allows for a reshuffling of genes in a population. Mutations also occur in sexually reproducing populations.

TEST SET 9-A

Name _____   Section _____

Choose the best answer to each of the following questions, and write the appropriate letter in the space provided.

Ans: d
pp. 222,
   223

_____ 1. In mitosis, daughter chromosomes separate from each other:
            (a) during prophase; (b) during metaphase; (c) in G2; (d) at the
            beginning of anaphase.

Ans: b
p. 223

_____ 2. Eukaryotes have complex chromosomes consisting of: (a) histone
            (Hl) and DNA; (b) several proteins and DNA; (c) several DNA
            strands; (d) DNA and RNA.

Ans: a
p. 224

_____ 3. Chromosomal movement (lining up for metaphase and during ana-
            phase) is directly dependent on: (a) spindle fiber activity;
            (b) DNA production; (c) centromere location; (d) nucleosomes.

Ans: a
pp. 225,
   226

_____ 4. All of the following statements are true of cytokinesis except:
            (a) Cytokinesis is part of mitosis. (b) Cytokinesis typically
            follows mitosis. (c) Cytokinesis in plants involves building
            a section of new cell wall. (d) Cytokinesis does not necessarily
            result in an equal division of cytoplasmic components.

Ans: b
p. 226

_____ 5. If a typically nondividing cell begins to divide again (as when
            malignant tumors grow): (a) it moves immediately into the Gl
            phase; (b) it moves into the S phase; (c) it moves immediately
            into the G2 phase; (d) it goes immediately into prophase.

Ans: d
p. 228

_____ 6. All of the following statements are true of meiosis except:
            (a) Meiosis occurs in sexually reproducing organisms. (b) Meiosis
            occurs in both plants and animals. (c) Meiosis results in haploid
            gametes. (d) One stage of meiosis is fertilization.

Ans: c
p. 229

_____ 7. In higher plants, meiosis gives rise directly to: (a) haploid
            gametes; (b) diploid gametes; (c) haploid spores; (d) multi-
            nucleate haploid bodies.

Ans: c
pp. 228,
   229

_____ 8. Sexual reproduction is important to species vitality and evolu-
            tion for all the following reasons except: (a) the combining of
            genetic information; (b) the shuffling of genetic information in
            the population; (c) the production of clones; (d) serving as a
            source of diversity.

Ans: b
p. 230

_____ 9. Homologous chromosome pairs: (a) carry genetically identical
            material; (b) are derived from two different individuals;
            (c) typically originate from a single parent; (d) pair (synapse)
            during mitosis.

Ans: a
p. 232

_____ 10. An interphase wherein each chromosome is replicated typically:
            (a) precedes meiosis I; (b) precedes meiosis II; (c) is part of
            mitosis; (d) follows metaphase I directly.

Ans: d
p. 232

_____ 11. Daughter chromosomes are likely to be different at all of the
            following times except: (a) after crossing over; (b) during
            metaphase I; (c) during anaphase I; (d) immediately before
            pairing.

Ans: c
p. 232

_____ 12. If chiasmata do not form: (a) homologues should all be geneti-
            cally identical; (b) nonhomologous chromosomes should all be
            genetically identical; (c) daughter chromosomes should be gene-
            tically identical; (d) crossing over probably occurred.

Ans: a
p. 237

_____ 13. If an XY male is actively forming sperm cells, how many of them are expected to carry the Y chromosome? (a) half; (b) 1/8; (c) 1/4; (d) all.

Ans: d
pp. 238,
    239

_____ 14. A prokaryotic cell: (a) lacks a chromosome; (b) has a nucleus with a nuclear membrane; (c) has a chromosome with complex protein attached; (d) has a ring-shaped chromosome.

Ans: b
p. 239

_____ 15. All of the following statements are true of bacterial asexual reproduction except: (a) The ring chromosome is attached to the plasma membrane. (b) A spindle with centrioles directs chromosomal movement. (c) DNA replicates. (d) Growth of new wall and membrane material separates the two attachment points of the two chromosomes following replication.

TEST SET 9-B

Name _____   Section _____

Choose the <u>best</u> answer to each of the following questions, and write the appropriate letter in the space provided.

Ans: a
p. 223

_____ 1. Telophase is <u>most</u> like: (a) the reverse of prophase; (b) the reverse of anaphase; (c) interphase; (d) metaphase.

Ans: d
p. 223

_____ 2. Proteins and DNA are arranged in a complex fashion to form the _____ of eukaryotes. (a) ribosomes; (b) tRNA; (c) mitochondria; (d) chromosomes.

Ans: b
pp. 222,
    223

_____ 3. Centromeres of daughter chromosomes: (a) repel each other to cause most of their separating movement from the metaphase plane to telophase polar areas; (b) attach to spindle fibers, and, as the spindle fibers move, the chromatids (daughter chromosomes) separate; (c) are predominantly involved in centriole activity; (d) are part of the spindle but guide chromosomal movement.

Ans: c
pp. 225,
    226

_____ 4. The pinching of microfilaments to divide the cytoplasmic components: (a) is part of karyokinesis; (b) usually occurs after prophase; (c) does not occur in plant cells that have cell walls; (d) causes an equal division of these cytoplasmic components.

Ans: c
p. 226

_____ 5. Molecules (other than DNA and protein for chromosomes) that are needed particularly for cell division are manufactured during: (a) Gap 1; (b) synthesis; (c) Gap 2; (d) cytokinesis.

Ans: a
p. 228

_____ 6. Asexual reproduction is accomplished: (a) without fertilization; (b) without cell division; (c) in animal cells only; (d) primarily in populations that cannot mutate.

Ans: b
p. 229

_____ 7. Spores produced from meiosis in higher plants give rise directly to: (a) diploid plants (sporophytes); (b) haploid plant forms (gametophytes); (c) triploid sporophytes; (d) diploid gametes.

Ans: d
p. 229

_____ 8. The greatest diversity in a population comes from: (a) asexual reproduction; (b) mitosis; (c) cloning; (d) sexual reproduction.

Ans: c
p. 230

_____ 9. All of the following statements may be true of the karyotype of a cell <u>except</u>: (a) It may be haploid. (b) It may be diploid. (c) It usually consists of chromosomes of similar size and shape. (d) It consists of specific numbers of chromosomes per cell type.

Ans: c
p. 232

_____ 10. DNA for all of meiosis is virtually all replicated: (a) in prophase I; (b) in interphase between meiosis I and meiosis II; (c) before meiosis I; (d) after crossing over occurs.

Ans: a
p. 232

_____ 11. Chiasmata are points of crossing over between: (a) homologues; (b) nonhomologous chromosomes; (c) centromeres and spindles; (d) adjacent bivalents (tetrad groups).

Ans: b
p. 237

_____ 12. If dyads do not separate until after gametes are fully formed, fertilization with a normal gamete would ordinarily result in: (a) diploid organisms; (b) triploid organisms; (c) tetraploid organisms; (d) haploid organisms.

Ans: d
p. 237

_____ 13. Assuming that a species has only two different kinds of chromo-somes (2n = 4) and that meiosis and fertilization are normal, how often do you expect an offspring would end up with all four chromosomes from the mother? (a) 1/4 of the time; (b) half the time; (c) 1/8 of the time; (d) none of the time.

Ans: a
p. 239

_____ 14. Circular DNA is characteristic of all the following except: (a) all eukaryotic nuclei; (b) all prokaryotic nuclei; (c) some viruses; (d) chloroplasts and mitochondria of eukaryotic cells.

Ans: b
p. 239

_____ 15. In bacteria, _____ separates the replicated chromosomes. (a) spindle microtubule movement; (b) growth of new wall and membrane material; (c) a nuclear membrane; (d) a mitochondrion.

CHAPTER TEN
MENDELIAN GENETICS

CHAPTER SUMMARIES AND OBJECTIVES

10-1 <u>Mendel's Strategy / "Experiment 1"</u> Mendel chose to work with the garden pea because the flower parts are so arranged that it is ordinarily self-pollinating, and he would know with certainty the parentage of any plants produced from cross-pollination. He also had available several pairs of true-breeding strains that exhibited contrasting traits (such as tall vs. dwarf). After cross-pollinating parental strains, he allowed the F_1 generation to self-pollinate, thus producing an F_2 generation. He also crossed offspring back with the true-breeding parental strains. Always, each type of progeny was counted and scored. These precise data were published with Mendel's conclusions in 1866, but the significance of his work was not recognized until 1900.

Mendel's "Experiment 1" was a monohybrid cross. He cross-pollinated parental strains (spherical seeds x dented seeds). The hybrid spherical seeds germinated and grew into plants (still known as F_1 hybrids) that self-pollinated and produced seeds that were 3/4 spherical and 1/4 dented. Mendel described the spherical trait as dominant because it was expressed to the exclusion of the dented (recessive) trait.

1. Support Mendel's choice of garden peas as a vehicle to study genetics.

2. In Mendel's strategy, explain why he made reciprocal parental crosses and then allowed the F_1's to self-pollinate.

3. Using a Punnett square and letters A and a to represent <u>any</u> dominant and recessive genes, respectively, show why monohybrid matings will always result in phenotypic ratios of 3:1 and genotypic ratios of 1:2:1.

10-2 <u>Segregation of Genes / The Test Cross</u> Mendel explained his data on the basis of discrete units (two for each trait). The fact that the dominant trait appeared to the exclusion of the recessive trait in the F_1's ruled out "blending" inheritance. True-breeding strains must have had two identical units controlling the inheritance of each trait he studied. When a dominant discrete unit (S = spherical) occurred with a recessive unit (s = dented) in the F_1's (S/s), the recessive unit would not be expressed. But when two recessive units (s/s) got back together in the F_2 generation, the recessive discrete unit was shown to have retained its identity.

Mendel's mathematical training led him to the conclusion that the ratios observed in the F_2 generation were nothing more than simple probabilities. The F_1 plants had both traits represented (S/s). When these peas "selfed," it was as though a female (ovary of the flower) of genotype S/s were mated with a male (stamen of the flower) of the same genotype. The discrete units of inheritance must separate (Mendel's Law of Segregation) when reproductive units are formed; pollination is necessary to restore two particulate units to each offspring. This restoration is done randomly: half of the eggs would carry the S unit, the other half, s. The same would be true of pollen for that same flower. The likelihood that two independent events will occur together in the offspring is nothing more than the product of the individual probabilities. For example, s/s offspring result from (1/2 egg carrying s) x (1/2 pollen carrying s) = 1/4 s/s.

Test crosses of the F_2 individuals helped support Mendel's conclusion. Individuals displaying dominant traits were crossed with true-breeding recessive (s/s) strains. He found that of the 3/4 showing the dominant trait, 1/3 of them produced offspring exhibiting the dominant trait only (the mating S/S x s/s), and 2/3 of them produced spherical and dented seeds in a ratio of 1:1 (the mating S/s x s/s).

4. Explain why F_1's would have been different in any of Mendel's experiments if "blending" inheritance had occurred.

5. Prepare a Punnett square for a hybrid mating. Then show how the same information may be obtained by calculating the products of the individual probabilities.

6. Explain why test crosses are useful in genetic research.

10-3 Incomplete Dominance / The Source of Differing Alleles More often than not, one allele (the alternative form of a given gene) is not completely dominant to another form. That "blending" does not occur, however, can be shown by matings that restore homozygous conditions; the heterozygote typically shows a form somewhere between dominant and recessive in incomplete dominance.

New or alternative forms of genes arise from mutations. Not only can one alternative form arise, but many examples of multiple alleles also exist. One example of a multiple allelic system is the ABO blood-grouping system in humans.

7. Explain how incomplete dominance is different from "blending" inheritance.

8. Assuming only one each allele for the A, B, and O blood factors, consider the six possible genotypes. Speculate about what would have happened to Mendel's work if he had worked with any multiple allelic systems.

10-4 Assortment of Genes / Linkage Crosses involving more than one gene pair were also done by Mendel. If two different genes were inherited together, the ratios of the offspring should be no different from those found for one gene. It was found, however, that for a dihybrid cross (two individuals heterozygous for two different genes), whenever one trait was completely dominant over its allele for both gene pairs and the genes were on different chromosomes, phenotypic ratios in the offspring were 9:3:3:1. Mendel recognized this as resulting from completely independent assortment of the gene pairs. For instance, a monohybrid cross would result in 3/4 of the offspring displaying the dominant trait. The chances that this trait will occur with a second dominant trait in a dihybrid cross should be 3/4 x 3/4, or 9/16. This is exactly what happened.

If genes are located on the same chromosome, there will be a preponderance of parental types and a shortage of recombinant types--how much of a shortage depends on how closely the genes are linked. The full set of genes found on a given chromosome constitutes a linkage group. Linkage groups are broken up by crossing over.

9. Explain why the ratios of offspring produced would be no different for two completely linked genes than for a single gene.

10. Explain why the phenotypic ratio of progeny of 9:3:3:1 is interpreted as having been produced by a dihybrid mating so long as true dominance/recessiveness exists.

10-5 <u>Sex Determination / Sex Linkage</u> Some diploid organisms, such as maize and earthworms, produce both eggs and sperm in the same individual (monoecious). Others--such as humans, most other animals, and some plants--produce male and female gametes in separate individuals (dioecious). Male drone honeybees are haploid; diploid honeybees are female, either workers or queens.

Most animals have sex chromosomes. Females of grasshoppers, fruit flies, and humans have two X chromosomes and males are X; females of birds, moths, and butterflies have only one sex chromosome and males have two. (Chromosomes that occur equally in all sexes are autosomes.) Some male animals, such as human males, have a Y chromosome along with the one X; the Y affects maleness. Other male animals may have a Y chromosome but be "male" with or without it (<u>Drosophila</u>). Grasshopper males have only one X chromosome.

Sex-linked traits are those carried on the X chromosome; therefore, in humans and other animals having XY males, the male is always hemizygous, having only one gene for each gene carried on the X chromosome. Y chromosomes carry very few genes. Eye color in <u>Drosophila</u> and hemophilia in humans are examples of sex-linked traits. A single dose of a recessive sex-linked trait in these males produces the full effect; only females can be heterozygous.

11. Answer: With reference to sex chromosomes, what is necessary in order for a Y chromosome to affect maleness?

12. Compare the sex-chromosomal complement of humans and <u>Drosophila</u> with that of birds and butterflies.

10-6 <u>The Ratios Are Averages, Not Absolutes</u> Statistical methods are used to determine whether the results of a given set of data fit the expected results. Sample size is very important if one is to evaluate data with confidence. A very small population may, by chance, seem to indicate unacceptable deviations from expected ratios.

13. Explain why a large sample is more statistically reliable than a small sample.

14. Consider the numbers of progeny scored by Mendel and whether his results would appear to be statistically reliable. Give one reason why his results have occasioned some criticism.

10-7 <u>Fruit Flies and Bread Molds / Recombination in Eukaryotes</u> <u>Drosophila</u> and maize have been studied so thoroughly that genetic maps have been constructed for their chromosomes. <u>Drosophila</u>, in particular, has been saved from "overstudy" obscurity by new studies in population genetics, genetics of development, and genetics of behavior.

The study of <u>Neurospora crassa</u> filled in the years between <u>Drosophila</u> and bacteria and viruses. The zygote is the only diploid cell of this bread mold, meiosis occurs immediately following formation of the zygote, and all meiotic products are lined up in order in a very narrow sac (ascus). Segregation, assortment, and recombination of genetic markers may be studied precisely.

Eukaryotes carry out meiosis in a very orderly fashion; this has led to much quantitative study on recombination. Crossing over breaks up linkage groups between homologous chromosomes at the four-strand stage. Any one exchange will involve only two chromatids, but multiple exchanges at other points in the same chromosomes may involve all four strands. The distance between marker genes determines the crossover frequency (and hence the recombinant frequency as well). Genes that are close together are less likely to have a crossover between them than genes that are farther apart. Recombination frequencies are converted to map units in the construction of genetic maps.

15. Explain why maize and <u>Drosophila</u> have been such good vehicles for the study of genetics.

16. <u>Neurospora crassa</u> offers something that corn and fruit flies do not offer. Describe its unique suitability for the study of genetics.

17. Diagram a pair of homologues at the four-strand stage, showing how three strands can be involved in crossing over, even though only two strands are involved at any one exchange.

10-8 Cytogenetics In order to correlate genetic maps derived from recombination frequencies with the physical structure of a chromosome, cytogeneticists have microscopically examined chromosomes. The cytogeneticist examines stained chromosome preparations of appropriate study material in order to become familiar with the banding pattern of each different chromosome. He must take advantage of exceptions to the normal banding pattern in order to correlate a particular band or bands with a particular gene. If a band is deleted in a given chromosome of a particular strain or stock of organisms, it must be correlated with a change in phenotype from the normal, and recombination frequencies must support the relative location of the genetic changes. An extensive inversion of a segment of a chromosome may be discerned through a change in banding patterns, and recombination studies should indicate a concomitant change in gene order.

In comparing physical and recombinational genetic maps, one sees that a map unit in one place on a chromosome may be different in measured length from a map unit at another place. Certain regions of chromosomes are capable of greater numbers of crossovers than other regions.

18. A major role for the cytogeneticist has been the preparation of physical maps for chromosomes. Describe what was necessary to come up with this type of map.

19. Explain why there may be a difference in actual distances between genes on physical genetic maps and on maps based on recombination frequencies.

10-9 From Genotype to Phenotype Some traits result from single gene pairs; other traits involve the interaction of more than one gene pair. Certain phenotypes may also depend on the physical environment.

One gene pair can mask the effect of other genes (epistasis). Wild-type coat color in mice is agouti, a dominant trait. Homozygous recessives are black. However, another gene pair determines whether any color at all will be produced. The color allele is dominant; homozygous recessives are albino. A dihybrid mating, therefore, will produce ratios of 9 agouti : 3 black : 4 albino. Here a/a is epistatic to B or b/b.

It is also possible for gene pairs both to be epistatic to each other, as in flower color of sweet peas. The dominant form of each of two genes must be present in order to produce purple flowers. If either gene pair is homozygous recessive, the flowers will be white. Color genes A and B are complementary to each other (mutually dependent). A dihybrid mating of purple sweet peas will produce offspring ratios of 9 purple : 7 white.

Some traits involve a number of pairs of genes for particular phenotypes, such as eye color in humans. One gene pair determines whether pigment will be produced, and other genes determine how much and where it will be deposited in the iris.

20. Describe epistatic gene interaction.

21. Compare single-gene inheritance patterns with multiple-gene inheritance patterns.

22. Explain how two apparently blue-eyed people can produce a brown-eyed child.

10-10 Quantitative Inheritance / Non-Mendelian Inheritance Some hereditary characters are controlled by many genes, each of which adds its effect to the total outcome. When this is the case, continuous variation is observed in the population. In humans, height, size, and skin color are all examples of this type of multiple-gene trait.

Environmental variables can also affect the expression of particular genes. Siamese cats have genes for the production of dark fur in all their cells, but dark fur is produced by skin in cooler parts of the anatomy only.

A population may have a given proportion of individuals with a particular geno-type. However, a lesser proportion may actually exhibit the expected phenotype. The lesser proportion is called the penetrance of the genotype.

The environment may also affect the degree to which a genotype is expressed (expressivity).

There are other examples of non-Mendelian inheritance, one of which is the control of traits through cytoplasmic DNA. Both chloroplasts and mitochondria contain circular DNA. In asexual reproduction, so long as each daughter cell receives some of each organelle, others can be made. Any particle inherited through the cytoplasm is usually maternally inherited, because it is the cytoplasm of the egg that becomes the cytoplasm of the zygote.

Pattern inheritance may not involve DNA or Mendelian inheritance. The sand shells constructed by the protist Difflugia corona that have teeth around the ventral opening are an interesting example of this type of inheritance.

23. Explain why multiple genes are necessary in order for a population to exhibit continuous variation in a trait.

24. Describe one way in which the environment may affect the degree to which a genotype is expressed.

25. Describe genotype penetrance.

26. Describe cytoplasmic maternal inheritance.

ARTICLE RESOURCES

Mendel's Strategy

Meinke, Peter. Mendel's law 1981. Science 81, 1981, 2(9), 75. A lovely poem about Mendel, inheritance, love, and the anticipation of a child to be born.

"Experiment I"

Duffy, Kevin. The legendary ostrich people. Science Digest, 1982, 90(6), 108. A description of dominant bidactyly, "lobster claw," found among some tribes in southern Africa. Nearly 25 percent of a tribe in Zambezi exhibit the trait. Apparently natural selection against the trait has not occurred.

Segregation of Genes

Keller, Evelyn Fox. McClintock's maize. Science 81, 1981, 2(8), 54-58. The life and work of Barbara McClintock of Cold Spring Harbor. Her work in cytogenetics predates that of Watson and Crick by nearly 15 years. It was based on chromosome mutations expressed in changing color patterns of corn kernels caused by movable genes.

The Test Cross

Incomplete Dominance

The Source of Differing Alleles

Assortment of Genes

Linkage

Sex Determination

Levy, Jerry. Sex and the brain. The Sciences, 1981, 21(3), 20-23 and 28. Evidence is presented to show that neurophysiological functions in males and females are partially biological in origin. Female hormones play a larger role in female development than the absence of male hormones.

Sex Linkage

The Ratios Are Averages, Not Absolutes

Fruit Flies and Bread Molds

Recombination in Eukaryotes

Cytogenetics

From Genotype to Phenotype

Bock, Walter J. The definition and recognition of biological adaptation. American Zoologist, 1980, 20(1), 217-227. Adaptation may be a causal factor in evolution. As discussed in this article, only phenotypic features of individuals are involved. Determination of adaptation depends on data gleaned from both laboratory and field studies.

Quantitative Inheritance

Non-Mendelian Inheritance

ESSAY QUESTIONS

1. Describe Mendel's strategy as he himself applied it to "Experiment 1."

Mendel cross-pollinated true-breeding garden peas (spherical seeds x dented seeds). He noted that all the F_1's produced spherical seeds. The F_1's were allowed to self-pollinate; therefore, he knew exact parentage of all generations. The F_2's were categorized as to spherical seeds or dented seeds; then they were counted and the numbers recorded. Back crosses were made of F_1's and F_2's, which exhibited spherical seeds with the dented seeded parental strain.

2. Compare the Punnett square method of calculating expected progeny classes with simple probability for a monohybrid mating.

All types of gametes are plotted on the Punnett checkerboard such that one symbolically makes all different fertilizations. Simple probability involves calculating the likelihood that two independent events will occur together by deriving the product of the individual probabilities of each event.

3. Discuss the rationale behind genetic mapping through the use of recombination frequencies.

Recombination frequencies are used to determine whether genes are linked in the first place. Then, the greater the recombination frequency, the farther apart the genes are located. Gene order is determined from the recombination frequencies in series of matings.

4. Compare Drosophila sex determination with human sex determination.

Drosophila	Humans
XX = female	XX = female
XY = male	XY = male
XO = male	XO = female
XXY = female	XXY = male
Y does not affect maleness.	Y positively affects maleness.

TEST SET 10-A

Name _____   Section _____

Choose the best answer to each of the following questions, and write the appropriate letter in the space provided.

Ans: d
pp. 244-
247

_____   1. Mendel's F_1 hybrids were allowed to self-pollinate instead of being cross-pollinated. Each of the following statements is true about why this was a valid procedure except: (a) Cross-pollination would have served no useful purpose in the production of the F_2 generation. (b) Self-pollination of garden peas did not require that Mendel physically manipulate each flower. (c) Female and male flower parts of the same flower have the same genetic information. (d) Different trends in the data would have occurred if F_1 hybrids had cross-pollinated.

Ans: a
p. 245

_____   2. In order to have been a ratio of exactly 3:1, one of Mendel's F_2 populations would have to have been _____ instead of 5474:1850. (a) 5550:1850; (b) 5398:1850; (c) 5474:1926; (d) 5474:5474.

Ans: b
p. 246

_____   3. The genetic constitution of an organism with respect to the traits being considered is its: (a) phenotype; (b) genotype; (c) physical appearance; (d) recessiveness.

Ans: c
pp. 246,
247

_____   4. Organisms that have like alleles for a given trait are: (a) dominant; (b) recessive; (c) homozygous; (d) heterozygous.

Ans: a
pp. 248,
249

_____   5. How will a test cross differ in progeny of a heterozygote from a homozygote (dominant)? (a) The heterozygote will produce half dominant and half recessive phenotypes. (b) There will be no difference in progeny. (c) The heterozygote will produce all dominant phenotypes. (d) The heterozygote will produce all recessive phenotypes.

Ans: b
p. 250

_____   6. In order for a cytologist to recognize whether or not a trait is incompletely dominant over another: (a) the cytologist must look at the chromosomes; (b) hybrids must exhibit a phenotype inter-mediate between the two contrasting homozygotes; (c) two con-trasting homozygotes must be allowed to self-pollinate; (d) heterozygotes must look like one parent or the other.

Ans: c
p. 250

_____   7. Mutant alleles occupy _____ wild-type alleles on homologous chromosomes. (a) different loci from; (b) gene sites beside those of; (c) the same locus as; (d) more loci than.

Ans: d
pp. 251,
252

_____   8. A phenotype ratio of 9:3:3:1 in the offspring of a mating of two organisms heterozygous for two traits can arise only when: (a) the two genes are linked; (b) garden peas are used in the mating; (c) each gene has mutated three times; (d) the gene pairs assort independently at meiosis.

Ans: c
p. 252

_____   9. In a dihybrid mating, what proportion of the offspring are expected to have a completely homozygous recessive genotype? (a) 1/4; (b) 3/16; (c) 1/16; (d) 1/8.

Ans: b
pp. 252,
   253

_____ 10. A test cross in <u>Drosophila</u> involving a completely wild-type female results in the following offspring proportions: 40% were completely wild type; 40% had purple eyes and vestigial wings; 10% had wild-type eyes and vestigial wings; 10% had purple eyes and wild-type wings. Each of the following statements is true <u>except</u>: (a) The wild-type female parent was heterozygous for <u>both</u> traits. (b) The two gene pairs assorted independently. (c) Parental types are wild/wild and purple/vestigial. (d) Recombinant types are wild/vestigial and purple/wild.

Ans: a
pp. 254,
   255

_____ 11. With respect to sex determination, each of the following statements is true <u>except</u>: (a) The Y chromosome plays no positive role in determining maleness in humans. (b) In many animals, males have only one sex chromosome. (c) In birds and butterflies, males have two sex chromosomes. (d) <u>Drosophila</u> males do not actually need the Y chromosome to be "male."

Ans: d
p. 257

_____ 12. When a hemophiliac man produces a family with a "carrier" female, _____ of their daughters are expected to have hemophilia. (a) none; (b) all; (c) 1/4; (d) half.

Ans: c
pp. 263,
   264

_____ 13. A change in gene order from ABCDEFGH to AFEDCGH represents what type of mutation? (a) inversion; (b) deletion; (c) inversion and deletion; (d) duplication.

Ans: b
p. 265

_____ 14. Assume that the genetics for the development of purple or white sweet pea flowers is not understood. A test cross of hybrid purple flower yields 3/4 white flowers (120) and 1/4 purple flowers (40) in the offspring. Then two hybrids are crossed, and 90 purple-flowered plants and 70 albino-flowered plants are produced. Each of the following statements is true of this situation <u>except</u>: (a) Two gene pairs are involved. (b) White is dominant to purple. (c) Either of two genes being homozygous recessive prevents color development. (d) Both matings give insight into the genetics of color inheritance in sweet peas.

Ans: a
p. 267

_____ 15. A particular trait is expected in a particular proportion of a population (the genes are known to be present); however, fewer individuals exhibit the trait than expected. (a) The penetrance of the genotype is less than its presence. (b) The expressivity of the genotype is less than its penetrance. (c) Maternal inheritance is probably involved. (d) Pattern inheritance is stronger than the genotype penetrance.

TEST SET 10-B

Name _____   Section _____

Choose the best answer to each of the following questions, and write the appropriate letter in the space provided.

Ans: c
p. 244

_____ 1. In order to get genes from each of two contrasting parental strains into an offspring, Mendel: (a) allowed self-pollination; (b) grafted the two strains together; (c) cross-pollinated; (d) provoked parthenogenesis.

Ans: b
p. 245

_____ 2. The ratio of 5550:1850 is the same as: (a) 1:1; (b) 3:1; (c) 2:1; (d) 4:1.

Ans: a
p. 246

_____ 3. The physical appearance of an organism with respect to a given character is its: (a) phenotype; (b) genotype; (c) ratio; (d) genetic constitution.

Ans: d
pp. 246,
247

_____ 4. Organisms that have two different forms of a gene for a given trait are: (a) homozygotes; (b) homologous; (c) recessive; (d) heterozygotes.

Ans: d
p. 249

_____ 5. How does one determine whether an organism is homozygous for a recessive allele? (a) One cannot make this determination without making a test cross. (b) The dominant trait will be apparent. (c) The organism will be intermediate in appearance between its true-breeding parental strains. (d) The organism's phenotype will be recessive.

Ans: c
p. 250

_____ 6. A green-flowering plant is crossed with a white-flowering one. All the progeny have light green flowers. Each of the following statements is true of the plants involved except: (a) Green flowers are apparently incompletely dominant over white. (b) The green-flowering plants were homozygous. (c) White flowers are dominant over green. (d) The white-flowering plants were of a true-breeding strain.

Ans: b
p. 250

_____ 7. An allele that occupies the same locus as a wild-type allele but codes for a different expression of the gene is: (a) not possible; (b) a mutation; (c) a multiple allele; (d) part of a multiple-gene system.

Ans: a
p. 251

_____ 8. Spherical, yellow garden peas are crossed with dented, green peas. All progeny are spherical and yellow. What traits are dominant? (a) spherical, yellow; (b) dented, green; (c) spherical, green; (d) It is not possible to tell without recovering F_2's.

Ans: a
p. 252

_____ 9. In a dihybrid mating, what proportion of the offspring are expected to have all four dominant alleles? (a) 1/16; (b) 1/4; (c) 3/16; (d) 1/8.

Ans: c
pp. 252,
253

_____ 10. A Drosophila test cross involving a female heterozygous for two traits results in only 10% recombinants and 90% parental types. The most likely reason for these results is that: (a) the two genes are located on nonhomologous chromosomes; (b) the recombinant types are carrying a lethal gene; (c) the two genes are linked; (d) there was not as much crossing over as there usually is.

Ans: d
pp. 254,
    255

_____ 11. Human males: (a) do not determine the sex of their children; (b) produce only one kind of gamete; (c) regularly produce three kinds of gametes: X-bearing, Y-bearing, and no X; (d) produce X and Y gametes in approximately equal numbers.

Ans: b
p. 257

_____ 12. When a hemophiliac man produces a family with a completely normal woman, _____ of their sons are expected to have hemophilia. (a) all; (b) none; (c) half; (d) 1/4.

Ans: a
pp. 263,
    264

_____ 13. When a homologue that has a deletion in the middle of it attempts to pair with its normal homologue in prophase I, what will occur? (a) The normal one will bulge out at the place where it has the gene and its homologue doesn't. (b) The normal one will just be longer than the one with the deletion and will stick out beyond the deletion chromosome. (c) There will be no pairing. (d) The deletion will prevent gene-for-gene pairing along the entire lengths of the two chromosomes.

Ans: d
p. 264

_____ 14. In coat color of mice, a/a (albino) is epistatic to B/- and b/b. B (agouti) is dominant to b/b (black). Two agouti mice mate (one is heterozygous for both genes, and the other is homozygous for both genes A and B). What should be produced? (a) Half black and half albino are expected. (b) All offspring are expected to be albino. (c) Half should be agouti and half black. (d) All are expected to be agouti.

Ans: b
p. 267

_____ 15. During sexual reproduction, maternal inheritance factors are usually passed on to offspring through: (a) the sperm cytoplasm; (b) the egg cytoplasm; (c) nuclear DNA; (d) protists.

CHAPTER ELEVEN
NUCLEIC ACIDS AS THE GENETIC MATERIAL

CHAPTER SUMMARIES AND OBJECTIVES

11-1 <u>What Does the Gene Control?</u> Genes cannot, by themselves, directly produce a phenotypic result such as eye color or flower color. The suggestion by Garrod in the early 1900s that some defects were caused by "inborn errors of metabolism" was a valuable insight. In the 1940s, Beadle and Tatum worked with the salmon-colored bread mold <u>Neurospora</u>. Wild-type <u>Neurospora</u> (prototrophs) can grow on a minimal medium, making all the chemical constituents of its cells from a few salts; a simple nitrogen source; a source of energy and carbon; and biotin, a vitamin. They caused an increase in the rate of mutation with ultraviolet light or X rays and then isolated various mutants. In almost every case, the nutritional requirement that had to be added to the medium for that mutant strain (auxotrophs) was a single compound. This led to the formulation of the one-gene, one-enzyme theory (one gene would code for one specific enzyme that was necessary for the mold to make a particular compound).

Further work--particularly that with sickle-cell hemoglobin--led to the refinement of this theory. All enzymes are proteins but not all proteins are enzymes; complex proteins are made of several smaller units. In order for DNA to code for all proteins, it could do so if one gene coded for one polypeptide.

1. Explain how considering alkaptonuria as an "inborn error of metabolism" could give any insight into gene action.

2. <u>Neurospora</u> was an excellent vehicle for the work Beadle and Tatum did. Describe the ways in which it is unique.

3. Compare the "one-gene, one-enzyme" and the "one-gene, one-polypeptide" theories by describing their similarities and the refinement that one has made in the other.

11-2 <u>What Is the Gene?</u> Recognition that the gene is made of DNA (deoxyribonucleic acid) came after a number of investigators contributed bits and pieces. Three key contributors were Griffith (1920s), Avery (1944), and Hershey and Chase (1952). Griffith, in studying the disease-causing properties of pneumococci, found two strains: S pneumococci had capsules and were virulent; R pneumococci were nonvirulent (with no capsules). R bacteria were transformed into S bacteria when heat-killed S pneumococci were inoculated into mice along with living R pneumococci. Cell-free extracts of S pneumococci could cause the same transformation: It must be a chemical transforming principle.

Avery tested all different types of purified, cell-free extracts from virulent bacteria. If the DNA in the sample was destroyed, transforming activity was lost. Everything else could be dispensed with.

Hershey and Chase labeled T_2 bacteriophage such that their protein coats and DNA inner cores were distinguishable. DNA was carried over from virus generation to generation, whereas protein was not. It followed, then, that the hereditary information of these viruses is contained in the DNA.

4. Relate the significance of transformation to the conclusion that DNA is the genetic material.

5. Describe the added points that Hershey and Chase contributed to the argument presented by Avery that hereditary information is contained in the DNA.

11-3 The Structure of the DNA Molecule / Structure of Ribonucleic Acid   Determination of the shape of DNA was possible because of the separate works of crystallographers Franklin and Astbury. X-ray diffraction revealed the molecules to be cylindrically spiral (helical). However, Watson and Crick made this interpretation instead of Franklin and Astbury. They also inferred that the helix was of two chains running in opposite directions (antiparallel) and twisted around each other. Also, using Chargaff's work (which showed that DNA, within any one species, has equivalent numbers of adenine and thymine, and that guanine is equal to cytosine), Watson and Crick constructed a model of what DNA must be like (1953). DNA is a polynucleotide; one nucleotide consists of a sugar (deoxyribose), a phosphate, and a nitrogenous base. Thymine and cytosine are pyrimidine bases, and adenine and guanine are purine bases. The two chains of polynucleotides are hooked together by hydrogen bonding between complementary base pairs. Adenine bonds to thymine, guanine to cytosine.

RNA is a polynucleotide. It contains ribose sugar instead of deoxyribose, and it contains uracil, a pyrimidine base, instead of thymine. Phosphate groups and the other three bases are the same and in similar locations, except that RNA is generally single-stranded. Under appropriate conditions or internal folding of the single strand, uracil is complementary to adenine.

6. Compare the structure of DNA with that of RNA.

7. Explain why Chargaff's work would be so critical to Watson and Crick's model building.

11-4 Implications of the Double-Helical Structure of DNA / Replication of the DNA Molecule   The nucleotide units are so repetitive that the linear sequence of nitrogenous bases--as the only thing different--has to be the code for genetic information. Further, it was noted that the proposed model of DNA would allow for reproduction of the molecule. If the two strands of the double helix unwound, by complementary base pairing the old strand could replace all missing nucleotides, thereby restoring the second chain. The proposed structure would also allow for occasional changes in the linear sequence of nucleotide pairs. This would explain some mutations.

Meselson and Stahl (1957) were able to show how DNA replicates by labeling the DNA of multiplying bacteria with "heavy" nitrogen (^{15}N). After labeling, the bacteria were transferred to media containing the more common--and lighter--form of nitrogen (^{14}N). Samples were taken after each generation to see how much heavy nitrogen was left. To do this quickly, Meselson and Stahl devised a density gradient using a 6M cesium chloride solution (this has a density similar to that of DNA). Upon centrifugation, all ^{15}N-^{15}N DNA should be at the bottom, ^{15}N-^{14}N in a higher band, and ^{14}N-^{14}N in a band above that. The DNA "generation" in which particular proportions occurred pointed to semiconservative DNA replication, wherein one "old" strand acts as a template for an entirely new strand.

8. After becoming thoroughly familiar with the complementary base pairs described in the previous section, determine what the two new strands containing ^{14}N would be. Begin with DNA strands immediately after complete labeling with ^{15}N. One strand is ATCGCCTAT. Determine the base sequence of the complementary strand before deriving the two new strands.

9. Distinguish semiconservative replication from conservative and dispersive replication.

11-5 Enzymes in DNA Synthesis / From DNA to Protein   Kornberg demonstrated that an enzyme complex, DNA polymerase, was necessary for the synthesis of DNA. DNA synthesis also requires nucleoside phosphates and intact DNA to function as a template. After Kornberg's work, certain experiments called DNA polymerase into question, but scientists are now accepting again its necessity for DNA synthesis. Kornberg also demonstrated that the addition of nucleotides to the growing chain always occurred at the same end: the one that has a free hydroxyl group on the number-three carbon of its terminal deoxyribose.

Because the two DNA strands of a helix are antiparallel, DNA synthesis is accomplished along one unwound strand in sequence, but in short reversed stretches that are promptly seamed together (with the DNA already made) along the other strand (which runs opposite to the first).

In making protein from DNA messages, DNA codes for RNA, RNA codes for the production of protein, and protein does not code for synthesis of protein, RNA, or DNA. This is the "central dogma" of molecular biology. Though Crick postulated adaptor molecules that would carry amino acids for construction of protein, Hoagland and Zamecnik actually found adaptor molecules. Transfer RNA (tRNA) recognizes the genetic messages and carries amino acids. They help translate the language of DNA into the language of proteins.

In 1960 Crick, Brenner, and Jacob postulated a messenger RNA (mRNA) as this code carrier from nucleus to cytoplasm, where protein synthesis occurs. The formation of a specific mRNA·under the control of specific DNA is called transcription. mRNA serves as a template upon which tRNA "adaptors" line up to bring amino acids in proper order into the growing polypeptide chain.

10. Determine the effect on DNA synthesis of antiparallel strands of DNA in the helical molecule.

11. State the "central dogma" of molecular biology in your own words.

11-6 Transcription / Transfer RNA   Transcription requires the enzyme RNA polymerase, nucleoside phosphates, and the DNA template. Only one of the DNA strands is transcribed. RNA polymerase catalyzes the continuous transcription of DNA in one direction, proceeding always in the same direction. The RNA transcript peels away as it is formed, allowing the DNA that has already been transcribed to be rewound. Some genes are transcribed from one strand, others from the opposite strand. Initiation sites tell the RNA polymerase where to attach and which strand to copy. Termination also requires specific base sequences in the DNA. Transcription is involved in the synthesis of mRNA, tRNA, and rRNA.

Genetic information transcribed in an mRNA molecule can be thought of as a series of three-letter "words." Each sequence of three nucleotides (a "word") specifies a particular amino acid. The "word" is a codon.

At one end of every tRNA molecule is a site to which the amino acid is attached. The charging of the tRNA is accomplished via specific activating enzymes for each amino acid. This uses one ATP at each conversion. At the other end of each tRNA is a group of three bases called the anticodon, which is the point of contact with mRNA. Each tRNA anticodon is distinctive. This is the essence of translation.

12. Cite the relationship between the codons of mRNA and the anticodons of tRNA.

13. Explain how transcription occurs in the making of mRNA, tRNA, and rRNA.

11-7 The Ribosome / Translation / RNA Viruses and the Central Dogma   Translation occurs at the site of the ribosome, which consists of a light subunit and a heavy subunit that are composed of rRNA and specific proteins. It begins with the formation of an initiation complex: a ribosomal light subunit, an mRNA strand bound to the light subunit at the mRNA starting point, and a charged tRNA (bearing a special first amino acid). The heavy ribosomal subunit then joins the complex. This first charged tRNA is at the A ribosome site. Then it moves over to the P ribosome site, and the ribosome moves down the mRNA one codon. Charged tRNAs always enter at the A site of the ribosome and by complementary base pairing, attach to the mRNA codon that is in place each time after the ribosome moves down mRNA the length of one codon. After being at the A site and attaching its amino acid to the growing polypeptide chain, each tRNA will move over to the ribosomal P site until a newly arriving charged tRNA frees it of its polypeptide chain.

As soon as enough of the mRNA chain has been translated and is available to attach to another ribosome, translation can begin again. The thread of mRNA, its attached ribosome, and the growing polypeptide chains are collectively known as polysomes or polyribosomes. Any one ribosome can serve as a synthesis site for all different types of protein. Whichever mRNA becomes associated with the ribosome confers the specificity for particular protein manufacture.

Many viruses have RNA instead of DNA. Tobacco mosaic virus (TMV) replicates its RNAs from RNA. Its host's cells provide the rest of the machinery for manufacturing its proteins. Rous sarcoma virus (causes a cancer in chickens) has RNA instead of DNA also; however, when it enters a host, the enzyme (reverse transcriptase) that it carries for the manufacture of DNA causes the host cell to make a DNA "transcript" of its viral RNA. The infected host chicken cells are permanently changed but do not lyse. The new DNA becomes part of the hereditary apparatus of the chicken cell.

14. Explain the following statement: Information does not flow from protein back to the nucleic acids.

15. Describe the roles of DNA, tRNA, mRNA, and rRNA in the manufacture of protein (translation).

11-8 The Genetic Code Logical thinking led to the conclusion that the sequence of bases of DNA determined amino acid sequence in proteins, but direct chemical methods were used to work out the actual "code book." Nirenberg's work in the early 1960s was of particular significance. He prepared artificial RNA, "poly(U)"; added it to a mixture of ribosomes, amino acids, activating enzymes, tRNAs, and other factors; and then recovered a polypeptide-polyphenylalanine. Other researchers working at the same time helped elucidate the code, too, using similar techniques. Nirenberg found that mRNA only three nucleotides in length (one codon) could bind to ribosomes which, in turn, would bind to the corresponding charged tRNA. Whatever artificial mRNA bound a particular tRNA and amino acid to the ribosome immediately made evident the codon for that amino acid.

There are 20 amino acids that must have codons and there are 4 bases. Hence 65 codons composed of 3 of the 4 bases each are possible. In actual fact, 3 of the 64 arrangements are chain terminators, and all but 2 amino acids have more than 1 codon. The code is considered degenerate because of this last fact. The code holds true for all prokaryotes and eukaryotes except for their mitochondria. The codons are typically written as letters representing the base sequence in mRNA. The master word of DNA, which has been transcribed into mRNA, may easily be determined by complementary base pairing.

16. Working out the codons without advanced nucleic acid chemistry required much ingenuity. List the significant events in this quest.

17. Describe the degeneracy of the genetic code.

11-9 Mutations / Point Mutations Mutations are heritable changes in the genetic information. The geneticist must pay attention to the exception--the only way to label a gene to discover its function is to find a mutation of that gene, a "marker." Extensive mutations may be rearrangements of large sections of chromosomes without genetic loss or deletions. Most mutations are point mutations involving single base pairs, and they typically occur in nature during replication of the DNA.

Nucleotide substitution mutations may be transition mutations (a purine, A or G, is replaced by the otherpurine, or a pyrimidine is replaced by a pyrimidine) or transversion mutations (a purine replaces a pyrimidine or vice versa). Abnormal pairing in DNA synthesis is often involved in transition mutations and may cause a significant or an insignificant change in a protein through the substitution of an amino acid. Point mutations may also involve the addition or deletion of single nucleotides. From the point of the change, the whole message will be read differently (frame-shift mutations).

18. Describe briefly the connection between mutations and evolution.

19. Explain the difference between extensive mutations and point mutations.

11-10 <u>Large-Scale Mutations</u> Deletion mutations completely remove part of the genetic message. They sometimes occur when two breaks heal at their broken ends with the omission of a segment; they also may occur when sister chromatids breaking at different points but rejoin, as in crossing over. This would result in one strand having a deletion and the other a duplication.

DNA rearrangement can occur through inversions--two breaks with the loose segment flipping over and then rejoining the rest of the chromosome. Translocations occur when a piece of one chromosome breaks off and rejoins a nonhomologous chromosome. The appearance of chromosomes during synapsis in prophase often shows what has occurred because of strange configurations. Homologous chromosome segments still pair gene for gene. Loops and cross-shaped arrangements often occur.

20. Describe how large-scale mutations differ from point mutations.

11-11 <u>A New Twist in the Study of DNA</u> Most DNA is of the right-handed double helix form (B-DNA). The last few years have seen the discovery of a zig-zag, left-handed twist DNA (Z-DNA). Though first discovered in small, synthetic DNA molecules, Z-DNA now appears to be naturally occurring in short stretches between B-DNA of the chromosomes of living organisms.

ARTICLE RESOURCES

<u>What Does the Gene Control?</u>

Conniff, Richard. "Supergene." <u>Science Digest</u>, 1982, 90(3), 62-65, 110-111. A small cluster of DNA in each cell on the sixth chromosome influences a wide range of biological processes. It plays a large role in immune function and has an effect on transplantation genetics.

<u>What is the Gene?</u>

Kornberg, Roger D., and Aaron Klug. The nucleosome. <u>Scientific American</u>, 1981, 244(2), 52-64. This article gives the flavor of how current research in molecular biology is carried out. The gene appears to lie coiled about a core of histones. These histones are chains of amino acids, in the core of which two units of each histone make up an octomer bead called a nucleosome.

<u>The Structure of the DNA Molecule</u>

<u>Structure of Ribonucleic Acid</u>

<u>Implications of the Double-Helical Structure of DNA</u>

<u>Replication of the DNA Molecule</u>

<u>Enzymes in DNA Synthesis</u>

<u>From DNA to Protein</u>

<u>Transcription</u>

<u>Transfer RNA</u>

Rich, Alexander, and Sung Hou Kim. The three-dimensional structure of transfer RNA. <u>Scientific American</u>, 1978, 238(1), 52-62. Transfer RNA acts to remove amino acids from the free state within the cell to an assembled protein molecule. Numerous diagrams and models clearly explain the process. Codons are clarified.

The Ribosome

Translation

Codebreaker. Scientific American, 1980, 242(2), 72-75. A report on the discrepancies found in the genetic code. There are 64 codons that can be represented in the chain of nucleotides but 20 words specifying the amino acids that make up a protein; 61 of the codons, each 3 letters long, specify an amino acid. Several different codons can specify the same amino acid with 3 of the codons acting as punctuation.

RNA Viruses and the Central Dogma

Butler, Jonathan, and Aaron Klug. The assembly of a virus. Scientific American, 1978, 239(5), 62-69. Tobacco-mosaic viruses are rods of a single protein encasing RNA or DNA. A simple virus consisting of a single strand of RNA encased in a rod of protein was studied. Once the virus is inside a living cell, the RNA is released and large numbers of viral particles are formed. The protein subunits will aggregate, depending on the pH and the ionic strength of the medium. Excellent diagrams and photographs are included.

The Genetic Code

Mutations

Cohen, Stanley, and James A. Shapiro. Transposable genetic elements. Scientific American, 1980, 242(2), 40-49. Transposable genetic elements (transposons) seem to be responsible for quantum jumps as well as small steps in evolution. This can account for ampicillin resistance and the transfer of this trait to other plasmids. A history of the discovery of transposable elements is given. Excellent diagrams.

Point Mutations

Large-Scale Mutations

A New Twist in the Study of DNA

Schimke, Robert T. Gene amplification and drug resistance. Scientific American, 1980, 243(5). Gene amplification is the duplication within the chromosomes of one or more genes. The amplified genes are capable of translation. This in turn can lead to more complex genomes. Methotrexate, a standard drug for the treatment of cancer, can lead to the cancer's resistance to the drug. This was studied in hamsters. Good photographs and clear diagrams.

ESSAY QUESTIONS

1. What does the gene control? Cite supporting evidence.

    Genes control the synthesis of polypeptides. Supporting evidence includes Tatum and Beadle's work with Neurospora, wherein mutations occurred, many of which caused only a single compound to have to be added to the growth medium, and Ingram's work, which showed that sickle-cell hemoglobin had one amino acid replaced in one of the two types of polypeptides.

2. Describe how Watson and Crick derived enough insight to build a model of DNA.

    Astbury's and Franklin's crystallographs convinced Watson and Crick that the DNA molecule was helical and helped determine certain measurements. Density measurements and trial models suggested two chains. The crystallographs pointed to antiparallel strands of DNA. Chargaff's work led to the pairing of A with T and of C with G.

3. Compare RNA with DNA.

    RNA   DNA

    single-stranded   double-stranded, helical
    uracil   thymine
    cytosine   cytosine
    adenine   adenine
    guanine   guanine
    ribose   deoxyribose
    phosphate groups   phosphate groups

4. Describe the relationship of the anticodon with the codon.

    DNA codes for both. One segment of DNA may be transcribed to mRNA, such as
    ATCGGT -- UAGCCA (two codons of three bases each). A DNA sequence of TAG would
    code for the first complementary anticodon of tRNA (AUC), and DNA of CCA would
    code for the second complementary anticodon of tRNA (GGU). The anticodon pairs
    complementarily with the codon when it delivers its amino acid to the ribosome.

TEST SET 11-A

Name _____ ' Section _____

Choose the best answer to each of the following questions, and write the appropriate letter in the space provided.

Ans: c
p. 272

_____ 1. Which of the following did Mendel and Garrod ("inborn errors of metabolism") have in common? (a) Both were cytogeneticists. (b) Both were pioneers in the mapping of chromosomes. (c) Both proposed major genetic insights that were not fully appreciated for a number of years. (d) Both were Austrian monks who worked with the genetics of garden peas.

Ans: a
p. 275

_____ 2. The finding that one of the two kinds of polypeptides differed by one amino acid between normal and sickle-cell hemoglobin led to the theory of: (a) one gene, one polypeptide, (b) one gene, one enzyme; (c) one gene, one phenotype; (d) one gene, one hemoglobin.

Ans: b
p. 275

_____ 3. For several years before it was determined that genes are particular sequences of DNA, the "gene material" was most widely thought to be: (a) fats; (b) protein, (c) carbohydrates: (d) starch.

Ans: d
p. 278

_____ 4. Hershey and Chase labeled T2 virus protein with ^{35}S and nucleic acid with ^{32}P. The results led to the conclusion that: (a) one gene codes for one polypeptide; (b) the genetic material is protein; (c) carbohydrates are enzymes; (d) DNA is the carrier of the hereditary information.

Ans: a
p. 279

_____ 5. Based on Chargaff's findings that A = T and C = G, Crick and Watson included which of the following concepts in their model? (a) Adenine paired with thymine. (b) Cytosine paired with thymine. (c) Guanine paired with a purine. (d) Thymine paired with a pyrimidine.

Ans: d
p. 280

_____ 6. Each of the following statements is true of RNA compared to DNA except: (a) RNA contains adenine, guanine, cytosine, and uracil. (b) RNA contains a ribose sugar. (c) RNA contains four bases. (d) RNA contains equal numbers of cytosine and guanine.

Ans: c
p. 285

_____ 7. Meselson and Stahl showed that the mechanism of DNA duplication is: (a) conservative replication; (b) modified replication; (c) semiconservative replication; (d) dispersive replication.

Ans: b
p. 285

_____ 8. _____ is needed for DNA synthesis. (a) RNA polymerase; (b) DNA polymerase; (c) A lipid; (d) Light as an energy source.

Ans: b
p. 287

_____ 9. Simply speaking, the _____ of molecular biology is that DNA codes for RNA and RNA codes for the production of protein. (a) peripheral dogma; (b) central dogma; (c) variant theme; (d) secondary interpretation.

Ans: a
p. 287

_____ 10. _____ translates the language of DNA into the language of proteins. (a) RNA; (b) DNA; (c) RNA polymerase; (d) DNA polymerase.

Ans: d
p. 288

_____ 11. Normally, RNA polymerase catalyzes the continuous transcription of DNA: (a) in both directions from the point of contact; (b) from the 3' end to the 5' end; (c) in the cytoplasm of the cell; (d) in one direction only.

Ans: c   _____   12. The enzyme that the Rous sarcoma virus carries for the manufac-
p. 294   ture of DNA, using its RNA as the template, is: (a) activating
                         enzyme; (b) RNA polymerase; (c) reverse transcriptase; (d) DNA
                         polymerase.

Ans: a   _____   13. UAA, UAG, and UGA are chain: (a) terminators; (b) initiators;
p. 296   (c) markers; (d) codes for methionine.

Ans: b   _____   14. Mutations that alter the function of genes are often referred to
p. 297   as: (a) duplications; (b) markers; (c) deletions; (d) inversions.

Ans: d   _____   15. Most mutations in nature occur during: (a) the transcription of
p. 297   RNA from DNA; (b) translation; (c) the formation of ribosomes;
                         (d) the copying of the genetic message.

TEST SET 11-B

Name _____   Section _____

Choose the best answer to each of the following questions, and write the appropriate letter in the space provided.

Ans: a
p. 274

_____ 1. Beadle and Tatum confirmed Garrod's insight into the reasons behind "inborn errors of metabolism" by showing that: (a) mutations are capable of altering the functions of enzymes so that they no longer do their jobs; (b) alkaptonuria is inherited in the Mendelian pattern of inheritance; (c) fruit flies change the color of their eyes in response to environmental change; (d) pneumococci can be transformed.

Ans: c
p. 275

_____ 2. The most widely accepted description of gene action today is: (a) one gene, one enzyme; (b) one gene, one phenotype; (c) one gene, one polypeptide; (d) one gene, one complex protein.

Ans: b
p. 276

_____ 3. The changing of bacteria from one strain to another in cell-free extracts without virus carriers is known as: (a) translation; (b) transformation; (c) transferral; (d) transduction.

Ans: c
p. 279

_____ 4. The success of Franklin's work in making crystallographs of DNA, depended on: (a) Astbury's support· (b) Chargaff's finding that A = T and C = G; (c) Wilkins uniform preparations of DNA; (d) Crick's model building.

Ans: d
p. 279

_____ 5. Crystallographs led Watson and Crick to all of the following conclusions except: (a) The DNA molecule was helical. (b) They measured certain of the distances within the helix. (c) The two DNA chains were antiparallel. (d) A purine and a pyrimidine always pair.

Ans: a
p. 280

_____ 6. RNA: (a) is single-stranded; (b) is a double helix; (c) contains thymine; (d) contains deoxyribose sugar.

Ans: b
p. 282

_____ 7. Three possible mechanisms of DNA duplication were seriously considered by scientists. Each of the following is one of them except: (a) conservative replication; (b) modified replication; (c) dispersive replication; (d) semiconservative replication.

Ans: d
p. 286

_____ 8. The fact that the DNA strands are _____ and the fact that nucleotides can be added to only one end cause one strand to be synthesized smoothly in one direction, but the other strand must be done in a series of short stretches that are promptly seamed together. (a) parallel; (b) helical; (c) twisted to the left; (d) antiparallel.

Ans: c
p. 287

_____ 9. Once genetic information has passed into _____ it cannot get out again. (a) lipids; (b) carbohydrates; (c) protein; (d) DNA.

Ans: a
p. 287

_____ 10. _____ serves as a template upon which "adaptor molecules" line up to bring amino acids in proper order into the growing polypeptide chain. (a) mRNA; (b) rRNA; (c) tRNA; (d) iRNA.

Ans: c
p. 290

_____ 11. _____, each specific for one amino acid, direct the "charging" of the tRNA molecule. (a) Anticodons; (b) Codons; (c) Activating enzymes; (d) Phospholipids.

Ans: b
p. 291

_____ 12. Translation begins with the formation of: (a) terminal polysomes; (b) an initiation complex; (c) a heavy ribosomal subunit; (d) an A site.

114   Chapter 11

Ans: d   _____   13. Poly(U) of the composition UUUUUU would code for_____
p. 294   phenylalanine(s). (a) only one; (b) six; (c) three; (d) two.

Ans: c   _____   14. Because eighteen amino acids are coded for by more than one
p. 296   sequence of three bases, the code is said to be: (a) regressing
                                 (b) advanced; (c) degenerate; (d) regenerate.

Ans: a   _____   15. Frame-shift mutations result from the addition or deletion of:
p. 299   (a) single base pairs; (b) 1-mm sections of DNA; (c) whole
                                 genes; (d) a sequence of three bases.

CHAPTER TWELVE
MICROBIAL GENETICS

CHAPTER SUMMARIES AND OBJECTIVES

12-1 <u>Mutations in Bacteria and Bacteriophage</u> The bacteria <u>Escherichia coli</u> can double
its population every 20 minutes under ideal conditions. If grown on a basic
nutrient agar, colonies that each arise from a single bacterium grow rapidly
until there is a complete "lawn" formed. If <u>E. coli</u> K and the bacteriophage T4
are mixed and plated out, a lawn forms with clear places in it (plaques). A
single T4 virus infects a single bacterium, the virus reproduces, and the bacterium lyses and releases hundreds of virus particles. The process continues
lysing bacteria until the plaque is visible even without magnification. If a
bacterial colony arises in the midst of a plaque, it is because a mutation has
occurred to make it resistant to T4 phage (<u>E. coli</u> K/4). A mixture of <u>E. coli</u>
K/4 and T4, plated out, will occasionally result in plaque formations caused by a
mutation in a T4 phage (T4h).

Many mutations exist in low frequency in populations. When the environment exerts
selection forces for a particular trait, the proportion of genotype in the population will shift. This is evolution.

   1. Determine why the one set of growth results caused the investigators to infer
       a bacterial mutation (<u>E. coli</u> K/4) and the other a viral mutation (T4h).

12-2 <u>A Sexual Process in Bacteria</u> Lederberg and Tatum demonstrated genetic recombination as a sexual phenomenon in <u>E. coli</u> K12 strain. Strain I required methionine
and biotin for growth (met⁻ bio⁻); strain II required threonine and leucine (thr⁻
leu⁻). When these strains were mixed and cultured on a medium having all four
amino acids and then washed and replated on minimal media (without the four amino
acids), a few colonies did appear on the plates. The colonies were from met⁺ bio⁺
thr⁺ leu⁺ bacteria. To explain this recovery of wild-type colonies, mutation was
ruled out (a double mutation of one of the types would have been necessary, and
the rates were too high to accept this); transformation was ruled out by Davis's
"U-tube" experiment, which demonstrated the necessity of cell-cell contact for the
change; conjugation was the only other possibility (sexual reproduction involving
genes from one strain being transferred to another).

   2. Support Lederberg and Tatum's reasoning that conjugation must have been the
       mode of introduction of the two genes into another host.

12-3 <u>Isolating Specific Bacterial Mutants</u> To isolate a new strain of bacteria carrying
a particular mutation, one begins with wild type, increases the mutation rate with
some agent such as U-V light, and grows them in a medium that contains the compound that will be needed by the desired mutant strain. Most of the cells will be
wild type; these must be eliminated to see whether any desired mutants exist. One
method of doing this is by "penicillin suicide." If one takes the mixture of wild

type and (hoped-for) mutants and grows them on a particular nutrient-deficient medium, the wild type will divide; the mutants that need the missing nutrient won't. Adding penicillin kills the growing cells. Culturing the remaining cells on a medium lacking penicillin but containing the particular nutrient will allow the desired mutants to grow if they exist.

To count the progeny of the various recombinant types from a cross between two strains of bacteria, replica plating may be done. After making a master plate, a sterile velvet on a block of wood is pressed onto the surface. The velvet is then touched to agar plates having various nutrients added or deficient. The locations of wild type colonies show up only on the minimal medium. Single mutants plus wild type will grow on the nutrient plates having a single additional nutrient (for whatever mutant one wishes to locate). Double mutants may be found on properly prepared plates (two added nutrients to the minimal medium) in different loci. Everything should grow on complete medium plates.

3. Outline the basic steps involved in isolating specific bacterial mutants.

4. Explain how replica plating leads to locating and counting the number of recombinant types from a particular cross.

12-4 The Bacterial Fertility Factor / Transfer of Male Genetic Elements / Sexduction
The Lederbergs found that genetic recombination in E. coli is a one-way process, involving a donor male strain and a female recipient strain. Other researchers confirmed the same phenomenon for other bacteria. Upon conjugating, the female becomes male due to the transfer of a fertility factor (F). The F factor is an extra piece of DNA that can replicate itself separately from the normal bacterial chromosome. One of the things controlled by genes on the F factor is the formation, on the surface of the male bacterium, of long, thin tubes with sticky ends (F pili), which attach to female cells and through which DNA may be transferred.

Hfr mutants were discovered in which the F factor is inserted into the bacterial chromosome. In these, the F factor is not usually transferred to the female, but there is a high frequency of recombination with the transfer of certain markers to the female strain during conjugation. By interrupted mating, researchers found that the number of markers transferred depended on the length of time allowed for conjugation, that the markers entered in a particular order, and that the F factor was almost never transferred from the Hfr male. Paying attention to the order in which the genetic markers were transferred allowed the chromosome to be mapped. It was also possible to determine that the chromosome had to be circular. Different strains were found to have the F factor inserted into the circular chromosome at different points; therefore, the circular chromosome would break open at different points--wherever the F factor was. The two strands separate from each other, and one is transferred. The opened chromosome moves through the pilus, with the F factor (at the tail end) not being included in the transfer. About half the markers are incorporated into the F⁻ chromosome, giving rise to recombinant bacteria. The rest are lost as the cell divides.

The F factor may separate from the chromosome--sometimes taking with it part of the chromosome--and behave as an episome. The modified F factor (F') contains alleles of the circular chromosome and may be used to study dominant-recessive traits in bacteria following transfer of the F' factor from male to female (sexduction or F-duction).

5. Cite the conditions necessary for the study of dominant-recessive relationships in normally haploid bacteria.

6. Explain why all recombinant progeny arise from a female infected by a male F factor.

12-5 Recombination in Viruses Hershey and Rotman demonstrated recombination in T2 bacteriophage by infecting E. coli with two strains simultaneously. A number of parental types were found--but also a significant number of recombinant types exhibiting both parental traits. Phage particles were later demonstrated to have a single circular chromosome.

7. Compare the chromosomes of bacteriophage with those of bacteria.

12-6 Lysogeny and the Disappearing Viruses Bacteriophage life cycles are not invariably lytic. Some phages are able to infect some bacteria and retreat temporarily
from being obvious, except that the host cell is immune to further infection from
the same phage strain. Bacteria harboring such an infection are lysogenic in
that, when they are combined with other bacteria sensitive to the phage, lysis
will occur in the sensitive cells. The virus (bacteriophage) DNA is incorporated
into the bacterial chromosome, is noninfective temporarily, and replicates along
with each bacterial division. U-V light and other mutagenic agents may induce
the prophage to separate from the bacterial chromosome, causing the production of
new phage particles and lysis of the host cell.

8. Compare prophage particles with F factor.

12-7 Transduction When the prophage escapes from the bacterial chromosome, it may take
with it some adjacent bacterial DNA. The new phage may introduce these bacterial
markers into other bacteria that they infect, resulting in genetic recombination
in the bacteria. This restricted transduction causes the newly infected bacteria
to become lysogenic.

It is possible, in the production of phage particles by the host cell, for bacterial DNA only to be incorporated into a phage protein coat. This particle is
infective for another host cell, but the only effect is the introduction of
bacterial DNA that may be incorporated into the new host chromosome. The new
host does not become lysogenic. This general transduction may involve any part
of the bacterial chromosome.

9. Compare restricted transduction with general transduction.

12-8 Episomes and Plasmids Episomes are nonessential genetic elements that may be
integrated into the main chromosome or may be independently replicating. They
are obtained from another bacterium or a virus. Plasmids are also nonessential,
but they exist only as free, independently replicating circles of DNA. (Episomes
are special types of plasmids.) In order for a plasmid to survive in a population, it must be a replicon; that is, it must be capable of independent replication so that it divides about the same time as the bacterial host.

10. Show how the episome is a special type of plasmid.

12-9 Recombinant DNA Technology Recombinant DNA is made up of connected segments from
mixed sources (from different species or even a combination of natural and synthetic DNA). Newly constructed genes may be inserted into host cells as plasmids
by transformation. Prokaryotes have been caused to make human products (such as
somatostatin) by such techniques. In the insertion of a gene into a circular
plasmid, restriction endonucleases are used to cut DNA at specific places, and
then sticky ends may be united permanently by DNA ligase.

11. Discuss the value of recombinant DNA technology to human beings.

ARTICLE RESOURCES

Mutations in Bacteria and Bacteriophage

Bishop, J. Michael. Oncogenes. Scientific American, 1982, 246(3), 80-92. Oncogenes are genes that cause cancers. They were first located in viruses. Their
cancerous potential has been demonstrated via genetic engineering. Cellular
oncogenes apparently direct the synthesis of a normal protein required for normal
growth. Cancerous growth occurs when abnormal amounts of this protein are formed.

A Sexual Process in Bacteria

Isolating Specific Bacterial Mutants

The Bacterial Fertility Factor

David, Bardell, Alexander Fleming and the age of antibiotics. The Science

<u>Teacher</u>, 1978, 45(3), 28-29. This biographical sketch of Alexander Fleming cites his research into lysozymes and pathogenic bacteria, as well as the discovery of penicillin.

Transfer of Male Genetic Elements

Sexduction

Recombination in Viruses

Lysogeny and the Disappearing Viruses

Transduction

Episomes and Plasmids

Novick, Richard P. Plasmids. <u>Scientific American</u>, 1980, 243(6), 102-122, 127. Plasmids appear to be subcellular organisms similar to viruses and occupying an ecological niche with other subcellular organisms such as viruses. Plasmids are found in bacteria. They have no protein coat and no extra cellular phase. They are present in all bacteria and appear to be present in most individual bacterial cells.

Recombinant DNA Technology

Baumel, Howard. "Genetic engineering" gains momentum. <u>The Science Teacher</u>, 1980, 47(4), 33-34. The accomplishments of researchers practicing genetic engineering are delineated. Simple and clear explanations of the process are given. Arguments for lifting of some of the governing standards are put forward.

Gilbert, Walter, and Lydia Villa-Komanoff. Useful proteins from recombinant bacteria. <u>Scientific American</u>, 1980, 242(4), 74-94. Recombinant DNA is used to produce inferon and insulin. Discusses methods of inserting DNA into bacteria. Well illustrated. Expands on the chapter treatment. Clear explanation of recombinant DNA technology.

Bauer, William R., F. H. C. Crick, and James H. White. Supercoiled DNA. <u>Scientific American</u>, 1980, 243(1), 118-132. Tumor viruses and the mitochondria of animal and human cells are supercoiled, as are the vectors of DNA in genetic engineering. The mathematics of supercoiling is discussed. Excellent diagrams to illustrate the mathematics of supercoiling are given.

Rensberger, Boyle. Tinkering with life. <u>Science 81</u>, 1981, 2(9), 44-49. Gene therapy as a direct result of the discovery of recombinant molecules. Genetic engineering and business have forged ties to market the new material for research purposes.

ESSAY QUESTIONS

1. Discuss evolution and selection as they pertain to the mutation of <u>E. coli</u> K to <u>E. coli</u> K/4 and T4 phage to T4h phage.

    These two types of mutations occur spontaneously in low frequencies all the time. The presence of T4 phage is an environmental selection factor for the enhancement of bacteria that are resistant to T4 infection--that is, <u>E. coli</u> K/4. K/4 bacteria cause the selection of T4h phage. The shift of genotype proportions in populations <u>is</u> evolution.

2. Describe sex in bacteria in terms of the factors that make one a male and another a female.

    Male bacteria have an F factor either as a free episome or as part of the bacterial chromosome. Transmission of the F factor to a recipient (female) causes the recipient to become male.

3. How could interrupted mating be used to map a bacterial chromosome?

    The bacterial chromosome breaks at the location of the F factor; one of the
    strands of DNA is then transferred linearly from the break point through the
    F pilus. Interruption of this transfer at various times allows investigators
    to check which marker genes enter the recipient in what order.

4. Compare sexduction with viral transduction.

    Sexduction is the carrying of bacterial chromosome DNA to a recipient as part
    of an F' factor during conjugation. Viral transduction--restricted--is the
    carrying of bacterial DNA with viral DNA from one host (which lyses) to a new
    host which is infected.

5. How are plasmids with spliced human genes transferred into bacterial cells such as
    E. coli?

    Modified plasmids with R factors in an appropriate solution and temperature
    will enter some bacterial cells by transformation. No carrier is necessary.
    Addition of the appropriate antibiotic kills all cells not carrying the
    appropriate R factor plasmid (with DNA from another source).

TEST SET 12-A

Name _____   Section _____

Choose the best answer to each of the following questions, and write the appropriate letter in the space provided.

Ans: d
p. 306
_____   1. Plaque formation depends on each of the following except:
(a) Viruses must infect bacteria for reproduction. (b) Bacteria must lyse. (c) Bacterial growth must be uniform enough for clearings to become visible. (d) Prophage must form.

Ans: b
p. 308
_____   2. In order for a mutation (a new allele) to become predominant in a population: (a) a large number of mutations must occur; (b) the environment must select for the mutation; (c) all wild-type organisms must die; (d) it must be recessive.

Ans: c
pp. 308,
    309
_____   3. If a U-tube were used with a fine filter separating the arms, killed bacteria were put in one side and live ones in the other, and the live bacteria picked up marker genes from the killed bacteria, _____ would have occurred. (a) transduction; (b) mutation; (c) transformation; (d) conjugation.

Ans: d
p. 311
_____   4. Each of the following statements is true about finding and isolating a new strain of bacteria carrying a particular mutation except: (a) It is more efficient to increase the rate of mutation. (b) The mutation would be expected to occur naturally at a low frequency. (c) One would have to get rid of the wild-type bacteria somehow. (d) Nutrients added to the medium or removed from it have little value in the isolation process.

Ans: a
p. 314
_____   5. DNA is transferred through the _____ into the _____ cell.
(a) F pilus, female; (b) F factor, male; (c) F pilus, male; (d) conjugation tube, male.

Ans: c
p. 314
_____   6. The Hfr F factor is ordinarily inserted into the: (a) male; (b) female; (c) bacterial chromosome; (d) T4 phage chromosome.

Ans: b
p. 315
_____   7. If gene B is always between marker genes A and C in interrupted mating studies, the most likely explanation is that: (a) gene A always leads the way; (b) the chromosome is circular; (c) gene C always leads the way; (d) the chromosome is linear, as in most higher organisms.

Ans: a
p. 317
_____   8. The carrying of marker genes with a separated F factor (F^1) from one bacterium to another occurs during: (a) conjugation; (b) translation; (c) meiosis; (d) mitosis.

Ans: a
pp. 318,
    319
_____   9. Bacteria harboring prophage are said to be: (a) lysogenic; (b) lysed; (c) immune to antagonistic bacteria; (d) sensitive cells.

Ans: d
p. 320
_____   10. The transduction of bacterial DNA only in a phage protein coat is: (a) conjugation; (b) sexduction; (c) restricted transduction; (d) general transduction.

Ans: c
p. 321
_____   11. Resistance factors (R factors) are carried on: (a) F factors; (b) episomes; (c) plasmids; (d) bacterial chromosomes.

Ans: c
p. 322
_____   12. Enzymes that cut DNA at specific places are: (a) DNA ligases; (b) DNA polymerase; (c) restriction endonucleases; (d) RNA polymerase.

Ans: a
p. 322

_____ 13. Somatostatin can be produced in great quantities now because of: (a) gene splicing and introduction of the human gene into E. coli; (b) its production in sea urchin eggs; (c) its usefulness in replacing insulin; (d) the discovery that it is made by fungi naturally.

Ans: b
p. 322

_____ 14. After being cut, bacterial DNA has sticky ends exposed that are capable of binding with _____ at low temperatures. (a) DNA polymerase; (b) complementary sticky ends; (c) DNA strands having the same base sequence as on each sticky end; (d) restriction endonucleases.

Ans: b
p. 324

_____ 15. Each of the following statements is true of recombinant DNA technology except: (a) Much of it is safe. (b) It cannot pose a health risk. (c) Some of it is hazardous. (d) Some of it carries appreciable risk.

TEST SET 12-B

Name _____   Section _____

Choose the best anwer to each of the following question, and write the appropriate letter in the space provided.

Ans: a
p. 306

_____ 1. Plaques in a bacterial lawn will not form unless: (a) viruses are actively reproducing; (b) viruses have mutations; (c) bacteria have mutations; (d) viruses lyse.

Ans: c
pp. 306,
    307

_____ 2. The mutation of E. coli K to E. coli K/4: (a) is the result of environmental selection; (b) made E. coli K/4 sensitive to T4 infection; (c) occurred spontaneously; (d) made T4 phage capable of infecting E. coli K cells.

Ans: d
p. 308

_____ 3. If a particular bacterium cannot grow without leucine, the genotype of the bacterium would be designated: (a) leu$^+$: (b) thr$^-$leu$^+$; (c) met$^+$thr$^-$leu$^+$; (d) leu$^-$.

Ans: b
p. 311

_____ 4. Sterilized velvet mounted on a wooden block is useful to the microbial geneticist; (a) in suspension dilution; (b) in replica plating; (c) in nutrient-deficient media preparation; (d) to grow Neurospora on.

Ans: a
p. 313

_____ 5. The most common manner in which sex change occurs in bacteria from male to female involves: (a) the loss of the F factor; (b) microsurgery; (c) the growth of F pili; (d) the addition of the F factor.

Ans: b
p. 314

_____ 6. Hfr males_____transmit their F factor to the female. (a) generally; (b) do not generally; (c) typically; (d) cannot.

Ans: c
p. 316

_____ 7. Chromosomes of Hfr males break at the location of the: (a) breakage gene; (b) plasmid; (c) F factor; (d) conjugation gene.

Ans: b
p. 317

_____ 8. Sexduction is remarkably like: (a) transformation; (b) viral transduction; (c) translation; (d) transcription.

Ans: a
p. 319

_____ 9. When a prophage leaves the chromosome: (a) the bacterium must then make new phages and lyse; (b) the bacterium will remain nonactive; (c) the virus makes bacterial DNA; (d) the bacterium makes viral DNA but not protein.

Ans: d
p. 320

_____ 10. The transduction of bacterial and viral DNA inside a phage protein coat is: (a) transcription; (b) sexduction; (c) general transduction; (d) restricted transduction.

Ans: c
p. 321

_____ 11. A plasmid that is capable of independent replication and is always free of the main chromosome is: (a) an episome; (b) a bacterial chromosome; (c) a replicon; (d) an F factor.

Ans: b
p. 321

_____ 12. Each of the following is an example of recombinant DNA except: (a) artificial DNA for the human somatostatin gene spliced into a plasmid; (b) a parental strand of DNA; (c) DNA spliced together from two different sources; (d) DNA that enters a new host by transformation and becomes part of the genome.

Ans: a
p. 322

_____ 13. Different restriction endonucleases recognize different _____ sequences in order to make specific "cuts." (a) base; (b) deoxyribose; (c) phosphate; (d) DNA ligase.

Ans: c   _____   14.   The uniting of sticky ends can be make permanent by an enzyme
p. 322   known as: (a) DNA polymerase; (b) RNA polymerase; (c) DNA
                          ligase; (d) restriction endonuclease.

Ans: a   _____   15.   In order to ascertain whether bacteria have taken up a desired
p. 324   plasmid (carrying an R factor and the spliced gene), investiga-
                          tors typically add: (a) a specific antibiotic relevant to the
                          R factor; (b) a specific amino acid; (c) a specific sequence
                          of DNA; (d) restriction endonuclease.

CHAPTER THIRTEEN
GENE EXPRESSION AND CELL DIFFERENTIATION

CHAPTER SUMMARIES AND OBJECTIVES

13-1 <u>The Control of Gene Expression / Control of Transcription in Prokaryotes</u> The overall strategies of some control mechanisms have become clear. By feedback inhibition, cells may decrease the activity of existing enzyme molecules in order to decrease the flow of a metabolite through a biochemical pathway, or they may switch off particular genes to stop enzyme production in order to accomplish the same thing. To increase the output of a metabolic pathway, cells may increase the number of genes for particular enzymes, speed up transcription (making mRNA), decrease the rate of mRNA breakdown, speed up translation, or decrease the rate at which enzymes are broken down. All five strategies result in an increase in the pool of enzyme molecules.

One of the best-understood strategies is the control of transcription in prokaryotes. Lactose induces transcription of mRNA and therefore production of the enzymes necessary for E. coli to utilize lactose. If lactose is removed, synthesis of the three specific enzymes ceases because the three structural genes are prevented from operating to make polycistronic messenger.

RNA polymerase attaches to DNA to begin transcription at particular sites known as promoters (one promoter per structural gene--cistron; one promoter per set of structural genes--polycistron). Efficient promoters precede structural genes for enzymes needed in large amounts; inefficient promoters control the synthesis of enzymes needed in tiny amounts.

1. Briefly summarize the overall strategies of the control of gene expression.

13-2 <u>Operons</u> An operon is a controllable unit of transcription and consists of a binding site for RNA polymerase (a promoter), a binding site for a specific repressor (an operator), and one or more structural genes. The repressor molecule is coded for by a separate gene site--a regulator gene. Since mRNA is made, the regulator gene must have a promoter. In the lactose inducible system, only a few repressor molecules are made, but they are made at a constant rate and are not subject to environmental control (they are constitutive, the promoter is inefficient, and there is no operator site for the regulator gene). When lactose or other β-galactosides are absent, repressor molecules attach immediately to the lactose operator. Lactose "induces" transcription of the structural genes by binding with the repressor molecules and making them incapable of continued attachment to the operator. With the repressor gone, RNA polymerase can attach to the promoter and transcription begins. The promoter is efficient and much mRNA is produced.

2. Diagram an operon.

3. Explain what occurs when lactose induces or activates transcription in prokaryotes.

13-3 Repressible Systems / Catabolite Repression / Eukaryotes and Positive Control
Repressible systems are capable of stopping production of the enzymes that are ordinarily used in biosynthesis when there are ample quantities of the product present. The repressor molecule is incapable of binding to the operator site by itself. The product of the biochemical pathway under consideration (such as tryptophan) acts as a corepressor and unites with the repressor molecule. This combination is capable of attaching to the operator which therefore turns off enzyme production.

Catabolite repression is a positive control process in contrast to the negative control of the inducible and repressible systems just discussed (their regulator molecules turn the systems off by attaching to operator sites). In catabolite repression, no operator site is involved. When glucose or some other excellent carbon source is present, there is no need for the cell to continue making the enzymes necessary to break down complex molecules to make more glucose available. When the glucose level is low, enzyme production occurs. A regulator protein (CAP) must bind to cAMP, the CAP-cAMP complex binds to the promoter, and then the promoter-CAP-cAMP complex binds RNA polymerase. A high glucose concentration causes catabolite repression by lowering the level of cAMP.

Functionally related gene clusters appear to be more common in prokaryotes than in eukaryotes. Positive control is more common in eukaryotes than in prokaryotes. The acidic chromosomal proteins of eukaryotes may be the positive regulatory elements that activate transcription of specific genes.

4. Compare and contrast control systems that turn the system off (negative control) with those that turn the system on (positive control).

5. Control systems make cell functions much more efficient. Make some brief explanatory statement about that phenomenon.

13-4 Control by Degradation / Control of Translation An enzyme may be stabilized by one of its substrates which, therefore, slows its rate of degradation. Continued synthesis of the enzyme would result in a rise in the overall level of enzyme presence. Elevated enzyme levels cause the substrate to be used up more rapidly, which then allows more rapid degradation of the enzyme.

The mRNA of bacteria lasts only minutes and translation begins while it is still being formed. In eukaryotes, the product of transcription is modified in the nucleus. It may be further modified in the cytoplasm prior to translation, and it lasts for hours or days. In the unfertilized egg of the sea urchin, mRNA is stored in a masked state (nonfunctioning). Following fertilization, the ribosomes are able to begin translation. Experimentation with Acetabularia mediterranea indicate that masked mRNA is stored in the stalk and later translated into cap proteins according to a prescribed biochemical schedule.

The immediate product of transcription in eukaryotes is heterogenous nuclear RNA (hnRNA). Excess materials (introns) are excised, and the remaining exons are spliced together to make the functioning mRNA. Other chemical changes and transport from nucleus to cytoplasm are also subject to regulation.

6. Explain how a reduced rate of degradation is involved in an overall rise of enzyme level.

7. Contrast the time schedule by which the products of transcription become ready to function in prokaryotes and in eukaryotes.

13-5 Repeated Genes and Gene Amplification / Chromosome Inactivation Several genes occur in multiple copies along the chromosomes of humans and other higher organisms (such as genes for histones and tRNA). Obviously these repeated genes allow more transcription than would occur if only one gene site for each were present. Genes responsible for rRNA synthesis (DNA of nucleoli) are thought to be amplified many times over in the oocytes of many animals. Prior to development of the fertilized egg, multiple copies of the nucleolar organizer are made, freeing numerous nucleoli in the nucleoplasm. This enables the large mass of oocyte cytoplasm to have enough rRNA to allow for protein synthesis during the entire period from fertilization to formation of the gastrula.

The abnormal presence of excess genetic material (as in Down's syndrome, resulting from trisomy-21) can cause some extremely critical effects. The extra number 21 is active, unlike multiple X chromosomes, all but one of which are inactivated. Inactivated X chromosomes form blobs of heterochromatin (Barr bodies) and do not produce mRNA. Normal human females have one Barr body; males have none. The Single Active X rule encompasses the findings of Lyon and Russell and explains why humans can survive with XO, XXX, XXY, and other combinations. To some extent, it also explains the similarity of men and women.

8. Explain the advantage to an embryo of having the oocyte begin with multiple nucleoli.

9. Excess genetic information may be disastrous, or the effects may be minimized. Illustrate both conditions.

13-6   Self-Assembling Molecules / Self-Assembly of a Bacteriophage   Though mRNA codes for the primary structure of proteins (linear arrangement of amino acids), the three-dimensional shapes of proteins depend on the spontaneous twisting, folding, and bonding between folds of the chain. These intricate shapes are caused by the sequence of amino acids themselves and by environmental (cellular) conditions. Once the subunits are synthesized, they aggregate spontaneously into functional proteins regulated by ion concentration, pH, oxidation-reduction potential, and concentration of the subunits. Enzymes may be disaggregated by changing one of these variables, and the restoration of original conditions may cause spontaneous reassembly of functional molecules.

Collagen is a good example of a self-assembled structural protein. The cell produces procollagen and secretes it into intercellular spaces, where it is partially broken down to form tropocollagen and two other polypeptides. In their final location, the tropocollagen molecules assemble into collagen fibrils that assume a precise orientation.

T4 bacteriophage is one of the most elaborately structured viruses. It has a head, neck and collar, a sheath and core, and an end plate with tail fibers. It does undergo self-assembly to an extraordinary degree of order and structure at the molecular level. Of particular note are the seventeen different proteins necessary for the end plate. This plate is formed by the step-by-step addition of each protein in a specific sequence--each one altering the structure, enabling it to add the next protein in sequence. The whole virus is more stable than the sum of its parts.

10. Self-assembling molecules reduce the need for extraordinary amounts of DNA. Briefly explain why this is so.

13-7   Assembly of an Organelle   Chloroplasts arise by division of existing chloroplasts in already green tissue. However, their beginning in nonphotosynthetic tissue is as small inpocketings of the plasma membrane. They become small proplastids containing prolamellar granules, which contain carotenoids and protochlorophyll a. In the light, protochlorophyll a is reduced to chlorophyll a, and the prolamellar granules develop into stacks of disks called grana. Other development leads to the mature organelle. Chloroplast DNA contains the genetic information for some of the chloroplast proteins; many others are in the DNA of the cell nucleus.

11. Relate the role of the cell nucleus to chloroplast development.

13-8   Cell Development / The Role of RNA   Differentiation of eukaryotic cells remains largely a mystery. Cellular differentiation is based on sequential changes in the amounts and kinds of proteins synthesized by the cell during its development. As the synthesis of one group of proteins slows or stops, the synthesis of others begins. The continuation of DNA synthesis and mitosis are not requirements for continued differentiation. In Acetabularia, neither nuclear DNA synthesis nor mitosis occurs until the cap begins to produce reproductive cysts. Ethel Harvey activated non-nucleated sea urchin egg fragments by treating them with a hypertonic salt solution. They cleaved and formed blastulas with blastocoels.

Obviously RNA synthesis was lacking, as were mitosis and DNA synthesis.

Messenger RNA must be synthesized before it can be translated into protein at the ribosomes. New protein, therefore, is preceded by the activation of a previously inactive gene and by the synthesis of a new kind of mRNA. A number of experiments point to the existence of long-lived mRNA in developing systems of eukaryotes (for example, mRNA for hemoglobin is clearly quite stable). In the early development of sea urchin eggs, mRNA inhibitors have no effect after protein synthesis has begun, but mRNA for proteins needed beyond the blastula state must be synthesized during blastula development. mRNA is masked prior to fertilization in sea urchin eggs. Fertilization permits the start of the protein synthesis required for development to the blastula stage.

12. Describe the relationship between gene activation, mRNA synthesis, and protein production.

13. Cite evidence supporting the existence of long-lived mRNA in eukaryotes.

13-9 Control of RNA Synthesis and Translation RNA synthesis (transcription) is regulated during development by several mechanisms--particularly gene amplification and gene activation. The same gene may become active in different tissues at different times in development. DNA in chromosomes is transcribed only when chromatin is in its extended state, as when puffs become visible. However, these are not the only bases of development; differentiation must involve changes in molecules or structures already present as well as effects on specific protein synthesis.

14. Give two means by which development is regulated through control of RNA synthesis.

13-10 Is Differentiation Reversible? In certain types of cells, differentiation is clearly irreversible--particularly in these cells that lose their nuclei, such as red blood cells and tracheids. Carrot and tobacco root callus cultures (Steward, Vasil, and Hildebrandt) showed clearly the retention of totipotency by these cells. Briggs, King, and Gurdon showed that amphibian cell nuclei retain the ability to direct the full development of enucleated eggs. Scientists must look further into aspects of totipotency, determination, and differentiation. Perhaps cytoplasmic influences on the nuclei may be involved.

15. Define totipotency.

ARTICLE RESOURCES

The Control of Gene Expression

Control of Transcription in Prokaryotes

Diener, I. O. Viroids. Scientific American, 1981, 244(1), 66-73. Viroids are strands of RNA and are the smallest known causes of infectious diseases in plants. Viroids are compared in size to bacteriophage, adenovirus, tobacco-mosaic virus, and E. coli. They are by far the smallest. Their discovery by gel electrophoresis is described.

Lake, James A. The ribosome. Scientific American, 1981, 245(2), 84-97. The ribosome is the organelle that binds messenger RNA and captures amino acids from the cytoplasm in the sequence coded by the nucleotides on the mRNA to form a protein. It is very small--only about 250 angstrom units at its largest dimension. Excellent photomicrographs of the ribosome, showing its basic structure, are included, and a model is deduced. The sites of protein synthesis are shown. Charts showing protein synthesis and the termination process whereby the protein and mRNA are released are very clear.

Operons

Repressible Systems

Catabolite Repression

Eukaryotes and Positive Control

Rodgers, Joann Ellison. Dr. Thymosin's remedy. Science 81, 1981, 72-76. Thymosin is a hormone that is secreted by the thymus and apparently triggers the body's immune response. Discovered in the 1960s by Abraham White and Allan L. Goldstein, it is a "natural" hormone that is causing quite a stir in the manipulation of immunity.

Control by Degradation

Control of Translation

Repeated Genes and Gene Amplification

Chromosome Activation

Self-Assembling Molecules

Self-Assembly of a Bacteriophage

Assembly of an Organelle

Cell Development

Woese, Carl R. Archaebacteria. Scientific American, 1981, 244(6), 98-122. Archaebacteria do not belong to the prokaryotes or to the eukaryotes. They have no nucleus and their biochemistry is different from that of eukaryotes and prokaryotes. Eukaryotes and prokaryotes are compared and contrasted. Archaebacteria include methanogens, thermoacidophilas, and extreme halophiles. These are explored in fascinating detail. A large chart compares archaebacteria, eubacteria, and eukaryotes.

The Role of RNA

Rich, Alexander, and Sung Hou Kim. The three-dimensional structure of transfer RNA. Scientific American, 1978, 238(1), 52-62. Transfer RNA are a class of small globular polynucleotide chains. These are 75-90 nucleotides long. They act in transferring amino acids within the cell into the assembled chain of protein. They act as an intermediary between the nucleic acids of the genetic code and the amino acids of the cell. Excellent diagrams and descriptions of tRNA.

Control of RNA Synthesis and Translation

Is Differentiation Reversible?

Holden, David J. Cloning: Learning to replay the genetic tape. The Science Teacher, 1979, 46(8), 27-29. Sorghum and tobacco clones were produced by removing the cell walls. These protoplasts fused as if they were gametes but had twice the number of chromosomes of the original cells. Gives a history of the discovery of cloning.

Shepard, James F. The regeneration of potato plants from leaf-cell protoplasts. Scientific American, 1982, 246(5), 154-166. Description of cloning in potato plants. Cloning was done to investigate the frequency of somatic variations in clonal populations. Only phenotypic variants were selected rather than varieties with altered chromosome numbers, which result in abnormal plants lacking vigor.

ESSAY QUESTIONS

1. Describe an operon of prokaryotes.

    An operon is a controllable unit of transcription having promoter, operator, and structural genes, in that order. RNA polymerase must attach to the promoter site for transcription of the structural genes to occur. Transcription may be blocked by the attachment of a regulator protein at the operator site.

2. Outline the steps in the lactose inducible system that prevent blockage of transcription.

Repressor molecules--coded for by regulator genes located elsewhere--attach to the operator site and prevent transcription of the structural genes for enzymes needed for lactose metabolism. Lactose presence is accompanied by lactose binding to the repressor molecule, making it incapable of continued attachment to the operator site. Transcription may then proceed.

3. What occurs in the tryptophan repressible system to block transcription?

The product tryptophan (corepressor) binds to the regulator protein; this combination is then capable of attachment to the operator site, thus turning off that operon.

4. What happens in eukaryotes to hnRNA before it can be transcribed?

Heterogenous nuclear RNA contains unusable segments (introns) that must be excised. The remaining extrons are spliced together to form workable mRNA.

5. Describe X-chromosome inactivation in humans and the effects of this phenomenon.

Only one X chromosome need function in any one cell to provide the products coded for by the X chromosome. Males ordinarily have only one X, so they would not have an inactive X--a Barr body. Females have two X's, and one in each cell is inactivated randomly. For any sex-linked heterozygous genes, both could be expressed but in different cells. One active X per cell helps explain the similarities between males and females.

6. Describe incidents which indicate that differentiation is reversible in some cells.

Steward (carrot) and Vasil and Hildebrandt (tobacco) produced whole plants from cultured callus cells derived from mature plants. Briggs, King, and Gurdon (amphibians) showed the development of enucleated eggs into which determined or differentiated nuclei were transplanted.

TEST SET 13-A

Name _____   Section _____

Choose the best answer to each of the following questions, and write the appropriate letter in the space provided.

Ans: c
p. 328

_____ 1. Changes in an organism with time, with an increase in order and complexity, constitute what is called: (a) determination; (b) organogenesis; (c) development; (d) totipotency.

Ans: a
p. 330

_____ 2. Glucose and galactose are: (a) breakdown products of lactose; (b) β-galactoside permeases; (c) β-galactosides; (d) disaccharides.

Ans: b
p. 331

_____ 3. Synthesis of an enzyme that will always be needed in tiny amounts is coded by a structural gene whose promoter: (a) must be amplified; (b) is inefficient; (c) is efficient; (d) can bind repressor protein.

Ans: d
p. 331

_____ 4. RNA polymerase attaches to the _____ of an operon. (a) operator; (b) regulator; (c) structural gene; (d) promoter.

Ans: a
p. 333

_____ 5. When the formation of an enzyme is turned off in response to an ample concentration of the product of the biochemical pathway in which the enzyme takes part, the enzyme is said to be: (a) repressible; (b) inducible; (c) a positive control process; (d) switchable.

Ans: b
p. 334

_____ 6. Blocking the synthesis of a family of enzymes through failure to synthesize a regulatory protein is characteristic of: (a) negative control; (b) positive control; (c) an inducible system; (d) a repressible system.

Ans. c
p. 335

_____ 7. The amount of enzyme within a cell depends on its rate of _____ and its rate of _____. (a) synthesis, production; (b) synthesis, transcription; (c) synthesis, degradation; (d) translation, transcription.

Ans: d
p. 337

_____ 8. In prokaryotes, translation begins while _____ is still being formed. (a) DNA; (b) hnRNA; (c) tRNA; (d) mRNA.

Ans: a
p. 338

_____ 9. RNA that contains sections of exons and introns is known as: (a) hnRNA; (b) tRNA; (c) rRNA; (d) extra RNA.

Ans: b
p. 340

_____ 10. An excess of the gene sites carried on chromosome 21 causes: (a) Barr bodies; (b) Down's syndrome; (c) hemophilia; (d) sickle-cell anemia.

Ans: c
p. 342

_____ 11. Arrange the following in the correct order for collagen self-assembly: I. tropocollagen; II. procollagen; III. collagen; IV. flagellin. (a) I, II, III, IV; (b) II, I, III, IV; (c) II, I, III (omit IV); (d) IV, III, II, I.

Ans: b
p. 345

_____ 12. Proplastids contain all of the following except: (a) protochlorophyll a; (b) nuclear DNA; (c) prolamellar granules; (d) carotenoids.

Ans: a
p. 347

_____ 13. The appearance of crystallin proteins in developing lens cells depends on: (a) synthesis of mRNA that codes for them; (b) synthesis of DNA just before their appearance; (c) the addition of actinomycin D; (d) transcription immediately prior to their appearance.

Ans: d   _____ 14. Fertilization and _____ are both major means of regulating
p. 348   protein synthesis (amounts and kinds) during development.
                             (a) mitosis; (b) DNA synthesis; (c) gene amplification;
                             (d) gene activation.

Ans: b   _____ 15. Briggs, King, and Gurdon are noted for their work with:
pp. 351,   (a) carrots; (b) amphibians; (c) snakes; (d) sea urchins.
    352

TEST SET 13-B

Name _____   Section _____

Choose the best answer to each of the following questions, and write the appropriate letter in the space provided.

Ans: a
p. 328

_____   1.   Differentiated cells are those: (a) whose fate has been achieved;
(b) that typically give rise to the whole organism; (c) blastula
cells that have become committed to a certain fate; (d) whose
DNA cannot ever again direct the formation of a whole organism
under any circumstances.

Ans: c
p. 330

_____   2.   Enzymes for lactose metabolism require _____ in order to be
formed. (a) three separate operons; (b) at least three
constitutive genes; (c) a polycistronic messenger molecule;
(d) three operator genes.

Ans: b
p. 331

_____   3.   An enzyme that will always be needed in large amounts is coded
by a structural gene whose promoter is: (a) inactive;
(b) efficient; (c) inefficient; (d) poor at binding RNA
polymerase.

Ans: b
p. 331

_____   4.   Repressor protein is able to bind to its specific _____ and
inducers. (a) promoter; (b) operator; (c) structural gene;
(d) RNA polymerase.

Ans: d
p. 333

_____   5.   Any gene that codes for mRNA must be preceded by: (a) an
operator; (b) efficient RNA polymerase; (c) a three-part
operon; (d) a promoter.

Ans: c
p. 334

_____   6.   In one catabolite repression system involving enzymes for making
glucose or other good carbon sources available to the cell, CAP
binds to cAMP; this complex binds to the appropriate promoter;
and RNA polymerase is able to bind to the promoter-CAP-cAMP
complex to begin transcription. What acts as the regulatory
molecule (its level is lowered in the presence of glucose)?
(a) CAP; (b) RNA polymerase; (c) cAMP; (d) glucose.

Ans: a
p. 336

_____   7.   When an enzyme is stabilized by one of its substrates, a high
substrate concentration tends to decrease the degradation rate
without affecting the rate of _____ for that system.
(a) synthesis; (b) substrate breakdown; (c) negative control;
(d) induction.

Ans: d
p. 337

_____   8.   Masked mRNA exists in unfertilized sea urchin eggs. What
unmasks it? (a) meiosis; (b) mitosis; (c) DNA synthesis;
(d) fertilization.

Ans: b
p. 339

_____   9.   The process of creating more genes of one kind to enhance
transcription is called: (a) gene repetition; (b) gene
amplification; (c) gene activation; (d) chromosome inactivation.

Ans: c
pp. 340,
341

_____   10.   The Single Active X rule means that: (a) three Barr bodies
will be formed in XO individuals; (b) males cannot be XXY;
(c) all but one X chromosome per cell are turned off; (d) only
one Barr body per cell can occur.

Ans: a
p. 343

_____   11.   If one were to mix headless and tailless T4 bacteriophage in a
test tube, what would be expected to occur? (a) formation of
completely normal viruses; (b) continued separation of headless
and tailless viruses; (c) nothing, because viruses need living
cells in which to reproduce; (d) continuation of tailless

viruses only, because the DNA is in the head.

Ans: a
p. 345
_____ 12. In the presence of light, protochlorophyll <u>a</u> is converted to:
(a) chlorophyll <u>a</u>; (b) chlorophyll <u>b</u>; (c) carotenoids;
(d) prolamellar granules.

Ans: c
p. 347
_____ 13. When activated with a hypertonic salt solution, sea urchin
non-nucleated egg fragments: (a) develop into sea urchins;
(b) progess through the gastrula stage; (c) cleave to form
"respectable" blastulas; (d) usually die.

Ans: b
p. 348
_____ 14. DNA in chromosomes is transcribed only when chromatin is in its
_____ state. (a) most contracted; (b) extended; (c) metaphase;
(d) condensed.

Ans: d
p. 351
_____ 15. _____ seem not to be so irreversibly differentiated as _____.
(a) animal cells, plant cells; (b) carrot cells, tobacco cells;
(c) South African clawed toad cells, carrot cells, (d) plant
cells, animal cells.

CHAPTER FOURTEEN
IMMUNE SYSTEMS AND DISEASE

CHAPTER SUMMARIES AND OBJECTIVES

14-1 <u>Specific Defenses Against Nonself Invaders / Cells of the Immune System</u> The functions of the vertebrate immune system are to recognize, selectively eliminate, and remember foreign invaders. There are two forms of immune responses in humans and many other vertebrates. The cellular immune response involves cells sensitive to nonself cells and molecules (most active against cancer cells, fungi, foreign tissue, multicellular parasites, and intracellular viral infections). The humoral immune response involves the making of plasma antibodies against bacteria and viruses that are not yet inside cells.

Cells that travel in the blood and lymph are involved in the immune system. Blood is 55-65% plasma; the remainder is red blood cells (erythrocytes), white blood cells (leukocytes), and platelets. Plasma minus clotting factor is serum. Lymph is in an accessory circulatory system and is essentially plasma and leukocytes that leave the blood capillaries. Lymph dumps back into the venous system in the lower neck region.

Leukocytes can move independently via pseudopods. Neutrophils and macrophages are key phagocytes, and lymphocytes are predominantly involved in the immune system. Lymphocytes originate in bone marrow. Ancestral cells give rise to B cell lymphocytes, which migrate to lymph nodes, and to T cell lymphocytes, which develop distinguishing properties in the thymus gland.

1. Distinguish between cellular and humoral immune responses.

2. Describe blood and lymph.

3. Name the key kinds of white blood cells and give their basic functions.

14-2 <u>The Immune Response</u> The humoral immune system is based on antibodies--also called immunoglobins (Ig)--that are produced by special lymphocytes (plasma cells). Each antibody recognizes and binds to the antigenic determinant site of a foreign body (the antigen). Some antibodies travel free in blood and lymph; others remain bound to lymphocyte membranes.

T cells of the cellular immune system have specific surface molecules (which may include parts of Ig molecules) that recognize and react to antigenic determinants on the surfaces of other cells (antigens). The specificity of B and T cells is in contrast to the nonspecific defense carried out by phagocytes, which engulf any foreign matter they encounter.

An animal does not require any previous encounter with a foreign body to mount an immune response to it. A person contains millions of distinct Ig specificities. The first encounter, however, is marked by a time lag of several days before the number of Ig's produced catches up with the number of invaders. Repeat exposures are marked by much greater and more rapid production of antibodies and activated lymphocytes due to immunological memory.

Immunological memory is the reason why immunization through vaccination is successful. Specific disease agents that are made harmless but retain their antigenic properties are introduced into the body in order to provoke a "first encounter" immune response. Exposure to the virulent disease organisms would then cause a swift immunological response. Under normal circumstances, the immune system distinguishes between self antigens and nonself antigens.

4. Distinguish between the antibodies of B cells and specific surface molecules of T cells.

5. An individual develops whooping cough but recovers. He encounters the disease organisms again the following year but does not again develop the disease. Briefly describe what happens in the immune system during the first encounter and during the second.

14-3 <u>Clonal Selection and Its Consequences</u> The clonal selection theory (Jerne and Burnet) holds that the individual animal possesses many different lymphocytes, each capable of producing one kind of antibody. It postulates a population of specific lymphocytes corresponding to each of the antigenic determinants to which an organism can respond. The arrival of the antigen merely activates the specific lymphocyte to multiply (producing a clone), and this results in an increase in Ig produced. Activated lymphocytes produce two kinds of cells: effector cells (B or T cells) and memory cells, which are long-lived and retain the ability to quickly produce more effector and memory cells when challenged with the appropriate antigen.

6. Compare the antibody-producing capabilities of one clone from an activated lymphocyte with those of another clone.

7. State the role of effector and memory cells.

14-4 <u>Self, Nonself, and Tolerance / Disorders of the Immune System</u> B and T cells that are not fully differentiated are very sensitive to antigen. Any of these lymphocytes that is capable of producing "anti-self" molecules will encounter "self" antigens which either inactivate or completely eliminate the anti-self lymphocytes. Thus, clones of these cells do not form (clonal deletion).

Before differentiation of immune specificities is complete (such as prior to birth), foreign antigen may be introduced into the individual. The individual will henceforth recognize that antigen as "self" (immunological tolerance).

Clonal deletion must occur throughout the life of the individual, because lymphocytes are constantly produced. If a given self protein were not produced for an extended period of time and then production commenced again, it might elicit a definite immune response (autoimmune disease). Examples of autoimmune diseases wherein one forms anti-self molecules are rheumatic fever, rheumatoid arthritis, and ulcerative colitis. The immune system may also overreact dangerously to a dose of antigen such as a bee sting.

Immune deficiency disorders may occur. For example, an individual may not form B cells, or B cells may lose the ability to give rise to plasma cells. A major line of defense against pathogens is therefore lost.

8. Explain what anti-self molecules are.

9. Describe the basic occurrences in clonal deletion.

10. Develop some understanding of an autoimmune disease.

11. Define immune deficiency disorder.

14-5 <u>The Immunoglobulins</u> The basic immunoglobulin molecule (a commonly occurring type) is a tetramer consisting of four polypeptides. There are two identical "light" chains and two identical "heavy" chains--held together by disulfide bonds--with each chain having a constant and a variable region. The variable regions give the specificity to Ig molecules. Each of the two arms is identical to the other, which sometimes leads to the formation of a complex network of antigen and antibody molecules.

12. Describe the basic similarities and differences between two different immuno-globulin molecules.

14-6 <u>Effectors of the Immune System / The Generation of Diversity</u> Following activation by an antigen, B and T cells produce effector and memory cells. The effectors produced by B cells are known as plasma cells, which are extremely efficient at making antibodies. The cells resulting from a clone of plasma cells all produce the same Ig and have identical antigen receptor sites on their surfaces. Through monoclonal antibody culturing techniques, it is possible to obtain large quantities of pure antibody.

Activated T cells produce several types of effectors. One is the killer T cell that recognizes and causes lysis of foreign cells or abnormal tumor cells; another is the helper T cell that must bind antigen before a B cell can respond to it.

The complement (effector) system assists in many defense reactions, though the approximately 11 complement proteins are not themselves immunoglobulins nor do they increase in concentration following immunization. Complement binds to the antigen-antibody complex. In one reaction, complement kills invading cells by damaging the plasma membrane; in another, it promotes phagocytosis; in still another, it triggers inflammation.

The germ-line hypothesis and somatic variation hypothesis both offer some sound ideas about the generation of the tremendous diversity that exists in the immune system. A considerable number of genes are involved in coding for Ig's; there is deletion, recombination, and splicing of DNA fragments during the differentiation of B cells; and the gene coding for one part of the light chain of Ig's undergoes an exceptional frequency of point mutations.

13. Describe the fate of activated B and T cells.

14. Describe the behavior of complement proteins.

15. Identify the source of diversity in the immune system.

14-7 <u>Transplants / Transfusions and ABO Blood Groups</u> Most transplants (other than those between identical twins and those that are done prior to development of immunological competence) fail due to the major histocompatibility complex. The multiple allelic genes of this complex code for specific proteins found on the surfaces of all cells of that animal. In transplants, the proteins are perceived as nonself and are rejected by the cellular immune system. The major histocompatibility complex of humans is known as the human leukocyte antigen (HLA) system. It apparently plays a great many roles in the immune system.

Karl Landsteiner worked out the blood compatibilities of the ABO multiple allelic system around 1900. Investigators later found just what the antigenic determinant of red blood cells is. Blood types A and B each have a different sugar attached to the ends of polysaccharide chains that occur on the surfaces of the RBC's. They each form a specific enzyme for the appropriate sugar to be attached. The I^O allele does not produce enzyme, so the last sugar is not added and no anti-O antibodies ever form. Blood type A individuals do produce anti-B antibodies, and blood type B individuals produce anti-A antibodies. Blood type AB individuals produce neither type of antibody. It would be disastrous if antibodies were formed against one's own RBC's. Type O individuals produce both anti-A and anti-B serum antibodies. Blood transfusions are a special kind of transplant and are best done between individuals of the same blood type. The plasma of the patient and the red blood cells of the donor must be compatible if mixing of (whole) un-matched blood types is ever done.

16. Describe the role of the major histocompatibility complex in transplant rejec-tion.

17. List the genotypes for each phenotype of the ABO system (types A, B, AB, and O).

18. Cite several reasons why an AB individual is sometimes identified as a universal recipient and a type O individual as a universal donor.

14-8 RH Incompatibility Among Caucasians, about 83% possess the dominant Rh^+ trait; their red blood cells form a surface antigen (the RH factor). Rh^+ individuals do not form antibody against the RH factor. Rh^- individuals do not produce the RH factor, the Rh^- condition behaves as a genetic recessive, and Rh^- individuals do form anti-RH antibody when Rh^+ red blood cells are introduced into their bodies.

An Rh^- woman who is pregnant with an Rh^+ child will form antibodies if fetal RBC's cross the placental barrier. When this occurs at birth, anti-RH antibody can be injected into the mother to clear out the foreign cells, if this step is taken before her immune system mounts a response (it must be within 72 hours). Once the mother makes anti-RH antibodies, they are capable of crossing the placenta. Then further pregnancies involving Rh^+ children would be affected in that the antibodies cause destruction of Rh^+ red blood cells.

19. Describe the genetics of inheritance of the RH factor.

20. Give the conditions under which a problem could arise concerning the RH factor.

14-9 Cancer The transformation of a normal cell into a cancer cell is marked by nuclear and cell membrane changes. They may be caused by mutation, selective activation of certain genes, or viruses (some oncogenic viruses are known). Cell surface molecules normally inhibit excessive growth of an organ; cancer cells, however, continue uncontrolled multiplication, forming tumors. Benign tumors grow in one place only, whereas malignant tumors invade surrounding tissues and spread to other parts of the body (metastasis) via the blood or lymph systems. The cellular immune system may have evolved principally as an immune surveillance system against cancers. Carcinogenic agents (such as sunlight and various chemicals) sometimes provoke the development of cancers, but most cancers develop in conjunction with old age as the immune system becomes less effective and growth can continue unchecked. Three principal lines of treatment are surgery, radiotherapy, and chemotherapy.

21. Describe how cancer cells differ from normal cells.

22. Discuss how the cellular immune system is related to the development of cancer.

14-10 Nonspecific Defense Mechanisms Against Pathogens Pathogens must face four major hurdles in order to be successful: they must arrive at the body surface of a potential host, get into the host, multiply there, and prepare to infect the next host. Potential hosts have several lines of defense against potential pathogens.

Once inside the host, the pathogen encounters the specific defenses of the immune system and such nonspecific defenses as competition for particular nutrients, antimicrobial proteins, phagocytosis, and inflammation. Neutrophils and macrophages are the main phagocytes. The engulfed pathogens are destroyed by the contents of lysosomes. Inflammation is a response to products released from damaged host cells (caused by infections, mechanical injuries, and burns). Redness and heat from the dilation of blood capillaries in the area, swelling from fluids released into tissues, and pain from the increased pressure and enzyme action are involved.

23. List the four major obstacles a pathogen must surmount in order to be successful.

24. Describe several nonspecific lines of defense.

ARTICLE RESOURCES

Specific Defenses Against Nonself Invaders

Simons, Kai, Henrik Garoff, and Ari Helenius. How an animal virus gets into and out of its host cell. Scientific American, 1982, 246(2), 58-66. For some years, the mechanisms whereby viruses enter the host cell were unknown. Now it appears that viruses enter via the same pathways as do many molecules required in cellular metabolism. The Semliki forest virus is studied because it contains only four protein molecules. Their order has been determined by the sequence of nucleotides in the viral RNA.

Cells of the Immune System

Koffler, David. Systemic lupus erythematosus. Scientific American, 1980, 243(1), 52-61. This is a disease of the immune system affecting many parts of the human body. Discusses the role of the immune system in protecting the body against infection. Discusses components of immune complexes and autoantibodies. Expands on material given in the text.

The Immune Response

Rose, Noel R. Autoimmune diseases. Scientific American, 1981, 244(2), 80-103. How the body rejects foreign substances. The errors in this system, however, are the subject of research by physicians and biologists. Autoantibodies have been discovered that can be used for diagnosing Grave's disease, juvenile diabetes, and myesthenia gravis. Excellent diagrams.

Clonal Selection and Its Consequences

Schloen, Lloyd Henry. Immortalizing immunity. The Sciences, 1980, 20(6), 14-17. Hybridoma technology is a new research tool that uses specific or monoclonal antibodies from individual hybridomas. A hybridoma results when a mouse mycloma cell is combined with a normal mouse spleen cell. It produces one particular type of antibody. These are monoclonal antibodies available to detect myocardial infarction, pregnancy, some types of cancer, and hepatitis. These monoclonal antibodies will also make tissue typing simpler.

Self, Nonself, and Tolerance

The Immunoglobins

Effectors of the Immune System

Leder, Philip. The genetics of antibody diversity. Scientific American, 1982, 246(5), 102-115. Antibodies are bifunctional bodies in that they have a variable region and a constant region. The variable region folds up to form an antibody-antigen site. The constant region serves the same function in every molecule of its type and determines how it will carry out its immunologic job in the body. These bodies offered the first clue to the genetic source of their diversity. Cloning of bacteria is discussed and diagrammed.

The Generation of Diversity

Transplants

Transfusions and ABO Blood Groups

Switching blood types. Science Digest, 1982, 90(4), 101. Brief note on blood types, their red blood cell coating, and how it is possible to use an enzyme from coffee beans to change type B cells to type O cells, leaving them functioning in a healthy manner.

How do our bodies make blood? Science Digest, 1982, 90(4), 103. Clear exposition of blood formation in simple terms. A good diagram is included.

Rh Incompatibility

Cancer

Henle, Walter, Gertrude Henle, and Evelyne T. Hennelti. The Epstein-Barr virus. Scientific American, 1979, 24(1), 48-59. This is a common virus causing infectious mononucleosis and is tied to two types of human cancer--Burkitt's lymphoma and nasopharyngeal carcinoma. Fascinating story of the search to link these three diseases to the Epstein-Barr virus. Detailed discussion of how the virus replicates.

Nonspecific Defense Mechanisms Against Pathogens

ESSAY QUESTIONS

1. Describe the basic activities of the humoral and cellular immune systems.

    The humoral system's B cells and plasma cells produce antibodies that are either free in plasma or attached to lymphocyte surfaces. It acts primarily against bacteria and viruses that are not yet in cells. Memory cells produced from activated B cells are responsible for immunization. The cellular immune system utilizes T cells, which do not secrete antibodies but have specific surface molecules that react with antigenic determinants. It acts primarily against cancer cells, fungi, foreign tissue, multicellular parasites, and intracellular viral infections.

2. What is blood? What does it consist of?

    Blood is a fluid tissue. About 55-65% is the watery matrix known as plasma. Plasma minus clotting factor is serum. About 35-45% of blood is composed of formed elements: red blood cells, white blood cells, and platelets.

3. What is the basic premise of the clonal selection theory?

    The individual makes enough different lymphocytes to account for its capability to respond to the tremendous number of antigenic determinants the organism may encounter. Such an encounter activates the specific kind of lymphocyte to produce a clone of effector cells and memory cells and results in specific antibodies being formed.

4. In discussing the body's immune system, what do we mean by "self" and "nonself"?

    Anything with a genetic basis that is characteristic of an individual is known as self. Anything foreign is nonself. The body distinguishes between the two by preventing the formation of cell clones that would be capable of making antibodies against "self." This is clone deletion.

5. Why do most transplants between genetically different individuals fail?

    The major histocompatibility complex codes for cell surface proteins on all cells of the individual. There are so many genes and multiple alleles in this complex that the occurrence of two totally compatible individuals is not likely. The cellular immune system of the transplant recipient interprets the transplant as foreign and attacks it.

6. What is the relationship between the immune surveillance system and cancer?

    The immune surveillance system ordinarily recognizes changed cell membranes of cancer cells as nonself and destroys them. When the body gets old, the surveillance system becomes less effective; therefore most cancers are diseases of old age.

TEST SET 14-A

Name _____   Section _____

Choose the <u>best</u> answer to each of the following questions, and write the appropriate letter in the space provided.

Ans: d
p. 356

_____ 1. The cellular immune response is most often involved in defense against each of the following <u>except</u>: (a) cancer cells; (b) fungi; (c) foreign tissue; (d) bacteria.

Ans: a
p. 357

_____ 2. Neutrophils play predominant roles in: (a) a first line of defense; (b) the cellular immune response; (c) oxygen transport; (d) the humoral immune response.

Ans: b
p. 358

_____ 3. B cells migrate from bone marrow directly to: (a) the thymus gland; (b) lymph nodes; (c) the thyroid gland; (d) other bones for storage.

Ans: c
p. 358

_____ 4. Antibodies are also called: (a) antigens; (b) antigenic determinants; (c) immunoglobulins; (d) lymphocytes.

Ans: c
p. 361

_____ 5. According to the clonal selection theory, each different population of lymphocytes produces: (a) millions of different antibodies; (b) B cells only; (c) a single kind of antibody; (d) both B and T cells.

Ans: a
p. 362

_____ 6. Preventing the formation of clones of anti-self lymphocytes is known as: (a) clonal deletion; (b) clonal survival; (c) antibody sensitivity; (d) immunological tolerance.

Ans: b
p. 364

_____ 7. A person who is born with a nonfunctioning thymus would have: (a) an autoimmune disease; (b) an immune deficiency disorder; (c) no lymphocytes; (d) T cells only.

Ans: d
p. 365

_____ 8. Activated T cells produce: (a) B cells and T cells; (b) plasma cells and memory cells; (c) plasma cells and effector cells; (d) effector cells and memory cells.

Ans: d
p. 367

_____ 9. Which of the following statements is <u>not</u> true of complement? (a) It is a protein. (b) It is not an immunoglobulin. (c) It does not increase in concentration following immunization. (d) It binds to free antigen.

Ans: a
p. 368

_____ 10. Specific proteins found on the surfaces of all the cells of the animal are coded for by the: (a) major histocompatibility complex; (b) ABO system; (c) immunoglobulin complex; (d) histone genes.

Ans: b
p. 370

_____ 11. A person of blood type AB forms which serum antibodies? (a) anti-A only; (b) none; (c) anti-A and anti-B; (d) anti-B only.

Ans: c
p. 370

_____ 12. The "universal donor" is identified as blood type: (a) A; (b) B; (c) O; (d) AB.

Ans: d
p. 370

_____ 13. Human anti-RH antibody is administered to an Rh^- mother within 72 hours following the birth of an Rh^+ child in order to: (a) neutralize fetal antibodies that crossed the placenta; (b) neutralize the mother's anti-RH antibodies; (c) destroy the mother's clone of lymphocytes that can make anti-RH antibodies; (d) destroy Rh^+ fetal red blood cells.

Ans: a
p. 372

_____ 14. A diffuse cancer (having no central location) would best be treated by: (a) chemotherapy; (b) respiratory therapy; (c) radiotherapy; (d) surgery.

Ans: b
p. 375

_____ 15. Inflammation is considered: (a) a specific defense response; (b) a nonspecific defense response; (c) an immune response; (d) part of antibody capacitation.

TEST SET 14-B

Name _____   Section _____

Choose the best answer to each of the following questions, and write the appropriate letter in the space provided.

Ans: a
p. 356

_____ 1. The humoral immune response acts primarily against: (a) bacteria; (b) cancer cells; (c) fungi; (d) intracellular viral infections.

Ans: c
p. 357

_____ 2. Which of the following are the predominant phagocytes? (a) macrophages and lymphocytes; (b) lymphocytes and neutrophils; (c) neutrophils and macrophages; (d) red blood cells.

Ans: b
p. 358

_____ 3. T cells develop their unique properties in the: (a) lymph nodes; (b) thymus gland; (c) thyroid gland; (d) bone marrow.

Ans: b
p. 358

_____ 4. Immunoglobulins recognize and bind to: (a) antibodies; (b) antigenic determinants; (c) whole antigens; (d) cell nuclei, ordinarily.

Ans: a
pp. 359,
    360

_____ 5. Vaccination against diphtheria and the development of a mild case of diphtheria are characterized by: (a) a time lag in the build-up of antibodies and activated lymphocytes; (b) immediate massive production of antibodies and activated lymphocytes; (c) entirely different immunological responses; (d) massive lymphocyte destruction.

Ans: d
p. 362

_____ 6. An immune response elicited against self proteins is: (a) impossible; (b) known as immunological tolerance; (c) clonal deletion; (d) an autoimmune disease.

Ans: c
p. 364

_____ 7. Which parts of common immunoglobulin molecules are virtually identical from one molecule to another? (a) the constant chains and the variable chains; (b) the variable regions; (c) the constant regions of both light and heavy chains; (d) the constant regions of the light chains only.

Ans: d
p. 365

_____ 8. Activated B cells produce: (a) B and T cells; (b) killer T cells and helper T cells; (c) T cells and plasma cells; (d) plasma cells and memory cells.

Ans: b
p. 365

_____ 9. A _____ cell must bind the antigen before a B cell can respond to the antigen. (a) killer T; (b) helper T; (c) different B cell; (d) complement.

Ans: a
p. 368

_____ 10. Transplants between individuals of different genetic constitution: (a) typically fail; (b) usually succeed; (c) have not been tried; (d) carry no risk for the recipient.

Ans: c
p. 370

_____ 11. A person of blood type O forms which serum antibodies? (a) anti-A only; (b) anti-B only; (c) anti-A and anti-B; (d) none.

Ans: d
p. 370

_____ 12. One's blood type is determined by: (a) white blood cell surface antibodies; (b) red blood cell surface antibodies; (c) white blood cell surface antigens; (d) red blood cell surface antigens.

Ans: b
p. 370

_____ 13. _____ can normally form anti-RH antibodies following the introduction of the appropriate antigen. (a) Rh^+ men; (b) Rh^- men; (c) Rh^+ women; (d) Rh^+ fetuses.

Ans: a
p. 371

_____ 14. The spreading of malignant cells to other parts on the body is known as: (a) metastasis; (b) benign intervention; (c) very uncommon; (d) chemotherapy.

Ans: c   _____   15.   Bacteria and fungi that typically live harmlessly on animal
p. 374   surfaces are <u>best</u> described as: (a) pathogens; (b) foreign
                                invaders; (c) normal flora; (d) toxic.

CHAPTER FIFTEEN
THE MULTICELLULAR WAY OF LIFE

CHAPTER SUMMARIES AND OBJECTIVES

15-1 <u>Advantages of Multicellularity</u> In truly multicellular organisms, individual cells differentiate, becoming better equipped to perform certain functions rather than having to do everything. Adjacent cells of the same type form tissues, cooperating tissues form organs, and cooperating organs form organ systems. Increased organization leads to increased efficiency of the whole organism.

Improved protection opened up different environments beyond aquatic habitats. Supporting tissues provide a stable location for other cells of complex shapes and functions. Multicellularity also allowed the development of larger organisms.

   1. Discuss the following concept: "Increased organization leads to increased efficiency."

15-2 <u>Effects of Scale</u> As they become larger without proportionate increases in sur- face area, individual cells lose considerably in the efficiency with which they can take up nutrients and eliminate wastes. The surface:volume ratio drops rapidly as the body becomes larger if there is no change in shape. Multicellu- larity is the most efficient means of increase in overall size. An increase in overall size must be accompanied by a proportionally greater increase in support tissue diameter. The science of biomechanics is the application of mechanical principles to the study of organisms. Size is an important factor in determining the performance of an organism as well as its overall structure and shape.

   2. Cite the basic principles explaining why a constant surface area typically would be less effective for exchanges as volume increases without basic changes in shape.

15-3 <u>Problems Associated with Multicellularity</u> New (and like) multicellular organisms must be reproduced--the pattern of development must be consistent each time. Specialized parts must develop for nutrition, a digestive system, and transport of materials to and from body cells. Gases must be exchanged with the environ- ment, and excretion must be taken care of. Larger organisms need structural support, and body parts must be coordinated in movement and function.

   3. Explain what problems are encountered when single cells (zygotes) give rise to whole multicellular organisms.

15-4 <u>Multicellular Fungi</u> Fungi all have filamentous structures (hyphae) collectively known as mycelia. The reproductive structures of fungi are called fruiting bodies and serve as the principal basis for fungal classification.

Slime molds are difficult to categorize as true fungi, protists, or separately in their two major groups, the plasmodial and the cellular slime molds.

4. Describe the general structure of multicellular fungi.

15-5 <u>Multicellular Plants</u> Multicellular plants are very diverse. The multicellular algae are virtually all aquatic and may be filamentous, two-dimensional, or three-dimensional in form. One line of land plants led to the bryophytes (such as mosses and liverworts), which do not have true leaves or true roots. Another line led to vascular plants, which have specialized xylem (water and mineral conduction) and phloem (manufactured food conduction). The ferns represent primitive vascular plants, and most have underground stems (rhizomes) from which true roots and fronds (leaves) arise.

Advanced vascular plants make seeds and do not require liquid water for fertilization (non-swimming sperm). One group of seed plants is the gymnosperms (such as pines and firs). The other group is comprised of the flowering plants (angiosperms). Part of the flower becomes the fruit. There are two main groups: the monocotyledons (monocots), which have one cotyledon (seed leaf), such as grasses, and the dicotyledons (dicots), which have two cotyledons, such as beans. Stems and leaves constitute the shoot system; and roots, with all their branches, make up the root system.

5. Describe the structural, functional, and reproductive diversity that exists within the algae, bryophytes, and vascular plants.

6. Explain how the reproductive strategies in plants changed in the transition from aquatic to dry environments.

15-6 <u>Animals</u> Sponges (phylum Porifera) are colonial associations of cells that may not have arisen from the same protist stock as truly multicellular animals. Primitive multicellular animals (Cnidaria) are represented by jellyfish, sea anemones, and hydroids. More advanced are the flatworms (Platyhelminthes) and roundworms (Nematoda). Segmented worms (Annelida), such as the earthworm, are quite successful. The phylum Mollusca is represented by snails, clams, and octopus.

The phylum Arthropoda is a very advanced, successful group; it includes insects, crabs, lobsters, mites, spiders, and centipedes. Adult echinoderms (Echinodermata) look simple but are an advanced group (examples include starfish, sea urchins, and sea cucumbers). The phylum Chordata includes a wide variety from humans to sea squirts and lancets. The most complex chordates are those with backbones (vertebrates). Classes of modern vertebrates include cartilaginous fishes, bony fishes, amphibians, reptiles, birds, and mammals.

7. Describe several representative animal groups, considering why terms such as <u>colonial</u>, <u>primitive</u>, and <u>advanced</u> are used to describe particular ones.

15-7 <u>Animal Structure</u> Each phylum is made up of organisms that have certain characteristics in common. <u>Cnidarians</u> have a mouth, digestive cavity, tentacles with stinging cells, a nerve net, and muscle fibers. <u>Annelids</u> have well-developed organ systems for many functions such as digestion, excretion, and locomotion. They typically have repeating, similar segments. <u>Mollusks</u> have a muscular foot and a mantle that secretes a shell in most representatives. Most have gills and a mouth rasping organ (radula). <u>Arthropods</u> have specialized, jointed appendages and chitinous exoskeletons that must be shed periodically for growth to occur.

Mammals have mammary glands for nursing their young, backbones encasing spinal cords (which have fluid-filled central canals), and a four-chambered heart in the area opposite the backbone. See Figure 19 for other organs of the human body.

8. Compare the body plans of each representative group described. For instance, describe some of the advances exhibited by arthropods over annelids.

15-8 <u>A Comparison of Animal Skeletal Systems</u> Annelid worms utilize muscles of the body wall and hydrostatic pressure to maintain their rounded body shape. Mollusks secrete hard shells that are primarily protective (and, for snails, supportive). Arthropods secrete chitinous exoskeletons with additional stiffening materials (except for joints) for added protection. Muscles attach to the exoskeleton. The exoskeleton does not grow but must be periodically molted, which leaves the soft body very vulnerable until the new exoskeleton is formed.

Endoskeletons are internally placed. Most vertebrate skeletons are bony (collagen plus calcium phosphate) and have articulating joints that allow movement in certain planes (muscles attach to the bones). Bony boxes protect certain internal organs, but not all soft tissues are protected by bone. Muscles tend to protect bones from fracturing. Bones are capable of growth.

9. Cite the basic problems of animals that are solved by skeletal formations.

15-9 <u>The Origin of Multicellularity</u> The transition from unicellularity to multicellularity occurred more than once. One step led to fungi. There may have been separate steps for bryophytes and vascular plants. Sponges are a separately derived group from true multicellular animals. Cnidarians may represent a separate line from other multicellular animals, and it is likely that other groups arose and became extinct.

There are two ways in which the transition is likely to have occurred. Once-independent unicellular organisms may have aggregated into colonies; then specialization occurred in the colonies and the specialized cells became unable to live independently. The second proposal is that a unicellular organism became multinucleate and divided into uninuclear cells that ultimately became differentiated.

10. Identify the two ways in which the transition from unicellularity to multicellularity is most likely to have occurred.

ARTICLE RESOURCES

<u>Advantages of Multicellularity</u>

<u>Effects of Scale</u>

Bakker, H. R., and A. R. Main. Condition, body composition, and total body water estimation in the Quokka, <u>Setonix brachyurus</u> (Macropodidae). <u>Australian Journal of Zoology</u>, 1980, 28(3), 395-406. A method for measuring the relationship between body weight and length is given. (The Quokka are wallabies.) Some of the difficulties involved in arriving at this method are discussed. Other methods for determining body weight from body measurements (such as snout-vent length in lizards and height, age, and sex tables in humans) are related. Statistical values are given.

<u>Problems Associated with Multicellularity</u>

<u>Multicellular Fungi</u>

Traub, Franz, Hans R. Hohl, and James C. Cavender. Cellular slime molds of Switzerland. II. Distribution in forest soils. <u>American Journal of Botany</u>, 1981, 68(2), 172-182. Large numbers of slime molds were found in the forest soils of Switzerland. Some are sole inhabitants of bog forests. Cellular slime mold diversity follows tree diversity. A soil supporting good bacterial growth is advantageous for slime molds.

<u>Multicellular Plants</u>

Gross, Katherine, L., and Judith D. Soule. Differences in biomass allocation to reproductive and vegetative differences of male and female plants of a dioecious, perennial herb, <u>Silene alba</u> (Miller) Krause. <u>American Journal of Botany</u>, 1981, 68(6), 801-807. Seed size was not a factor in differences between male and female plants. Male flower production was 3-4.2 times that of similar-sized females. The male flower, however, weighs less than half as much as the female flower. The female plant is larger than the male.

<u>Animals</u>

Animal Structure

Garcia Bellido, Antonio, Peter A. Lawrence, and Gines Movata. Compartments in animal development. _Scientific American_, 1979, 241(1), 102-110. The folding and thickening of two-dimensional sheets of cells is the fundamental way in which all animals develop. This gives rise to a question. How does the chromosomal information control the structure and organization of cell layers? The development of D. melanogaster is explored. There is enough information on the mouse to indicate that it, too, has compartments that generate large, well-defined regions of the adult.

Wolpert, Lewis. Pattern formation in biological development. _Scientific American_, 1978, 239(4), 154. Pattern formation and cell differentiation are two important processes in embryonic development. The position of the cell in the developing embryo may well determine how, and into what, it will develop. Pattern formation and regulation are similar in different systems. This apparently is not dependent on DNA.

Todd, James T., Leonard S. Mark, Robert E. Shaw, and John B. Pittenger. The perception of human growth. _Scientific American_, 1980, 242(2), 132. Mathematical analysis of human growth, especially craniofacial morphology. Examines the effects of stress on various tissues. Expands on the text. Useful for the instructor.

A Comparison of Animal Skeletal Systems

The Origin of Multicellularity

ESSAY QUESTIONS

1. Why is multicellularity the most efficient way to increase size?

    The surface:volume ratio drops rapidly as the body becomes larger. A cell would be severely limited in growth unless it were to increase its surface tremendously by many infoldings and outpocketings. The non-pocketed membrane could not function efficiently for nutrient, gas, and waste exchange between the cell interior and its environment. For instance, a ten-fold increase along one edge of a cube results in a thousand-fold increase in volume. Multicellularity provides the greatest surface area while allowing development of a larger organism.

2. Describe the anatomy of multicellular fungi.

    They all have multicellular filamentous structures--hyphae--that are collectively known as a mycelium. The fungal body may be mostly a collection of diffuse hyphae (bread mold) or it may develop elaborate reproductive structures called fruiting bodies (mushroom, puffball). All the reproductive structures are known as fruiting bodies.

3. Make one descriptive statement about each of the following plant groups: algae, bryophytes, primitive vascular plants, gymnosperms, and angiosperms.

    Many possible answers, including the following: Algae are mostly aquatic; bryophytes are small without true roots or leaves; primitive vascular plants do not make seeds; gymnosperms produce seeds without the assistance of flowers; and angiosperms are the flowering plants and produce seeds.

4. What were some of the significant structural advances of arthropods over annelids?

    Arthropods have specialized, jointed appendages; annelids' appendages are not specialized or jointed. The segments of arthropods exhibit some specialization; those of annelids are more similar and repetitive. Arthropods secrete chitinous exoskeletons; annelids depend on a "hydrostatic skeleton."

5. What are some of the key differences and similarities between the two ways in which multicellularity is most likely to have originated?

One assumes aggregation of cells into a colony; the other assumes the division of a multinucleate structure. Both then follow with cell specialization and loss, through differentiation, of the ability to live independently.

TEST SET 15-A

Name _____ Section _____

Choose the best answer to each of the following questions, and write the appropriate letter in the space provided.

Ans: b
p. 383

_____ 1. _____ allowed the development of large organisms. (a) The transition to dry land; (b) The development of multicellularity; (c) The development of unicellularity; (d) The development of non-swimming sperm.

Ans: a
p. 385

_____ 2. The surface:volume ratio _____ as the body becomes larger. (a) drops rapidly; (b) increases rapidly; (c) remains the same; (d) increases slowly.

Ans: d
p. 386

_____ 3. Any one cell of a multicellular organism must accomplish each of the following functions except: (a) to take in nutrients; (b) to carry out excretion; (c) to exchange gases with the cellular environment; (d) to reproduce the entire organism.

Ans: c
p. 386

_____ 4. The totality of the filaments comprising a fungus is known as: (a) the hypha; (b) the fruiting body; (c) the mycelium; (d) the cellular mode.

Ans: c
p. 388

_____ 5. _____ are vascular plants that do not make seeds. (a) Mosses; (b) Gymnosperms; (c) Ferns; (d) Grasses.

Ans: a
p. 388

_____ 6. Each of the following has xylem and phloem except: (a) liverworts; (b) ferns; (c) monocots; (d) gymnosperms.

Ans: d
p. 390

_____ 7. _____ are more colonial than truly multicellular. (a) Platyhelminthes; (b) Nematoda; (c) Cnidaria; (d) Porifera.

Ans: b
p. 393

_____ 8. Of the following, _____ are the most complex and advanced. (a) echinoderms; (b) vertebrates; (c) sea squirts; (d) roundworms.

Ans: c
p. 392

_____ 9. A starfish is a(an): (a) arthropod; (b) mollusk; (c) echinoderm; (d) chordate.

Ans: d
p. 393

_____ 10. Each of the following is characteristic of a jellyfish except: (a) It has a mouth. (b) It has tentacles with stinging cells. (c) It has a digestive cavity. (d) It is segmented.

Ans: a
p. 393

_____ 11. Each of the following is characteristic of an earthworm except: (a) It has a muscular foot and mantle. (b) It has a digestive system. (c) It has an excretory system. (d) It has repeating, similar segments.

Ans: b
p. 393

_____ 12. Each of the following is characteristic of a clam except: (a) It has gills. (b) It has stinging cells. (c) It has a muscular foot. (d) Its mantle secretes a shell.

Ans: c
p. 394

_____ 13. Each of the following is characteristic of arthropods except: (a) They molt. (b) They have specialized appendages. (c) They have five pairs of hearts. (d) They have jointed appendages.

Ans: b
pp. 394,
    396

_____ 14. In order for significant growth to occur in arthropods, the _____ must molt. (a) endoskeleton; (b) chitinous exoskeleton; (c) calcium carbonate-hardened bivalved shell; (d) tentacles.

Ans: a
p. 397

_____ 15. Multicellularity probably originated: (a) many times; (b) once only; (c) twice only; (d) most directly from viruses.

TEST SET 15-B

Name _____   Section _____

Choose the best answer to each of the following questions, and write the appropriate letter in the space provided.

Ans: d
pp. 382,
    383

_____   1.   Each of the following is true of differentiated cells except: (a) They can form large organized structures. (b) They specialize and perform certain functions efficiently. (c) They increase the efficiency of the organism as a whole. (d) They retain their embryonic totipotency.

Ans: b
p. 384

_____   2.   The volume of a cube 100 microns on an edge is _____ as(than) that of a cube 10 microns on an edge. (a) the same; (b) 1000 times greater; (c) 10 times greater; (d) 1,000,000 times smaller.

Ans: a
p. 386

_____   3.   A system that hydrolyzes complex foods to their component parts and absorbs them is best known as: (a) a digestive system; (b) a transport system; (c) a support system; (d) an excretory system.

Ans: b
p. 388

_____   4.   The plasmodium body form is characteristic of one subdivision of: (a) mushrooms; (b) slime molds; (c) cellular molds; (d) hyphae.

Ans: c
p. 388

_____   5.   Of the following, the only group that has true roots or leaves is the: (a) algae; (b) bryophytes; (c) ferns; (d) mosses.

Ans: a
p. 389

_____   6.   Seed coats form from flower parts in: (a) angiosperms; (b) gymnosperms; (c) bryophytes; (d) algae.

Ans: c
p. 390

_____   7.   Of the following, the phylum _____ is the most primitive group. (a) Chordata; (b) Arthropoda; (c) Porifera; (d) Cnidaria.

Ans: d
p. 392

_____   8.   A snail is a(an): (a) arthropod; (b) echinoderm; (c) chordate; (d) mollusk.

Ans: a
p. 393

_____   9.   Backbones are characteristic of each of the following except: (a) crabs; (b) bony fishes; (c) reptiles; (d) mammals.

Ans: b
p. 393

_____   10.   Annelids have: (a) stinging cells; (b) similar, repeating segments; (c) jointed appendages; (d) a muscular foot and a mantle.

Ans: c
pp. 392,
    394

_____   11.   Chitinous exoskeletons are characteristic of: (a) clams; (b) earthworms; (c) insects; (d) humans.

Ans: a
p. 394

_____   12.   Most _____ have a mouth rasping organ (radula). (a) mollusks; (b) annelids; (c) cnidarians; (d) mammals.

Ans: b
p. 393

_____   13.   A characteristic that humans share with snakes is: (a) a mantle; (b) a backbone; (c) jointed appendages; (d) functioning gills.

Ans: d
p. 396

_____   14.   The supporting framework of an earthworm may be best known as: (a) an endoskeleton; (b) an exoskeleton; (c) a muscle mantle; (d) a "hydrostatic skeleton."

Ans: c
p. 397

_____   15.   Each of the following is thought to be true of the origin of multicellularity from unicellularity except: (a) It occurred in the origin of fungi. (b) It occurred in the origin of sponges. (c) All groups in which it occurred exist today. (d) There may have been separate transition steps for bryophytes and vascular plants.

CHAPTER SIXTEEN
PLANT DEVELOPMENT

CHAPTER SUMMARIES AND OBJECTIVES

16-1 <u>Embryo Formation in Flowering Plants / Seed Germination</u> The embryo sac in the pistil contains several nuclei; one is the egg nucleus. Fertilization occurs here; the zygote, sac, and surrounding tissue give rise to the seed (which contains the embryo). The first mitotic division of the zygote gives rise to two cells: one becomes the suspensor, the other the embryo proper. The suspensor becomes filamentous and, at first, the embryo is a ball of cells. The covering layer (protoderm) will give rise to the epidermis. In dicots, the formation of cotyledons and axis elongation cause the embryo to resemble a torpedo externally. Internally, differentiation begins. The region below the cotyledons is the hypocotyl. The region between the cotyledons (above the hypocotyl) is the shoot apex; the opposite end is the root apex. Much mitosis occurs in the apical regions.

Seed germination begins with imbibition of much water. The seed expands and certain metabolic changes occur: preformed enzymes are activated, RNA and new enzymes are synthesized, cell respiration rate increases, and other metabolic pathways become activated. DNA synthesis coincides with the later emergence of the radicle through the seed coat.

Until photosynthesis is possible, food reserves from the endosperm are used--taken in through the cotyledons. The cotyledons may have absorbed all the reserves and be quite plump (peanuts) or they may remain thin and draw on endosperm reserves as needed (castor bean). Many plants store reserves in seeds as fats or oils; others store starch. The seed contains amino acid reserves and nucleotide polymers. These reserves must be digested so that the breakdown products may enter the cotyledons.

1. Describe early plant embryonic growth from zygote to torpedo stage.

2. Describe what happens to promote growth after the seed takes up the water of imbibition.

3. Cite the relationship between endosperm and cotyledons.

16-2 <u>Gibberellins</u> Embryonic activity is accompanied by secretion of gibberellins. These diffuse from the embryo through the endosperm to the aleurone layer of grass seeds. A crucial series of events is triggered and aleurone grains break down yielding amino acids. The amino acids are used to assemble digestive enzymes, such as amylases, proteases, and ribonucleases). These and other enzymes return to the endosperm to catalyze hydrolysis of reserve polymers for embryonic use.

A growth-promoting substance (gibberellin) was described in 1925 as the cause of "foolish seedling" disease. It was produced by a fungus and caused excessive seedling growth--they are spindly and die before producing seed. It was later found that there was more than one gibberellin and that higher plants also produce them. Mutant dwarf corn plants are apparently unable to produce their own gibberellins. When gibberellins are applied to normal (tall) plants, there is virtually no effect; treated dwarf plants grow as tall as their normal relatives.

Gibberellins also cause fruit growth from unfertilized flowers, increase the rate of fruit growth, and promote seed germination in some species. Plant growth substances typically have multiple effects within the plant.

4. Describe the role of gibberellins in seed germination.

5. Describe the genetic insight afforded by the normal growth of dwarf corn plants when gibberellin was added.

16-3 Seed Dormancy / Phytochrome Some seeds can germinate as soon as they are produced, but many are initially dormant (which may last from weeks to decades). The mechanisms of dormancy are numerous and diverse. (1) An impermeable seed coat may prevent the entry of water or oxygen; abrasion of the seed coat must occur. (2) The seed coat may merely mechanically restrain the embryo. Softening of the seed coat must occur, as, perhaps, by soil microorganisms. (3) Chemical inhibitors of germination such as abscisic acid may be involved. Increasing concentrations of growth promoters may overcome the inhibitor, or it may be leached out or changed by scorching. (4) A prolonged period of cold temperatures (freezing) may be required to break dormancy. This may decrease the content of germination inhibitors. (5) Some seeds remain dormant until they have dried extensively (tomato, lima bean). Dormancy may ensure germination at the right place and time.

Some seeds must be exposed to light to germinate. Blue light or red light is most effective to induce germination; far-red light reverses the effect of these blue and red light bands. If different wavelengths of light are administered in close succession, the seeds respond to the final exposure only. This phenomenon is due to light effects on a bluish pigment known as phytochrome. The red-light-absorbing form (P_r) is converted to P_{fr} upon absorbing one photon of red light. P_r is stable in darkness, but P_{fr} gradually disappears once formed. If red light exposure occurs, germination occurs; far-red light continues dormancy. P_{fr} apparently also affects overall plant form, flowering, and leaf movement.

6. List several mechanisms of dormancy.

7. Expand on the idea that dormancy may ensure germination at the right place and time.

8. Describe the relationship between red and far-red phytochrome.

16-4 Early Seedling Growth The radicle (embryonic root) escapes the seed coat first; later the shoot must escape the coat. Germination typically occurs in the dark in soil, and the plant must reach the surface to begin photosynthesis. Etiolated seedlings of flowering plants exhibit several adaptive features that help in this process. (1) They do not use resources to make chlorophyll. (2) The shoot elongates rapidly: a coleoptile protects the delicate tip of grasses as it elongates; the hypocotyl of most dicots elongates, carrying the cotyledons to the surface; the epicotyl of other dicots (pea) elongates, leaving the cotyledons in the soil. (3) Dicot seedlings form hooks at the shoot apex (which protect them). (4) No leaf expansion occurs.

Exposure to light converts red-absorbing phytochrome (P_r) to P_{fr}, which in turn initiates chlorophyll synthesis, slows shoot elongation, straightens the apex hook, and begins leaf expansion.

9. Explain what adaptations etiolated seedlings exhibit which help them reach the soil surface to begin photosynthesis.

10. Describe the effects of light on red-absorbing phytochrome (P_r) when the seedling first becomes self-established.

16-5 <u>Auxin / Auxin and Cell Walls</u> Charles Darwin and his son Francis investigated the effects of auxin (indoleacetic acid) using canary grass seedlings. One of the things they studied was phototropism. Other investigators found that the "eye" of the plant was, in fact, a chemical message. Frits Went isolated the growth substance from oat coleoptiles in 1926. Auxin moves primarily from the biological growing tip toward the biological base--polar movement. Should light come from one side only, however, auxin at the stem-growing tip moves toward the shaded side and then down the axis in a polar fashion, causing more elongation on that side. Thus it results in a coleoptile or petiole bending toward the light (phototropism). Should a plant be tipped on its side, more auxin is transported to the lower side and causes elongation upward (geotropism) against gravity. The actual bending occurs a few millimeters below the tip.

In many species, the rooting of cuttings can be sharply improved by dipping the cut surfaces into an auxin solution. Abscission of leaves in nature is determined in part by the transport of auxin, which is produced in the blade, through the petiole; abscission is delayed by auxin. Apical dominance is maintained by auxin. If the stem apex is removed, lateral buds are released from inhibition and develop.

In higher plants, the principal strengthening component of the cell wall is cellulose. Cellulose molecules tend to associate with others, forming crystalline regions (micelles). Bundles of cellulose molecules (including many micelles) constitute microfibrils. Rigidity of the wall is based on a network of cellulose microfibrils embedded in a jelly-like matrix of other, smaller polysaccharides. For the cell to expand, this wall must loosen. The predominant transverse weave of most of the microfibrils allows expansion primarily along the long axis of the cell. Eventually, longitudinal molecular reorientation limits this expansion pattern, too.

If cell walls are bent but snap back, they are elastic. If the bending is prolonged and molecular reorientation of cellulose occurs to prevent total unbending, this is plasticity. Auxin increases plasticity. There is some evidence that the weave of cell walls may be loosened by auxin to allow plastic bending. Apparently auxin-induced hydrogen ion secretion is related to growth; the wall-loosening factor seems to be nothing but hydrogen ions. The H^+ ions may activate some enzyme that digests certain cellulose linkages. Gibberellins, too, can loosen the cell wall.

11. Trace the experiments that led to the discovery of auxin and its mode of action.

12. Cell elongation is a primary effect of auxin. Describe what happens in stem elongation, phototropism, geotropism, and cell wall molecular loosening.

13. Describe the relationship between auxin and (a) the rooting of cuttings, (b) abscission, and (c) apical dominance.

16-6 <u>The Growing Regions of the Plant</u> Plants grow in meristem regions instead of all over. Shoot apical meristems extend the stem lengths and form buds at their flanks, into which are built leaf and lateral bud primordia. Roots grow at root apical meristems. Woody plants grow in diameter by cell division in vascular cambium. Cork cambium forms bark, which becomes impregnated with suberin and protects the plant from drying.

14. List the various meristematic areas of plant growth.

16-7 <u>Differentiation and Organ Formation / Cytokinins</u> Pith, an undifferentiated plant tissue, is easily grown in tissue culture. Notching the culture growth and inserting a stem or applying auxin and coconut milk (which is a rich source of growth substances) causes tissue differentiation (conductive cells) below the notch. Tobacco pith cultures are stimulated by auxins to form roots and by cytokinins to form buds and shoots.

Cytokinins are growth substances that (1) stimulate the formation of buds and shoots, (2) promote cell division in cultured tissues, (3) cause germination of some seeds that ordinarily germinate only after exposure to light, and (4) cause lateral swelling of stems and roots (especially the fleshy roots of radishes and some other plants). Cytokinins are formed in roots and transported to other parts of the plant.

15. Explain the role of auxins and cytokinins in differentiation.

16. A primary function of cytokinins is promoting cell division. Describe how this role is related to bud formation, germination, and growth of fleshy roots.

16-8 Bud Dormancy and Abscisic Acid   Perennial plants of the temperate zones are adapted for minimizing the damage caused by rigorous winters. The onset and termination of winter dormancy are positive events in the life of the plant. Deciduous tree buds form bud scales that are covered with wax and may form cottony modified leaves in winter bud development that act as insulation. The leaves of these trees fall off, and the scars are sealed with a corky layer. Trunk growth ceases and the freezing point of sap is lowered.

The onset of winter dormancy is cued by night length. Leaves carry out a response to night length (photoperiodism). The growth substance, abscisic acid, is produced; it inhibits stem elongation and is present in elevated concentration in dormant buds and some dormant seeds.

The breaking of winter dormancy generally does not occur until the buds have undergone a period of winter cold. This leads to a decrease in abscisic acid and an increase in growth promoters (particularly gibberellins).

17. Cite some advantages to the plant of winter dormancy and leaf abscission.

18. Describe the roles of night length and chilling in the onset and breaking of winter dormancy.

16-9 Senescence / Growth Substances and Senescence   Senescence is deteriorative changes with aging. It is illustrated by leaf deterioration and abscission in the autumn and by death after flowering and setting fruit. Before the leaves die and are shed, their proteins are hydrolyzed to yield amino acids, which are stored in the stems. Death after flowering and fruit set in some species allows the plant to have put all its energy into seeds to produce more offspring. This is an adaptation for species survival.

Roots produce cytokinins, which appear to delay senescence of leaves. They cause a redistribution of biologically active materials from one part of a plant to another. Treatment of a leaf with cytokinin results in movement of materials into the treated leaf.

The gas ethylene, produced by all parts of the plant, promotes senescence. Auxin delays leaf abscission; ethylene promotes it. Ethylene promotes fruit ripening, and it is active in other stages of plant growth. It inhibits stem elongation in general and specifically maintains the apical hook of dicot seedlings through an asymmetric production of the gas, inhibiting elongation on the inner surface of the hook. Ethylene promotes lateral swelling of stems (as do cytokinins) and causes stems to lose their sensitivity to geotropic stimulation.

19. Describe the conservation of resources before leaf death and abscission.

20. Explain why ethylene has been called the "senescence hormone."

ARTICLE RESOURCES

Embryo Formation in Flowering Plants

Seed Germination

Egli, D. B., Joanna Fraser, J. E. Leggett, and C. G. Poneliet. Control of seed growth in Soya beans. Annals of Botany, 1981, 48(2), 171-176. Since the seed is the primary source of photosynthesis during reproductive growth, it is necessary to understand the mechanisms controlling the rate of seed growth. Growth rate correlates highly with seed size. Cell numbers per seed cotyledons also correlated with seed size.

Gibberellins

Seed Dormancy

Thoradgill, Paul F., Jerry M. Baskin, and Carol C. Baskin. Dormancy in seeds of Frasera caroliniensis (Gentianeae). American Journal of Botany, 1981, 68(1), 80-86. In the seeds of this plant the embryos are under-developed and dormant. They were chilled at 5°C for 205 days; 87% of the seeds germinated. A discussion of morphological and physiological dormancy is given.

Phytochrome

Early Seedling Growth

Auxin

Hayes, Alice B., and James A. Lippencott. The timing of and effect of temperature on auxin-induced hypnostic curvature of the bean primary leaf bud. American Journal of Botany, 1981, 68(3), 305-311. It is important that the growth of cells in the upper portion of the leaf blade must be correlated with the cells of the lower portion so that a planar structure may be maintained during this period of active expansion. The regulation of pinto bean leaf planar growth as affected by time, removal of auxin, second auxin applications, and temperature are discussed.

Auxin and Cell Walls

The Growing Regions of the Plant

Erickson, Ralph O., and Wendy Kuhn Silk. The kinematics of plant growth. Scientific American, 1980, 242(5), 134-151. Plant roots, stems, and leaves all exhibit growth rates that are analogous to the mathematical models found in fluid dynamics. Growth rates within plant organs, specifically spatial variation, are difficult to find in plant literature. Hence this well-illustrated article is important.

Differentiation and Organ Formation

Cytokinins

Davey, J. E., and J. Van Staden. Cytokinins in spinach chloroplasts. Annals of Botany, 1981, 48(2), 243-246. Cytokinins delay senescence in detached leaves kept in the dark. This is an indication that cytokinins have an effect on chloroplast metabolism. They also affect the maturation and replication of plastids. When applied in the light, cytokinins reduce chlorophyll degradation and promote chloroplast protein synthesis. Chloroplasts were extracted from spinach leaves kept in the light and from spinach leaves kept in the dark. Cytokinins were found in the chloroplasts of the leaves kept in the light, indicating that cytokinin function and light are closely correlated.

V. Jaiswal, S. J. H. Rizui, D. Mukerji, and S. N. Mathur. Cytokinins in root nodules of Phaseolus mungo. Annals of Botany, 1981, 48(3), 301-305. For the instructor, a discussion of how to extract cytokinins from root nodules.

Bud Dormancy and Abscisic Acid

Senescence

Growth Substances and Senescence

Dyche, Kevin E. Allelopathy--Chemical warfare in the plant kingdom. The Science Teacher, 1978, 45(6), 40-41. Discusses allelopathy--the inhibitor of plant growth, population, or behavior via chemicals from another plant. Walnut, mint, and sage are three common sources of allelopathic chemicals.

ESSAY QUESTIONS

1. Describe the three component processes involved in the development of the plant from zygote to mature plant.

    Cell division is necessary to achieve multicellularity. Cell expansion is necessary for growth in size. Differentiation is the modification and specialization of cells, resulting in particular structures and enabling cells to perform specific functions.

2. What role do cotyledons play in early plant growth?

    They are part of the embryo and are the gateway into the growing embryo for nutrients from the endosperm. They are known as "seed leaves" and may become the first photosynthesizing structures of the seedling if the hypocotyl expands, carrying them to the soil surface, as occurs in many dicots.

3. List the meristems of plants and describe their predominant activities.

    Shoot apical meristems extend the lengths of stems and on their flanks form lateral bud and leaf primordia. Root apical meristems extend the lengths of of roots. Cork cambium produces new cells primarily in the outward direction; they become impregnated with suberin. This outer bark protects against excessive drying in some climates. Vascular cambium is a lateral meristematic region, forming new wood from the inner cells and phloem from the outer cells.

4. How does leaf abscission benefit deciduous trees of the temperate zone?

    When branches are brittle from freezing temperatures, they would be more likely to break in wind if leaves were present. Prior to abscission, nutrients are moved into the trunks and branches and conserved. Leaves would freeze and become nonfunctional anyway. Leaf drop prevents much water loss during winter.

5. How might a gardener most effectively induce a new hedge to branch more in order to fill in open spots?

    Trimming the growing tips regularly should release the lateral buds from apical dominance, as controlled by the production of auxin in the tips. Lateral bud development provides new branches that also begin producing auxin, so trimming must be repeated.

TEST SET 16-A

Name _____ Section _____

Choose the <u>best</u> answer to each of the following questions, and write the appropriate letter in the space provided.

Ans: c
p. 401

_____ 1. The first division of the zygote produces two cells; one will become the embryo proper, the other the: (a) endosperm; (b) cotyledons; (c) suspensor; (d) proembryo.

Ans: b
p. 402

_____ 2. Endosperm is used as a source of food reserves until: (a) germination begins; (b) photosynthesis begins; (c) fruit set; (d) the suspensor forms.

Ans: a
p. 403

_____ 3. Nucleotides are required by the growing embryo to synthesize: (a) RNA and DNA; (b) amino acids; (c) fats; (d) glucose.

Ans: d
p. 406

_____ 4. Dwarf corn plants are apparently unable to make: (a) nucleotides; (b) proteases; (c) auxin; (d) gibberellin.

Ans: d
pp. 403,
404

_____ 5. Amino acids available from the breakdown of aleurone grains are used to assemble: (a) ribosomes; (b) nucleotides; (c) gibberellins; (d) digestive enzymes.

Ans: b
pp. 406,
407

_____ 6. In some seeds, the level of abscisic acid does not decline during germination; the strategy employed to allow germination is to overcome the abscisic acid through gradually increasing the concentrations of: (a) growth inhibitors; (b) growth promoters; (c) nucleotides; (d) amylases.

Ans: a
p. 408

_____ 7. The form of phytochrome that absorbs red light is: (a) P_r; (b) P_{fr}; (c) P_{lac}; (d) P_i.

Ans: c
p. 408

_____ 8. The _____ escape(s) the seed coat first. (a) hypocotyl; (b) cotyledons; (c) radicle; (d) shoot apex.

Ans: c
pp. 409,
410

_____ 9. Each of the following statements is true of etiolated seedlings in soil <u>except</u>: (a) The shoot elongates rapidly. (b) Dicot seedlings have a hook at the apex. (c) The shoot is green. (d) No leaf expansion occurs.

Ans: b
p. 411

_____ 10. Frits Went is credited with isolating: (a) cytokinins; (b) auxin; (c) gibberellin; (d) ethylene.

Ans: a
pp. 415,
416

_____ 11. Cellulose wall molecular loosening is known to be caused by gibberellins and: (a) auxin; (b) growth inhibitors; (c) hydroxide ions; (d) chloride ions.

Ans: d
p. 416

_____ 12. Stem lengths are extended by cell division in the: (a) root apical meristem; (b) cork cambium; (c) vascular cambium; (d) shoot apical meristem.

Ans: d
p. 418

_____ 13. Each of the following statements is true of cytokinins except: (a) They stimulate the formation of buds and shoots. (b) They promote cell division. (c) They cause lateral swelling of stems and roots. (d) They prevent germination in the presence of red light.

Ans: a
p. 419

_____ 14. The onset of winter dormancy is cued by: (a) night length; (b) a slow temperature decline; (c) atmospheric CO_2; (d) the first frost.

Ans: c   _____   15.   Fruit ripening (as in apples) is promoted by: (a) short nights;
p. 421   (b) increased rainfall; (c) ethylene gas; (d) cytokinins.

TEST SET 16-B

Name _____   Section _____

Choose the best answer to each of the following questions, and write the appropriate letter in the space provided.

Ans: b
p. 401

_____ 1. Epidermis develops from the: (a) suspensor; (b) protoderm; (c) interior of the proembryo; (d) endosperm.

Ans: d
pp. 402,
   403

_____ 2. More energy is available to the seed if food reserves are stored as: (a) starch; (b) glucose; (c) proteases; (d) fats or oils.

Ans: a
p. 403

_____ 3. The growing embryo requires _____ as building materials for its myriad proteins. (a) amino acids; (b) starches; (c) lipids; (d) nucleotides.

Ans: c
p. 404

_____ 4. Before complex molecules of food reserves can be transported into the growing embryo, they must be: (a) changed to aleurone grains; (b) converted to gibberellin; (c) hydrolyzed; (d) reconstituted by condensation.

Ans: c
p. 404

_____ 5. The gibberellin that caused "foolish seedling" disease of rice was produced by: (a) algae. (b) the rice seedlings. (c) a fungus. (d) Penicillium notatum.

Ans: a
p. 406

_____ 6. The most common chemical inhibitor of seed germination is: (a) abscisic acid; (b) gibberellin; (c) auxin; (d) ethylene.

Ans: b
pp. 407,
   408

_____ 7. A seed that requires light to germinate is exposed to red light for five minutes, then exposed to far-red light for five minutes, and then returned to darkness. (a) The seed will germinate. (b) The seed will remain dormant. (c) The embryo will undergo senescence. (d) The seed coat will crack.

Ans: d
pp. 409,
   410

_____ 8. Etiolated seedlings (in soil): (a) form chlorophyll; (b) of grasses form a hook at the apex; (c) are inhibited from elongation; (d) do not expand their leaves.

Ans: a
p. 409

_____ 9. When cotyledons are carried up past the soil surface, elongation of the _____ causes it. (a) hypocotyl; (b) epicotyl; (c) radicle; (d) coleoptile.

Ans: c
p. 410

_____ 10. Another name for auxin is: (a) abscisic acid; (b) stress hormone; (c) indoleacetic acid; (d) senescence hormone.

Ans: d
pp. 412,
   413

_____ 11. Auxin delays leaf abscission and maintains: (a) the hook of etiolated seedlings; (b) continuous fruit ripening; (c) continuous lateral bud development; (d) apical dominance.

Ans: b
p. 417

_____ 12. In lateral growth, new wood is produced by cell division in: (a) cork cambium; (b) vascular cambium; (c) shoot apical meristems; (d) root apical meristems.

Ans: a
p. 418

_____ 13. Cytokinins are formed in _____ and transported to other parts of the plant. (a) roots; (b) leaf tips; (c) petioles; (d) flowers.

Ans: d
pp. 419,
   420

_____ 14. Dormant winter buds are characterized by each of the following except: (a) Bud scales form. (b) They are coated with a waxy material. (c) They have an elevated concentration of abscisic acid. (d) There is an increase in gibberellin concentration.

Ans: b
p. 420

_____ 15. Deciduous tree leaf senescence and abscission in the temperate zones are considered: (a) liabilities; (b) adaptive in nature; (c) not useful but not harmful either; (d) infrequent events.

CHAPTER SEVENTEEN
REPRODUCTION IN FUNGI AND PLANTS

CHAPTER SUMMARIES AND OBJECTIVES

17-1 <u>Asexual and Sexual Reproduction in Fungi and Plants</u> The least complicated way to reproduce is asexually. Making spores containing at least one full complement of DNA and packaging them for survival under diverse conditions is a primary means of asexual reproduction. Plants employ several types of vegetative reproduction: horizontal stems (runners), underground stems (rhizomes), modified fleshy leaf bases (bulbs), and modified stems (corms) can all root and give rise to new plants.

Asexual reproduction leads to offspring genetically identical to the parent cell. Sexual reproduction provides for genetic diversity through recombination. Most higher organisms reproduce sexually; most plants and all fungi reproduce asexually as well.

Some plant varieties can reproduce asexually only. The navel orange has arisen only once in history, apparently, in a plantation on the Brazilian coast. One seed gave rise to one tree that had aberrant flowers; seedless fruit were formed. Every navel orange in the world comes from a tree derived by bud grafting from another, and so on, back to this original Brazilian tree. Strawberries can reproduce either sexually or asexually but, once a genetic strain is developed for a desired fruit, asexual reproduction via runners is typically employed. Annual crops (such as grain crops) usually employ sexual reproduction and are grown from seed.

1. Describe the advantages of sexual reproduction and those of asexual reproduction.

2. Cite the kinds of plants that can reproduce by asexual means only.

17-2 <u>Alternation of Generations</u> Alternation of generations is a life cycle pattern in which a multicellular sporophyte (2n) form is followed by a multicellular gametophyte (n) form. Sporophyte and gametophyte may look alike (isomorphic, as in <u>Ulva</u>, the sea lettuce) or look different (heteromorphic, as in fungi and many algae). Meiosis occurs in sporocytes of the sporophytes to produce haploid spores. The spore is the first cell of the gametophyte which later produces gametes by mitosis. The gametes may all look alike (isogamous, as in some algae and fungi), or the gametes of each mating type may look different (anisogamous, as in <u>Ulva</u>). The fusion of two gametes (one from each of two mating types) produces a diploid zygote (the first cell of the sporophyte). The intermediate life cycle just described may be contrasted with a haplontic or to a diplontic life cycle. In the haplontic cycle, the only diploid cell is the zygote, which immediately undergoes meiosis to produce haploid spores (these become multicellular gametophytes, as in <u>Ulothrix</u>, a filamentous alga). In the diplontic life cycle, meiosis of sporocytes directly produces haploid gametes which fuse to form 2n zygotes (these become multicellular sporophytes, as in some algae).

The anisogamous gametes of Ulva can also function as zoospores, giving rise to gametophytes when they germinate without undergoing fertilization. This is not so common, however, as the direct production of zoospores by mitosis in either gametophyte or sporophyte (asexual reproduction of many fungi and algae).

3.   Describe alternation of generations.

4.   Explain the genetic effect in terms of the numbers of sets of chromosomes on the multicellular adult form (a) when meiosis occurs in the zygote, and (b) when meiosis produces gametes directly.

17-3   Heterokaryon Formation in the Fungi   The life cycle of Neurospora crassa typifies heterokaryon (different nuclei) formation of advanced fungi. Each gametophyte mycelium of Neurospora is of one mating type or another. The tips of hyphae of compatible mating types touch, enzymes digest the cell walls, and cytoplasm plus nuclei (n) from one type invade the other. The nuclei divide mitotically; the heterokaryon filament develops into a heterokaryon mycelium which forms fruiting structures. The fruiting structures contain elongated sacs (asci) into which two nuclei (one from each parental mating type) have been segregated. Prior to fusion of these two nuclei, the cells of the heterokaryon mycelium are n + n instead of 2n or n. After the zygote forms, meiosis occurs in that diploid nucleus (the only truly diploid cell in the cycle). The four haploid products divide mitotically to form eight spores per ascus.

Heterokaryosis is considered a significant genetic peculiarity of the fungi. Also, fungi have developed a mode of nuclear exchange in sexual reproduction that does not require liquid water for gamete transport.

5.   Define heterokaryon.

6.   Describe heterokaryon formation in Neurospora crassa.

7.   Explain the significance of sexual reproduction without the need of liquid water for gamete transport.

17-4   Multiple Hosts in the Fungal Life Cycle   Symbiotic fungi--particularly parasitic ones--may be very specific as to host organisms. Some use different hosts for different stages of the life cycle. The wheat rust fungus, Puccinia graminis, illustrates the utilization of two different hosts, the extent of heterokaryosis, and the sheer complexity of some fungal life cycles.

Summer heterokaryon hyphae form orange uredospores in wheat, each of which contains two nuclei--one each derived from each parental mating type (+ and -). These infect other wheat plants throughout the summer. In late summer, dark brown teliospores form, which constitute the over-wintering stage. Each teliospore contains two truly diploid cells, each diploid nucleus being derived from fusion of + and - nuclei (they are not heterokaryons). In the spring, each cell of the teliospore develops into a reproductive structure, each producing four basidiospores by meiosis (+ and - spores are formed). Wind dissemination gets these spores to the intermediate host, the barberry bush. The + or - basidiospores germinate into + or - hyphae. Flasks develop on the upper leaf surface in which + or - pycniospores are formed. Then + pycniospores fuse with - hyphae in the flasks after being transferred from one flask to another by insects. The hyphae extend through the leaf and form aecia on the lower leaf surface. The nucleus from the pycniospore divides repeatedly. In the aecia, heterokaryon cells form once more (+ and - nuclei); these become aeciospores which are wind-borne to wheat plants. When these spores germinate, they produce the summer hyphae. Wheat rust may be controlled by removal of the barberry plants or, as in more recent years, by development of rust-resistant strains of wheat.

8.   Explain the value of a detailed description of the life cycle of Puccinia graminis toward an understanding of symbiotic fungi.

9.   Give two logical reasons for the use of methods of controlling wheat rust that have been successful.

17-5 <u>Mosses and Ferns</u> Mosses and vascular plants exhibit adaptations for life on land and a progressive independence from a watery environment. Mosses have a predominant gametophyte--the leafy, small green plants that form mats of growth. At the tops of the upright gametophytes, antheridia produce sperm or archegonia produce eggs. These multicellular reproductive structures illustrate a trend toward protecting developing eggs and sperm. Sperm must swim to the eggs, but water droplets from dew or rain are sufficient. The zygote develops into a sporophyte which is totally dependent on the gametophyte. It remains attached by its foot to the female gametophyte, and a stalk supports a capsule in which sporophytes undergo meiosis, producing haploid spores.

Ferns are primitive plants that possess true vascular tissue (xylem and phloem). They, too, exhibit alternation of generations and swimming sperm, but the sporophyte is the predominant body form. They produce sporangia on the underside of frond leaflets where haploid spores are produced. The gametophytes are very small, inconspicuous, and independent. The fertilized egg begins development in the archegonium, but the development of roots allows sporophyte independence. Fern sporophytes can be very large and long-lived.

10. Compare mosses and ferns, and explain why ferns can grow larger than mosses.

17-6 <u>The Seed Plants / The Flower</u> The most advanced plants produce seeds: gymnosperms (pines) and angiosperms (flowering plants). The alternation of generations in these plants is characterized by dominant sporophyte forms and dependent gametophytes.

Modified leaves become the sporangia of seed plants, forming male or female sporocytes in cones or flowers. Sporangia are separate for the production of male or female spores. Following meiosis, the male gametophyte from each spore at least begins development in the sporangium, forming pollen grains. The female gametophyte develops completely within the sporangium.

The most typical pattern of female gametophyte development in angiosperms is as follows: Only one of the products of meiosis functions. It divides mitotically three times, forming eight nuclei. This is the mature female gametophyte. One nucleus becomes the functional egg nucleus and will be fertilized by a sperm nucleus; two nuclei will fuse with a second sperm and give rise to the triploid endosperm--the food reserves of the seed. Seed coats develop around the female gametophyte from surrounding sporophyte (flower) tissue. The zygote begins early embryonic development as the seed is forming. By the time the seed is ready to be shed, it typically becomes dormant and remains so until germination.

Pollen grains develop in the male structures of the flower. When one lands on a female sporangium, a pollen tube grows toward the female gametophyte, digesting sporophyte tissue in its pathway. One of the pollen nuclei will divide to form two sperm nuclei, which take part in the double fertilization just described.

Seeds are much sturdier than one-celled spores. They provide much protection for the enclosed embryo. The possession of seeds is a major reason for the enormous evolutionary success of seed plants.

Flower parts include sepals (outermost ring), petals (inside the sepals), stamens (inside the petals), and pistil (center of the flower). Most flowers have both pistil and stamens; some have separate flowers for pistils and stamens. Many flowers do not have sepals or petals. Stamens (male sex organs) have a supportive filament and an anther where pollen is made. Pistils have a stigma (catches pollen), style, and ovary. The pistil forms from the fusion of modified leaves (carpels).

11. Describe the change in gametophyte/sporophyte dependence and/or independence from ferns to flowering plants.

12. Explain how gametophytes are protected during development in seed plants.

13. Cite the survival advantages to species that form seeds over those plants that form spores only.

17-7 <u>Pollination</u> The mechanisms of pollen transport are legion. Self-pollination occurs prior to the opening out of the flower parts. Outcrossing (cross-pollination) is characteristic of most angiosperms and gymnosperms. Wind pollination occurs typically in those plants with sticky or feathery receptive surfaces on their pistils. Animal pollinators have often co-evolved with the plants they pollinate, benefitting from the nectar or pollen. Hummingbirds are often associated with tubular, large, bright, odorless flowers that produce much nectar. Bats pollinate flowers that open at night and have characteristic odors. Odors are generally important to insect-pollinated flowers.

14. Describe the relationship between flower and pollinator.

17-8 <u>Fruit</u> Fruits arise from floral parts; a fruit may consist of a mature ovary and its seeds, or it may include other flower parts of structures closely related to it. Many strategies have evolved for seed dispersal, many of them related to fruits. Buoyant fruits may be wind dispersed (maple samaras, dandelions); sticky or barbed fruits hitch animal rides easily; fleshy fruits tend to be eaten and their seeds dropped elsewhere with feces.

Fruit development typically occurs after fertilization, so pollination is important to fruit formation. The pollen tube is aided in finding the female gametophyte by growth substances produced in the pistil and by an attractant released from the ovary. Pollen tube growth triggers pistil growth in some species.

Fruit growth involves auxin and gibberellins. Seedless grapes can be sprayed with gibberellins and will grow as large as seeded varieties. In many species, treatment of an unfertilized ovary with auxin or gibberellins will cause parthenocarpy. Parthenocarpy occurs spontaneously in some plants. The fleshy part of strawberries is a modified stem called the receptacle. It is induced to grow by auxin that is produced by the dry fruits (achenes) that become embedded in the receptacle.

15. Describe the relationship between a fruit and floral parts.

16. List three seed dispersal strategies.

17. Describe the role of gibberellins and auxin in fruit growth.

17-9 <u>Photoperiodic Control of Flowering / Importance of Night Length / Circadian Rhythms and the Biological Clock</u> Species that all flower at about the same time, even if planted days apart, exhibit photoperiodism. Short-day plants (SDP) are those that must have day lengths of a certain critical time period or shorter (Maryland Mammoth tobacco). Long-day plants (LDP) must have daylight of so many hours or longer (spinach, clover). Short-long-day plants (SLDP) flower during the long days before midsummer; they need short days followed by long days. Long-short-day plants (LSDP) flower after the long days of summer are followed by shorter ones (fall blooming). There are more day-neutral plants than short-day and long-day plants.

Plants actually measure the length of darkness instead of daylight. Cocklebur (SDP) is especially sensitive to night length. It requires 9 hours or more of darkness. Experiments were done wherein LDP and SDP were exposed to dark-interrupted days and light-interrupted nights. Only SDP were affected by one of the sets of conditions: light-interrupted nights. SDP must have long nights of a critical length (or longer). Even brief interruptions effectively produced shorter night lengths, which allows flowering in LDP but not in SDP. Phytochrome is somehow involved. Red-light interruption of darkness was most effective and its effect could be reversed by exposure to far-red light. Other evidence points to the interaction of night length with an internal timer (biological clock).

Circadian rhythms are exhibited by events that happen at about the same time daily. Examples of such "events" are peak activity periods of insects or flying squirrels; temperature fluctuations of hamsters; and enzyme level in humans, such as alcohol dehydrogenase being lowest in the early morning and highest near five in the evening. Plant sleep movements and spore formation in <u>Neurospora</u> are rhythmic occurrences in plants and fungi.

Peaks of activity periods in circadian rhythms are not temperature-dependent, but the amplitude of fluctuation may be reduced by lowered temperatures. Circadian rhythms persist even in the absence of light-dark cueing (unless required for expression of the activity). Circadian rhythms can be entrained within limits. Phase shifts may be caused by establishing something other than a 24-hour day-night cycle or by interrupting dark periods with light exposure.

18. Restate the conditions necessary for short-day plants in terms of night length.

19. Restate the conditions necessary for long-day plants in terms of night length.

20. Describe circadian rhythms.

17-10 A Flowering Hormone?/ Vernalization and Flowering Each leaf is apparently capable of timing night length. The following evidence suggests that a chemical messenger--a flowering hormone--is produced. (1) If induced leaves are removed from the plant too quickly, flowering will not occur. (2) If cocklebur plants are grafted together but only one is induced, both will flower. (3) A single leaf of the SDP Perilla was induced and then grafted onto a noninduced Perilla. It caused flowering. The elusive hormone has been named florigen and is thought to be the same for SDP and LDP because one may be induced, the two grafted together, and both will flower.

In wheat and rye, there are two categories of flowering behavior. Spring wheat flowers that very growing season after being planted in the spring. Winter wheat must go through a period of cold or it will not flower the following growing season. Winter wheat is high-yielding and a preferred crop. Lysenko of Russia induced flowering via "winter" temperatures and called the process "vernalization." Once accomplished, vernalization is a stable condition.

21. Summarize the evidence for a flowering hormone, florigen.

22. Describe vernalization.

ARTICLE RESOURCES

Asexual and Sexual Reproduction in Plants

Alternation of Generations

Heterokaryon Formation in the Fungi

Multiple Hosts in the Fungal Life Cycle

Strobel, Gary A., and Gerald N. Lanier. Dutch elm disease. Scientific American, 1981, 245(2), 56-66. A discussion of the devastating plant disease caused by a fungus, Ceratocystis ulmi, and carried by beetles, Hybergopinus rufipes and Scolytus multistriestus. The life cycle of the fungus is diagrammed. Inhibition of the fungus by an antagonistic bactericum, Pseudomorras syringae, is treated.

Mosses and Ferns

Pietropaolo, Patricia. Plant tissue culture. The American Biology Teacher, 1981, 43(9), 508-512. Asexual reproduction of Boston ferns. Step-by-step procedures are outlined.

The Seed Plants

The Flower

Pollination

Pettitt, John, Sophie Ducker, and Bruce Knox. Submarine pollination. Scientific American, 1981, 244(3), 134-143. Examines pollination in plants that flower under water. These include 12 genera with about 50 species. Eelgrass, surf grass, pondweed, frogbit, sea wrack, and turtle grass are common examples. The greatest concentrations are in the Caribbean and Indo-Malaysian regions. Nine of the 12 genera are dioecious, which is most unusual. The pollen grains produced are bouyant, and the pistils have receptive stigmas to capture the pollen grains floating by.

Fruit

Rick, Charles M. The tomato. Scientific American, 1978, 239(2), 76-87. New production technology and genetic modification have made the tomato a major food crop. The twelve chromosomes of the tomato have been studied extensively. Diagrams of the flower and fruit are given. Production practices are shown. An interesting but brief history of the use of tomatoes in the diet is included.

Beard, Benjamin H. The sunflower crop. Scientific American, 1981, 244(5), 150-161. The sunflower is a valuable source of plant oil. Its life cycle is described, as are methods of planting and harvesting. This is an interesting plant because it has two different types of flowers--a ray flower, and a disk floret, developing on the head of the plant. A discussion of hybridization is given.

Solomon, Stephen. Amaranth, ancient food of the future. Science Digest, 1982, 90(6), 36-39, 102. Amaranth, a plant used by ancient Aztecs, has exceptional nutritional properties. Its seeds are rich in protein and the essential amino acid lysine, which is missing from rice, corn, and wheat. This plant can grow in soils and under conditions in which most food plants would die.

Photoperiodic Control of Flowering

Abori-Haidar, San S., and David W. Burger. Floral induction does not alter the growth parameters of fuschia. American Journal of Botany, 1981, 68(9), 1278-1281. Forty terminal cutlings of Fuscia hybrida, a long-day plant, were taken from stock plants grown under short-day conditions. Short-day conditions were provided during the rooting period. Eight hours of daylight were provided. Long-day induction was started 17 days after the plants were placed in the growth chamber. Growth rate was found to be unchanged and leaf initiation rates were unaltered by the transition to flowering or bud formation.

Downs, R. J., and J. M. Bevington. Effect of temperature and photoperiod on growth and dormancy of Betula papyrifera. American Journal of Botany, 1981, 68(6), 795-801. Catkins of paper birch from Alaska, Ontario, Montana, Vermont, New Hampshire, and West Virginia were gathered. They were air-dried and the seeds stored at 4°C until used. The seeds, which are light-requiring, were sown on a peatlite-sand substrate and placed in a mist chamber lighted 20 hours a day. Germination was completed and the seedlings well established within 3 weeks. The seedlings were then placed in conditions of long or short photoperiods and in conditions of cold or warm temperatures. The results showed different responses of plants from the various geographic locations.

Importance of Night Length

Circadian Rhythms

A Flowering Hormone?

Vernalization and Flowering

ESSAY QUESTIONS

1. Draw a diagram to illustrate the alternation of generations.

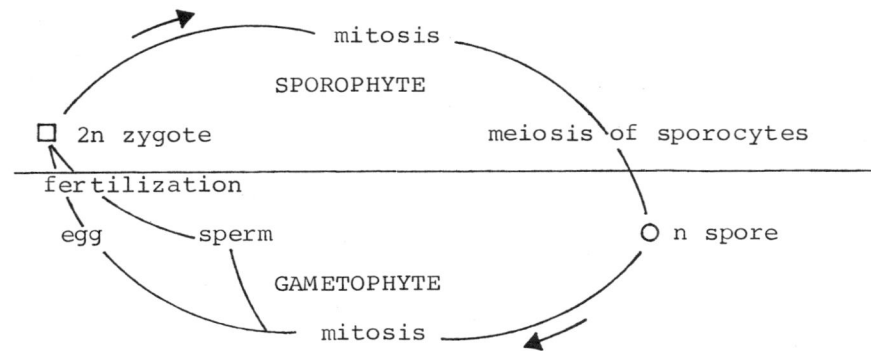

2. Compare mosses, ferns, and seed plants in terms of dependent vs. independent gametophytes and sporophytes and the dominant body form.

	Gametophyte	Sporophyte	Dominant body form
Mosses	independent	dependent	gametophyte
Ferns	independent	independent	sporophyte
Seed Plants	dependent	independent	sporophyte

3. Diagram a flower, showing the typical locations of the floral parts.

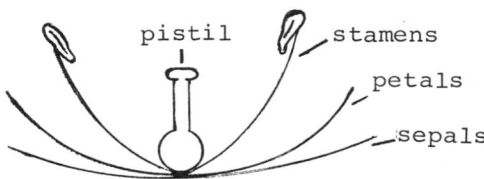

4. What are some of the dispersal strategies observed in seeds?

    Wind dispersal requires development of buoyancy. Hitchhiking with animals requires some means of hanging on. Being carried elsewhere and dropped with feces requires being edible and being palatable at the right time of year.

5. Describe the evidence suggesting that florigen is a chemical messenger.

    If induced leaves are removed too quickly, flowering will not occur. If cocklebur plants are grafted together but only one is induced, both will flower. When a single leaf of Perilla is induced and then grafted onto noninduced Perilla plants in succession, it causes all of them to flower.

TEST SET 17-A

Name _____   Section _____

Choose the best answer to each of the following questions, and write the appropriate letter in the space provided.

Ans: a
p. 425

_____ 1. Strawberries reproduce asexually via: (a) runners; (b) rhizomes; (c) bulbs; (d) corms.

Ans: c
pp. 426,
    427

_____ 2. The first cell of the sporophyte is the: (a) egg; (b) sperm; (c) zygote; (d) spore.

Ans: b
p. 427

_____ 3. The production of unlike gametes is known as: (a) isogamy; (b) anisogamy; (c) isomorphism; (d) heteromorphism.

Ans: d
p. 430

_____ 4. Heterokaryosis is a distinctive trait of: (a) algae; (b) higher animals; (c) higher plants; (d) fungi.

Ans: d
p. 431

_____ 5. Removal of _____ will control Puccinia graminis infection. (a) fences; (b) cedar apples; (c) lilac bushes; (d) barberry bushes.

Ans: c
p. 430

_____ 6. Basidiospores of P. graminis germinate in the: (a) fall; (b) winter; (c) spring; (d) summer.

Ans: a
pp. 432-
    434

_____ 7. Mosses and ferns both have: (a) swimming sperm; (b) vascular tissue; (c) rhizomes; (d) flowers.

Ans: a
p. 434

_____ 8. A pollen grain is: (a) a male gametophyte; (b) a male sporophyte; (c) diploid; (d) a female gametophyte.

Ans: b
p. 435

_____ 9. Seed coats develop around the: (a) male gametophyte; (b) female gametophyte; (c) pistil; (d) pollen.

Ans: c
p. 436

_____ 10. Most angiosperms and gymnosperms: (a) self-pollinate; (b) have co-evolved with each other; (c) cross-pollinate; (d) wind-pollinate.

Ans: b
p. 437

_____ 11. The fleshy part of fruits typically develops from: (a) the embryo; (b) floral parts; (c) the female gametophyte; (d) the stem.

Ans: d
pp. 439,
    440

_____ 12. _____ require only long nights of a critical length or longer in order to flower. (a) LDP; (b) LSDP; (c) Day-neutral plants; (d) SDP.

Ans: a
p. 441

_____ 13. Interruption of darkness with exposure to red light effectively: (a) shortens the night; (b) lengthens the night; (c) makes cocklebur flower; (d) prevents long-day plants from flowering.

Ans: c
p. 444

_____ 14. If cocklebur plants are grafted together but only one is induced, _____ will flower. (a) neither; (b) only the induced one; (c) both; (d) both die before they.

Ans: b
p. 445

_____ 15. A biennial plant requires _____ year(s) to develop. (a) one; (b) two; (c) 100; (d) several.

TEST SET 17-B

Name _____   Section _____

Choose the best answer to each of the following questions, and write the appropriate letter in the space provided.

Ans: c
p. 426

_____ 1. Navel orange trees are produced by _____ only. (a) rhizomes; (b) corms; (c) bud grafting; (d) seeds.

Ans: d
pp. 426, 427

_____ 2. The first cell of the gametophyte is the: (a) egg; (b) sperm; (c) zygote; (d) spore.

Ans: a
pp. 427, 428

_____ 3. In _____ life cycle, the only diploid cell is the zygote, which immediately undergoes meiosis to produce haploid spores. (a) a haplontic; (b) a diplontic; (c) Ulva's (sea lettuce); (d) a sporophytic.

Ans: b
p. 430

_____ 4. The best description of the genetic constitution of heterokaryon mycelia is: (a) 2n + n; (b) n + n; (c) n; (d) 2n.

Ans: a
p. 430

_____ 5. Each of the following statements is true of Puccinia graminis except: (a) It utilizes only one host. (b) It exhibits haploidy during part of its life cycle. (c) It has a complex life cycle. (d) It exhibits heterokaryosis during part of its life cycle.

Ans: d
p. 430

_____ 6. Uredospores of P. graminis form during: (a) fall; (b) winter; (c) spring; (d) summer.

Ans: c
pp. 432, 433

_____ 7. Ferns have _____, but mosses do not. (a) swimming sperm; (b) multicellular reproductive structures; (c) vascular tissue; (d) spores.

Ans: b
pp. 434-436

_____ 8. In flowering plants, sperm are transported by _____ to reach the egg. (a) swimming; (b) a pollen tube; (c) petals; (d) the filament.

Ans: b
p. 436

_____ 9. Most flowers have both: (a) sepals and petals; (b) stamens and pistil; (c) petals and stamens; (d) petals and pistil.

Ans: a
p. 436

_____ 10. Rotting flesh odors or dung odors produced by flowers would be most likely to attract _____ as pollinators. (a) flies; (b) bees; (c) moths; (d) bats.

Ans: c
p. 438

_____ 11. In many species, treatment of an unfertilized ovary with auxin or gibberellin will cause: (a) fruit ripening; (b) flower death; (c) parthenocarpy; (d) defective pollen development.

Ans: d
p. 439

_____ 12. _____ require only days of a critical length or shorter in order to flower. (a) LDP; (b) LSDP; (c) SLDP; (d) SDP.

Ans: b
pp. 441, 442

_____ 13. Recurring enzyme level peaks and troughs in a 24-hour period in humans are an example of: (a) phytochrome control; (b) a circadian rhythm; (c) vernalization; (d) heterokaryosis.

Ans: a
p. 445

_____ 14. Florigen, if it exists, is thought to be _____ for SDP and LDP. (a) the same; (b) different; (c) phytochrome; (d) an electrical impulse.

Ans: c
p. 445

_____ 15. The induction of flowering through producing tolerable winter conditions in the laboratory is known as: (a) a circadian rhythm; (b) spring wheat induction; (c) vernalization; (d) biennial stimulation.

CHAPTER EIGHTEEN
NUTRITION OF PLANTS AND FUNGI

CHAPTER SUMMARIES AND OBJECTIVES

18-1 <u>Mineral Nutrients</u> All mineral elements required by higher plants are derived from soil or rock. Essential elements are those that are needed for growth and repro- duction, cannot be replaced by any other element, and must be used directly. Some elements fulfill multiple roles. Some nutrients are needed in greater quantities, but <u>all</u> essential nutrients are still essential. Macronutrients are those that are present in plant tissue in concentrations of at least 1 milligram per gram of dry matter; micronutrients are present in concentrations of 100 micrograms or less per gram of dry matter.

To determine whether a mineral is essential, plants may be grown in solutions that are deficient in a selected element. Before mineral-deficient plants die, they usually produce characteristic deficiency symptoms. There have been some problems associated with the techniques used (including impure salts, atmospheric contami- nants, and unsuspected sources of particular minerals). If any further essential nutrients are found, more sophisticated techniques probably will be required.

1. Describe requirements for labelling a mineral as an essential element.

2. Explain why the recognition of mineral deficiency symptoms would be useful to the horticulturist.

18-2 <u>Soils</u> Soils are important to plants for their mineral nutrients and their water supply, for soil oxygen, and for mechanical support. Soil harbors bacteria that are beneficial to plants and bacteria that are harmful. Soils are constantly changing in any one area, and they vary considerably among different geographical areas due to different circumstances during soil formation. All soils can be seen to consist of two or more horizons (recognizable horizontal layers). Generally minerals from the top horizon are leached and carried to a deeper layer.

Mineral depletion through leaching or crop removal may be counteracted by ferti- lizers. The three elements most commonly added to agricultural soils are nitrogen (N), phosphorus (P), and potassium (K). A 5-10-5 fertilizer is common: 5% N, 10% P, and 5% K. Organic fertilizers improve the physical characteristics of the soil as well as their mineral content, whereas inorganic fertilizers are more precise for specific nutrients only.

3. Describe the relationship between soil horizons and fertilizers.

18-3 <u>The Scope of Photosynthesis</u> In photosynthesis, light energy is used to raise an electron to an excited state and then to trap this energy in usable form. During photosynthesis, water is split, CO_2 is fixed, energy is bound, and O_2 is released to the atmosphere. An amount of oxygen equivalent to that present in the earth's entire atmosphere is produced by photosynthesis every two years.

4.   Explain the significance of O_2 production to the living world as it exists today.

18-4 <u>Nitrogen-Fixing Organisms</u> Nitrogen fixers (monerans) are capable of converting atmospheric N_2 to NH_3 (ammonia). Without the nitrogen fixers, no other known organisms would survive. Nitrogen fixers are identified in an array of photosynthetic, aerobic, anaerobic, and symbiotic bacteria and in photosynthetic blue-green algae. Blue-green algae are predominant nitrogen fixers in aquatic ecosystems. Root nodule symbiotic bacteria fix most of the nitrogen that becomes available to soil-growing organisms. Other symbiotic, nodule-living bacteria are critical to the success of pioneer species such as alder, western mountain lilac, and eastern sweet gale.

5.   Explain the importance of nitrogen fixation to the rest of the living world.

18-5 <u>Nitrogen Fixation / Symbiotic Nitrogen Fixation / Denitrification</u> In biological nitrogen fixation, dinitrogen (N_2) molecules are progressively reduced by the addition of hydrogen atoms until the three bonds between the nitrogens are cleaved. The reactants are bound to nitrogenase for each of the reactions, and much ATP is used in the process. Nitrogenase is extremely sensitive to oxygen. Many nitrogen fixers are anaerobic; those that are not must decrease their internal oxygen levels drastically. Most nitrogen is fixed biologically; a very small amount is fixed by lightning, volcanic eruption, and forest fires.

Some bacteria live mutualistically with leguminous plants--both benefit. Nodules form on the roots and house members of the <u>Rhizobium</u> species. Neither can fix nitrogen while living separately. Root hairs release an attractant; the bacteria multiply and release a growth substance that causes the cell wall of the root hair to invaginate. When the infection thread invaginates to the cortex tissue, 4n plant cells are commonly encountered, and they are stimulated to divide rapidly. The infection thread bursts, releasing <u>Rhizobium</u> into the cortical cell cytoplasm. <u>Rhizobium</u> cells increase about tenfold in size and develop elaborate, folded, internal membranes (bacteroid stage). Before they can fix nitrogen, the bacteroids become surrounded with hemoglobin, which apparently acts as an oxygen trap to keep the bacteroids anaerobic.

Nitrogen gas (N_2) is continually returned to the atmosphere by aerobic bacteria (<u>Bacillus</u> and <u>Pseudomonas</u>) under anaerobic conditions (denitrification). Nitrate is used as a terminal electron acceptor instead of oxygen. N_2 gas and H_2O are produced.

6.   Describe briefly the role of nitrogenase in nitrogen fixation.

7.   Explain the value of the relationship of mutualistic bacteria with legumes.

8.   Describe the replenishing of the atmosphere with N_2 gas.

18-6 <u>Industrial Applications of Nitrogen Fixation / Nitrification</u> Production of nitrogen-containing fertilizer is the most energetically expensive aspect of crop production in the United States. The Haber process combines N_2 and H_2 gases to make ammonia (NH_3). Research is being done on the insertion of nitrogenase genes into bacterial plasmids and crop plants. Perhaps the greatest problem is that of excluding free oxygen and providing strong reducing agents.

Nitrogen fixers release ammonia (NH_3, toxic to plants) and ammonium ions (NH_4^+) into the soil. Plants may safely take up NH_4^+ but will preferentially take up nitrates (NO_3^-) instead under more acidic conditions. Soil bacteria accomplish nitrification (the oxidation of ammonia to nitrate). <u>Nitrosomonas</u> and <u>Nitroso-coccus</u> convert ammonia to nitrite ions (NO_2^-), and <u>Nitrobacter</u> oxidizes nitrites to nitrates. These chemosynthetic bacteria use energy released in their oxidation of ammonia or nitrite to power their life activities.

9. Explain what the Haber process essentially does.

10. Define nitrification.

11. Describe the role of soil bacteria in nitrification.

18-7 Nitrate Reduction / Sulfur Metabolism After plants take in nitrates, they reduce them back to ammonia before using them further. The steps from NO_2^- to NH_3 occur in chloroplasts. NH_3 is then used in the synthesis of amino acids and other nitrogen-containing compounds. All the nitrogen in the animal world gets there by way of the plant kingdom.

All living things require sulfur. It is a constituent of two amino acids, cysteine and methionine, which are in nearly all proteins. Plants and fungi take up sulfate ions (SO_4^{2-}), sulfur's most oxidized state. Plants reduce SO_4^{2-} and incorporate it into cysteine; then they make all other sulfur-containing compounds from metabolism of cysteine. Animals must obtain cysteine and methionine from plants. Certain chemosynthetic bacteria oxidize hydrogen sulfide (H_2S) to sulfate and use the energy to make ATP and fix CO_2.

12. Describe the significance to the animal world of nitrogen incorporated into organic molecules by plants.

13. Outline the relationship between sulfur, cysteine and methionine, plants, and animals.

18-8 Heterotrophic Seed Plants / Nutrition of Fungi A few higher plants are parasitic. Indian pipe does not carry on its own photosynthesis but obtains nutrients via a fungal bridge between its roots and those of nearby actively photosynthesizing plants. Others (such as mistletoe) carry on photosynthesis but not enough to support full normal growth.

Some plants are carnivorous: Venus's-flytrap, sundew, and pitcher plants all trap and digest insects. They all grow well without insects but grow faster and are a darker green when insects are part of their diet.

Fungi are all heterotrophic; many are saprophytic, absorbing organic nutrients from dead organic matter and causing decomposition. Those that are not sapro- phytic are parasitic. Facultative parasites can be grown on a defined medium, but obligate parasites will grow on particular hosts only. Fungi do not fix nitrogen, and most must absorb thiamine (vitamin B_1) and biotin from their environment.

Some fungi function as active predators. Some trap microscopic protists or animals in sticky mycelia. A few form constricting rings that swell and trap their prey as they attempt to crawl through the ring.

Lichens are composite plants of symbiotic associations between a fungus and an alga. Some fungi live symbiotically with the roots of certain plants, forming mycorrhizae. Some leaf-cutting ants farm fungi: they feed the fungi, remove extraneous species, and eat their fungus crops. Scale insects may be covered by fungal hyphae, which protect most of the scale colony, but the fungus actually infects (parasitizes) some of the insects.

14. Describe nonphotosynthetic nutrition of seed plants and fungi.

ARTICLE RESOURCES

Mineral Nutrients

Walker, C. D., and J. F. Loneragen. Effects of copper deficiency on copper and nitrogen concentrations and enzyme activities in aerial parts of vegetative subterranean clover plants. Annals of Botany, 1981, 47(1), 65-73. Shoot tips as well as young and mature leaves had their copper concentrations reduced by 35-65 percent when copper deficiencies were permitted. Nitrogen concentration were depressed by 20 percent in the leaves. All shoot tip enzymes were depressed by 70-90 percent.

Guttridge, C. G., E. G. Bradfield, and R. Holder. Dependence of calcium transport into strawberry leaves on positive pressure in the xylem. Annals of Botany, 1981, 48(4), 473-480. Leaf tip burn in strawberries is caused by a local deficiency of calcium. Calcium, apparently, is moved at night when guttation occurs. An apparatus was set up whereby strawberry plants could be grown in different solutions during daytime and nighttime hours. The same experiment done on tomato plants gave the same results.

Soils

The Scope of Photosynthesis

Bardell, David. Bacterial photosynthesis without chlorophyll. The American Biology Teacher, 1982, 44(5), 278-279, 313. Halobacterium halobrium, a species found in Great Salt Lake and the Dead Sea, where high concentrations of salt are to be found, is capable of photosynthesis. It contains neither bacterial nor plant chlorophyll. The light-sensitive molecule rhodopsin is located in the cell membrane. This molecule is also present in vertebrates with retinal eyes and in some invertebrates.

Nitrogen-Fixing Organisms

Nitrogen Fixation

Peat, J. R., F. R. Minchin, B. Jeffcoat, and R. J. Summerfield. Young reproductive structures promote nitrogen fixation in soya bean. Annals of Botany, 1981, 48(2), 177-182. Nitrogen fixation occurs rapidly during the transition from the vegetative to the reproductive phase of development in the soya bean. Removal of flower buds and other reproductive structures at various times in the reproductive cycle caused a marked change in nitrogenase activity. This was reflected in the reduced nitrogen content of the plants when compared with control and vegetative plants.

Van Raalte, Charlene. Nitrogen fixation by nonleguminous plants. American Biology Teacher, 1982, 44(4), 229-232, 254. Discusses the role of such plants as alder trees and shrubs, sweet-fern, bayberry, and sweet gale. Good box describing nitrogen fixation. Lists nonleguminous plants known to form nodules on roots with actinomycete bacteria. Uses of actinomycete-nodulated plants are shown.

Symbiotic Nitrogen Fixation

Denitrification

Industrial Applications of Nitrogen Fixation

Nitrification

Nitrate Reduction

Sulfur Metabolism

Heterotrophic Seed Plants

Heslop-Harrison, Yolande. Carnivorous plants. Scientific American, 1978, 238(2), 104-114. Carnivorous plants live in a nutrient-poor environment. By adopting the habit of capturing and digesting their prey, they have overcome this obstacle. Their prey is usually small insects, though some instances of small tree frogs, mice, and small fish have been recorded. These animals supply the plant with nutrients that are scarce. The carnivorous plants are to be found in bogs, marshes, and impoverished soil in forest openings, and on crumbly soil associated with weathered limestone. Electron micrographs illustrate the glands used for capturing prey and those used for prey digestion.

Nutrition of Fungi

ESSAY QUESTIONS

1. What are the characteristics of an essential element?

    The element must be essential for normal growth and reproduction. The element cannot be replaced by another element. The requirement must be direct, not the result of some indirect effect.

2. Describe the primary uses of organic and inorganic fertilizers.

    Organic fertilizers contain materials that improve the physical properties of the soil, providing spaces for gases, root growth, and drainage. Inorganic fertilizers provide a rapid increase in soil nutrients and can be formulated to meet the requirements of a particular soil and a particular crop more precisely.

3. Describe briefly what occurs in nitrogen fixation, nitrification, and denitrification.

    Nitrogen fixation: $N_2 \rightarrow NH_3$ or NH_4^+ by reduction

    Nitrification: $NH_3 \rightarrow NO_3^-$ by oxidation

        Nitrosomonas and Nitrosococcus--$NH_3 \rightarrow NO_2^-$

        Nitrobacter--$NO_2^- \rightarrow NO_3^-$

    Denitrification: $NO_3^- \rightarrow N_2^-$ gas

4. Briefly describe the nutrition of fungi.

    The fungi are all heterotrophs that obtain their food by absorption. Saprophytes obtain their nutrients from dead organic matter. Parasites live on live organisms.

TEST SET 18-A

Name _____   Section _____

Choose the best answer to each of the following questions, and write the appropriate letter in the space provided.

Ans: d
p. 450

_____ 1. If a nutrient is ordinarily used by a plant but can be replaced by a different element, it is considered: (a) essential; (b) deficient; (c) a micronutrient; (d) nonessential.

Ans: a
p. 453

_____ 2. A 5-10-5 N-P-K fertilizer contains _____% nitrogen. (a) 5; (b) 10; (c) 15; (d) 20.

Ans: d
p. 451

_____ 3. Plants are benefitted by soil components in all of the following except: (a) soil water; (b) mechanical support; (c) mineral nutrients; (d) CO_2 gas.

Ans: b
p. 453

_____ 4. In photosynthesis, light energy raises a(an) _____ to an excited state. (a) proton; (b) electron; (c) neutron; (d) CO_2 molecule.

Ans: c
p. 453

_____ 5. Without _____, no other known organisms would survive. (a) aerobic bacteria; (b) blue-green algae; (c) nitrogen fixers; (d) parasites.

Ans: a
p. 455

_____ 6. In the organic reduction of N_2, the reactants are bound to: (a) nitrogenase; (b) proteases; (c) amylases; (d) nitrate ions.

Ans: b
p. 457

_____ 7. Rhizobium invades the cytoplasm of plant _____ cells. (a) xylem; (b) cortical; (c) phloem; (d) meristem.

Ans: c
p. 457

_____ 8. Nitrogen gas is returned to the atmosphere via: (a) nitrification; (b) nitrogen fixation; (c) denitrification; (d) assimilation.

Ans: d
p. 458

_____ 9. In the Haber process, _____ are combined to make ammonia. (a) water and carbon dioxide; (b) water and nitrogen gas; (c) nitrates and water; (d) nitrogen gas and hydrogen gas.

Ans: a
p. 458

_____ 10. _____ oxidizes nitrite to nitrate. (a) Nitrobacter; (b) Nitrosococcus; (c) Nitrosomonas; (d) Pseudomonas.

Ans: b
p. 459

_____ 11. After plants take in nitrates, they reduce them back to _____ before using them further. (a) amino acids; (b) ammonia; (c) nitrogen gas; (d) water and nitrogen.

Ans: c
p. 459

_____ 12. Cysteine and methionine are sulfur-containing: (a) fatty acids; (b) ATP; (c) amino acids; (d) phospholipids.

Ans: a
p. 460

_____ 13. The Venus's flytrap is: (a) a carnivorous plant; (b) a fungus; (c) a parasite; (d) a saprophyte.

Ans: d
pp. 461,
    462

_____ 14. Each of the following may be used to describe certain fungi except: (a) saprophytic; (b) parasitic; (c) predatory; (d) photosynthetic.

Ans: b
p. 462

_____ 15. Mycorrhizae are associations of plant roots and: (a) certain algae; (b) certain fungi; (c) certain lichens; (d) predatory fungi.

TEST SET 18-B

Name _____   Section _____

Choose the best answer to each of the following questions, and write the appropriate letter in the space provided.

Ans: c
pp. 450,
    451
_____ 1. Micronutrients are _____ than macronutrients. (a) less essential; (b) more essential; (c) needed in smaller quantities; (d) needed in greater quantities.

Ans: d
pp. 452,
    453
_____ 2. Inorganic industrial fertilizer commonly contains all of the following except: (a) nitrogen; (b) phosphorus; (c) potassium; (d) auxin.

Ans: a
p. 452
_____ 3. Minerals usable to crop plants are those available in the: (a) upper horizons; (b) lower horizons; (c) parent bedrock; (d) undecomposed surface litter.

Ans: d
p. 453
_____ 4. Photosynthesis does each of the following except: (a) release O_2 to the atmosphere; (b) fix CO_2; (c) split H_2O; (d) synthesize amino acids.

Ans: b
p. 454
_____ 5. Planting alfalfa or clover every so often in a field is done to increase the useful _____ content of the soil. (a) oxygen; (b) nitrogen; (c) carbohydrate; (d) potassium.

Ans: c
pp. 455,
    457
_____ 6. Most nitrogen is fixed: (a) by volcanic action; (b) by forest fires; (c) biologically; (d) aerobically.

Ans: a
p. 457
_____ 7. Legumes and Rhizobium can live separately, but neither can _____ while doing so. (a) fix nitrogen; (b) carry out respiration; (c) make amino acids; (d) burn glucose.

Ans: b
p. 457
_____ 8. Which of the following statements it not true of Pseudomonas and Bacillus? (a) They carry out denitrification. (b) They fix free nitrogen. (c) They use NO_3^- as a terminal electron acceptor under anaerobic conditions. (d) They are normally aerobic.

Ans: a
p. 458
_____ 9. The Haber process is used to make: (a) ammonia; (b) amino acids; (c) nitrates; (d) nitrites.

Ans: d
p. 458
_____ 10. _____ oxidizes ammonia to nitrite. (a) Nitrobacter; (b) Bacillus; (c) Pseudomonas; (d) Nitrosomonas.

Ans: c
p. 459
_____ 11. All the nitrogen of the animal world gets there by way of the _____ world. (a) fungal; (b) animal; (c) plant; (d) bacterial.

Ans: b
p. 459
_____ 12. Cysteine and methionine are _____ -containing amino acids. (a) potassium; (b) sulfur; (c) DNA; (d) phosphorus.

Ans: c
p. 460
_____ 13. Indian pipe is identified as: (a) a saprophyte; (b) a fungus; (c) a parasite; (d) a carnivorous plant.

Ans: a
p. 461
_____ 14. Fungi are all: (a) heterotrophs; (b) saprophytes; (c) parasites; (d) carnivorous.

Ans: d
p. 462
_____ 15. Fungal parasites that cannot be grown on any available defined medium are said to be: (a) nonexistent; (b) facultative; (c) anaerobic; (d) obligate.

CHAPTER NINETEEN
PLANTS AND THEIR TERRESTRIAL ENVIRONMENT

CHAPTER SUMMARIES AND OBJECTIVES

19-1 The Plumbing of a Vascular Plant Vascular plants have three principal regions: leaves, stem, and root system. All stems and leaves comprise the shoot system. Vascular tissues distribute water, minerals (through xylem), and manufactured nutrients or food (through phloem). Nutrients travel from sources (where they are manufactured) to sinks (where they are used or stored).

All vascular plants have tracheids in their xylem. These die before assuming their ultimate vascular function; they are spindle-shaped and interconnected by numerous pits in the cell walls. Xylem vessels (also dead cells) occur almost exclusively in flowering plants. They are larger in diameter than tracheids and lose their end walls so as to form continuous tubes from one element to another. Thin-walled parenchyma cells store food in xylem; fibers (dead cells) provide support in the xylem of angiosperms. In gymnosperms vessels and fibers are absent.

Phloem consists primarily of living cells. Sieve tube elements of flowering plants arranged end-to-end form tubes. The elements develop small holes in their ends, which give a sieve appearance. In the maturation of these elements, the central vacuoles lose their membranes and the vacuole solution (plus some cytoplasm) forms a continuous fluid phase from one element to another. Nuclei break down, but a layer of cytoplasm remains as a lining to the cell wall. Companion cells adjacent to the sieve tube elements retain all their organelles and may regulate the performance of the elements. Storage cells and dead fibers are also part of phloem.

1. Explain why vascular plants need vascular tissue.

2. Describe the types of cells in xylem.

3. Describe the types of cells in phloem.

19-2 The Root Many gymnosperms and dicots exhibit tap root systems with less prominent secondary roots. Monocots and some dicots have fibrous root systems that hold soil very well. A protective root cap is at the tip of each root, sloughed off cells are replaced by the meristem, and the root cap apparently detects the pull of gravity. The root apical meristem just above the root cap provides all cells for growth in root length. Above the meristem is a region of cell elongation and then a region of cell differentiation.

The outer layer of cells differentiates into epidermis. These cells in the region of differentiation produce long, delicate root hairs--extensions of individual epidermal cells. Root hairs take up water and minerals. Internal to the epidermis is the cortex layer (several cells thick). In many trees, epidermal and sometimes cortical cells become associated with a fungus, forming mycorrhizae (increases absorption of water and minerals). Proceeding inward, a single layer of thicker walled cells is encountered: the endodermis. Waxy Casparian strips containing suberin (around endodermal cells) are placed so as to form waterproof seals. Water and dissolved minerals must pass through the cytoplasm of endoderm cells to get to the central vascular cylinder: pericycle, xylem, and phloem. Pericycle cells are relatively undifferentiated; branch roots arise in the pericycle. The very center is occupied by xylem in a multiple pointed pattern. Phloem occurs between the tips of the points and the pericycle.

4. Compare a tap root system with a fibrous root system.

5. Name the areas of a root tip that are encountered longitudinally.

6. Name the different cell types, in order, as they are encountered in a root cross section just above the region of differentiation.

19-3 The Stem Vascular tissues run continuously from root through stem, although the spatial arrangements change from one to the other. The stem may be green; it bears leaves; and, in each axis between leaf and stem, there is a lateral bud that may develop into a branch. At the tip of each stem is an apical bud that contains a shoot apical meristem and leaf primordia. Some stems are highly modified: The potato tuber is part of a stem, and lateral buds are at the "eyes." And the runners of strawberry plants, Bermuda grass, and iceplant are horizontal stems that root at intervals.

The vascular bundles of dicots form a ring in the interior of the stem; they are scattered throughout the stem of monocots with pith cells in between. Internal to the ring of vascular bundles is the pith center of dicots, while the cortex is exterior to the bundles. The outermost cell layer of the young stem is the epidermis. Stems bear leaves and flowers, conduct water and other materials to and from the flowers and leaves, store foods, and minimize water loss in their functions.

7. Contrast the tissue found in the center of dicot stems with roots.

19-4 Growth in Diameter of Stems and Roots Thickening of stems and roots of woody plants (gymnosperms and dicots) is due to a lateral meristematic tissue called vascular cambium. Vascular cambium begins as the single layer of cells between xylem and phloem. Cortical cells between vascular bundles of stems also begin to divide and are continuous with the original vascular cambium; thus an entire ring of vascular cambium develops. Complete cylinders of xylem form interior to the cambium, with phloem forming exerior to the cambium.

Continuous laying down of these interior layers causes the epidermis and cortex to crack and ultimately to be lost. Cells derived from the outermost phloem begin to divide and produce layers of cork that are waterproofed with suberin. The wood of a tree is xylem; the bark is phloem and cork. Vascular cambium gives rise to all the types of phloem and xylem cells we have discussed. The living storage cells of xylem and phloem must have their nutrient supplies maintained. To do this, cells in the vascular cambium give rise to rows of living cells, which are perpendicular to the xylem vessels and phloem sieve tubes, forming vascular rays.

8. Describe lateral stem and root growth.

9. Describe bark formation.

10. Explain the value of vascular rays.

19-5 The Leaf Leaves carry on most of the photosynthesis, produce food, release oxygen, reduce nitrate to ammonia, and are responsible for photoperiodic measurements.

The photosynthesizing cells (those that have chloroplasts) of a plant are mesophyll cells--typically arranged in two layers. They are serviced by a network of vascular tissue (veins). Each leaf has an epidermis that completely covers it and reduces the rate of water loss. Many plants are described as C_3 photosynthesizing plants--the CO_2 level must be maintained at a particular level to support net photosynthesis. The C_4 plants, however, can fix CO_2 at a level below the C_3 plants. They have mesophyll arranged in concentric layers around the veins, and a bundle sheath surrounds the xylem/phloem unit. C_4 chloroplasts are different from C_3 chloroplasts.

11. Describe the primary functions of leaves.

12. Identify the kinds of cells that can carry out photosynthesis.

19-6 <u>Stomata: A Compromise / Reversed Stomatal Cycle in the Crassulaceae</u> Epidermis prevents water loss--the waxy cuticle facilitates this. The cuticle is impermeable to CO_2 as well as to H_2O; therefore, the evolution of stomata was an elegant compromise. The stomatal complex consists of the hole in the epidermis (stoma) and the two guard cells that control the opening. Stomata typically close at night and when water is being lost at too great a rate.

The transport of K^+ into guard cells from surrounding epidermis and the concomitant influx of H_2O causes turgidity of the guard cells and opening of the stoma because of their structural characteristics. Closure of the stomata is caused by K^+ being actively transported out of the guard cells and the concomitant loss of H_2O. A low level of CO_2 inside the air spaces of stomata apparently provokes stomatal opening. If the water potential of guard cells is too negative, they release abscisic acid. They then release K^+ and close the stomata.

Many succulent plants, those that live in dry areas, and those that live in marine-associated habitats exhibit a backward stomatal cycle: Their stomata are open at night and closed during the day. At night, CO_2 diffuses into leaf spaces and reacts with PEP to form organic acids such as malic acid and aspartic acid. This allows CO_2 to be stored as organic acids in higher concentrations than would occur with normal diffusion of atmospheric CO_2 (0.03% of normal atmosphere).

With daybreak, stomata may then close to prevent excessive water loss, the organic acids are broken down to release CO_2 behind these closed doors, and photosynthesis may proceed for a period of time instead of minutes, as would occur if the CO_2 were in gas form only.

13. Explain why there would be less water loss at night with stomata open than during the day.

14. Describe the relationship between K^+ level in guard cells and turgidity.

19-7 <u>Special Adaptations of Leaves</u> Plants living in particularly dry areas may exhibit one or more of the following adaptations: heavy cuticle layer, stomata in sunken cavities, fleshy leaves for water storage, leaf production when water is abundant, spines rather than leaves with photosynthesis occurring in the stems and leaves that roll up during dry periods.

Aquatic angiosperms face the opposite problem, effecting gaseous exchange when water is so abundant. Water lilies and duckweeds have stomata on the upper epidermis instead of predominantly on the lower leaf surfaces as most other plants do.

15. Describe how plants cope with the environmental stresses of too much or too little water.

19-8 <u>Transport in the Xylem / Transport in the Phloem</u> Water enters root hairs or mycorrhizae osmotically. Water evaporates from mesophyll cells of leaves, establishing a pulling effect on the continuous water columns of xylem due to the tension in the water column (because of cohesion of water molecules). Mineral ions in solution rise passively with the water column--they may be redistributed via phloem. All of the xylem transport involves no work on the part of the plant.

Phloem transport appears to be by bulk flow of phloem sap. Sugars and solutes are actively transported into the sieve tubes at the sources, then actively removed at the sinks. In effect, the hydrostatic pressure at the source end would squeeze the fluid contents toward the sink end. Some of the features of this "pressure-flow" model are still debatable.

16. Describe the cohesion-tension explanation of xylem water/mineral movement.

17. Describe the "pressure-flow" model of phloem transport.

19-9 <u>Strengthening Tissues in the Shoot</u> Nonwoody plants depend on cell turgidity, sclerenchyma cells, and collenchyma cells for support. Woody plants are supported by wood: Fibers have the thickest walls and provide the most support. The sclerenchyma cells of young trees and nonwoody plants are cells that die when mature--they are supportive. Fibers are a type of sclerenchyma cell that is greatly elongated. Some sclerenchyma cells are not so elongated yet do provide strength. Collenchyma cells remain alive after laying down thick walls--they are flexible in contrast to the brittleness of sclerenchyma cells.

18. Describe how a herbaceous (nonwoody) plant stands upright.

ARTICLE RESOURCES

The Plumbing of a Vascular Plant

The Root

Ellmore, George S. Root dimorphism in <u>Ludwigia peploides</u> (Onagraceae) structure and gas content of mature roots. <u>American Journal of Botany</u>, 1981, 68(4), 557-568. <u>Ludwigia peploides</u> is an amphibious plant that grows prostrate in water-logged soil. It gives rise to two distinct root types, an upward-growing and a downward-growing root. They both emerge from nodes and have cell pattern lineages that are similar. Well illustrated, with five pages of photomicrographs.

The Stem

Growth in Diameter of Stems and Roots

Schwarz, Jack, On the up and up. <u>Science 81</u>, 1981, 2(2), 56-60. Monocultures of forest trees may be the solution to a dwindling supply of wood for industry. Cloning is discussed.

The Leaf

Kemp, Paul R., and Gary L. Cunningham. Light, temperature, and salinity effects on growth, leaf anatomy, and photosynthesis of <u>Distichlis spicata</u> (L.) Greene. <u>American Journal of Botany</u>, 1981, 68(4), 507-516. Growth rates of this grass in controlled environment chambers were reduced significantly by low temperature or high substrate salinity when the grass was grown in low light. The growth rates at high light were not affected by salinity or temperature. There was no correla-tion between net photosynthesis, and growth salinity has little effect on cell morphology in this plant.

Stomata: A Compromise

Rugenstein, Seanna R., and Nils R. Larsten. Stomata on seeds and fruits of <u>Bauhinia</u> (Leguminosae: Caesalpinioideae). <u>American Journal of Botany</u>, 1981, 68(6), 873-876. Seeds do not ordinarily have stomata, but they have been found on about 23 monocot and dicot families. Excellent photomicrographs. The authors are not sure whether or not these seed stomata are functional but speculate that they may be after the seed is shed.

Revised Stomatal Cycle in the Crasslaceae

Special Adaptations of Leaves

   Schrock, Gould F. A laboratory exercise to assess transpiration. The American
   Biology Teacher, 1982, 44(4), 242-245, 252. Sunflower seedlings are employed in
   this method of determining the rate of transpiration by gravimetric means. A count
   of the stomata on the adaxial and abaxial surface is made.

   Watterlond, Michael. Stopping by woods. Science 81, 1981, 92-93. Formation of
   fall colors in leaves after cessation of photosynthesis. Chlorophyll is no longer
   produced, allowing other pigments to become visible.

Transport in the Xylem

Transport in the Phloem

Strengthening Tissues in the Shoot

   Carlquist, Sherwin. Types of cambial activity and wood anatomy of Stylidium
   (Stylidiaceae). American Journal of Botany, 1981, 68(6), 778-785. Excellent
   photomicrographs of xylem and no-xylem parenchyma. A periderm does not develop
   from the outer cortical cells. This plant has a wood of heightened mechanical
   strength, leading to a shrubby growth form. The stems do not exceed 1 cm in
   diameter.

ESSAY QUESTIONS

1. Describe a root tip in terms of the regions and functions of cells encountered
    longitudinally.

    The root cap covers and protects the growing tip. The root apical meristem
    replaces sloughed off root cap cells and provides all the cells for root
    growth in length. The region of cell elongation is encountered next (cells
    elongate). Above cell elongation, the region of cell differentiation occurs,
    where epidermal cells develop root hairs, and vascular tissue and other special
    cell types differentiate.

2. What is the significance of the Casparian strips of endodermal cells in plants?

    The strips contain a waxy suberin, and they are oriented to prevent leakage
    of any absorbed materials between the cells. Instead, water and minerals
    must go through the cytoplasm of endodermal cells to reach the interior
    vascular tissue.

3. Describe the development of a vascular cambium cylinder in woody stems.

    At first vascular cambium occurs between phloem and xylem of vascular bundles
    only, which are arranged in a ring. Then cortical cells at the same depth
    as the vascular bundle cambium begin to divide, joining the existing cambium
    and forming a continuous ring. This makes possible the formation of the
    familiar annual rings of woody plants.

4. Why are stomata considered a compromise?

    Plants must be protected against excessive water loss--an epidermis with a
    waxy cuticle does just that. But plants must also take in gaseous CO_2 and
    give off O_2 and enough water vapor to keep xylem operating correctly. The
    "compromise" that evolved is the stomatal complex, which can be closed and
    opened under the appropriate conditions.

5. What are some of the special adaptations that allow conservation of water by plants
    that live in particularly dry areas?

    These adaptations include secreting a heavier cuticle, locating stomata in
    sunken cavities, storing water in fleshy leaves, producing leaves only when

water is abundant, forming spines instead of leaves, and rolling leaves up during dry periods.

6. Water movement in xylem requires no work on the part of the plant. Why is this so?

Water enters the roots via osmosis, and it leaves as vapor through the stomata. Water molecules have such cohesiveness that each molecule of water that leaves establishes a tension on the water column in the xylem tube, pulling water upward. None of these activities requires the expenditure of plant energy.

TEST SET 19-A

Name _____   Section _____

Choose the best answer to each of the following questions, and write the appropriate letter in the space provided.

Ans: c
p. 467

_____ 1. The stems and leaves comprise the _____ of a plant. (a) root system; (b) meristematic regions; (c) shoot system; (d) seedling.

Ans: b
pp. 468, 469

_____ 2. When mature, each of the following is a dead cell except: (a) xylem vessel; (b) phloem sieve tube element; (c) sclerenchyma; (d) tracheid.

Ans: a
p. 470

_____ 3. Monocots typically have: (a) fibrous root systems; (b) taproot systems; (c) woody tissue; (d) netted leaf venation.

Ans: d
p. 472

_____ 4. Branch roots arise in the: (a) xylem; (b) endodermis; (c) vascular cambium; (d) pericycle.

Ans: d
p. 472

_____ 5. Each of the following is part of the central vascular cylinder of roots except the: (a) pericycle; (b) xylem; (c) phloem; (d) endodermis.

Ans: b
p. 474

_____ 6. Apical buds always contain: (a) root apical meristems; (b) leaf primordia; (c) root caps; (d) cork.

Ans: a
p. 474

_____ 7. The vascular bundles of monocot stems: (a) are scattered; (b) form a ring; (c) have epidermal cells between them; (d) are mostly pith.

Ans: c
p. 475

_____ 8. Each of the following is part of bark except: (a) cork; (b) companion cells; (c) xylem; (d) phloem.

Ans: c
p. 475

_____ 9. The tissue responsible for producing new phloem and xylem in the growth in diameter of woody stems and roots is known as: (a) vascular bundles; (b) an apical meristem; (c) vascular cambium; (d) cork cambium.

Ans: a
p. 477

_____ 10. The plant structure responsible for photoperiodic measurements is the: (a) leaf; (b) stem; (c) flower; (d) root.

Ans: c
p. 479

_____ 11. CO_2 enters leaf cells: (a) by diffusing through the cuticle; (b) primarily through xylem transport; (c) by diffusing into mesophyll cells from the spaces inside stomata; (d) the same way water gets in.

Ans: b
p. 479

_____ 12. Stomatal opening is preceded by the transfer of _____ into guard cells. (a) magnesium ions; (b) potassium ions; (c) calcium ions; (d) starch.

Ans: b
p. 479

_____ 13. Plants with a reversed stomatal cycle typically: (a) open their stomata during daylight hours; (b) open their stomata during the night; (c) open their stomata in the morning and evening; (d) cannot open their stomata.

Ans: a
p. 482

_____ 14. Water movement in xylem: (a) is all passive; (b) requires active transport; (c) is dependent on the minerals dissolved in it; (d) proceeds from leaf to root in direction.

Ans: d
p. 484

_____ 15. Nonwoody plants depend on each of the following for support except: (a) cell turgidity; (b) sclerenchyma; (c) collenchyma; (d) wood fibers.

TEST SET 19-B

Name _____   Section _____

Choose the best answer to each of the following questions, and write the appropriate letter in the space provided.

Ans: a
p. 467

_____ 1. The site of production of organic molecules by a plant is known as the: (a) source; (b) mineral depot; (c) sink; (d) root hair.

Ans: d
pp. 468,
    469

_____ 2. Fibers occur in each of the following except: (a) flowering plants; (b) xylem; (c) phloem; (d) gymnosperms.

Ans: c
pp. 469,
    474

_____ 3. Each of the following statements is true of dicots except: (a) Most have taproot systems. (b) Some have fibrous root systems. (c) None is woody. (d) Many are herbaceous.

Ans: b
pp. 471,
    472

_____ 4. In young roots, the cells next to the epidermis are _____ cells. (a) endodermal; (b) cortical; (c) pericycle; (d) xylem.

Ans: c
p. 472

_____ 5. Seepage between cells of the endodermis is controlled by: (a) the deposition of cork; (b) a barrier composed of fibers; (c) Casparian strips; (d) lignin deposits.

Ans: a
pp. 474,
    475

_____ 6. The phloem of dicot stem vascular bundles is most commonly _____ with reference to xylem placement. (a) more toward the exterior; (b) more toward the interior; (c) at the same depth; (d) alternating.

Ans: d
pp. 472,
    474

_____ 7. _____ are in the same spatial location in both roots and stems of herbaceous dicots. (a) Epidermis and xylem; (b) Xylem and phloem; (c) Cortex and xylem; (d) Epidermis and cortex.

Ans: b
pp. 475,
    476

_____ 8. Nutrients are distributed from the periphery of stems to living cells in the interior of xylem in woody plants via: (a) vascular cambium; (b) vascular rays; (c) xylem vessels; (d) phloem.

Ans: d
pp. 475,
    476

_____ 9. Year after year, the laying down of wood will cause all of the following results except: (a) Early on, the epidermis will crack. (b) The cortex will crack. (c) Growth rings will occur. (d) Production of functional xylem will cease.

Ans: b
p. 477

_____ 10. Cells of leaves that carry out photosynthesis are _____ cells. (a) epidermal; (b) mesophyll; (c) phloem; (d) companion cells.

Ans: c
p. 479

_____ 11. The stomatal complex is composed of: (a) guard cells and mesophyll; (b) mesophyll and intercellular air spaces; (c) guard cells and stoma; (d) stoma and mesophyll.

Ans: a
p. 479

_____ 12. Stomata close when guard cells: (a) become flacid; (b) become turgid; (c) take up potassium ions; (d) carry out photosynthesis.

Ans: b
p. 479

_____ 13. Plants with reversed stomatal cycles store CO_2 as: (a) sugars; (b) organic acids; (c) calcium carbonate; (d) starch.

Ans: d
p. 481

_____ 14. Duckweeds typically have their stomata: (a) on the upper and lower epidermis; (b) on the lower epidermis; (c) on their stems; (d) on the upper epidermis.

Ans: c
p. 482

_____ 15. Water _____ causes tension due to cohesion of water molecules in the water column of xylem. (a) condensation; (b) ionization; (c) evaporation; (d) hydrolysis.

CHAPTER TWENTY
REPRODUCTION IN ANIMALS

CHAPTER SUMMARIES AND OBJECTIVES

20-1 <u>An Overview of Asexual Reproduction / Asexual Reproduction in Animals</u> In asexual reproduction, offspring are produced by a single parent and are genetically identical to the parent. Asexual reproduction allows for the production of large numbers of offspring, but it does not help to produce genetic diversity in a population. Such a population, therefore, does not have a variety of individuals on which natural selection can act.

There are several modes of asexual reproduction. Budding is the process whereby a new individual develops as an outgrowth of the parent. Many organisms can reproduce by regeneration, in which a new organism is grown from a portion of an older organism. Parthenogenesis, in which eggs develop without fertilization, occurs in several animal groups.

1. Explain the advantages and disadvantages of asexual reproduction.

2. Give some examples of asexual reproduction methods in animals.

20-2 <u>An Overview of Sexual Reproduction / Sexual Reproductive Systems of Animals / Vertebrate Reproductive Systems</u> In sexual reproduction, half the genome from each parent is pooled in the zygote (fertilized egg). Thus a great degree of genetic diversity can occur in a sexually reproducing population, due to the constant reshuffling of traits. Sexually reproducing animals exist as either males, which produce sperm, or females, which produce eggs. Organisms that possess both male and female reproductive systems in the same individual are said to be hermaphroditic. Such organisms frequently have mechanisms for avoiding self-fertilization. Fertilization may be external (occurring in the environment) or internal (occurring within the body of the female).

The eggs of vertebrates are produced in the ovaries of the female and typically contain yolk, stored food for the embryo. In mammals, the egg leaves the ovary and passes through the oviduct to the uterus. The terminal end of the uterus is the vagina, which receives the penis of the male. Sperm are produced in the testes and move via the vas deferens to the urethra, a tube leading through the penis.

3. Trace the path of an egg and a sperm through the reproductive tract of female and male mammals, respectively, and be familiar with the structure of the reproductive systems of other types of animals.

4. Explain why sexual reproduction is advantageous to a population.

20-3 <u>Gamete Production (Gametogenesis) / Spermatogenesis / Oogenesis</u> In gametogenesis, diploid germ cells give rise to haploid gametes. Gametogenesis occurs in three steps: mitotic proliferation of germ cells, reduction of the diploid number of chromosomes by meiosis, and maturation of the haploid cells into sperm and eggs.

Spermatogenesis is the process of sperm formation. Spermatogonia decrease in size to produce primary spermatocytes, which undergo the first division of meiosis to become secondary spermatocytes. Secondary spermatocytes undergo the second meiotic division to become spermatids. Spermatids, of which four are produced for every primary spermatocyte, mature into sperm. The sperm consists of a head, which contains the genetic material, a midpiece, which contains centrioles and mitochondria, and a tail.

Eggs are formed by the process of oogenesis. Oogonia enlarge to form primary oocytes. These complete the first division of meiosis to produce one secondary oocyte and the first polar body. The secondary oocyte gives rise, through the second meiotic division, to an ootid and a second polar body. The ootid matures into an ovum, or egg. Formation of the polar bodies results from asymmetrical division of the cytoplasm during meiosis, allowing most of the yolk to remain in the single cell that is destined to become the ovum.

5. Trace the steps in gametogenesis for both male and female animals.

6. Explain what is meant by the "polarity" of an ovum.

20-4 <u>Juxtaposing Egg and Sperm / Fertilization Events / Location of Mates and Courtship Among Invertebrates / Vertebrate Mate Location and Courtship</u> Fertilization is the coming together of egg and sperm. There are a number of means by which this is accomplished in the animal kingdom. Spawning involves the broadcasting of great numbers of gametes into the environment; fertilization occurs by chance. In amplexus, the male grips the female and stimulates her to release eggs, which he simultaneously fertilizes with his sperm. Sperm may also be placed by the male in a spermatophore, which is later found by the female and transferred to her body, where fertilization occurs internally. In mammals, sperm is transferred via the male's penis into the vagina of the female during the process of copulation. Copulation occurs in other animals as well.

In mammals the events involved in fertilization are as follows: Many sperm are generally introduced into the female reproductive tract. Sperm secrete hyaluronidase, which breaks up the cluster of cells surrounding the egg. The acrosomal reaction results in the release by the sperm of enzymes which digest substances covering the surface of the egg, allowing the plasma membranes of the gametes to contact each other. Contact between the plasma membranes causes them to fuse, and the nucleus of the sperm is drawn into the egg. Fusion of the plasma membranes also activates the egg, which begins to divide and develop into an embryo.

Among invertebrates, mate location is achieved by many channels of communication. These include: sound, chemical attractants, touch, and visual cues. Once a mate has been located, a period of courtship often ensues, the purpose of which in invertebrates is usually to ensure that the potential mate is of the correct species.

Vertebrates employ the same channels of communication to locate mates as are employed by invertebrates. Courtship may involve very complex behavior. In some mammals, ovulation may not occur until courtship and mating have taken place.

7. Describe the events involved in the fertilization of a mammalian egg.

8. Give several examples of mate location and courtship behavior in animals. At least one of your examples should be drawn from among the vertebrates.

20-5 <u>Protection of the Embryo / Nurture of the Young / Dispersal of the Young</u> The developing embryo may be protected in a variety of ways. In many insects, eggs are provided with a tough outer coating that retains moisture and shelters the developing embryo. The amniotic egg of birds and reptiles is a sealed container in which the embryo can develop in the fluid-filled amniotic cavity. Gas exchange occurs through the shell. Attached to the embryo are the yolk sac, chorion, amnion and allantois, membranes that participate in nourishment, gas exchange, and waste disposal.

In mammals, as well as in a number of other animals, the young are retained within the body of the mother until their development is largely completed. In marsupial mammals, the young are born at a very immature stage and continue their development in the mother's marsupium, or pouch. In placental mammals, including humans, the young develop within the uterus. The chorion of the embryo comes into contact with the uterine wall to form, along with the allantois and part of the uterus itself, the placenta. Juxtaposition of the blood vessels of mother and embryo within the placenta allows for the exchange of foodstuffs, gases, and waste materials.

Increasing the space colonized by an organism's offspring increases the chance that the organism's genes will be preserved in the new generation. Dispersal occurs most frequently by means of eggs or immature stages in most animals.

9. Describe each of the following methods for protecting developing young: amniotic egg, marsupium, placenta.

20-6 Reproductive System of the Human Male / Reproductive System of the Human Female / Sexual Responses in Humans The reproductive biology of humans is typical of mammals in general. The testes of the male are enclosed within a sac, the scrotum, and contain seminiferous tubules in which sperm are produced. Erection of the penis facilitates its insertion into the vagina of the female. Friction with the vaginal wall stimulates ejaculation, discharging semen into the vagina.

In the female, ovaries produce eggs, the vagina receives sperm from the male, and the uterus carries the developing embryo. The production of eggs (ovulation) follows a precise pattern that is under hormonal control. In the human female the uterine wall is first prepared to receive an embryo. This is followed by ovulation. If fertilization does not occur, the uterine wall breaks down and menstrual flow occurs. In other mammals there is usually no menstrual flow. During the time of ovulation, however, the females of many mammal species become sexually receptive. Human females may be sexually receptive throughout the ovarian cycle.

In both male and female humans, the sexual response consists of four phases: excitement, plateau, orgasm, and resolution. The responses of females differ from those of males in that females generally can experience multiple orgasms, whereas males generally cannot. Also, there is in males a refractory period following orgasm, during which neither erection nor orgasm can be achieved.

10. Describe the anatomy of both male and female human reproductive systems.

11. Answer: What are the similarities and differences between the sexual responses of men and women?

12. Outline the major events in the human ovarian cycle. Cite two ways in which the human cycle differs from that of most other mammals.

20-7 Pregnancy / Childbirth / Menopause Pregnancy begins with fertilization, which can occur within half an hour of ejaculation. Implantation of the zygote into the uterine wall occurs within a few days. The placenta forms within the first few weeks of fetal development, connected to the embryo by the umbilical cord. Fetal membranes also form early in development.

Pregnancy ends in childbirth, the onset of which is the various stages of labor. In the first stage, dilation of the cervix and contractions of the uterus begin. In the second stage, the baby's head becomes visible, and the infant is soon eased from the womb to begin an independent existence. The final stage of labor consists of the expulsion of the placenta and fetal membranes from the uterus.

Menopause is the cessation of menstruation and is one of the changes associated with the loss of reproductive ability in women from 45 to 60 years of age. Hormonal changes also occur at this time, as frequently do psychological problems.

13. Outline the events involved in a normal pregnancy, beginning with fertilization and ending with expulsion of the placenta.

ARTICLE RESOURCES

An Overview of Asexual Reproduction

Asexual Reproduction in Animals

An Overview of Sexual Reproduction

Sexual Reproductive Systems of Animals

Vertebrate Reproductive Systems

Gamete Production (Gametogenesis)

Spermatogenesis

Oogenesis

Juxtaposing Egg and Sperm

Fertilization Events

Location of Mates and Courtship Among Invertebrates

Vertebrate Mate Location and Courtship

   Wiley, R. Haven, Jr. The lek mating system of the sage grouse. Scientific American, 1978, 238(5), 114-125. In a lek system a small percentage of males mate with a large percentage of females. The males engage in an unusual courtship display that they perform in a territory they stake out. This territory is important. If it is within or near the mating center of the lek, the male occupying it has an excellent opportunity for multiple copulations.

Protection of the Embryo

   Rahn, Hermann, Amos Ar, and Charles V. Paganelli. How bird eggs breathe. Scientific American, 1979, 240(2), 46-55. How gases are exchanged and water is lost through the shells of bird's eggs. Details the growth of the embryo within the egg. The shape and size of egg shell pores is explored for various species of birds. The rate of diffusion of gases is explored. The gas conductance of eggs increases in proportion to the size of the egg. Good charts and photomicrographs.

   Randal, Judith. Breeding the perfect cow. Science 81, 1981, 2(9), 86-92. Surrogate cows can produce as many as 89 calves by means of embryo transfer. Reproduction engineering has become a major factor in animal husbandry.

Nurture of the Young

   Wolpert, Lewis. Pattern formation in biological development. Scientific American, 1978, 239(4), 154-164. The positioning of cells determines the pattern of development of the embyro. How this is accomplished is yet to be discovered.

Dispersal of the Young

Reproductive System of the Human Male

Reproductive System of the Human Female

   Sloane, Ethel. Dispelling myths about female potential. The Science Teacher, 1980, 47(7), 14-21. A summary of what is known about female reproductive physiology. Also a history of myths surrounding the female potential from early

Egyptian times to the present. A clear presentation of the menstrual cycle.

Sexual Response in Humans

White David. Pursuit of the ultimate aphrodisiac. Psychology Today, 1981, 15(9), 9-12. Discussion of human pheromones. Research from England and the United States indicates that more questions have been raised than have been answered.

Pregnancy

Williams, Douglas, Jon Hendriks, and Thomas Mertens. Forging the rights of the unborn. The Science Teacher, 1978, 45(8), 24-25. Brief but lucid article on fetal research. Distinguishes between therapeutic and nontherapeutic research. Poses some very real ethical questions. Guidelines issued by the National Commission for Protection of Human Subjects of Biomedical and Behavioral Research, 1975, are given.

de St. Jorre, John. The morning sickness drug controversy. The New York Times Magazine, 1980, 113, 124-128 (Oct. 12). A controversy has arisen over bendectin, a drug used for nearly a quarter of a century to control morning sickness. Studies have shown that bendectin may cause birth defects, though not on the order of thalidomide.

Winick, Myron. The web of hunger: Food and the fetus. Natural History, 1981, 90(1), 76-81. A discussion of physiological changes during pregnancy that are affected by nutrition. Improper nutrition will reduce the necessary increase in maternal blood volume as well as placental blood flow. More energy is expended during lactation, which is prepared for during pregnancy.

Beaconsfield, Peter, George Birdwood, and Rebecca Beaconsfield. The placenta. Scientific American, 1980, 243(2), 94-102. Origins of the placenta and its importance during pregnancy are stressed. Its complementary functions to both mother and embryo are featured. This makes it a unique organ for investigating many subjects ranging from drug effects, genetic code, normal growth, organ transplant, and hormone synthesis.

Childbirth

Menopause

Reproduction Summarized

Fleagle, John J. In the beginning.
Murphy, Cullen. A survey of the research.
Dick Howard, A. E. The sexes and the law.
    The Wilson Quarterly, 1982, 6(1), 51-95. A series of three articles devoted to sexual differences viewed from the perspectives of evolution, cultural development, and the law. Research in sexual differences with regard to behavior, development, and learning is explored. The law continues to view male and female differently. Good for both instructor and student.

Caplan, Arnold I. Embryological development: Evolutionary history, genetic bias, and cellular environment control the flow of developmental events. Parts I and II. Journal of College Science Teaching, 1981, x(4 and 5), 226-230, 289-293. A lengthy two-part article on embryology in which the extracellular connections between neighboring cells, tissues, and organs provide information for evolutionary and genetic potential. This expands on the material in the text and can be of use to instructor and student.

ESSAY QUESTIONS

1. What are some of the pros and cons of asexual and sexual methods of reproduction?

    Asexual reproduction allows for the production of large numbers of offspring but provides no opportunity for recombination of the genetic material within a population. The only source of genetic diversity in a species that reproduces asexually is mutation. Sexual reproduction, on the other hand, requires a substantial investment of energy for producing gametes, bringing them together, and (frequently) caring for the offspring. Balancing this cost, however, is the accrual of a great variety of individuals within a sexually reproducing population. Each individual has a different genetic makeup on which natural selection can act.

2. Briefly outline the events involved in the formation of sperm and egg cells. Include the concept of polarity in your answer. How do the polar bodies sometimes function in animals that do not discard them?

Male germ cells are called spermatogonia. These develop into primary spermatocytes. Primary spermatocytes undergo the first meiotic division to become secondary spermatocytes, which in turn undergo the second meiotic division to become spermatids. Spermatids differentiate into sperm cells, developing a head, midpiece, and tail. Primary oocytes develop from oogonia in the female. The first meiotic division produces a secondary oocyte and the first polar body. Cytoplasmic distribution is asymmetrical, leaving the secondary oocyte with most of the yolk. The secondary oocytes of mammals do not undergo the second meiotic division until after entrance of the sperm. The secondary oocyte divides meiotically to produce an ootid, which matures into an ovum, and a second polar body. As the ovum matures, it developes polarity, a distribution of cytoplasmic substances throughout the egg such that different parts of the egg give rise to different parts of the embryo. In animals that do not discard the polar bodies, one polar body recombines with the ovum to produce a diploid cell that develops into an embryo without fertilization by a sperm.

3. Give three examples of means, other than copulation, employed by animals to ensure that sperm and egg find each other.

Amplexus, which occurs in frogs, for example, ensures that male and female will release gametes into the environment at the same time and place. The palolo worm accomplishes the same thing by releasing body segments filled with gametes at a precise time of the year. The body segments swarm on the surface of the ocean, releasing gametes. In many terrestrial arthropods and some salamanders, the male deposits sperm in a spermatophore. This may be found at a later time by the female, which picks up the spermatophore and introduces sperm into her body, with fertilization occurring internally.

4. For each of the channels of sexual communication discussed in the text (sound, chemical, touch, and visual) give two examples, one drawn from among the invertebrates and one drawn from among the vertebrates.

Sound: Male mosquitos are attracted to the hum of the female. Male songbirds sing to encourage females to enter their territories to mate. Chemical: Gypsy moth females secrete disparlure, which the males "home in on" to arrive in the female's vicinity. Female dogs in heat secrete a scent that attracts male dogs from long distances away. Touch: Males of some species of spiders must pluck the strands of the female's web in a highly ritualized manner to inform her that he is a male spider of her species, rather than a food item. Ovulation in some mammals (rabbits, for example), does not occur unless the female has been stimulated by insertion of the penis. Visual: Male fireflies flash rhythmically and are answered by females on the ground. The males land and mate with females that emit the appropriate pattern of flashing. Mating in the three-spined stickleback involves a complicated exchange of visual cues between the male and female fishes. For example, the swollen abdomen of a ripe female stimulates the male to begin his zigzag dance, which initiates the courtship procedure. The female, in turn, responds to the red underbelly of the male fish.

5. Name two ways in which the sexual response of women differs from that of men. Name two ways in which the ovarian cycle of humans differs from that of most other mammals.

    Women generally can experience several orgasms in rapid succession, whereas men generally cannot. And in men there is a refractory period following ejaculation during which another orgasm is not possible, whereas there is no refractory period in females. Most female mammals do not experience menstrual bleeding as human females do, and the females of most mammal species are sexually receptive only during a certain time of the ovarian cycle or at a particular season of the year. Human females may be sexually receptive at any time.

6. Sketch your own reproductive system and label at least six of the parts.

    The sketch should correctly and clearly show at least the following structures: (males) testis, scrotum, epididymis, vas deferens, urethra, penis; (females) ovary, oviduct, uterus, cervix, vagina, clitoris.

7. What is the most probable explanation of the complex series of events that must precede the fusion of the nuclei of sperm and egg? List these events in their correct order.

    Fertilization events, including capacitation, the release of hyaluronidase, the acrosomal reaction, and the formation of the fertilization membrane all serve to (1) ensure the entry of only sperm produced by a male of the correct species and (2) prevent the entry of more than one sperm into a single egg.

TEST SET 20-A

Name _____   Section _____

Choose the best answer to each of the following questions and write the
appropriate letter in the space provided.

Ans: c
p. 489

_____ 1. The development of an unfertilized egg is called: (a) zygotic
maturation; (b) regeneration; (c) parthenogenesis; (d) sexual
reproduction.

Ans: a
p. 491

_____ 2. An individual possessing either a female or a male reproductive
system is termed: (a) dioecious; (b) parthenogenic; (c) hermaphroditic; (d) monoecious.

Ans: a
p. 493

_____ 3. Proliferation of germ cells in males occurs by a process called:
(a) mitosis; (b) meiosis; (c) spermatogonia; (d) parthenogenesis.

Ans: b
p. 495

_____ 4. The function of the midpiece of a sperm is to: (a) carry the
genetic information; (b) provide energy for movement; (c) carry
the enzymes that facilitate fertilization; (d) orient the pattern
of microfilaments for motility.

Ans: a
p. 496

_____ 5. The two meiotic divisions in oogenesis are unique because the:
(a) cytoplasmic division is asymmetrical; (b) first meiotic
division produces ootids; (c) second meiotic division produces
secondary oocytes; (d) reduction division is the second meiotic
division.

Ans: a
p. 498

_____ 6. The enzyme hyaluronidase affects the process of fertilization
by disrupting the: (a) cells surrounding the egg; (b) fertilization membrane; (c) acrosome; (d) bindin attachment site.

Ans: b
p. 499

_____ 7. Species specificity during fertilization in invertebrates is
maintained by the: (a) specific activity of hyaluronidase
enzymes; (b) specific bindin-receptor molecule reaction;
(c) formation of several fertilization membranes; (d) second
meiotic division of the ovum.

Ans: a
p. 502

_____ 8. Courtship in stickleback fishes is initiated via: (a) visual
cues; (b) audible cues; (c) bioluminescence; (d) chemical
attractants.

Ans: c
p. 505

_____ 9. The membrane that serves to "waterproof" the contents of the
amniotic egg is the: (a) amnion; (b) allantois; (c) chorion;
(d) placenta.

Ans: b
p. 505

_____ 10. Which of the following membranes is part of the embryonic
contribution to placenta formation? (a) amnion; (b) chorion;
(c) uterine membrane; (d) fertilization membrane.

Ans: a
p. 509

_____ 11. At what age does cell division in potential ova cease?
(a) prenatal; (b) preadolescence (1-12 years); (c) adolescence
(12-19 years); (d) menopause (40-50 years).

Ans: d
p. 508

_____ 12. The function of the Cowper's glands is to: (a) store mature
sperm; (b) initiate spermatogenesis; (c) lower the temperature
to the optimum required for sperm maturation; (d) aid in semen
formation.

Ans: a
p. 512

_____ 13. _____ cross(es) the placenta by diffusion. (a) Oxygen;
(b) Vitamins; (c) Hormones; (d) Antibodies.

Ans: b
p. 509

_____ 14. In human females, ovulation occurs: (a) every 14 days; (b) once a month; (c) every 24 hours; (d) only during the menstrual flow.

Ans: c
p. 515

_____ 15. Cessation of menstruation in mature women is called: (a) menses; (b) meninges; (c) menopause; (d) meningitis.

TEST SET 20-B

Name _____   Section _____

Choose the best answer to each of the following questions, and write the
appropriate letter in the space provided.

Ans: a
p. 489

_____ 1. A recombination of genetic information within a species occurs
           in the: (a) zygote; (b) process of parthenogenesis; (c) pro-
           duction of gemmules; (d) process of fragmentation.

Ans: c
p. 491

_____ 2. An individual possessing both a female and a male reproductive
           system is termed: (a) dioecious; (b) parthenogenic;
           (c) monoecious; (d) conjugated.

Ans: d
p. 492

_____ 3. Eggs from females with seminal receptacles: (a) develop par-
           thenogenetically; (b) develop within the gonads; (c) are
           fertilized externally; (d) are fertilized internally.

Ans: c
p. 492

_____ 4. In humans, spermatogenesis occurs in the: (a) vas deferens;
           (b) epididymis; (c) seminiferous tubules; (d) seminal vesicles.

Ans: c
p. 495

_____ 5. The function of the acrosome of a sperm is to: (a) carry the
           genetic information; (b) provide energy for movement; (c) carry
           the enzymes that facilitate fertilization; (d) orient the pattern
           of microtubules for motility.

Ans: c
p. 496

_____ 6. The process of oogenesis produces one ovum and: (a) one polar
           body; (b) two polar bodies; (c) three polar bodies; (d) four
           polar bodies.

Ans: a
p. 500

_____ 7. During oogenesis in mammals, the second meiotic division occurs:
           (a) after the entry of a sperm; (b) after fusion of sperm and
           egg nuclei; (c) before the egg is released by the ovary;
           (d) after implantation.

Ans: b
p. 501

_____ 8. Fireflies locate mates via: (a) rhythmic touch patterns;
           (b) bioluminescence; (c) audible cues; (d) chemical attractants.

Ans: d
p. 499

_____ 9. The uptake of calcium ions by a sperm ultimately aids in the:
           (a) meiotic division of the sperm nucleus; (b) production of
           hyaluronic acid by the sperm; (c) stabilization of the acrosome
           membrane; (d) formation of the fertilization membrane.

Ans: a
p. 504

_____ 10. The membrane that surrounds the developing embryo in an amniotic
            egg is the: (a) amnion; (b) allantois; (c) chorion; (d) placen-
            ta.

Ans: c
p. 505

_____ 11. The _____ is part of the maternal contribution to placenta
            formation. (a) amnion; (b) chorion; (c) uterine membrane;
            (d) fertilization membrane.

Ans: d
p. 510

_____ 12. One of the major differences between the sexual response cycles
            in males and females is the: (a) increase in blood pressure in
            males; (b) increase in heart rate in females; (c) presence of a
            refractory period in females after orgasm; (d) presence of a
            refractory period in males after orgasm.

Ans: c
p. 514

_____ 13. _____ cross(es) the placenta by active transport. (a) Oxygen;
            (b) Urea; (c) Hormones; (d) Antibodies.

Ans: b   _____   14.   In humans, fertilization occurs in the: (a) ovaries; (b) ovi-
p. 511   ducts; (c) vagina; (d) uterus.

Ans: b   _____   15.   Menopause in women is accompanied by the cessation of all:
p. 515   (a) ovarian functions; (b) ovulation; (c) vaginal functions;
                            (d) clitoral functions.

CHAPTER TWENTY-ONE
ANIMAL DEVELOPMENT

CHAPTER SUMMARIES AND OBJECTIVES

21-1 <u>Embryos and Animal Affinities / Preformation versus Epigenesis</u> An embryo is an animal or plant in an early stage of development, that directly or indirectly obtains food from its mother. A larva is an immature form of an animal, which feeds independently and differs in appearance from the adult. Development occurs at all levels of organization. Differentiation is the phenomenon whereby cells become specialized. Morphogenesis involves the establishment of particular shapes or arrangements of tissues. Development also involves growth, the increase in mass from cell division and cell expansion.

Before the publication of the Darwinian theory of evolution, Karl von Baer published the first major presentation of developmental biology. Baer's law states that the features that are shared among a group of organisms appear earlier in embryonic development than do those features that are unique to a particular species. When the theory of evolution was adopted, von Baer's statements came to be re-interpreted as "otongeny recapitulates phylogeny." This is a statement of the "biogenetic law." There are many exceptions to this generalization, but it helps illuminate relationships between embryos that are not apparent in adult organisms.

Early developmental biologists were divided into two schools of thought concerning the way in which a spherical zygote produces an adult typical of the species. The earliest theory, the theory of preformation, held that a miniature adult is present within the zygote. Later, the alternative theory of epigenesis, which held that the adult was built from materials present in the zygote, was proposed. Modern views contain elements of both theories: epigenetic development operates on the basis of the genes, a set of preformed instructions.

1. Summarize early theories on the nature of development, and explain how the modern view embodies elements of both early ideas.

2. Discuss the experiments of Roux and Driesch.

21-2 <u>Establishing Multicellularity / Formation of the Blastula</u> The zygote gives rise to hundreds of cells through the process of cleavage. The pattern of cleavage is determined by a number of factors. One factor is the distribution of yolk within the egg. Another is the degree of specificity of early blastomeres. In some animals, the fates of the individual blastomeres may be fixed very early, whereas in other animals, loss of some blastomeres in early development is compensated for by the other cells. The former pattern is called mosaic development; the latter pattern is called regulative development.

Despite many differences, the process of cleavage is similar in all animals with regard to several of its aspects. Mitosis is rapid at this stage, resulting in the formation of a hollow ball or disk of cells, the blastula.

3. Outline the early steps in the formation of an embryo from a zygote.

21-3 The Formation of Embryonic Germ Layers / Gastrulation in Chordate Embryos / Sheet to Tube to Brain: Forming the Nervous System Gastrulation, the formation of a gastrula from a blastula, involves the movement of cells from the surface of the embryo to the interior. Gastrulation results in the formation of a two-layered embryo at first, with an inner layer of endoderm and an outer layer of ectoderm. Later, a layer of mesoderm forms between the endoderm and ectoderm. Invagination during gastrulation forms a new cavity, the archenteron, which opens to the exterior through the blastopore.

In chordate embryos, the notochord forms from a portion of the mesoderm layer, establishing the anterior-posterior axis of the body. In fishes and birds the blastula is disk-shaped, and gastrulation results in the formation of the primitive streak.

The nervous system of the frog may be used as an example of the development of an organ system from the gastrula. By the time the process of neurulation begins, the gastrula has three embryonic germ layers, and the anterior-posterior axis is established. Neurulation gives the embryo left and right sides. The dorsal side of the embryo thickens to form the neural plate, the edges of which begin to fold inward to form the neural groove. As development proceeds, the edges of the neural groove come together, forming a closed neural tube. After formation of the neural tube, the anterior end becomes segmented into compartments that will later form parts of the brain.

4. Trace the development of an embryo from blastula through the neural tube stage.

5. Define the terms gastrula, gastrulation, endoderm, ectoderm, mesoderm, archenteron, primitive streak, and neurulation.

21-4 Prospective Fate and Prospective Potency / Induction and the Organizer / Instructive versus Permissive Induction It is possible to determine which structures in the embryo will give rise to which structures in the adult. The range of possible developmental behaviors of blastoderm cells is their prospective potency, whereas the prospective fate of these cells is determined by their location in the embryo--that is, their relationships to other cells. By the late gastrula stage, tissue has a prospective potency identical with its prospective fate. That is, the fate of the tissue has been determined by the late gastrula stage.

Experiments have shown that determination of the fate of a specific tissue is the result of interaction with other specific tissues. For example, tissue from the dorsal lip of the blastopore in an amphibian embryo induces the formation of a neural tube in the ectodermal tissue with which it is in contact. Similarly, contact of the optic vesicles of the developing forebrain with the overlying ectoderm causes the formation of a lens placode in the overlying ectoderm. The lens placode eventually forms the lens of the eye. Experimental evidence indicates that inductions such as these are due to the transfer of chemicals, called inducers, from one tissue to the other. The exact chemical nature of these inducers has not been determined.

Induction may be either instructive or permissive. In instructive induction, the inducer determines which of several paths the target tissue is to follow. In permissive induction, the inducer merely triggers a preset developmental pattern in the target tissue.

6. Differentiate between the members of each of the following pairs: prospective fate / prospective potency, instructive induction / permissive induction.

7. Briefly describe an experiment demonstrating the existence, action, or nature of the process of embryonic induction.

21-5 Asymmetry in the Egg and Zygote Polarity, a difference between one end and the other, develops early, even in the egg itself. An unequal distribution of materials in the cytoplasm of the egg is maintained in the zygote. In the process of cleavage, these materials become asymmetrically distributed among the blastomeres. Such cytoplasmic differences mediate some embryological events. For example, it can be shown experimentally that normal development in the sea urchin requires material from both the animal and vegetal poles of the embryo. Unequal distribution of materials between these two halves is present even in the unfertilized egg. Similarly, the material in the gray crescent of amphibian eggs is required for normal development. Forcing the egg to divide in such a way that one blastomere contains all of the gray crescent and the other one produces one normal embryo (from the part with the gray crescent) and an unorganized mass of cells.

8. Explain how the experiments of Hörstadius and of Spemann demonstrated that part of embryonic development is attributable to unequal distribution of materials in the egg cytoplasm.

21-6 Aggregation Behavior in Embryonic Cells Cellular affinities may underlie many of the phenomena of gastrulation. When embryonic cells are separated in the laboratory and allowed to reaggregate, they sort themselves out into endoderm, ectoderm, and mesoderm, with all of the tissues in their proper orientation. Apparently, cells of the same type preferentially associate with each other.

9. Outline the experiments Holtfreter conducted in elucidating the nature of affinities between embryonic cells. How could such affinities account for the phenomena associated with gastrulation?

21-7 Postembryonic Development / Larval Development and Metamorphosis / Regeneration / Senescence and Death Development does not, of course, cease with the development of the embryo. Many changes, including further differentiation and morphogenesis, as well as extensive growth, mark postembryonic development. Rates and patterns of growth vary from one type of organism to another, and different parts of an organism may grow at different rates.

Many species go through a larval stage. The dramatic changes associated with the maturation of the larva into an adult are called metamorphosis. The change from a tadpole to a frog is an example. Complete metamorphosis in insects is a familiar pattern of larval development.

Some animals have the ability to replace lost parts with new ones. This phenomenon is called regeneration. The process seems to be under the control of the nervous system. The ability of a cut stump to regenerate a lost limb is evidence that the cells of the stump have not undergone irreversible changes in their nuclei during the developmental process.

The life history of some animals includes a period of senescence, or age-associated decline in the individual's physical capacities. Senescence in humans is frequently referred to as "old age." A growing body of evidence suggests that the events of senescence and death are programmed into the individual's genome.

10. Trace the various pathways of postembryonic development mentioned in the text. These include larval development and metamorphosis, patterns of growth, regeneration, and senescence.

ARTICLE RESOURCES

Embryos and Animal Affinities

Preformation versus Epigenesis

Establishing Multicellularity

Formation of the Blastula

The Formation of Embryonic Germ Layers

Gastrulation in Chordate Embryos

Sheet to Tube to Brain: Forming the Nervous System

Meier, Stephen, and Patrick P. L. Tam. Metameric pattern development in the embryonic axis of the mouse. I. Differentiation of the cranial segments. Differentiation, 1982, 21(2), 95-108. Scanning electron microscope analysis of the mouse embryo during early stages of development shows the presence of somitomeres at the cranial end of the primitive streak and on either side of Hensen's node. The reading material is difficult, but the photomicrographs are spectacular.

Tam, Patrick P. L., Stephen Meier, and Anton G. Jacobson. Differentiation of the metameric pattern in the embryonic axis of the mouse. II. Somitomeric organization of the presomitic mesoderm. Differentiation, 1982, 21(2), 109-122. Scanning electron microscope analysis of the mouse embryo aged 8-11 days post coitum. The tail portion of the embyro was isolated and cut into lateral halves, thus exposing the presomitic mesoderm. Pages of photomicrographs illustrate the discussion. The captions are well done.

Cowan, Maxwell W. The development of the brain. Scientific American, 1979, 241(3), 112-133. Major regions of the brain can be traced by labeling various areas of the neural plate at a very early embryonic stage. The development of the nervous system from the ectoderm is traced. Migration of nerve cell nuclei in epithelial tissue forming the wall of the embryo is shown. Good drawings and photographs.

Prospective Fate and Prospective Potency

Induction and the Organizer

Instructive versus Permissive Induction

Asymmetry in the Egg and Zygote

Aggregation Behavior in Embryonic Cells

Postembryonic Development

Winick, Myron. The web of hunger: Food and the fetus. Natural History, 1981, 90(1), 76-81. Physiological adjustments to pregnancy that are affected by nutrition are discussed. The required increase in maternal blood volume and placental blood flow are discussed. A mother who is inadequately nourished will have a fetus whose growth has been impaired. Underweight pregnant women must gain an average of 25 pounds during pregnancy in order to produce an infant of normal size.

Larval Development and Metamorphosis

Nijhut, H. Frederick. The color patterns of butterflies and moths. Scientific American, 1981, 245(5), 140-151. Pigments as well as small structural features of the scales of the wings of butterflies and moths are responsible for the patterns appearing on the wings. The rules of design were discovered in the 1920s. About 75 percent of the color patterns of moths and butterflies can be derived from these rules of design. Positional information seems to be the clue to the appearance of circles and rings. Cauterization and implantation are the two methods employed for determining the sources of positional information.

Youngqren, Newell A., and Mac E. Hadley. Thyroid hormones and amphibian metamorphosis. The American Biology Teacher, 1981, 43(1), 21-25. Amphibian metamorphosis is initiated and controlled by a number of complex hormones of the thyroid and pituitary glands. A clear chart showing these relationships is given. Descriptions of experiments to demonstrate metamorphic change, quantitative aspects of hormonal action, and the effect of antithyroid drugs such as thiocarbamides on metamorphosis.

Regeneration

Senescence and Death

Fries, James F. Aging, natural death, and the compression of morbidity. New England Journal of Medicine, 1980, 303(3), 130-135. The period of senescence will decrease, as will extensive health care for the elderly. This is the prediction of Dr. Fries. In this century alone the life span has gone from 47-73 years, but the maximum life span has not shown a comparable increase. This is because we have eliminated a great many of the causes of infant death. A fleeting bout with senescence after a life of rigorous physical, emotional, and intellectual activity is envisioned.

ESSAY QUESTIONS

1. Outline the major features of animal development from the time of fertilization until the completion of neural tube formation.

    Cleavage begins with the first mitotic division of the zygote and continues until hundreds or thousands of embryonic cells, or blastomeres, are produced. These cells are organized into a hollow ball or disk-shaped structure called the blastula. Mitosis during cleavage is very rapid, and the total volume of the blastula is similar to that of the zygote. The blastula consists of a sheet one cell thick, called the blastoderm, surrounding an internal cavity, or blastocoel. Invagination of cells of the blastula, beginning at the dorsal lip of the blastopore in amphibians, produces a two-layered gastrula. The internal cavity of the gastrula, the archenteron, is lined by endoderm cells and is open to the outside through the blastopore. The outer layer of the gastrula is composed of ectoderm cells. Between these two layers, the third embryonic germ layer, the mesoderm, forms. The process of neurulation begins when cells from the dorsal lip of the blastopore induce a thickening in the ectoderm at the anterior end of the embryo. The ectoderm begins to fold upward, forming a neural groove, the edges of which eventually come together to form the neural tube. By the neural tube stage, the embryo has a definite anterior-posterior axis, a gut cavity (the archenteron) open to the outside through what will become the anus (the blastopore), and a definite left and right side (as a result of neurulation).

2. What is meant by the term embryonic induction? Explain how induction plays a role in the formation of the eye in frog embryos. Identify the biologist who first conducted experiments on eye formation in the frog, and cite experimental evidence suggesting that inducers are chemical in nature.

    Embryonic induction is the process whereby the differentiation of specific embryonic tissues is influenced by reciprocal contacts with other specific tissues. Lewis's experiments with eye development in frogs, for example, demonstrated that a complex series of induction events take place. Swellings of the anterior portion of the neural tube form optic vesicles, which come into contact with ectoderm of the head region. The optic vesicle induces formation of a lens placode in the overlying ectoderm. The lens placode develops into a structure that will ultimately become the lens of the eye. The lens, in turn, induces the formation of an optic cup from the optic vesicle vesicle and also induces the ectoderm overlying the lens to form the cornea of the eye. Lewis found that placing an impenetrable barrier between the lens placode and the optic cup prevented the development of the lens, suggesting that the induction is mediated by chemicals exchanged between the inducing tissue and the target tissue.

3. Cite experimental evidence for instructive induction in the differentiation of wing ectoderm in the chick embryo.

    Wing ectoderm, which would normally produce wing feathers, can be caused to produce thigh feathers, or even scales or claws, simply by transplanting it to overlie appropriate mesodermal tissue. That is, wing ectoderm transplanted to a region over thigh mesoderm will produce thigh feathers. This experiment demonstrates that the chemical signals transmitted from mesoderm to ectoderm are specific in nature.

4. Contrast the theories of preformation and epigenesis, and describe the work of Roux and Driesch in resolving this controversy. How does the modern view of development encompass both these early theories?

The theory of preformation suggested that all features manifested in the adult were present in a preformed condition in the zygote. Some biologists claimed to be able to see a tiny homunculus, or miniature human, within human eggs or sperm. The alternative theory, epigenesis, suggested that the adult organism is constructed of materials present within the zygote. Roux attempted to resolve this controversy by destroying one of the cells of a sea urchin embryo at the two-celled stage. The result was an abnormal embryo. It was concluded that the theory of preformation was supported. Three years later, Driesch repeated Roux's experiments but was careful to separate the two cells completely, rather than simply destroying one and leaving it attached to the other. This resulted in the development of a normal embryo, and the theory of epigenesis was upheld. The modern view of development embodies elements of both these ideas. Development is epigenetic; there is no preformed individual within the zygote. However, the set of instructions by which development is directed <u>are</u> preformed. They are the genes inherited by the zygote from its parents.

5. What evidence can you offer to support the idea that human senescence and death may be genetically programmed?

    The suggestion that some sort of "timer" runs throughout the life of an individual is supported by two lines of evidence. First, there are a number of genes that are not expressed until relatively late in life, long after maturity. Second, the experiments of Hayflick on cultured human cells indicate that a nucleus cannot continue to divide forever. The older the person from whom cells are cultured, the fewer cell divisions occur before the culture dies. Cells from human embryos can divide about 50 times, whereas those from middle-aged persons can divide only about 20 times. Transplantation of young nuclei into older cells increases the number of divisions that the cells can undergo before dying.

6. Do you think it is theoretically possible for a human being to regenerate an amputated limb? Why?

    The evidence suggests that all cells in the body are totipotent. That is, the changes that occur during differentiation are not irreversible. It further appears that regeneration in other vertebrates is under control of the nervous system. Perhaps electrical stimulation of cells at the stump of an amputated limb could be used to stimulate cell division and consequent regeneration of the limb.

TEST SET 21-A

Name _____   Section _____

Choose the best answer to each of the following questions, and write the appropriate letter in the space provided.

Ans: a
p. 519

_____   1. Which of the following statements expresses Baer's law?
(a) Shared features of a large group of animals appear early in embryonic development. (b) Shared features of a large group of animals appear late in embryonic development. (c) Unique features of individuals are predominantly expressed early in embryonic development. (d) Unique features of individuals are expressed only in the zygotic stage.

Ans: b
p. 520

_____   2. Homologous structures are formed from: (a) similar zygotes; (b) similar embryonic structures; (c) different embryonic tissues; (d) different patterns of hormonal activity in young adults.

Ans: c
p. 521

_____   3. The experiments of Driesch using sea urchin eggs showed that: (a) differentiation had occurred at the two-cell stage; (b) differentiation had not occurred at the two-cell stage; (c) preformation was a viable hypothesis for embryonic development; (d) epigenesis was not a viable hypothesis for embryonic development.

Ans: d
p. 523

_____   4. In frog eggs, the gray crescent forms: (a) at the vegetal pole; (b) between the animal and vegetal poles; (c) opposite the site of sperm entry; (d) at the site of sperm entry.

Ans: a
p. 523

_____   5. The development of identical twins in humans is an example of: (a) regulative development; (b) mosaic development; (c) differentiation in the zygote; (d) asymmetrical cleavage in the zygote.

Ans: d
p. 525

_____   6. The outer germ layer in a gastrula is termed the: (a) blastoderm; (b) endoderm; (c) mesoderm; (d) ectoderm.

Ans: d
p. 527

_____   7. In chordates, the blastopore will ultimately become the: (a) digestive tract; (b) nervous system; (c) mouth; (d) anus.

Ans: a
p. 530

_____   8. Neurulation is the result of differentiation in the: (a) ectoderm; (b) mesoderm; (c) endoderm; (d) archenteron.

Ans: d
p. 531

_____   9. The process of determination is nearly complete by the: (a) early blastula stage; (b) late blastula stage; (c) early gastrula stage; (d) late gastrula stage.

Ans: b
p. 532

_____   10. The embryonic organizer is the: (a) ectoderm; (b) dorsal lip of the blastopore; (c) neural tube; (d) primitive streak.

Ans: a
p. 533

_____   11. Lewis's studies on lens formation in the frog eye showed that: (a) optic vesicles induce ectoderm to form a lens placode; (b) lens placodes induce the formation of optic vesicles in ectoderm; (c) optic vesicles differentiate into lens placodes; (d) lens placodes differentiate into optic vesicles.

Ans: d
p. 537

_____   12. Studies by Holtfreter on dissociation and reaggregation of embryonic tissue showed that: (a) the ectoderm is influenced by the mesoderm to form ectoderm-mesoderm aggregates; (b) the endoderm is influenced by the mesoderm to form endoderm-mesoderm aggregates; (c) the three germ layers will reaggregate in inverse order after dissociation; (d) the three germ layers will reaggregate in the correct order after dissociation.

Ans: b _____ 13. Hayflick's studies on human cells grown in tissue culture showed
p. 541   that: (a) the cells have an infinite life span; (b) the cells
              have a finite life span; (c) the senescence patterns are directed
              by cytoplasmic events; (d) the senescence patterns are directed
              by the contact with other cells.

Ans: c _____ 14. Imaginal disks first appear in the insect: (a) adults; (b) pupa;
p. 539   (c) larva; (d) zygote.

Ans: a _____ 15. Programmed cell death may be the result of: (a) gene expression;
p. 540   (b) an increase in the total amount of DNA; (c) induction of
              embryonic tissue; (d) regeneration.

TEST SET 21-B

Name _____   Section _____

Choose the best answer to each of the following questions, and write the appropriate letter in the space provided.

Ans: a
p. 520

_____ 1. One of the criteria for classifying animals into different phyla is the number of similar features shared by the: (a) embryos; (b) adolescent organisms; (c) young adult organisms; (d) mature adult organisms.

Ans: b
p. 520

_____ 2. "Ontogeny recapitulates phylogeny" is a statement of: (a) Baer's law; (b) the biogenetic law; (c) the genetic law; (d) the morphogenetic law.

Ans: d
p. 522

_____ 3. The initial activation of the zygote results in: (a) a series of meiotic divisions; (b) a large increase in total mass as a result of a series of cell divisions; (c) a single meiotic division followed by a series of mitotic divisions; (d) cleavage of the cytoplasm.

Ans: b
p. 523

_____ 4. The initial plane of cleavage in frog eggs depends on the position of the: (a) blastomeres; (b) gray crescent; (c) vegetal pole; (d) zygote.

Ans: b
p. 523

_____ 5. Differentiation in the echinoderm zygote results in a _____ pattern of development. (a) mosaic; (b) regulative; (c) biogenetic; (d) symmetrical.

Ans: c
p. 525

_____ 6. The middle germ layer in a gastrula is termed the: (a) blastoderm; (b) endoderm; (c) mesoderm; (d) ectoderm.

Ans: b
p. 527

_____ 7. In chordates, the digestive cavity is formed from the: (a) blastocoel; (b) archenteron; (c) blastopore; (d) endoderm.

Ans: d
p. 530

_____ 8. Initial thickening of the ectoderm results in the formation of the: (a) archenteron; (b) neural tube; (c) neural groove; (d) neural plate.

Ans: a
p. 532

_____ 9. Transplanting cells from one part of the chordate blastula to another results in: (a) the rearrangement of tissues in the gastrula; (b) the formation of two embryos; (c) no significant change in embryonic development; (d) the death of the embryo.

Ans: b
p. 532

_____ 10. The embryonic organizer induces the formation of a(an): (a) archenteron; (b) neural tube; (c) digestive cavity; (d) primitive streak.

Ans: c
p. 533

_____ 11. The development of feathers in a chicken is determined by the type of inducing: (a) ectoderm; (b) endoderm; (c) mesoderm; (d) blastopore.

Ans: b
p. 539

_____ 12. Replacement of liver cells after injury is: (a) metamorphosis; (b) regeneration; (c) differentiation; (d) impossible.

Ans: d
p. 536

_____ 13. Spemann's studies on frog embryogenesis showed that: (a) the mosaic pattern is established with the first cleavage; (b) the gray crescent maintains the vegetal pole; (c) the gray crescent results from differentiation in the zygote; (d) the gray crescent is required for normal embryogenesis.

Ans: d   ____   14. Grasshopper nymphs are formed during the process of: (a) blas-
p. 539   tula formation; (b) gastrulation; (c) complete metamorphosis;
                              (d) gradual metamorphosis.

Ans: d   ____   15. Which of the following cell combinations has the greatest
p. 541   longevity? (a) an old nucleus in old cytoplasm; (b) an old
                              nucleus in young cytoplasm; (c) a young nucleus in old cyto-
                              plasm; (d) cancer cells.

CHAPTER TWENTY-TWO
ANIMAL NUTRITION

CHAPTER SUMMARIES AND OBJECTIVES

22-1 <u>Nutrient Requirements / Sources of Organic Nutrients / Mineral Nutrients / Vitamins / Other Nutritional Deficiency Diseases in Humans</u> All animals need a source of energy. They also need a source of carbon skeletons, which are used as building blocks for the synthesis of specific organic compounds. The primary sources of both energy and carbon skeletons are carbohydrates, lipids, and proteins.

Extracellular digestion is necessary, especially in the case of proteins, because protein molecules are usually too large to pass through cell membranes and because "foreign" proteins inside cells would trigger immune responses.

Plants can synthesize all of the amino acids. Animals can synthesize all but eight; they must obtain these by eating meat or suitable plant material. Certain lipids and fatty acids are also essential in the diet. Essential minerals include calcium (bones, nerve function); phosphorus (DNA-RNA, ADP-ATP); sulfur (a component of some proteins); iron (hemoglobin); and copper (hemocyanin and certain proteins).

Most vitamins cannot be synthesized and must be ingested in small amounts. Vitamins are characterized as water-soluble (C, B complex) or fat-soluble (A, D, E, and K). B vitamins act as coenzymes for cellular respiration; C is necessary for the formation of collagen; A is essential in vision; D (which can be synthesized by the skin) is involved in the absorption of calcium; and K is essential in blood clotting. Nutritional deficiency diseases include scurvy (vitamin C), beriberi (thiamine), pellagra (nicotinamide), rickets (vitamin D), one form of pernicious anemia (B_{12}), kwashiorkor (essential amino acids), and goiter (iodine).

1. Describe the significance of carbon skeletons.

2. Explain why extracellular digestion of proteins is necessary.

3. Distinguish between essential and non-essential amino acids.

4. Explain how a vegetarian can avoid protein deficiency.

5. List the names and functions of five minerals that are important in the diet.

6. Indicate which of the following vitamins are water-soluble and which are fat-soluble, and give their functions: C, thiamine, nicotinamide, D, K, B complex.

22-2 <u>Nutrient Procurement: The Act of Predation</u> Techniques of carnivorous predation include aggressive hunting (stalking, chasing, or use of tentacles) and killing (with jaws and teeth or by injection of toxins), followed by swallowing (of small pieces after tearing up the prey, of the entire prey, or of softened tissue following predigestion). Herbivores are predators of plants. Deposit feeders sort through debris to obtain organic material. Parasites are fluid feeders that prey on their hosts. Other fluid feeders attack their prey only temporarily and remove fluids for use as food. Filter feeders filter organic material from the water in which they live.

7. Distinguish among carnivores, herbivores, and omnivores and give examples of each.

8. Distinguish between deposit feeders and filter feeders and give examples of each.

9. Define the following terms: autotroph, heterotroph, saprophyte, parasite, and predator.

22-3 Food Processing / Digestion in Tubular Guts / The Vertebrate Gut Once food is captured, it is passed through a mouth into the gut. Here digestion occurs (the breakdown of macromolecules into soluble micromolecules). The products of digestion then pass into the body through the walls of the gut.

A simple sac-like gut occurs in coelenterates. The single opening serves as both mouth and anus. In flatworms, the sac is branched; in some the pharynx is extendable and helps in gathering food. In higher metazoans the gut is tubular with openings at both ends. The gut varies among species but may include the following: mouth, buccal (oral) cavity, esophagus, gizzard, crop, stomach, midgut or small intestine, hindgut or large intestine, muscular rectum, and anus. The hindgut may contain bacteria that may help in digestion and in vitamin production. The portion of the gut in which most absorption occurs is often modified to allow greater surface area. For example, the lining of the small intestine of vertebrates has finger-like projections called villi. The wall of the gut is composed of several layers: mucosa (produces mucus, enzymes, and other secretions), submucosa (contains blood vessels, lymphatic vessels, and nerves), muscularis externa (moves food by peristalsis), and serosa. The gut is covered by a membrane called the peritoneum, which lines the abdominal cavity.

10. Describe how Hydra and Paramecium capture and digest prey.

11. Describe the location and function of villi.

12. List and describe the tissue layers that make up the wall of the gut.

13. Explain how ingested material moves from one end of the gut to the other.

22-4 Principles of Digestion / Mammalian Chewing and Swallowing / The Stomach / Intestines / Enzymatic Activity in the Small Intestine Digestion involves the enzymatic cleavage of macromolecules into monomers, which can pass through cell membranes and which can later be assembled into appropriate polymers. This cleavage is accomplished by hydrolysis, the addition of water molecules across chemical bonds. To prevent self-digestion, this enzymatic activity occurs outside cells; food inside the gut is really outside the body, for it has not yet passed through cell membranes. Many digestive enzymes are produced in inactive forms called zymogens, which are activated by necessary conditions outside the cell. The gut lining is protected by mucus or (in insects) by chitin.

In the mammalian digestive tract, the mouth contains teeth of various types: incisors for biting, canines for tearing, premolars and molars for crushing and grinding. Each tooth consists of a crown and root and is made of dentine covered with enamel. In the mouth, food is exposed to a starch-digesting enzyme called salivary amylase. Swallowing moves the bolus through the pharynx and into the esophagus, and peristaltic waves push it into the stomach. Meanwhile the glottis is closed to keep food out of the respiratory tract.

In the stomach, food is exposed to a low pH and to an endoprotease called pepsin, and it is mixed by muscular contractions. The resulting liquid gradually enters the small intestine, which has a basic pH. The first part is called the duodenum, where most digestion occurs. Next is the jejunum and finally the ilium, where most absorption of nutrients occurs. Pancreatic enzymes (three proteases, amylase, and lipase) enter the duodenum through the pancreatic duct. The intestinal wall produces three disaccharases and two proteases. Bile from the liver facilitates fat digestion. Absorption involves both diffusion and active transport. Once absorbed, monosaccharides and amino acids pass directly into the blood. Fatty acids and glycerol are formed back into fats, which pass first into the lymphatic vessels and later into the blood.

Peristaltic waves move the digestive residue from the small intestine into the large intestine where additional water and some minerals are absorbed, forming the semi-solid feces.

14. Cite the functions of the process of digestion.

15. Outline the digestion of fats.

16. Explain why cells aren't digested by the proteases they produce.

17. Describe the structure of a tooth, and name and give the functions of the four types of teeth present in the human mouth.

18. Outline the digestion of carbohydrates.

19. Describe the function of pepsinogen and the role of HCl.

22-5 Hormonal Controls The assembly-line process of digestion requires coordination so that each step proceeds in an orderly fashion. Both nervous and hormonal controls are involved. Hormonal control is accomplished by three hormones: secretin, cholecystokinin, and enterogastrone. Secretin and cholecystokinin are produced by cells in the walls of the duodenum when material enters from the stomach. Both stimulate the release of pancreatic secretions (bicarbonates by secretin and enzymes by cholecystokinin) and the release of bile by the gall bladder. Enterogastrone is produced by the duodenum in the presence of fatty acids, causing a slowdown of stomach activity. This results in less acid production and causes food to remain in the stomach longer.

20. Explain why both nervous and hormonal control of digestion are necessary.

21. Describe the functions of secretin, cholecystokinin, and enterogastrone.

22. Answer: Why is it advantageous for stomach function to slow down during fat digestion?

22-6 Herbivores and Carnivores The problem in eating plant material is the necessity of digesting cellulose. Only a few animals produce a cellulase; most make use of bacteria or other starch-digesting microorganisms in their gut instead.

Ruminants have a specialized food-storage organ that consists of four chambers: the rumen, reticulum, omasum, and abomasum. The first two work like a fermentation vat in which microorganisms digest cellulose. Regurgitated contents of the rumen form the cud. Cud chewing aids digestion by breaking down plant tissue. Water reabsorption occurs in the omasum. All three chambers are modifications of the esophagus. The abomasum is the true stomach, in which the microorganisms are killed and the proteins they contain are digested. Digestion from this point on is comparable to that described for other mammals. Some mammalian herbivores have an enlarged caecum that contains microorganisms that aid digestion. Coprophagy is a means of utilizing nutrients produced in the caecum.

Compared to herbivores, carnivores eat food that has a higher nutrient content and (since it lacks cellulose) is more easily digested. Thus a carnivore has a shorter gut with no special microorganism-containing chambers. A tapeworm is a highly specialized parasitic "carnivore" that absorbs predigested nutrients directly from the gut of its host.

23. Describe the digestive problem faced by most animal herbivores.

24. Describe the structure, function, and origin of the ruminant digestive tract.

25. Answer: What is the function of cud chewing?

26. Indicate the significance of an enlarged caecum and of coprophagy.

27. Given an anatomical description of an animal's gut, indicate whether this animal is an herbivore or a carnivore.

ARTICLE RESOURCES

Nutrient Requirements

Sources of Organic Nutrients

Mineral Nutrients

Dwyer, Johanna. Nutritional requirements for adolescents. Nutrition Reviews, 1981, 39(2), 56-69. This is a grey area in nutrition. We have studies on infants, young children, and adults but none on adolescents. Complicating the issue is the fact that nutritional needs do not rise at a constant rate in adolescence. There are also sex differences to be accounted. This issue of Nutrition Reviews presents papers given at an international conference on "Nutrition in Adolescents." It contains papers on anorexia nervosa, acne, obesity, food habits, and social development.

Rhodes, Martha. The "natural" food myth. The Sciences, 1979, 19(5), 30. The popular reasons for assuming that natural foods are better are given, as are the reasons why foods with additives do more to promote health. Ninety nine percent of the food we eat is natural. There are still some toxic substances present, but they are in such minute quantities that they are not hazardous to people. A great safety hazard would arise if all additives would be removed from the food supply. Food-borne illnesses caused by bacteria are prevented from becoming a public health problem.

Mauer, Allan. Hot dogger or hamburger? Omni, 1980, 2(10), 36. The choice of hot dogs or hamburgers for lunch may reveal something about the personality of the one making the choice.

Vitamins

Other Nutritional Deficiency Diseases in Humans

Iatrogenic kwashiorkor in California. Nutrition Reviews, 1981, 39(11), 397-399. An unsigned review in clinical nutrition by the editors of Nutrition Reviews. Four children in California suspected of having a milk protein sensitivity were prescribed a milk substitute of high-fat, low-protein nondairy creamers and soft diets. Within 1.5 to 6 months all of the children developed severe edema and other symptoms of kwashiorkor. The Committee on Nutrition of the American Academy of Pediatrics recommended breast feeding. Nondairy creamers apparently lack the required amounts of vitamin A, vitamin C, thiamine, riboflavin, protein, calcium, and iron--as well as other minerals. They should never be substituted for milk in infant diets.

Steinman, Linda. Rhythm and blahs. Omni, 1981, 3(5), 117-118. The jet lag experienced by travelers can be upsetting. Altering the diet to give different time cues can overcome jet lag. Meal composition is monitored. Social activity is timed and caffeine intake is controlled.

Nutrient Procurement: The Act of Predation

Jones, Jack Colvard. The feeding behavior of mosquitoes. Scientific American, 1978, 238(6), 138-148. Contrary to popular belief, the mosquito does not insert a hypodermic-like needle into the skin in order to feed. The anatomy of the proboscis is described. Male and female mosquitoes feed on different things. The male prefers sugar water. The female prefers human blood. The life cycle of the mosquito is described, as are some of the diseases they transmit.

Food Processing

Digestion in Tubular Guts

The Vertebrate Gut

Principles of Digestion

Mammalian Chewing and Swallowing

The Stomach

Intestines

Enzymatic Activity in the Small Intestine

Moog, Florence. The lining of the small intestine. Scientific American, 1981, 245(5), 154-162, 166, 171, 174-176. The small intestine plays a very active role in the digestion of food. It not only breaks down nutrients but absorbs them as well. This is done through a variety of transport systems. Excellent drawings and photographs accompany the clear text.

Hormonal Controls

Herbivores and Carnivores

ESSAY QUESTIONS

1. Consider the following statement from your text: "The processes of nutrition lie at the boundary of the animal world."
   a. What are nutritional processes?

      Basically, all nutritional processes involve the procurement of energy and the modification of stored energy into forms in which it can be absorbed and utilized. All nutrients are important either for the energy they contain or for the role they play in the release, storage, or use of energy for carrying on the processes of life.

   b. If the "boundary" referred to in this statement is the boundary between living and nonliving, how do nutritional processes make life possible?

      Life requires energy, and no life was possible until some means were evolved by which energy available in the environment could be obtained and utilized.

   c. If the "boundary" referred to is the boundary between animals and plants, how do animal and plant nutritional processes differ? How are they similar?

      Plants obtain energy directly from the sun; animals obtain energy indirectly. The methods of obtaining energy, then, are quite dissimilar, but the methods of utilization (as in cellular respiration) are often quite similar.

2. The terms "nutrients" and "food" are sometimes used interchangeably, but in the nutritional sense food is a more general term.
   a. Distinguish between nutrients and food, and give an example of a non-nutrient food.

      Food is anything that is ingested. Nutrients are the components of food that are essential to life. Saccharin is an example of a food with no nutrient value.

   b. Distinguish between "building block" nutrients and nutrients that are energy sources.

      "Building block" nutrients supply carbon skeletons for the synthesis of other substances. Other nutrients are especially important as energy sources. Some nutrients serve both functions.

c. Distinguish between mineral nutrients and vitamin nutrients.

Mineral nutrients are elements that play important roles in various physiological process and in the structural composition of certain tissues and
organs. Vitamins are compounds that are important in many metabolic
processes. Several serve as coenzymes or as parts of coenzymes.

d. Could water be considered a nutrient?

Water is also an essential component of the diet because of its role as the
primary solvent in biological systems.

3. Generally animals are predators of plants, and not vice versa.
a. How do plants obtain their nutrients?

Plants are autotrophs and obtain energy directly from the sun.

b. It is theoretically possible for a plant to supplement its nutritional needs
by occasionally preying on animals. Which of the nutrients that make up the
body of an animal might be of some use to a plant?

Most everything in the body of an animal could be used by a plant--either
as building blocks or for energy--if it were in the proper form.

c. Describe some possible techniques of carnivorous predation that might be
utilized by a plant in preying on animals.

There are numerous possibilities, including pitcherlike or traplike leaves
for entrapment and/or vinelike branches for entanglement.

d. Why would it be necessary for such a plant to be able to produce digestive
enzymes? What enzymes would be required? Would this digestion be extracellular or intracellular? In what forms would the animal nutrients probably
be absorbed into the plant cells?

All nutrients must be converted to monomers that can be absorbed, and all
would require the same enzymes used by animals for digestion. All digesting
would be done extracellularly.

e. Do carnivorous plants actually exist? Do some plants also prey on other
plants?

Yes, in both cases.

TEST SET 22-A

Name _____   Section _____

Choose the <u>best</u> answer to each of the following questions, and write the appropriate letter in the space provided.

Ans: b
pp. 544-
546

_____   1. Carbohydrates, lipids, and proteins are major classes of nutrients that are said to serve a "dual role" in nutrition. This dual role results because these nutrients: (a) provide both carbon and nitrogen, both of which are necessary for the synthesis of all essential organic compounds; (b) can provide both carbon skeletons and energy; (c) have both nutrient and nonnutrient functions; (d) contain the carbon skeletons necessary for the synthesis of both enzymes and vitamins.

Ans: c
pp. 543-
545

_____   2. All of the following statements about modes of nutrition are true <u>except</u>: (a) All photosynthetic plants are autotrophs. (b) All animal parasites are carnivores. (c) All predators are carnivores. (d) All animals are heterotrophs.

Ans: a
p. 546

_____   3. All of the following statements about cholesterol and unsaturated fatty acids are true <u>except</u>: (a) Cholesterol is produced in the human liver in sufficient quantities and should be totally avoided in the diet. (b) Unsaturated fatty acids cannot be synthesized in sufficient quantities by the body and must be included in the diet. (c) Unsaturated fatty acids are essential components of cell membranes. (d) A dietary deficiency of unsaturated fatty acids may result in impaired milk production and in other reproductive difficulties.

Ans: d
pp. 546-
548

_____   4. Which of the following statements about the importance of minerals is true? (a) Iron is important in muscle contraction and also in the electrical balance in cells. (b) Magnesium, manganese, zinc, and cobalt ions are particularly important in the osmotic balance of tissues. (c) Potassium, sodium, and chlorine play important roles as cofactors for enzymes. (d) Iron is the oxygen-binding atom in both hemoglobin and myoglobin, and it undergoes redox reactions in cellular respiration.

Ans: a
pp. 552,
560,
563

_____   5. Mucus plays an important role in the procurement and processing of nutrients. All of the following statements are true <u>except</u>: (a) Mucus consists of several digestive enzymes, the most important of which is a protease called mucin. (b) Mucus lubricates the intestinal lining and protects it from abrasion. (c) Mucus shields the gut lining from being digested by activated enzymes. (d) Mucus is used by certain stationary filter feeders to extract food particles from the water in which they live.

Ans: b
pp. 553,
554,
556

_____   6. All of the following statements about guts are true <u>except</u>: (a) All animal groups that are evolutionarily more advanced than flatworms have a gut which is basically tubular. (b) Vertebrates are unique because they are the only animals that have a gut with two openings: a mouth and an anus. (c) Cnidarians have a saclike gut with a single opening that serves as both a mouth and an anus. (d) A <u>Paramecium</u> utilizes food vacuoles in which digestion occurs.

Ans: b
p. 556

_____ 7. All of the following conditions provide an increase in the surface area of the gut for absorption except: (a) A long dorsal infolding of the intestine occurs in the earthworm, and a spiral valve occurs in sharks. (b) Extensive branching of the intestinal tract, between stomach and anus, occurs in most coelenterates. (c) In many vertebrates, the gut is greatly folded. (d) Many tiny fingerlike projections, called villi, occur on the inner intestinal wall in many vertebrates.

Ans: d
pp. 557-
558

_____ 8. Peristalsis: (a) is a partial paralysis of gut mobility caused by too much enterogastrone; (b) results from the simultaneous contraction of all the muscles of the gut wall, followed by the simultaneous relaxation of these same muscles; (c) results from rapid contraction of muscle segments occurring in the submucosa, causing a spasm and restricting the passage of food; (d) results from the successive contractions of adjacent muscles that ring the gut, causing a moving wave of constriction.

Ans: a
pp. 559-
560

_____ 9. Protein-digesting enzymes don't digest the animal that produces them. All of the following conditions help to prevent self-digestion except: (a) Proteases are able to distinguish "self" proteins from ingested foreign proteins. (b) The digestion of proteins takes place outside the body of the animal. (c) Proteases are produced in inactive form. (d) Mucus shields the gut lining from being digested by enzymes.

Ans: c
p. 563

_____ 10. Pepsin: (a) is an inactive zymogen that acts as an exopeptidase; (b) is produced by the pancreas and functions as a dipeptidase under high-pH conditions; (c) is derived from pepsinogen under acidic conditions; (d) is a carbohydrase that continues the digestive action of the salivary enzymes.

Ans: b
pp. 563-
565

_____ 11. Carbohydrates are digested by enzymes produced by all of the following except: (a) salivary glands; (b) the liver; (c) the pancreas; (d) the small intestine.

Ans: d
p. 565

_____ 12. All of the following statements about the absorption of the products of lipid digestion are true except: (a) Fatty acids and glycerol are resynthesized into fats immediately after absorption. (b) After digestion, the products of lipid digestion enter the lymphatic vessels of the intestinal wall. (c) After absorption, the products of fat digestion eventually enter the blood. (d) Unlike most macromolecules, lipids are readily absorbed (as such) through the cell membranes of the cells of the gut, and it is not necessary that they be converted into monomers.

Ans: a
p. 566

_____ 13. Secretin is a hormone that: (a) is produced by the duodenal lining; (b) causes the stomach to decrease its muscular activity and rate of secretion; (c) causes the pancreas to secrete enzymes; (d) causes the gall bladder to empty.

Ans: b
pp. 566-
567

_____ 14. The grass eaten by ruminants is a poor source of proteins. These animals obtain much of their protein by: (a) eating insects present on the grass; (b) digesting the microorganisms (present in the first two chambers of the "stomach") that are able to synthesize proteins from inorganic nitrogen present in the grass; (c) synthesizing all of the essential amino acids from carbon skeletons obtained from carbohydrates; (d) using carbon dioxide and methane gas (produced by fermentation in the rumen) to synthesize the necessary proteins.

Ans: c
p. 567

_____ 15. All of the following statements are true of both cud chewing and coprophagy except: (a) Both increase the efficiency of the digestive process. (b) Both occur in certain mammalian herbivores. (c) Both occur regularly in ruminants and involve the regurgitation of material from the first two chambers of the "stomach." (d) Both occur in animals that utilize microorganisms to assist in the digestive process.

TEST SET 22-B

Name _____   Section _____

Choose the best answer to each of the following questions, and write the appropriate letter in the space provided.

Ans: a
p. 545

_____   1.   Proteins are digested "outside" the body (extracellularly). This occurs for all of the following reasons except: (a) ingested proteins are first digested into amino acids, which are quite different from those that occur inside cells and would be recognized as foreign. (b) Proteins are macromolecules that are too large to pass readily through plasma membranes. (c) If taken directly into cells, ingested proteins would be attacked by the immune system. (d) The same enzymes that digest ingested proteins would also digest the protein components of cells.

Ans: c
pp. 545,
546

_____   2.   The essential amino acids are essential because: (a) they are the only amino acids that can be synthesized into proteins; (b) they cannot be synthesized by any living organisms, including plants, even when an adequate source of nitrogen is available; (c) they cannot be synthesized by humans, and all must be obtained in the diet; (d) they can be used to produce all of the lipids that are essential in the body.

Ans: a
pp. 548-
550

_____   3.   Which of the following statements about vitamins is true? (a) Water-soluble vitamins are essential in all animals. (b) Vitamin C (acetic acid) is a fat-soluble vitamin that regulates the absorption and metabolism of calcium. (c) Vitamin K is a water-soluble vitamin that plays an important role in cellular respiration. (d) Vitamin D is important in vision and is considered a vitamin because it is synthesized in the human skin under conditions of bright light.

Ans: a
pp. 545,
546,
550,
551

_____   4.   A strict vegetarian diet: (a) will eventually lead to a protein deficiency, because the essential amino acids can be obtained only from meat, eggs, or dairy products; (b) will eventually lead to pernicious anemia, because plants don't produce vitamin B_{12};

(c) will eventually result in a vitamin deficiency causing scurvy; (d) will maintain good health, but only if the water-soluble vitamins are stored in the body in sufficient quantities prior to beginning the diet.

Ans: c
pp. 563-
564

_____   5.   If the stomach contents of an animal should have a low pH, it is reasonable to predict that: (a) the stomach has no enzymatic digestive function; (b) food will pass rapidly through the stomach and will be delivered to the midgut in essentially the same form in which it was ingested; (c) the midgut will have a higher pH (neutral or somewhat basic) and that most digestion and absorption will take place here; (d) the stomach will be an important site for the absorption of the products of digestion and will contain large colonies of microorganisms.

Ans: a
p. 557

_____   6.   The wall of the gut consists of four layers. These include: (a) the mucosa--the innermost layer that produces mucus and digestive enzymes; (b) the submucosa--the fibrous outermost layer; (c) the gastrovascularis--the layer containing the nerves, blood vessels, and lymph vessels that serve the gut; (d) the serosa--a layer consisting mostly of muscles.

Ans: c
pp. 559-
    560

7. Digestion involves all of the following processes except: (a) the breaking down of macromolecules into their constituent monomers; (b) hydrolysis, in which water is added across chemical bonds; (c) the conversion of soluble polymers into insoluble monomers; (d) the modification of ingested nutrients into forms in which they can be absorbed.

Ans: a
p. 563

8. All of the following statements about human saliva are true except: (a) Human saliva is produced by numerous cells that line the inside of the mouth. (b) Human saliva contains only one type of digestive enzyme, a carbohydrase. (c) Human saliva contains antimicrobial agents. (d) Human saliva helps to compact food into a bolus.

Ans: b
p. 563

9. All of the following statements about the stomach are true except: (a) The stomach begins the digestion of proteins. (b) The stomach is the part of the gut that plays a major role in the absorption of nutrients. (c) Periodic contractions of powerful compressor muscles in the wall of the stomach serve to churn and mix the stored food with the gastric juices. (d) The activity of the stomach is controlled by enterogastrone.

Ans: a
p. 565

10. Digestion of fats: (a) occurs entirely in the small intestine; (b) results in part from the effects of fat-digesting enzymes present in the bile from the gall bladder; (c) occurs primarily as a result of lipases produced by cells that line both the large and small intestines; (d) does not occur in most non-Caucasians beyond about four years of age.

Ans: a
p. 565

11. Active transport always involves: (a) the use of ATP; (b) the movement of dissolved substances from areas of high concentration to areas of low concentration; (c) the movement of small molecules by diffusion; (d) systems that are highly specific so that a system which moves amino acids will move only specific types of amino acids.

Ans: a
p. 565

12. Once absorbed, monosaccharides and amino acids pass directly into the blood. On the other hand, fatty acids and glycerol pass: (a) first into lymph vessels and into the lymph; (b) directly into the cells in which they will be stored as fat; (c) by diffusion through the interstitial fluid against the concentration gradient; (d) passively from cell to cell by active transport.

Ans: b
p. 564

13. All of the following occur in the large intestine except: (a) absorption; (b) digestion; (c) vitamin synthesis; (d) formation of undigested residue into feces.

Ans: d
p. 566

14. Enterogastrone: (a) is produced as a result of the stimulatory effects of acidic foods in the stomach; (b) causes the gall bladder to release bicarbonate; (c) converts trypsin into trypsinogen; (d) decreases the activity of the stomach in order to slow the rate at which food enters the duodenum.

Ans: b
p. 567

15. Coprophagy occurs: (a) most commonly in those herbivores that lack a caecum; (b) as a means of increasing the efficiency of the digestive process; (c) in the upper portions of the large intestine and facilitates the digestion of cellulose by cellulases produced by the caecum; (d) when food is regurgitated from the rumen and rechewed.

CHAPTER TWENTY-THREE
GAS EXCHANGE AND TRANSPORT IN ANIMALS

CHAPTER SUMMARIES AND OBJECTIVES

23-1 <u>Problems of Gas Exchange / Rate of Oxygen Consumption / Oxygen Availability / Size</u>
Animals obtain energy from food by oxidizing it to CO_2 and water. The process requires oxygen, which must be dissolved in water and transported throughout the body. Similarly, the resulting CO_2 must leave solution. Diffusion is adequate for gas exchange in unicellular aquatic organisms, but larger animals require specialized structures. Respiration is influenced by three factors: rate of O_2 consumption, availability of O_2, and body size. Gills and lungs are among the major gas-exchange structures.

1. Describe the basic gas-exchange problems faced by all but unicellular aquatic organisms.

2. Name the three major factors influencing respiration.

23-2 <u>Special Respiratory Mechanisms / Insect Tracheae / Lungs</u> All gas-exchange systems share one principle: diffusion across a wet membrane. Gills are an aquatic adaptation to increase surface area; covered gills require appropriate structures or behavior to circulate water through them. Lungs are a terrestrial adaptation to protect respiratory surfaces and reduce water loss. Amphibians respire directly through the skin and buccal cavity; most also have lungs. A few aquatic reptiles also breathe partly through the skin and/or cloaca. Mammals and birds, which need large amounts of oxygen, have elaborate lungs. To obtain the energy needed for flight, birds have air sacs and other structures that enable the lungs to work more efficiently. Insects obtain their oxygen through tubes called tracheae, which open to the outside through spiracles. Larger insects also have air-filled sacs and perform ventilating movements.

3. Describe the functions of gills, of lungs, and of insect tracheae.

4. Explain how a bird's respiratory system differs from that of a mammal.

5. Name structures other than lungs that are used for respiration in amphibians; in reptiles.

23-3 <u>Transport in Animal Bodies / Open and Closed Systems / The Heart</u> Animals need transport systems to deliver nutrients and oxygen to cells and to remove wastes. Blood serves these functions in all animals. In many, it also contains oxygen-carrying pigments. Transport systems may be open (blood fills all spaces between organs) or closed (blood is pumped through vessels). The heart is an organ that collects and pumps blood; in some organisms there are "booster" hearts that aid circulation. Vertebrates other than fish have two circulatory pathways: pulmonary and systemic. Amphibians and most reptiles have a three-chambered heart that allows some mixing of oxygenated and deoxygenated blood in the single ventricle. Mammals and birds have more efficient four-chambered hearts.

6. Describe the functions of blood.

7. Differentiate between open and closed circulatory systems.

8. Compare the transport systems of fish, amphibians, and mammals.

23-4 <u>The Human Circulatory System / The Human Heart / Blood Movement and Pressure / The Capillary System</u> Blood is circulated in humans as follows: right ventricle to alveoli in lungs via pulmonary artery, arterioles, and capillaries; oxygenated blood from alveoli to left atrium via capillaries, venules, and pulmonary veins; left atrium to left ventricle; systemic circulation via aorta, coronary arteries, and other major arteries; deoxygenated blood to right atrium via major veins (including hepatic portal vein) and superior and inferior venae cavae. The heart has four valves that move blood in the proper direction. The heartbeat is myogenic (initiated within the muscle itself) and is regulated by two nodal centers, the S-A node and the A-V node. The heartbeat rate is influenced by nerves and hormones. Systole is the contraction of the heart; diastole is its relaxation. The return of venous blood to the heart is aided by muscular activity, which squeezes the veins. When blood enters capillaries, blood plasma (containing oxygen and nutrients) is forced out by relatively high pressure. At the venous end, pressure is lower; and CO_2 and wastes enter by diffusion, while water follows by osmosis.

Nerves and hormones influence sphincters, which in turn affect blood flow through capillaries.

9. Describe the path followed by blood in the human circulatory system.

10. Differentiate between myogenic and neurogenic heartbeats.

11. Describe the process of gas and nutrient exchange in capillaries.

23-5 <u>Blood / Blood Pigments / Carbon Dioxide Transport</u> Blood consists of plasma, erythrocytes, leukocytes, and platelets. Plasma consists of water, nutrients, hormones, and antibodies and other proteins, waste products, and ions. Erythrocytes transport oxygen; those of mammals lack nuclei. Leukocytes are larger and less numerous, have nuclei, and are capable of amoeboid movement. They defend the body against foreign materials. Platelets are fragments of giant bone marrow cells; they aid in blood clotting. Hemoglobin is the oxygen-carrying pigment in vertebrates. The uptake or release of oxygen by hemoglobin depends on partial pressure (P_{O_2}). Animals with higher metabolic rates tend to have hemoglobin that gives up O_2 more readily; conversely, the hemoglobin of high-altitude mammals must take up oxygen more readily. Myoglobin stores oxygen in muscles, releasing it only at low P_{O_2}. Other oxygen-bearing pigments are found in invertebrates. Most CO_2 is carried in the plasma as HCO_3^- ions, which form carbonic acid. This lowers pH and causes hemoglobin to release oxygen more readily.

12. Name and give functions of the major blood components.

13. Explain the significance of partial pressure.

14. Explain why organisms need oxygen-carrying pigments.

23-6 <u>The Lymphatic System</u> Some fluid from capillaries enters the lymphatic system, which consists of progressively larger vessels ending in the thoracic duct. Lymph (water, solutes, and white blood cells) is returned to the blood via the superior vena cava. Lymphatic vessels, like veins, have valves. Circulation is aided by skeletal muscles or, in some animals, by lymph hearts. Lymph nodes produce lymphocytes and are sites for the removal of foreign materials by phagocytosis. The rate of lymph formation is influenced by blood pressure, blood protein content, and leakiness of capillaries.

15. Describe the structure and function of the lymphatic system.

ARTICLE RESOURCES

Problems of Gas Exchange

Rate of Oxygen Consumption

Oxygen Availability

Size

Special Respiratory Mechanisms

Lungs

> Diamond, Stuart. Breathing facts. Omni, 1981, 3(5), 38. Few studies on human breathing have been made. Six liters of air are breathed each minute. Particles inhaled remain in the lungs for varying periods of time. Smoking apparently worsens the damage done by pollutants, particularly coal dust and asbestos.

Insect Tracheae

Transport in Animal Bodies

> Blakeslee, Anton. Cold noses. Omni, 1980, 2(11), 39. The cold nose of the dog acts as a heat exchanger for the brain. In conjunction with panting, the cold nose of a dog can cool the blood two to three degrees before it reaches the brain. The nose has 20 times the evaporative surface of the tongue.

Open and Closed Systems

The Heart

The Human Circulatory System

The Human Heart

> Johansen, Kaj. Aneurysms. Scientific American, 1982, 247(1), 110-125. Repetitive hydraulic stress is generated in the arteries by the pumping of the heart. The artery wall is usually able to withstand this stress because of the structure of the wall itself. A balloonlike dilation, an aneurysm, occurs when the structure of the wall gives way. The formation of aneurysms increases with age and in direct proportion to age. However, better diagnostic techniques and methods of repair have been found recently.

Blood Movement and Pressure

The Capillary System

Blood

> Block, Irvin. Microwar. Science Digest, 1982, 90(7), 86-88. Role of blood components in healing wounds. Micrograph of blood clotting showing red blood cells, white cell, and fibrin. Discusses phagocytes, macrophages, histocytes, and granulocytes as well as histamines and lymphocytes.

> Doolittle, Russel F. Fibrinogen and fibrin. Scientific American, 1981, 245(6), 126-135. Fibrinogen is a plasma protein that converts into fibrin to form a blood clot which in time breaks down. A history of the controversy over the structure of fibrinogen is given. An analysis of the alpha-chain sequence of amino acids is given. Supplements the discussion in the text.

> Allen, T. D., and T. M. Dexter. Ultrastructural aspects of erythropoietic differentiation in long-term bone marrow culture. Differentiation, 1982, 2(2), 86-94.

Erythroid stages are beautifully illustrated using scanning electron microscopy. Pages of photomicrographs are enhanced by clear captions. These are so well done that it is not essential to read the entire article to grasp the sequence of events in erythropoietic differentiation.

Zucker, Marjorie B. The functioning of blood platelets. Scientific American, 1980, 242(6), 86-90, 96, 103. Excellent photomicrographs of blood platelets are given. These platelets play a critical role in the stoppage of blood flow. Platelets have no nuclei and die within about 10 days, yet their metabolism is active. It is carried out by enzymes in the cytoplasm and mitochondria. There are well-defined organelles within the cytoplasm of platelets.

Blood Pigments

Carbon Dioxide Transport

The Lymphatic System

ESSAY QUESTIONS

1. Describe some mechanisms used to increase gill efficiency in cephalopod mollusks and in fish.

    Cephalopods have gill hearts. Fish exhibit increased surface area and counter-current devices and water is forced through the gill cavity.

2. Why do you think people and animals tend to stretch and yawn upon arising from a night's sleep?

    Perhaps it is to help return venous blood to heart and circulate oxygenated blood, following relative inactivity.

3. Describe all major structures and functions involved in the breathing cycle of a typical bird and of a mammal. How might the respiratory system of a large, flightless bird (such as an ostrich) differ from that of a typical songbird?

    For the structures and functions in birds and mammals, see Chapter 23, page An ostrich is large and flightless, so its metabolic rate is probably lower, its breathing slower, its air sac system less extensive, and its bones heavier. (Remember that the object is to have the student speculate about likely adaptations on the basis of knowledge he or she has acquired. Whether this speculation corresponds to reality in ostriches makes little difference.)

4. Diagram and describe the sequence of events involved in blood clotting after tissue damage occurs.

    See text page 594.

TEST SET 23-A

Name _____   Section _____

Choose the best answer to each of the following questions, and write the appropriate letter in the space provided.

Ans: a
p. 574
_____ 1. Why must warm-blooded aquatic animals obtain their oxygen from the air rather than from the water? (a) They need more energy than fish do. (b) When a structure (such as a gill) is lost in the course of evolution, it can never be regained. (c) To maintain their body temperature, they must live fairly near the surface where the water contains little oxygen. (d) All animals obtain their oxygen from the air.

Ans: c
pp. 572-
573
_____ 2. Why do many fishes tend to prefer cool water? (a) It contains more insects. (b) Actually they prefer shady areas, which are safer and only incidentally are cooler. (c) It contains more oxygen. (d) They become sluggish at higher temperatures.

Ans: d
p. 573
_____ 3. Why do air-breathing fishes tend to occur in the tropics? (a) Primitive organisms live where conditions have been stable for a long time. (b) Only humid air can be taken in by a gill. (c) They have become extinct elsewhere due to human activities. (d) Warm, stagnant water, which is often found in the tropics contains little oxygen.

Ans: a
p. 571
_____ 4. Must all gas-exchange surfaces be wet? (a) Yes, always. (b) No, only gills must be wet. (c) No, only if the partial pressure of oxygen is low. (d) No, a terrestrial salamander has dry skin.

Ans: a
p. 572
_____ 5. Which of the following mammals has the highest rate of oxygen uptake? (a) shrew; (b) raccoon; (c) human; (d) hippopotamus.

Ans: d
p. 586
_____ 6. Why is the left ventricle the strongest of the heart's four chambers? (a) It forces blood through the lungs. (b) It contains the most blood. (c) It has no valves to assist in directing the flow of blood. (d) It forces blood through the systemic circuit.

Ans: c
p. 588
_____ 7. The heartbeat in humans: (a) is determined solely by the SA node; (b) is determined solely by the SA and AV nodes; (c) is affected by many factors, such as exercise and hormones; (d) is neurogenic, meaning that it can be influenced by sympathetic and parasympathetic nerves.

Ans: c
p. 587
_____ 8. All of the following statements about heart murmurs are true except: (a) They can be present from birth. (b) They can be caused by disease. (c) They are always harmful, reducing heart efficiency. (d) They can often be cured surgically.

Ans: b
p. 596
_____ 9. The hemoglobin of a human fetus: (a) is the same as that of the adult; (b) has a higher affinity for oxygen than that of the adult; (c) has only two protein subunits instead of four; (d) has a lower affinity for oxygen than adult hemoglobin, because oxygen is supplied by the mother.

Ans: c
p. 597
_____ 10. Myoglobin: (a) has only two protein subunits, like fetal hemoglobin; (b) has a lower affinity for oxygen than hemoglobin; (c) causes steak to be red; (d) is a blood pigment found in certain marine invertebrates.

Ans: b
p. 595
_____ 11. Hemoglobin: (a) occurs only invertebrates; (b) has four protein subunits; (c) is the only known blood pigment; (d) is the same in all vertebrates.

Ans: b   _____   12.   In mammals, lymph returns to the circulatory system via:
p. 598   (a) lymph hearts; (b) pressure from skeletal muscles; (c) diffu-
                            sion; (d) active transport.

Ans: d   _____   13.   Oxygen requirements are generally independent of: (a) activity
p. 572   level; (b) size; (c) air temperature; (d) taxonomic group.

Ans: a   _____   14.   The oxygen content of water: (a) influences the efficiency of
pp. 572-   gills; (b) falls as the water temperature falls; (c) is unrelated
   573   to salt content; (d) is higher in stationary water than in running
                            water.

Ans: d   _____   15.   Aquatic insects have: (a) gills; (b) lungs; (c) chloroplasts;
p. 580   (d) tracheae.

TEST SET 23-B

Name _____   Section _____

Choose the best answer to each of the following questions, and write the appropriate letter in the space provided.

Ans: a
pp. 582-
    583

_____ 1. A _____ has an open circulatory system. (a) lobster; (b) earthworm; (c) squid; (d) salamander.

Ans: b
p. 585

_____ 2. All of the following have a four-chambered heart except the: (a) alligator; (b) turtle; (c) rabbit; (d) duck.

Ans: b
p. 573

_____ 3. The net effect of bacteria in water is to: (a) increase oxygen by photosynthesis; (b) decrease oxygen by respiration; (c) increase CO_2 by photosynthesis; (d) decrease CO_2 by respiration.

Ans: a
p. 573

_____ 4. What is meant by the "diving reflex" of some mammals? (a) Blood flow is diverted to the brain during diving. (b) Breathing automatically stops while underwater. (c) The animal dives into the nearest body of water to escape predation. (d) The heart rate increases to supply blood to the muscles used in swimming.

Ans: d
p. 591

_____ 5. Harvey described all major features of the human circulatory system except the: (a) heart; (b) arteries; (c) veins; (d) capillaries.

Ans: a
p. 577

_____ 6. What is meant by a countercurrent distribution system? (a) Fluid in two structures moves in opposite directions. (b) Materials in solution move against the current. (c) Fluids in two structures move in the same direction. (d) Heat is transferred from a warmer system to a cooler system.

Ans: d
p. 578

_____ 7. How do frogs normally fill their lungs? (a) They swallow oxygenated water. (b) They use a suction-pump mechanism. (c) They have no lungs--they breathe through their skin. (d) They force a mouthful of air into their lungs.

Ans: a
p. 578

_____ 8. Aquatic reptiles use all of the following structures for breathing except: (a) gills; (b) lungs; (c) skin; (d) cloaca.

Ans: d
p. 579

_____ 9. What is the function of capillary networks in the air sacs of birds? (a) gas exchange; (b) to warm the air to make it lighter; (c) waste removal; (d) There are no such networks in the air sacs of birds.

Ans: a
p. 579

_____ 10. Parabranchi differ from alveoli in that they: (a) are open on both ends; (b) have valves to control their opening and closing; (c) retain some "dead" air after each breathing cycle; (d) lack capillary networks.

Ans: d
p. 581

_____ 11. Insects are very numerous and successful, yet they have a simple, largely passive method of gas exchange. Why do vertebrates need more elaborate gas-exchange structures? (a) They are more active. (b) They have a higher rate of oxygen consumption. (c) They occupy a wider range of habitats. (d) They are larger.

Ans: c
p. 582

_____ 12. How does the open transport system of a nematode worm differ from that of an insect? (a) In insects, oxygen is transported primarily by the blood. (b) Nematodes have hemoglobin; insects have hemocyanin. (c) Insects have a heart; nematodes do not. (d) Nematodes have blood vessels; insects do not.

Ans: a   _____   13.   The term "pacemaker" refers to: (a) the SA node; (b) the AV
p. 587   node; (c) the bundle of His; (d) an artificial device.

Ans: c   _____   14.   What is the significance of the Purkinje fibers? (a) They
p. 588   contract to drive blood through the atrium. (b) They initiate
                                       the heartbeat in insects. (c) They transmit signals from the AV
                                       node to the ventricle. (d) They result from cholesterol deposits
                                       in the arteries.

Ans: d   _____   15.   Why don't the atria and ventricles contract at the same time?
p. 588   (a) The beat of the atrium is initiated by sympathetic nerves,
                                       whereas that of the ventricle is initiated by parasympathetic
                                       nerves. (b) The atrium is controlled by the SA node, whereas
                                       the ventricle is controlled solely by the AV node. (c) They do
                                       contract at the same time; if they get out of synchrony, the
                                       result is a heart attack. (d) It takes time for a signal to
                                       pass from the SA node to the AV node.

CHAPTER TWENTY-FOUR
EXCRETION AND OSMOREGULATION

CHAPTER SUMMARIES AND OBJECTIVES

24-1 The Excretion of Nitrogenous Wastes The vertebrate kidney, like other excretory systems, maintains homeostasis by waste excretion and by control of water and salt balance. Ingestion of excess proteins and nucleic acids necessitates removal of nitrogenous wastes as ammonia, urea, or uric acid. Ammonia is highly toxic but also highly soluble, so it is excreted primarily by aquatic animals. Urea is less toxic, but energy is expended to produce it. Uric acid is still less toxic and requires even more energy for its formation; being relatively insoluble in water, it limits water loss.

   1. List the three forms in which nitrogenous wastes can be excreted, and explain the advantages and disadvantages of each.

   2. Explain what is meant by homeostasis.

24-2 Water, Salts, and the Environment / Isotonic Organisms and Osmoregulating Fishes / Water Balance in Terrestrial Organisms Organisms must maintain water and salt balance. Environments are categorized as hypotonic (osmotic potential greater than that of the organism), hypertonic (osmotic potential less than that of the organism), terrestrial (land), and isotonic (osmotic potential the same as that of the organism). Species fall into two categories: euryhaline (tolerating a wide range of salt concentration) and stenohaline (lacking such tolerance). Within each group, some species control their internal salinity (osmoregulators), whereas others allow it to vary (osmoconformers). Marine bony fishes drink large amounts of sea water and excrete excess salts. Freshwater fishes produce very dilute urine and do not drink. Sharks and related fish retain urea so that water tends to enter the body. Adaptations for reducing water loss in terrestrial animals include relatively impermeable body surfaces, replacement of lost water and salts, and internal fertilization.

   3. Name the four major types of environments in terms of water and salt balance.

   4. Explain how animals maintain water and salt balance in each environment.

24-3 Types of Excretory Systems / Simple Protonephridia / Malpighian Tubules / Ultrafiltration in the Crustaceans / Metanephridia Excretory systems remove from the urine those substances that need to be reabsorbed. Protonephridia, in aquatic invertebrates, are simple tubules that move fluids out of the body while selectively reabsorbing certain materials. Terrestrial arthropods, including insects, have similar Malpighian tubules, which add solutes selectively to the urine. Crustaceans have another type of protonephridium, called a green gland; it is similar to the vertebrate kidney in that ultrafiltration is followed by selective reabsorption. Annelids and mollusks have metanephridia: tubules open at both ends that actively transport selected materials from coelomic fluid into capillaries.

5. Differentiate between Malpighian tubules and crustacean green glands.

6. Differentiate between protonephridia and metanephridia.

7. Explain the general functions of all excretory systems.

24-4 The Vertebrate Excretory System / The Mammalian Nephron / Urine Formation / Final Processing of Urine / The Cortex, The Medulla, and Water Retention / What the Kidney Can and Cannot Do In vertebrates, the excretory tubule is called a nephron. One end forms a capsule, surrounding a glomerulus (knot of capillaries), while the other end empties into a collecting duct. Many nephrons form a kidney, which empties into a ureter. Urine may be released directly from the ureter (as in freshwater fishes) or it may first accumulate in a bladder (as in mammals).

Blood enters the mammalian kidney via the renal artery, which divides to form arterioles and the capillaries of the glomerulus, which is surrounded by Bowman's capsule. Blood leaves the nephron by way of venules and the renal vein. The kidney consists of an outer cortex (glomeruli, part of nephrons) and an inner medulla (loops, collecting ducts). From the nephrons, urine enters collecting ducts leading to the renal pelvis and ureter. Water, with solutes of low to medium weight, is squeezed from the glomerulus into the nephron by blood pressure. This fluid then changes in composition as it passes through the nephron. In the proximal convoluted tubule, glucose and sodium are actively reabsorbed; chloride ions and some water follow. Several substances, including drugs, are actively secreted here. Next, in the loop of Henle, sodium is removed and a salt gradient is created. Potassium, hydrogen ions, and ammonia are secreted into the distal convoluted tubule. Water is lost here, and more sodium is reabsorbed; these two processes are regulated by ADH and aldosterone. More water is lost in the collecting duct. Nephrons are designated as cortical (within the cortex) or juxtamedullary.

8. Explain the function of the loop of Henle.

9. Explain how the vertebrate nephron differs from protonephridia and from metanephridia.

10. Describe how the urine changes as it passes through the nephron.

ARTICLE RESOURCES

The Excretion of Nitrogenous Wastes

Water, Salts, and the Environment

Wintsch, Susan. Beating the heat. Science 81, 1981, 2(5), 80-82. Discusses the role of water in combatting overheating and dehydration in athletes. Isotonic drinks such as Gatorade are compared with water, and their effects on athletes are noted.

Quayle, Robert, and Fred Dochring. Heat stress: A comparison of indices. Weatherwise, 1981, 34(3), 120-124. Ways of measuring heat stress in humans are discussed. The two factors contributing most to heat stress are temperature and humidity. Heat stress accounted for about 20,000 deaths between 1936 and 1975. Anyone who has gone through a heat wave with high humidity accompanying it will be interested in this article.

Isotonic Organisms and Osmoregulating Fishes

Water Balance in Terrestrial Organisms

Types of Excretory Systems

Simple Protonephridia

Malpighian Tubules

Ultrafiltration in the Crustaceans

Metanephridia

The Vertebrate Excretory System

Frömter, E. Electrophysiological analysis of rat venal sugar and amino acid transport. Pflüger's Archiv European Journal of Physiology, 1982, 393(2), 179-189. Glucose and/or amino acids are transported across rat venal proximal tubular epithelium. This is associated with two electrical events: (1) depolarization of the tubular cell membranes and (2) a reduction of the resistance of the brushborder membrane. Excellent discussion for the instructor.

The Mammalian Nephron

Urine Formation

Final Processing of Urine

The Cortex, the Medulla, and Water Retention

Lindstedt, Stan L., and James H. Jones. Desert shrew. Natural History, 1980, 89(1), 46-53. This desert animal is the smallest animal to maintain a constant body temperature. This it does despite adverse desert conditions of high temperature and little water. It conserves what water it has by restricting most of its activities to night. It also reclaims a great deal of water that would otherwise be lost through exhaling. Urine is highly concentrated; hence little if any water is lost by this means.

What the Kidney Can and Cannot Do

ESSAY QUESTIONS

1. Discuss the principle of countercurrent distribution as it applies to the function of the loop of Henle.

   See Box B in Chapter 24 of the text. (page 616)

2. Draw and label a diagram of a human nephron. For each major structure, indicate the principal changes that occur in the composition of the urine at that point.

   See Figure 12-17. (page 613)

3. Diagram the excretory tubules of an insect and those of a crustacean, labeling major parts. Explain the differences and similarities between these two types of protonephridia.

   See Figure 7 for insects and Figure 8 for crustaceans. (page 609)

4. Name the three major forms of nitrogenous waste and the groups of organisms that excrete them. Explain how a group's principal waste product is related to its mode of life.

   See Table II in Chapter 24 of the text. (pages 602-603)

TEST SET 24-A

Name _____   Section _____

Choose the best answer to each of the following questions, and write the appropriate letter in the space provided.

Ans: c
p. 605
_____   1. Why is active transport necessary in an isotonic environment?
(a) Excess salt tends to enter cells and must be pumped out.
(b) Excess water tends to enter cells and must be pumped out.
(c) Body fluids have a different composition from that of sea water. (d) Diffusion alone is too slow, except in small organisms.

Ans: c
p. 603
_____   2. What is the main source of ammonia in the diapers of human infants? (a) For a few weeks, the principle excretory product of the newborn infant is ammonia, because the newborn has been living in an aquatic environment. (b) No ammonia should be present unless bacteria have produced it. (c) All humans excrete some ammonia. (d) Infants ingest more nucleic acids than do adults.

Ans: b
p. 614
_____   3. How does vertebrate urine differ from blood? (a) They are the same except that the urine normally contains no blood cells.
(b) Urine contains less glucose and more ammonia. (c) Urine contains less sodium and more calcium. (d) Urine contains more amino acids and fewer ketones.

Ans: d
p. 602
_____   4. Why don't plants need excretory systems? (a) They generate no waste products. (b) They have no fluid transport system.
(c) They retain their nitrogenous wastes to conserve water.
(d) The substances they need are obtained in simple forms.

Ans: b
p. 603
_____   5. Humans are categorized as: (a) ammonotelic; (b) ureotelic;
(c) uricotelic; (d) telocentric.

Ans: d
p. 603
_____   6. The least toxic form of nitrogenous waste is: (a) urea;
(b) ammonia; (c) nitric acid; (d) uric acid.

Ans: a
p. 616
_____   7. Which of the following statements about the loop of Henle is true? (a) In the ascending limb, Na^+ and/or Cl^- ions are pumped out. (b) In the descending limb, Na^+ and/or Cl^- ions are pumped out. (c) In the descending limb, water moves in by osmosis.
(d) In the ascending limb, water moves out by osmosis.

Ans: d
p. 618
_____   8. All of the following statements about ADH are true except:
(a) Its production is inhibited by alcohol. (b) The higher the ADH level, the more water is reabsorbed. (c) It helps to maintain homeostasis. (d) It is produced in the adrenal cortex.

Ans: a
p. 618
_____   9. Bowman's capsule is located in: (a) the renal cortex; (b) the loop of Henle; (c) the glomerulus; (d) the collecting duct.

Ans: b
p. 618
_____   10. The majority of nephrons in humans are: (a) medullary; (b) cortical; (c) juxtamedullary; (d) juxtacortical.

Ans: a
p. 603
_____   11. Reptiles may do well in arid climates partly because they:
(a) excrete uric acid; (b) have long loops of Henle to increase water reabsorption; (c) can live on metabolic water derived from dry food; (d) can absorb water from the atmosphere on dry days.

Ans: d   _____   12. Which structure is absent from the nephron of a freshwater fish?
p. 612   (a) glomerulus; (b) capsule; (c) collecting duct; (d) loop of
                         Henle.

Ans: a   _____   13. The sequence followed by fluid passing through a mammalian
p. 614   nephron is: (a) descending limb of loop, ascending limb, collec-
                         ting duct, renal pelvis; (b) loop, collecting duct, capsule,
                         glomerulus; (c) glomerulus, capsule, loop, collecting duct;
                         (d) ascending limb of loop, descending limb, ureter, collecting
                         duct.

Ans: c   _____   14. Cortical nephrons include: (a) capsule and glomerulus only;
p. 618   (b) capsule, glomerulus, and loop of Henle; (c) capsules,
                         glomeruli, loop of Henle, and collecting duct; (d) capsules,
                         glomeruli, loop of Henle, collecting duct, and ureter.

Ans: a   _____   15. The glomeruli of marine fishes, compared with those of freshwater
p. 612   fishes, are: (a) fewer and smaller; (b) more numerous but
                         smaller; (c) fewer but larger; (d) more numerous and larger.

TEST SET 24-B

Name _____   Section _____

Choose the best answer to each of the following questions, and write the appropriate letter in the space provided.

Ans: d
pp. 608-
  609

_____ 1. All of the following are protonephridial structures except:
(a) solenocytes; (b) green glands; (c) Malpighian tubules;
(d) nephrons.

Ans: b
p. 610

_____ 2. Which of the following organisms has a metanephridial excretory system? (a) crab; (b) snail; (c) butterfly; (d) lamprey.

Ans: d
p. 611

_____ 3. All vertebrate excretory systems include a: (a) urinary bladder;
(b) loop of Henle; (c) Bowman's capsule; (d) glomerulus.

Ans: c
p. 618

_____ 4. The human kidney performs all of the following functions except:
(a) controlling water balance; (b) disposing of toxic wastes;
(c) secreting aldosterone; (d) preventing loss of glucose.

Ans: b
p. 603

_____ 5. Ammonia requires less energy to produce than other nitrogenous wastes, so why aren't all animals ammonotelic? (a) They lack the necessary enzymes. (b) Ammonia is highly toxic. (c) Ammonia tends to form kidney stones. (d) All animals are ammonotelic (which means "able to excrete some ammonia").

Ans: a
p. 601

_____ 6. Artificial kidneys are inferior to natural ones in that they:
(a) are very large; (b) do not remove waste products; (c) do not control water balance; (d) do not adjust mineral ion levels.

Ans: a
p. 611

_____ 7. Nephrons are found: (a) only in vertebrates; (b) only in terrestrial vertebrates; (c) only in warm-blooded terrestrial vertebrates; (d) only in mammals.

Ans: d
p. 603

_____ 8. Uric acid crystals in human urine can result from any of the following except: (a) drinking coffee; (b) taking the "memory pills" (DNA and RNA) sold in health-food stores; (c) eating meat; (d) drinking excess water.

Ans: c
p. 612

_____ 9. Marine birds and turtles have a special gland in the head that excretes salt. Marine mammals do not have a salt gland. What does this difference suggest? (a) Because mammals excrete primarily urea, their urine is dilute enough to carry away excess salt. (b) Because mammals, unlike birds and turtles, bear live young, excess salts are consumed by the growing fetus. (c) The kidneys of birds and reptiles concentrate urine less efficiently, and the salt gland compensates. (d) Mammals get rid of excess salt by perspiring instead.

Ans: a
p. 603

_____ 10. When a tadpole becomes a frog, its excretory system changes:
(a) from ammonotelic to ureotelic; (b) from ureotelic to uricotelic; (c) from ureotelic to ammonotelic; (d) from protonephridial to metanephridial.

Ans: c
p. 605

_____ 11. In an isotonic environment: (a) water has the same composition as the organisms that live in it; (b) water has the same osmotic potential as the organisms that live in it; (c) water has the same osmotic potential as some of the organisms that live in it; (d) active transport is unnecessary.

Ans: a
p. 604

_____ 12. A euryhaline organism: (a) tolerates a wide range of salt concentration; (b) tolerates a narrow range of salt concentration;
(c) controls its internal salt concentration; (d) allows its
internal salt concentration to vary.

Ans: d
p. 605

_____ 13. Both hypertonic and hypotonic environments present serious
problems. As a result, fishes: (a) are not very numerous in
species and may be on the way to extinction; (b) are absent from
all but the most hospitable bodies of water; (c) must be anadro-
mous; (d) have evolved numerous adaptations for maintaining water
and salt balance.

Ans: a
p. 606

_____ 14. The gills of freshwater fishes are used not only for breathing,
but also for: (a) ingesting salt; (b) excreting salt; (c) ingesting water; (d) excreting water.

Ans: b
p. 606

_____ 15. Unlike other marine fishes, sharks: (a) do not urinate; (b) do
not drink; (c) do not have true nephrons; (d) do not use active
transport.

CHAPTER TWENTY-FIVE
ANIMAL HORMONES

CHAPTER SUMMARIES AND OBJECTIVES

25-1 <u>General Properties of Hormones</u> Animal hormones are produced in one part of the body, released into the bloodstream, and transported to target cells where they induce specific effects. Hormones control growth, reproductive development, and many other aspects of physiology. Hormonal communication is slow, yet minute amounts of hormone can evoke dramatic responses in the target cells.

   1. Explain what determines how a given cell will be affected by a particular hormone.

25-2 <u>Glands and Hormones / Chemistry of Hormones</u>   The endocrine system is composed of the endocrine tissues and glands, by which hormones are synthesized and released. Several types of molecules may function as hormones: proteins, steroids, amines, and amino acid derivatives. Most hormone molecules are large enough to be recognized by the target cells but small enough to be readily transported by the bloodstream.

   2. Answer: What is a hormone? What types of substances may act as hormones? How are hormones similar among various species of animals? How are they different?

25-3 <u>Developmental Hormones: A Vertebrate Example / Developmental Hormones in Moth Larvae / Developmental Hormones in a Bloodsucking Bug</u> The hormonal control of growth and development in animals is precisely coordinated. In the frog, for example, the pituitary secretes a hormone that induces secretion of thyroxin by the thyroid. Thyroxin triggers the changes associated with metamorphosis.

Molting and metamorphosis in invertebrates are also under hormonal control. In silkworm moths, brain hormone secreted by specialized brain cells causes the production of ecdysone by the prothoracic glands. The corpora allata secretes juvenile hormone. Ecdysone induces molting, while the amount of juvenile hormone determines the form of the new exoskeleton.

Through experiments with <u>Rhodnius</u>, a bloodsucking bug, Wigglesworth established much of what we know about hormonal control of insect development. The interaction of hormones may bring about developmental changes that are quite different from those induced by one hormone acting alone.

   3. Explain how developmental processes may be controlled by hormones, giving a specific vertebrate or invertebrate example.

   4. Answer: What is the relationship between insect hormones and the defense mechanisms of certain plants?

25-4 <u>The Anterior Pituitary / Pituitary and Hypothalamus / Growth Hormone</u> The pituitary gland is a small, round organ located at the base of the brain and consists of an anterior and a posterior lobe. The anterior pituitary secretes tropic hormones, which control the development and function of other endocrine organs. The posterior pituitary is an extension of the hypothalamus. The hypothalamus produces vasopressin and oxytocin, which are stored in the posterior pituitary. Oxytocin induces uterine contractions during childbirth and the milk let-down reflex during nursing. Vasopressin acts as an antidiuretic hormone and produces **an increase in blood pressure. The brain controls the anterior pituitary by means of the hypothalamus. The hypothalamus secretes special hormones that induce or inhibit the secretion of tropic hormones by the anterior pituitary.**

Defects in the endocrine system can result in gross developmental errors. Excessive production of growth hormone by the anterior pituitary in juveniles can lead to gigantism, whereas inadequate production leads to dwarfism. As might be expected, pituitary hormones are precisely regulated by positive and negative feedback mechanisms.

5. Explain how positive or negative feedback mechanisms work to regulate the secretion of pituitary hormones.

6. Define neurosecretion. Explain how the hypothalamus acts as a mediator between the forebrain and the pituitary gland.

25-5 <u>Human Reproductive Hormones / Hormones and the Menstrual Cycle / Hormones and Embryonic Sexual Differentiation</u> Puberty is under the control of the hypothalamus and pituitary gland. In males, releasing factors trigger the secretion of gonadotropic hormones by the pituitary. LH stimulates the secretion of androgens, including testosterone, by the testes. FSH induces spermatogenesis. LH and testosterone affect body maturation and sexual potency. **Androgens are responsible for** the development of secondary sex characteristics. In the female, FSH and LH are also secreted from the anterior pituitary in response to releasing factors from the hypothalamus. They induce the ovaries to secrete estrogen and progesterone, which initiate the development of secondary sex characteristics.

The menstrual cycle is a series of hormone-controlled changes in the mature female. The cycle lasts about 28 days and is marked by ovulation and the preparation of the uterus to receive a fertilized egg. If an egg is fertilized, the uterine wall is maintained and nourishes the embryo. If fertilization does not occur, the uterine lining is shed in the process of menstruation, and the cycle begins again. FSH, LH, estrogen, and progesterone are all involved in the menstrual cycle, and are regulated by positive and negative feedback.

In the mammalian fetus, hormones induce alternative paths of development. As gonads differentiate, androgens are produced by normal XY individuals, leading to male development. In XX individuals, androgens are not produced and female development occurs.

7. Explain how puberty, with the concomitant appearance of secondary sex characteristics, occurs in human males and females.

8. Trace the hormonal changes associated with the menstrual cycle in cases wherein an egg <u>is</u> and in cases wherein an egg is <u>not</u> fertilized.

9. Describe how hormones produced in differentiating gonads influence fetal sexual differentiation.

25-6 <u>The Adrenal Medulla / The Adrenal Cortex</u> The adrenals are located on top of each kidney and consist of two parts. The adrenal medulla secretes adrenalin and noradrenalin in response to stress; the adrenal cortex secretes corticoids. Corticoids are involved in prolonged adjustments of the body to stress. ACTH causes the release of corticoids by the adrenal cortex.

10. List three ways in which adrenalin affects the "fight or flight" response.

11. Name and describe the function of two types of corticoids.

25-7   Some Other Glands and Their Hormones   The parathyroid glands are located in or near the thyroids. These glands secrete parathormone, which induces an increase in blood calcium. The thyroid glands produce calcitonin, which reduces blood calcium. Negative feedback links the levels of these two hormones.

   12. Explain why calcium is vitally important in the animal body.

25-8   Peptide Hormones and the Second Messenger / Steroid Hormone Action   Protein-like hormones act by means of a second messenger, cyclic AMP (cAMP). When the hormone reaches the target cell, it attaches to a specific receptor molecule in the cell membrane. This activates the enzyme adenylate cyclase, which converts ATP to cAMP within the cell. The cell's response to the hormone results from a series of reactions activated by cAMP.

   Steroid hormones are fat-soluble and readily pass through the cell membrane. Inside the target cell, steroid hormones combine with specific receptor molecules and stimulate the synthesis of new proteins.

   13. Explain the mechanisms by which protein-like hormones and steroid hormones exert their effects on target cells.

25-9   Controlling Hormone Levels   Hormone levels are regulated in several ways. Rates of synthesis and secretion are controlled. Hormones may be excreted or inactivated by the liver. Some hormones become active only in the target cells.

   14. Speculate on the consequences resulting from an inability to regulate hormone levels within the body.

25-10 Prostaglandins   Prostaglandins are specialized lipids that are produced in all parts of the body. They are involved in several aspects of reproductive physiology, in the control of hormone secretion, in inflammation reactions, in blood clotting, in smooth muscle tone, and in pain reactions.

   15. Describe three roles played by prostaglandins in the reproductive physiology of humans.

25-11 Discovery of "Natural Opiates" / Endorphins and Hormones--and a Common Precursor / Activities of Endorphins   Endorphins are peptides occurring in the brain that resemble opiates such as morphine and opium. These "natural opiates" have a high binding affinity for opiate receptors in the brain.

   Endorphins are "coded for" in fragments of the large protein β-lipotropin, which is itself contained within a larger protein, pro-opiomelanocortin. Thus a single segment of DNA codes for several hormones and neurotransmitters.

   Endorphins function both as hormones and as neurotransmitters. They produce an analgesic effect and have been implicated in acupuncture and the placebo effect. In addition, endorphins seem to be involved with various other physiological responses, including hibernation.

   16. Explain why endorphins constitute a major link between the endocrine and nervous systems of animals.

25-12 Evolution of Regulatory Systems   The endocrine and nervous systems probably had common origins in the earliest marine organisms, which exchanged chemical messages via pheromones. As multicellular plants and animals evolved, these compounds took on highly specialized, integrative functions. Some came to act rapidly over short distances (nervous system), while others came to act slowly over longer distances (endocrine system).

   17. Explain how and why the regulatory systems of animals probably evolved.

ARTICLE RESOURCES

General Properties of Hormones

Glands and Hormones

Chemistry of Hormones

Developmental Hormones: A Vertebrate Example

Developmental Hormones in Moth Larvae

Developmental Hormones in a Bloodsucking Bug

The Anterior Pituitary

Pituitary and Hypothalamus

Crews, David. The hormonal control of behavior in a lizard. Scientific American, 1979, 241(2), 180-187. Both behavior patterns and the function of the pituitary gland are modulated by hormones acting on the central nervous system. Steroid-concentrating areas were found in the preoptic area, the anterior and basal hypothalamus, and the limbic system of the brain. The behavior of the male influences the reproductive physiology of the female, and the presence of females has a strong influence on the testicular growth and sperm development in males.

Heller, H. Craig, Larry I. Crawshaw, and Harold T. Hammel. The thermostat of vertebrate animals. Scientific American, 1978, 239(2), 102-113. Chordate animals have a limited temperature range. At the top level for survival is 45°C, at which temperature proteins become inactivated and denature. The lower level is about 0°C, when intercellular water forms ice crystals and ruptures and kills cells. Each chordate has an optimal temperature at which it operates efficiently. Any change in this may interrupt the rates of chemical reactions that are temperature-dependent. The hypothalamus is very sensitive to temperature changes and is the primary activating signal for thermoregulation activities in small animals.

Growth Hormone

Human Reproductive Hormones

Youcha, Geraldine. The quest for the male pill. Science Digest, 1982, 90(3), 33, 116. Attempts to answer the question of why scientists have not found a perfect contraceptive for males. Apparently, drugs that act as effective spermicides may attack other organs and will suppress libido. At present, male response to steroids is only about 76 percent, whereas 98-99 percent of women respond to them.

Hormones and the Menstrual Cycle

Hormones and Embryonic Sexual Differentiation

The Adrenal Medulla

The Adrenal Cortex

Some Other Glands and Their Hormones

Notkins, Abner Lewis. The causes of diabetes. Scientific American, 1979, 241(5), 62-73. In the United States, diabetes is the third most common cause of death. The history of the discovery of the role of the pancreas and of insulin is discussed. There are two types of diabetes that can be distinguished clinically: maturity-onset diabetes, which develops in some obese persons, and the juvenile-onset diabetes, which is an acute disease. Ways of treating the two types are given.

Peptide Hormones and the Second Messenger

Steroid Hormone Action

Controlling Hormone Levels

Prostaglandins

Rosenfeld, Albert. Science letter: The body's own wonder drug. Saturday Review, 1979, Oct. 13, 16-18. Prostaglandins are manufactured by the cells and come in sixteen known kinds. They control a variety of life processes. Five thousand research papers a year on this topic alone are being published.

Discovery of "Natural Opiates"

Benefits of the big sleep. Science Digest, 1982, 90(2), 103. Comparison of causes of hibernation in warm-blooded and cold-blooded animals. Effects of hibernation are listed.

Endorphins and Hormones--and a Common Precursor

Activities of Endorphins

Evolution of Regulatory Systems

Deerden-Smith, Jo. How to win the mating game--by a nose. Next, 1980, 1(5), 85-89. Human pheromones have been investigated primarily in the sexual area, but there is evidence that they may be a factor in mother-baby bonding, aggressive crowd behavior, stress, and madness. Current research on human pheromones is described. Reflections on future applications as well as the use and abuse of such findings.

ESSAY QUESTIONS

1. Explain how juvenile hormone regulates the growth of larval insects.

    If juvenile hormone is present in great quantities during molting, subsequent exoskeletons that are formed will be of the larval type. As the larva grows after each molt, less and less juvenile hormone is produced. Eventually there is not enough juvenile hormone left to induce a larval exoskeleton. A pupal exoskeleton is formed instead, and the adult emerges after the next molt.

2. Explain the functioning of the pituitary gland, mentioning at least three of the hormones it secretes.

    The pituitary gland is a small, round organ at the base of the brain. The anterior lobe of the pituitary secretes tropic hormones in response to releasing factors produced by the hypothalamus. Tropic hormones such as thyrotropin, ACTH, FSH, and LH control the development and function of other endocrine tissues. Oxytocin and vasopressin, which are secreted by the hypothalamus, are stored in, and released from, the posterior lobe of the pituitary. When released into the bloodstream, oxytocin and vasopressin travel to their target cells, where they induce specific responses.

3. Contrast the mechanisms whereby protein and steroid hormones, respectively, exert their effects in appropriate target cells.

    Protein hormones bind with specific receptor molecules on the surface of their target cells, activating adenylate cyclase. Adenylate cyclase catalyzes the formation of cyclic AMP from ATP. The cell's specific response to the hormone results from a series of reactions activated by cyclic AMP. Steroid hormones pass readily through the membranes of cells. Inside their target cells, steroid hormones bind with specific receptor molecules. The hormone-receptor complex then enters the nucleus, where it stimulates the synthesis of messenger RNA.

4. List three ways in which adrenalin affects the "fight-or-flight" response.

    Adrenalin induces an increase in heart rate and blood pressure. It also produces an increase in blood sugar concentration. And it inhibits digestive and reproductive functions.

5. List three ways in which prostaglandins are involved with the reproductive physi-ology of humans.

    Prostaglandins initiate uterine contractions at the onset of childbirth. They may also cause menstrual cramps. And they promote contractions of the uterus during sexual intercourse, thereby helping sperm travel through the female reproductive tract.

6. Explain how birth control pills work, in terms of hormonal control of the menstrual cycle.

    FSH and LH, secreted by the pituitary, stimulate the growth and maturation of ovarian follicles, which secrete estrogen. Rising estrogen levels stimulate the production of gonadotropin-releasing factors by the hypothalamus, in turn causing the pituitary to step up FSH and LH production. As the LH level peaks, ovulation occurs. The follicle cells left behind after ovulation form the corpus luteum under stimulation from LH. The corpus luteum secretes proges-terone and estrogen, which help to prepare the uterine lining to receive the fertilized egg and to maintain the lining during pregnancy. Progesterone and estrogen also have a negative feedback effect on the hypothalamus. Lowered output of releasing factors from the hypothalamus in turn causes the pituitary to reduce LH production, inhibiting ovulation. "The Pill" takes advantage of this negative feedback effect of progesterone and estrogen. The synthetic progesterone- and estrogen-like compounds in birth control pills "trick" the hypothalamus into turning off secretion of gonadotropin-releasing factors, thus turning off LH secretion by the pituitary and inhibiting ovulation.

TEST SET 25-A

Name _____   Section _____

Choose the best answer to each of the following questions, and write the appropriate letter in the space provided.

Ans: c
p. 626

_____   1. The specificity of a cellular response to a hormone depends on: (a) the structure of the hormone; (b) the activity of the hormone; (c) the type of target cell; (d) the type of nervous stimulation.

Ans: b
p. 630

_____   2. The presence of ecdysone in silkworm larvae causes: (a) pupa formation; (b) molting; (c) formation of additional larvae; (d) imaginal disk formation.

Ans: a
p. 631

_____   3. Which of the following is a tropic hormone? (a) ACTH; (b) insulin; (c) estrogen; (d) testosterone.

Ans: d
p. 633

_____   4. Releasing factors produced by the hypothalamus stimulate the secretion of hormones from the: (a) adrenal cortex; (b) adrenal medulla; (c) gonads; (d) anterior pituitary.

Ans: b
p. 635

_____   5. _____ are gonadotropic hormones. (a) Oxytocin and vasopressin; (b) LH and FSH; (c) Estrogen and testosterone; (d) Insulin and glucagon.

Ans: c
p. 635

_____   6. In males, luteinizing hormone (LH) acts to stimulate production of: (a) progesterone; (b) spermatozoa; (c) testosterone; (d) FSH.

Ans: a
p. 636

_____   7. High levels of progesterone result in: (a) low LH production; (b) high LH production; (c) low estrogen production; (d) high ACTH production.

Ans: b
p. 638

_____   8. Anticipation of a "fight or flight" situation causes the adrenal medulla to release a greater amount of: (a) norepinephrine; (b) epinephrine; (c) aldosterone; (d) cortisone.

Ans: a
p. 638

_____   9. Which of the following hormones is not produced by the adrenal cortex? (a) parathormone; (b) cortisone; (c) aldosterone; (d) sex hormones.

Ans: b
p. 639

_____   10. The level of calcium in the blood may be raised by: (a) stimulating the production of calcitonin; (b) stimulating the production of parathormone; (c) inhibiting the production of glucagon; (d) stimulating the production of insulin.

Ans: d
p. 641

_____   11. Which of the following hormones does not use cAMP as a second messenger? (a) thyrotropic hormone; (b) glucagon; (c) insulin; (d) estrogen.

Ans: c
p. 641

_____   12. Steroid hormones act by: (a) binding to cell surface receptors; (b) altering the activity of proteins that are present in the cell; (c) stimulating new protein synthesis; (d) increasing the levels of cAMP and ATP.

Ans: b
p. 646

_____   13. The physiological changes that occur prior to hibernation in animals can be induced by: (a) prostaglandins; (b) endorphins; (c) pheromones; (d) glucocorticoids.

Ans: b   _____   14. Aspirin relieves pain by suppressing the formation of: (a) pher-
p. 643   omones; (b) prostaglandins; (c) endorphins; (d) gonadotropins.

Ans: c   _____   15. The glucose concentration in the blood is raised by the action
p. 628   of: (a) insulin; (b) aldosterone; (c) glucagon; (d) vasopressin.

TEST SET 25-B

Name _____   Section _____

Choose the <u>best</u> answer to each of the following questions, and write the appropriate letter in the space provided.

Ans: d
p. 626

_____   1. The response of a system to hormonal stimulation is determined by: (a) the structure of the hormone; (b) the activity of the hormone; (c) signals from the autonomic nervous system; (d) specific target cell receptors.

Ans: d
p. 630

_____   2. The absence of juvenile hormone in a silkworm results in: (a) molting; (b) formation of additional larvae; (c) inhibition of the imaginal disks; (d) pupa formation.

Ans: b
p. 631

_____   3. ACTH, LH and FSH are all: (a) ovarian hormones; (b) tropic hormones; (c) steroid hormones; (d) testicular hormones.

Ans: c
p. 633

_____   4. Oxytocin and vasopressin are produced by the: (a) posterior pituitary; (b) anterior pituitary; (c) hypothalamus; (d) adrenal cortex.

Ans: d
p. 634

_____   5. Human growth hormone can be used to treat dwarfism in: (a) puppies less than one year of age; (b) adult pygmy horses; (c) post-adolescent humans; (d) pre-adolescent humans.

Ans: a
p. 635

_____   6. In females, FSH acts to stimulate the: (a) growth of ovarian follicles; (b) corpus luteum; (c) production of testosterone; (d) production of glucocorticoids.

Ans: c
p. 635

_____   7. The production of _____ is stimulated during the luteal phase of the menstrual cycle. (a) LH; (b) oxytocin; (c) progesterone; (d) ACTH.

Ans: a
p. 638

_____   8. Actual participation in an aggressive encounter induces the release of a large amount of: (a) norepinephrine; (b) epinephrine; (c) aldosterone; (d) cortisone.

Ans: c
p. 638

_____   9. _____ works to reduce inflammation in body tissues. (a) Aldosterone; (b) Epinephrine; (c) Cortisone; (d) Norepinephrine.

Ans: b
p. 639

_____   10. The level of calcium in the blood may be lowered by stimulating the production of: (a) parathormone; (b) calcitonin; (c) glucagon; (d) insulin.

Ans: a
p. 640

_____   11. Epinephrine keeps the level of cAMP high by: (a) stimulating the activity of adenyl cyclase; (b) stimulating the production of ATP; (c) inhibiting the activity of phosphodiesterase; (d) inhibiting the production of ATP.

Ans: c
p. 640

_____   12. Peptide hormones act by: (a) stimulating new protein synthesis in a cell; (b) binding to cytoplasmic receptor molecules; (c) altering the activity of proteins present in the cell; (d) activating specific genes in the DNA.

Ans: d
p. 644

_____   13. The opiate receptors in the brain bind: (a) pheromones; (b) glucocorticoids; (c) prostaglandins; (d) endorphins.

Ans: c
p. 642

_____ 14. Individuals with testicular feminization (XY) have the external anatomy of a woman. The problem is: (a) too much testosterone; (b) too much dihydrotestosterone; (c) lack of cytoplasmic receptors for dihydrotestosterone; (d) lack of testes.

Ans: d
p. 633

_____ 15. Vasopressin is responsible for: (a) lowering the blood glucose level; (b) raising the blood glucose level; (c) increasing the volume of urine produced; (d) decreasing the volume of urine produced.

CHAPTER TWENTY-SIX
NEURONS AND NERVOUS SYSTEMS

CHAPTER SUMMARIES AND OBJECTIVES

26-1 <u>Major Divisions of the Nervous System / Cells of the Nervous System / Reflex Arcs</u> Data from the environment enter the body via the sensory nervous system and pass to the central nervous system (CNS). Actions triggered by this input are accomplished through the peripheral nervous system, which is divided into the autonomic nervous system, serving smooth muscles and glands, and the somatic system, serving skeletal muscles.

Nervous tissue consists of glial cells and neurons. Neurons are composed of the cell body, dendrites, axons, and terminal boutons. The junction between neurons, or between a neuron and a muscle cell, is termed a synapse. Glial cells may serve a structural, phagocytic, or nutritive role in nervous tissue.

A reflex is an automatic action in which a motor response follows rapidly after a sensory stimulus. A sensory neuron carries the impulse from the stimulus to the spinal cord, where a synapse with a motor neuron causes the impulse to travel to the appropriate muscle for the reflex action.

1. Describe the major divisions of the nervous system, giving the function of each part.

2. Define axon, dendrite, terminal bouton, synapse, and reflex arc.

26-2 <u>The Resting Potential / Electrotonic Potentials</u> Neurons have an electrical gradient across their cell membranes. The interior of the cell carries a negative charge relative to the exterior. This membrane potential is generated by the unequal distribution of ions across the cell membrane. Resting neurons have a membrane potential of about -75 mV, the resting potential. The membrane of the neuron is impermeable to all ions except K^+. Because of concentration differences, K^+ tends to diffuse out of the cell, causing a net buildup of positive charge on the outside. Diffusion stops when the external positive charge is sufficient to repel any additional K^+ ions trying to diffuse out.

Changes in membrane potentials can be induced experimentally by applying an electric current across the membrane. The result is either a hyperpolarization or a depolarization of the membrane. These changes are called electrotonic potentials. Upon removal of the current, the normal resting potential of the neuron is restored.

3. Explain how the resting potential is established in a neuron.

26-3 <u>Action Potentials / Mechanism of the Action Potential / The Refractory Period /</u>
<u>Propagation of the Action Potential</u> When a nervous impulse reaches a particular
point on an axon, depolarization occurs. Within a few milliseconds, the cell
returns to the resting potential. This positive polarization of the neuron, or
action potential, constitutes the nervous impulse. The impulse travels down the
axon with no change in magnitude. The intensity of a given stimulus is not reflected in the magnitude of the action potential, but rather in the number of
action potentials propagated along the axon in a given length of time.

With depolarization, the membrane of the neuron suddenly becomes permeable to
sodium ions (Na^+), allowing them to rush into the cell. The influx of positive
ions is enough to cause a reversal of the membrane potential. When V_m becomes
positive, impermeability to Na^+ and permeability to K^+ return. Efflux of K^+
returns the membrane to resting potential.

With the restoration of the resting potential, there is a refractory period
during which the neuron cannot undergo another action potential. This is because
the sodium channels cannot immediately reopen. The original ionic balance of the
neuron is restored by active transport of Na^+ and K^+.

Movements of positive ions during an action potential result in depolarization of
adjacent areas of the axon. In this way, the action potential is propagated
along the axon, as adjacent depolarization causes Na^+ channels to open and a new
action potential to arise. More rapid conduction occurs on neurons that are
myelinated. On these neurons, action potentials leap between Schwann cells, from
node to node, rather than traveling over each bit of the axon.

4. Diagram the ionic movements occurring in a neuron (1) at resting potential,
   (2) during an action potential, and (3) during restoration of the resting
   potential.

5. Explain how nervous impulses are transmitted along the axon. How is transmission improved by Schwann cells?

26-4 <u>Synaptic Transmission / Neurotransmitters / Clearing the Synapse of Neurotrans-</u>
<u>mitters</u> The junction between two neurons is called a synapse. Transmission of
a nervous impulse from one neuron to another may be by electrical or chemical
means. Electrical transmission occurs via a physical bridge between the neurons.
This bridge is called a connexon. Chemical transmission across the synapse
occurs as follows. Terminal boutons of the presynaptic neuron contain numerous
synaptic vesicles, which release neurotransmitters into the synaptic cleft when
the impulse reaches the terminal boutons. The neurotransmitters diffuse across
the cleft and bind with neuroreceptor proteins on the membrane of the postsynaptic neuron. Binding of neurotransmitters with neuroreceptor proteins allows Na^+
to diffuse through the membrane, which, in turn, triggers an action potential in
the postsynaptic neuron.

Each chemical synapse uses a specific neurotransmitter, of which eight are known,
including acetylcholine, noradrenalin, serotonin, and glycine. Acetylcholine is
the most important neurotransmitter for motor neurons arising from the spinal
cord and for all nerve-to-skeletal-muscle synapses. The other neurotransmitters
function mainly in brain synapses.

Neurotransmitters are quickly removed from the synaptic cleft to prevent their
constant binding to the neuroreceptors and the continuous triggering of action
potentials that would result. Neurotransmitter may be allowed to diffuse away
from the cleft, is broken down by enzymes, or is reabsorbed from the cleft.

6. Explain the process of chemical transmission of a nervous impulse across a
   synapse.

26-5 <u>Events in the Postsynaptic Neuron / Summation</u> A synapse may be either excitatory
or inhibitory; that is, it may serve to trigger an impulse in the postsynaptic
neuron or to make it less likely to fire. The nature of the synapse is determined by the nature of the neuroreceptors in the membrane of the postsynaptic

neuron.

Neurons usually receive input from many synapses, some excitatory, others inhibitory. If active at the same time, the effects of excitatory and inhibitory synapses tend to cancel each other. Stimulation of the postsynaptic neuron can occur if enough excitatory synapses fire at the same time (spatial summation) or if a single excitatory provides rapid, repeated stimulation (temporal summation).

7. Explain what determines whether a postsynaptic neuron will fire (undergo depolarization).

26-6 Nervous Integration Nervous integration is the controlled interaction of many neurons to produce complex patterns of behavior. A single neuron may synapse with many others, some excitatory and some inhibitory. As a result, stimulation of a single neuron can have many and varied effects throughout the entire organism. Organized clusters of neurons, called ganglia, exist in many invertebrates. In the ganglion, each neuron has a particular function.

8. Define nervous integration, and construct a simple model to account for a variety of responses arising simultaneously from a single stimulus.

26-7 The Autonomic Nervous System / Cephalization / Structures of Vertebrate Brains The autonomic nervous system (ANS) is made up of those neurons that control smooth muscles and glands and consists of two subsystems. The sympathetic system produces effects related to the "fight-or-flight" response, whereas the parasympathetic system produces a general "slowing-down" reaction.

Cephalization refers to the gathering of sense organs at one end of an organism and the formation of ganglia to organize sensory input. In higher animals, the anterior ganglia have been further developed into a brain, which has specific regions to carry out specific functions.

The vertebrate brain can be subdivided into the hindbrain, midbrain, and forebrain. The hindbrain, or medulla, contains axons of both sensory and motor neurons and, in part, forms the cerebellum. The midbrain controls eye muscles in all vertebrates and processes optic input in all but the mammals, in which it processes input from the ears. The forebrain is further subdivided into the diencephalon and telencephalon. The former contains the hypothalamus, posterior pituitary, and thalamus. The telencephalon is the center for processing olfactory information in fish, but it reaches its greatest complexity in higher mammals, where the bulk of it constitutes the cerebrum.

9. Describe the functioning of the autonomic nervous system and its subdivisions. How do the two subdivisions differ, with respect to both structure and to function?

10. Outline the major subdivisions of the vertebrate brain, and, for humans, associate functions with each subdivision.

26-8 The Cerebrum / The Hypothalamus / The Cerebellum / The Limbic System and the Emotions / The Reticular Formation One sees an increase in the size and development of the cerebrum as one progresses from the lower to the higher vertebrates. In humans the cortex of the cerebrum is large and heavily convoluted and is the site of the most complex brain functions. Specific regions of the cortex are associated with specific sensory or motor functions. A large area of the frontal lobes of the cortex is responsible for complex social senses. The cerebral cortex is also the center for the processing of all sensory input and for memory.

The hypothalamus plays a vital role in body maintenance, monitoring, and triggering responses to changes in body temperature, blood glucose, water balance, blood pressure, and fat and carbohydrate metabolism. It also contains centers for producing complex behaviors, such as those to satisfy hunger, thirst, or other appetites.

The cerebellum modifies instructions from the cerebral cortex in light of constant sensory input about the position and orientation of body parts. This modification is important in coordination and in maintaining equilibrium and balance during movement.

The limbic system is thought to be the neural basis of emotions. It may work in concert with the hypothalamus and other structures of the brain. The reticular formation is a diffuse net of neurons running through the brain stem. It is responsible for attention, arousal, and concentration on particular sensory inputs or muscular activities.

11. State the location and function of each of the following: cerebrum, cerebellum, hypothalamus, reticular formation, and limbic system.

26-9 Sleeping and Dreaming / Memory Sleep has two main stages, each with different associated brainwave activity. In light sleep, muscles maintain tone, but they relax suddenly when one enters deep sleep. Deep sleep is also known as REM (rapid eye movement) sleep. Dreaming takes place during REM sleep, as do some limb movements.

Memory, both short-term and long-term, is a function of the cerebral cortex. Changes in particular synapses that lead to a specific pattern of neuron firing are part of the establishment of long-term memory, but they are not thought to be its prime basis. Long-term memory establishment seems to be a two-step process. The first step provides a transient memory trace that can be made more permanent in a second step involving protein synthesis. The fixation of long-term memory is accomplished largely in the hippocampus.

12. Explain how EEG measurements can be used to characterize the stages of sleep.

13. Cite experimental evidence for the contention that memory is contained on the surface of the cortex and that the establishment of long-term memory involves protein synthesis.

ARTICLE RESOURCES

Major Divisions of the Nervous System

Hubel, David H. The brain. Scientific American, 1979, 241(3), 44-53. The history of neuroanatomy is given. Comparative size and complexity of brains from fish to cats to humans. (This issue is devoted to the brain. Eleven excellent articles are included.)

Cells of the Nervous System

Stevens, Charles F. The neuron. Scientific American, 1979, 241(3), 54-65. Well-illustrated article on neurons. Anatomy types and physiology are discussed. The sodium pump is explained in detail but clearly. The relationship between membrane proteins and neuron function is explored.

Iverson, Leslie L. The chemistry of the brain. Scientific American, 1979, 241(3), 134-150. Chemical transmitters form a system overlaid on the neuronal circuits of the brain. There are thirty or more of these substances, each of which has a characteristic excitatory or inhibitory effect on neurons. Excellent charts, drawings, and chemical models.

Reflex Arcs

The Resting Potential

Electrotonic Potentials

McAuliffe, Kathleen. I sing the body electric. Omni, 1980, 3(2), 70-73, 98-99. Electrical signals are used by living organisms to regulate development, growth, and repair. Bone healing via electric coils has been explored. Regeneration of amphibian lost parts by the use of electric currents is illustrated. Constant current in the peripheral nerves appears to give rise to the field patterns displayed by all animals.

Action Potentials

Mechanisms of the Action Potential

The Refractory Period

Propagation of the Action Potential

Synaptic Transmission

Neurotransmitters

Clearing the Synapse of Neurotransmitter

Events in the Postsynaptic Neuron

Summation

Nervous Integration

   Nauta, Walle J. A., and Michael Feirtag. The organization of the brain. Scientific American, 1979, 241(3), 88-111. Excellent diagrams and models illustrate this discussion of the mammalian central nervous system. Each section is studied morphologically and physiologically as well as embryologically. The interconnectedness of the brain is shown.

Autonomic Nervous System

Cephalization

Structures of Vertebrate Brains

The Cerebrum

The Hypothalamus

The Cerebellum

The Limbic System and the Emotions

   Guerin, Joel. Chemical feelings. Science 80, 1979, 1(1), 28-36. Peptides in the brain play a key role in influencing feelings and behavior. This article catalogs the brain peptides and summarizes the complexity of the systems involved. There is evidence that these large molecules may be used as neurotransmitters.

   Cytowic, Richard E. The long ordeal of James Brady. New York Times Magazine, 1981, 109 (Sept. 27), 27-31. This article describes the extent of James Brady's injuries sustained in the assassination attempt on President Reagan. It also tells of the surgery performed on him and the factors that played a role in his recovery. The frontal lobes of his brain were injured, and 20 percent of the right frontal lobe was surgically removed. Injury to the frontal lobes can affect personality, mood, and judgment. The patient's mental attitude plays a role in recovery, and here it is crucial.

The Reticular Formation

Sleeping and Dreaming

   Hooper, Judith. Dreaming. Omni, 1979, 2(3), 46. Universal themes seem to appear in everyone's dreams. These include loss, falling, being chased, and flying. Speculations about their meaning are advanced.

Memory

ESSAY QUESTIONS

1. Explain the mechanism whereby an action potential is generated and propagated.

    At rest, there is a potential of -75 mV across the cell membrane of the neuron. The membrane is freely permeable to K^+ ions but to no others. K^+ diffuses out, following the concentration gradient, until the buildup of positive charge on the outside of the cell prevents further diffusion. When the cell is stimulated, the membrane suddenly becomes permeable to Na^+ ions, which rush in, reversing the membrane potential (depolarization). Ion pumps soon restore the resting potential. Depolarization of one region of the axon stimulates the adjacent region, producing another action potential. As a result, the action potential is propagated along the axon in a series of depolarization events.

2. Compare and contrast the two types of synaptic transmission with regard to structure and function.

    In an electrical synapse, cytoplasmic links called connexons provide a physical connection between two neurons. The nerve impulse may travel directly from one neuron to another. In a chemical synapse, on the other hand, there is no physical connection between the neurons. The presynaptic neuron has terminal boutons containing synaptic vesicles, which in turn contain neurotransmitters. When an impulse reaches the synapse, the neurotransmitters are released into the cleft between the neurons. The postsynaptic neuron has in its cell membrane specific neuroreceptors with which the neurotransmitter molecules bind. Such binding causes the membrane of the postsynaptic neuron to depolarize, triggering an action potential. An electrical synapse can, at least theoretically, operate in both directions; a chemical synapse is unidirectional, because only the presynaptic neuron releases neurotransmitters and only the postsynaptic neuron possesses neuroreceptors.

3. Outline the major divisions of the vertebrate brain, and give the function of each in humans.

    The hindbrain consists of the medulla and cerebellum. The medulla contains the axons of sensory neurons (except those for smell and vision) and axons that provide input to motor neurons. The cerebellum is involved with the coordination of body movements and with equilibrium. The midbrain controls external eye muscles and processes auditory input. The forebrain is subdivided into the diencephalon and the telencephalon. The former contains the hypothalamus, which controls physiological and behavioral responses to maintain homeostasis; the posterior pituitary (see Chapter 25); and the thalamus, which integrates sensory input and controls arousal. The telencephalon is composed largely of the cerebrum, which is the site of most complex brain activities, including association, memory, and learning; much sensory input and motor control; and complex social senses, such as the concepts of good and evil. The limbic center, forming a double loop around the center of the brain, is the seat of emotions. The reticular activating system, a diffuse net of neurons in the brainstem, controls attention, arousal, and concentration.

4. What are the major functional and structural differences between the sympathetic and parasympathetic divisions of the autonomic nervous system?

    The parasympathetic system is involved with excitatory reactions, such as those of the fight-or-flight response, whereas the sympathetic system has the opposite function, that of mediating "slowing-down" responses. At their target organs, the two systems differ in neurotransmitters used. The sympathetic system releases noradrenalin or adrenalin, whereas the parasympathetic system releases acetylcholine. The other major structural difference lies in the placement of the synapses between the motor neurons. In the sympathetic system, the cell bodies of presynaptic neurons lie in the spinal cord and synapse with the postsynaptic neuron in a ganglion near the spinal cord. Presynaptic neurons of the parasympathetic system have their cell bodies in the medulla, and their axons extend all the way to the target organ. Thus the parasympathetic synapses lie in a ganglion very near the target organ, and the postsynaptic neuron is quite short.

5. What is cephalization, and what adaptive advantages does it confer on animals that exhibit cephalization?

Cephalization is the gathering of the sense organs in the head, or anterior end, of the animal. This has the advantage of placing the sensory apparatus at the leading end of the animal, ready to probe the environment in the region lying in the direction of the animal's movement. This aids in both detection of food and avoidance of danger. Cephalization also refers to the development of ganglia or a brain. In a ganglion, individual neurons can be specialized for different functions, increasing efficiency and making possible the triggering of a variety of behavioral responses by a single stimulus. In a brain, entire regions, consisting of many neurons each, can become specialized for particular tasks. This allows the integration of a variety of sensory inputs and the coordination of a variety of motor responses. In short, complex behavior becomes a possibility.

6. It has been suggested that much of what we call "normal behavior" consists in maintaining certain neurons in the hypothalamus in favorable electrochemical states. Cite some information from this chapter that would tend to support this proposition.

Nervous activity is essentially electrochemical in nature, with the activity of any particular neuron determined by the excitatory and inhibitory impulses it receives from other neurons with which it synapses. Furthermore, it is clear that the hypothalamus contains the neural centers that monitor and control the internal environment and mediate the behavior associated with the satisfaction of hunger, thirst, and other appetites. Some evidence even points to the existence of pleasure and punishment centers. Through the pituitary gland, the hypothalamus is also involved in the physiological activities (and resulting behavior) associated with anger, aggression, and fear (fight-or-flight reaction).

TEST SET 26-A

Name _____   Section _____

Choose the best answer to each of the following questions, and write the appropriate letter in the space provided.

Ans: d
p. 650

_____   1.   In the central nervous system, the most abundant cell type is the: (a) motor neuron; (b) sensory neuron; (c) interneuron; (d) glial cell.

Ans: b
p. 651

_____   2.   The _____ carries a nerve impulse toward a synapse. (a) glial cell; (b) presynaptic neuron; (c) postsynaptic neuron; (d) synaptic neurotransmitter.

Ans: c
p. 654

_____   3.   In the stimulation of a simple reflex arc, the muscular response is swift because: (a) the nerve impulse by-passes the central nervous system; (b) the brain processes some nerve impulses very rapidly; (c) the nerve impulse passes almost directly from the motor neuron to the sensory neuron; (d) the nerve impulse passes almost directly from the sensory neuron to the motor neuron.

Ans: d
p. 657

_____   4.   Which of the following statements describes an action potential? (a) The size of the action potential decreases along the axon. (b) Action potentials increase with an increase in stimulating current. (c) Action potentials become electrotonic potentials as the impulse travels along the axon. (d) All action potentials in a single neuron are of the same magnitude.

Ans: d
p. 659

_____   5.   The V_m of a nerve cell membrane is restored to a "resting" value by an: (a) efflux of Na^+; (b) influx of Na^+; (c) influx of K^+; (d) efflux of K^+.

Ans: c
p. 659

_____   6.   A neuron cannot be immediately restimulated to another nerve impulse level. The short interval of time during which restimulation is not possible is called the: (a) action potential; (b) electrotonic potential; (c) refractory potential; (d) resting period.

Ans: a
p. 659

_____   7.   Nerves that have myelin sheaths are able to: (a) increase the nerve impulse conduction rate; (b) decrease the nerve impulse conduction rate; (c) increase the action potential for each nerve impulse; (d) decrease the action potential for each nerve impulse.

Ans: d
p. 661

_____   8.   Neurotransmitter molecules are produced by: (a) connexons; (b) enzymes in the synaptic cleft; (c) postsynaptic neurons; (d) presynaptic neurons.

Ans: d
p. 663

_____   9.   A synapse may be either excitatory or inhibitory depending on: (a) the type of neurotransmitter; (b) the presynaptic terminal boutons; (c) the size of the synaptic cleft; (d) the postsynaptic neurotransmitter receptors.

Ans: b
p. 663

_____   10.   A postsynaptic neuron is brought to threshold level by simultaneous signals from several excitatory synapses. This process is called: (a) temporal summation; (b) spatial summation; (c) neurotransmitter summation; (d) neutralization.

Ans: c
p. 666

_____   11.   Increased blood pressure, accelerated heart rate, and dilation of the pupils in the eyes are produced by the excitation of the: (a) somatic nervous system; (b) parasympathetic nervous system; (c) sympathetic nervous system; (d) voluntary nervous system.

Ans: a
p. 668

_____   12. The neurotransmitter that is released by postsynaptic neurons
in the parasympathetic nervous system is: (a) acetylcholine;
(b) epinephrine; (c) norepinephrine; (d) serotonin.

Ans: b
p. 669

_____   13. The part of the brain that is more complex in mammals than in
amphibians is the: (a) midbrain; (b) cerebrum; (c) cerebellum;
(d) medulla.

Ans: d
p. 674

_____   14. The phase of rest known as light sleep is controlled by the:
(a) RAS; (b) limbic system; (c) somatic system; (d) raphe.

Ans: b
p. 675

_____   15. Your memory of this material for an exam one week from now will
be accomplished via the activity of the: (a) raphe; (b) hippocampus; (c) locus ceruleus; (d) limbic system.

TEST SET 26-B

Name _____   Section _____

Choose the best answer to each of the following questions, and write the appropriate letter in the space provided.

Ans: d
p. 649

_____ 1. The cell that carries the nerve impulse is called a: (a) glial cell; (b) nerve fiber; (c) phagocyte; (d) neuron.

Ans: c
p. 651

_____ 2. A _____ carries a nerve impulse away from a synapse. (a) glial cell; (b) presynaptic neuron; (c) postsynaptic neuron; (d) synaptic neurotransmitter.

Ans: d
p. 651

_____ 3. The movement of a nerve impulse through a nerve cell is: (a) axon to cell body to dendrite; (b) axon to dendrite to cell body; (c) cell body to dendrite to axon; (d) dendrite to cell body to axon.

Ans: b
p. 657

_____ 4. A very strong stimulus applied to a nerve cell causes: (a) an increase in the magnitude of each action potential; (b) an increase in the frequency of the action potentials; (c) an increase in the conversion of action potentials to electrotonic potentials; (d) a decrease in the magnitude of the electrotonic potentials.

Ans: a
p. 658

_____ 5. Depolarization of the nerve cell membrane results in: (a) an increased permeability to Na^+; (b) a decreased permeability in Na^+; (c) a decreased permeability to K^+; (d) a nerve impulse with no appreciable movement of Na^+.

Ans: b
p. 659

_____ 6. The nerve impulse travels along an axon in one direction because the _____ prevents a "backing-up" movement. (a) cell body; (b) refractory period; (c) myelin; (d) electrotonic potential.

Ans: c
p. 663

_____ 7. The binding of a neurotransmitter to a neuron in an inhibitory synapse results in: (a) depolarization of the cell membrane; (b) generation of an action potential; (c) little or no change in the resting potential; (d) increased permeability of the membrane to Na^+.

Ans: b
p. 668

_____ 8. The neurotransmitter released by postsynaptic neurons in the sympathetic nervous system is: (a) acetylcholine; (b) epinephrine; (c) acetylcholinesterase; (d) serotonin.

Ans: a
p. 674

_____ 9. The period of sleep known as REM sleep is controlled by the: (a) locus ceruleus; (b) occipital cortex; (c) raphe; (d) limbic system.

Ans: a
p. 672

_____ 10. Many of the internal processes involved in the maintenance of homeostasis are monitored by the: (a) hypothalamus; (b) cerebrum; (c) cerebellum; (d) telencephalon.

Ans: b
p. 673

_____ 11. Your fear of the outcome of this examination is controlled by the: (a) RAS; (b) limbic system; (c) somatic system; (d) raphe.

Ans: a
p. 663

_____ 12. A postsynaptic neuron is brought to threshold level by rapid repeated stimulation through a single excitatory synapse. This process is called: (a) temporal summation; (b) spatial summation; (c) neurotransmitter summation; (d) neutralization.

Ans: d   _____   13.   Some studies indicate that the establishment of long-term
p. 675   memory requires: (a) interaction of the cerebrum with the limbic
                           system; (b) processing of signals by the raphe; (c) processing
                           of signals by the locus cereuleus; (d) protein synthesis.

Ans: d   _____   14.   The function of acetylcholinesterase is to: (a) facilitate the
p. 663   uptake of acetylcholine by postsynaptic cells; (b) produce
                           acetylcholine in a synaptic cleft; (c) pass a nerve impulse
                           through a connexon; (d) inactivate acetylcholine in a synaptic
                           cleft.

Ans: c   _____   15.   The hypothalamus is located in the: (a) hindbrain; (b) midbrain;
p. 669   (c) diencephalon; (d) telencephalon.

CHAPTER TWENTY-SEVEN
SENSORY SYSTEMS

CHAPTER SUMMARIES AND OBJECTIVES

27-1 <u>Information Filtering in Sensory Systems / Types of Sensory Receptors / Receptor Cells</u> Data about the internal and external environments of an organism are picked up by sensory receptors and transmitted to the central nervous system. All data are not treated with equal importance. As a result, the animal is not conscious of every bit of information in the world around it. The information detected is that which is necessary or useful for survival or well-being, and it was selectively processed by the nervous system.

To obtain different types of information, different types of receptors are required: photoreceptors detect light; thermoreceptors detect temperature; chemoreceptors detect scent, taste, and internal chemical concentrations; mechanoreceptors detect touch, vibration, and related sensations; electroreceptors measure changes in electric fields. Organisms have evolved in different types of environments and have developed different types of sensory receptors for the needs of their differing lifestyles. Messages from sensory receptors are transmitted by action potentials to different parts of the brain.

Receptor cells are neurons, modified neurons, or neuron-like cells that pass information to sensory neurons by way of a synapse. The central nervous system is alerted to the intensity of a stimulus by the rate at which the action potential is produced.

1. List at least five types of sensory receptors and describe their functions.

2. Explain how a sensory receptor passes information to the central nervous system and how the CNS determines the intensity of a stimulus.

27-2 <u>Visual Systems</u> The simplest visual system is a patch that is sensitive to light intensity. Animals that possess only such simple structures do not detect images. Animals with true eyes may also have simple eyes (ocelli) for light detection. Arthropods have compound eyes made up of many units called ommatidia. Each ommatidium is composed of two structures that focus light onto a rhabdom and its surrounding retinula cells. Because the image formed is a composite of information from each ommatidium, compound eyes do not allow for sharp images, but they do allow the detection of rapid movements.

3. Describe the structures that form the arthropod eye.

27-3 <u>Vertebrate and Cephalopod Eyes</u> Light striking the vertebrate eye first encounters the protective cornea, then passes through an opening in the retractable iris called the pupil, and finally strikes the retina. Photoreceptor cells are located in the retina. The fovea is an area on the retina where reception of the most distinct portion of the visual image takes place. Sensory information that strikes the photoreceptors is passed to the brain via the optic nerve. The eyes of cephalopods are similar to those of vertebrates but lack a "blind spot." In vertebrates, the optic nerve interferes with light reception in the area where it leaves the retina. In cephalopods, the nerve forms behind the photoreceptors.

4. Identify each of the following structures: rods, cones, cornea, iris, pupil, retina, optic nerve, bipolar cells, ganglion cells, fovea. Tell how each functions in vision.

27-4 Photoreceptors of the Vertebrate Eye / Events in the Photoreceptors / Color Vision
Rods are responsible for light absorption in dim light, but the image formed under these circumstances lacks color. Cone cells, which function in brighter light, are responsible for color vision. The sharpness of the image focused on the fovea is due to the concentration of cones in this area. Peripheral vision, which is not sharp but is highly light-sensitive, is due to a concentration of rods in the periphery of the eye.

A chemical substance (rhodopsin) is responsible for the photosensitivity of the photoreceptor cells of most animals. Rhodopsin consists of a protein (opsin) and a derivative of vitamin A, retinaldehyde. The isomer of this compound found in unstimulated rhodopsin is 11-cis retinaldehyde. When stimulated by light, 11-cis retinaldehyde changes shape to the all-trans form. Bleaching occurs when further stimulation causes rhodopsin to separate into opsin and retinaldehyde. This separation causes a change in the polarization of membranes (generator potential), which can trigger action potentials in neighboring neurons and the optic nerve.

There are three types of cone cells, each with a different pigment, that respond to different wavelengths of light. Colorblindness is the result of the absence of the class of cone cells with the pigment that absorbs a particular color. Most mammals are at least partially colorblind. Color pigments all have 11-cis retinaldehyde, but they have different opsin components.

5. Explain the changes that occur on the molecular level as a photoreceptor cell receives light stimulation.

6. Explain how colors are perceived.

27-5 Information Processing in an Invertebrate Eye Some processing occurs in the eye before the visual information is passed on to the brain. Neurons in the compound eye of the horseshoe crab, which carry visual messages to the CNS, also form synapses with neighboring neurons. Synapses with the optic nerve are excitatory, but synapses with neighboring neurons are inhibitory. Inhibition of output from surrounding cells is called lateral inhibition. Lateral inhibition enables the animal to detect contrast at edges between light and dark objects, thereby improving resolution of moving objects.

7. Explain how lateral inhibition in Limulus improves the resolution of moving objects.

27-6 Information Processing in the Vertebrate Retina / Visual Information Processing in the Vertebrate Brain In vertebrates, the possession of circular areas on the retina, called receptive fields, and the presence of two kinds of neurons, horizontal cells and amacrine cells, allow for processing of visual information in the retina before messages are sent to the brain. "On"-center receptive fields have an inner core of receptors that excite a ganglion cell when illuminated. Cells in the outer ring of the field inhibit the ganglion cell. Ganglion cells are inhibited when light strikes the center of an "off"-center field and are excited when the outer ring of the field is illuminated. Both "off"- and "on"-center fields allow vertebrates to perceive contrast between light and dark images. Horizontal cells receive messages from receptor cells. Because of synapses with bipolar cells, other horizontal cells, and the photoreceptors, horizontal cells can also sharpen contrast through lateral inhibition. Amacrine cells do not receive direct messages from photoreceptors. Instead, they receive messages from bipolar cells and other amacrine cells, and they respond to changes in light intensity, increasing their ability to detect moving objects.

The optic nerve in mammals goes directly to the lateral geniculate body in the brain. The cells of the lateral geniculate body all respond to different small areas of the retina and behave like "on"-center and "off"-center fields. Data are transmitted from the lateral geniculate body to the visual cortex, where different receptor cells respond to different types of visual input. There are receptors for lines of light oriented in certain directions (horizontal, vertical, and so on). Other receptors detect edges where bright and dark areas meet. There are particular "edge detectors" for receiving stimuli from edges with the light and dark areas oriented in a specific way.

8. Explain how information processing in the vertebrate eye is similar to that described for Limulus.

9. Identify "on"-center fields, "off"-center fields, amacrine cells, and horizontal cells.

10. Name and describe the functioning of two areas of the brain in which visual information is processed.

27-7 Auditory Systems In humans, sound waves traveling through the air are funneled through the outer ear and strike the tympanic membrane. Vibrations of the tympanic membrane are then transmitted to the middle ear via three ossicles and finally strike the oval window of the cochlea. The cochlea is separated into two fluid-filled chambers by a basilar membrane. Sound receptors embedded in the tectorial membrane are normally in contact with the basilar membrane. When sound waves strike the oval window, the resulting pressure waves in the fluid of the cochlea cause the basilar membrane to bulge. This, in turn, causes deformation of receptors, depolarization, and an action potential in nerve endings surrounding the receptors. Waves resulting from different sound frequencies are detected in specific areas of the organ of Corti, and the brain determines pitch by the location of the sensitive cells that are generating action potentials.

11. Diagram a human ear, clearly label the parts, and explain the pathway that auditory information takes before reaching the brain.

27-8 Chemoreception / Olfaction / Chemoreception in Arthropods Chemoreception gives animals the ability to taste (gustation), which is important in feeding; the ability to smell (olfaction), which is also important in feeding and in the sexual behavior of many animals; and the ability to monitor the concentrations of chemicals in their internal environment. Chemoreceptors are usually naked endings of specialized bipolar sensory neurons.

In vertebrates, the receptors detecting airborne chemicals are sensory neurons in the epithelium at the back of the nose. It has been suggested that, when a specific chemical is present, a macromolecule in the neuron's cell membrane combines with the chemical. The resulting change in the shape of the macromolecule causes an increase in the permeability of the neuron's cell membrane and thereby creates a generator potential.

Male silkworm moths have chemoreceptors located on their antennae. Females of this species release a chemical pheromone that the male detects. When the chemical diffuses into the base of sensory hairs on the antennae of the male, and is detected by at least 200 hairs per second, the male will search for the female by following the chemical "trail."

12. Describe one way in which generator potentials may be created in an olfactory chemoreceptor.

13. Explain how the male silkworm moth locates his mate.

27-9 Mechanoreception / Proprioception / Other Receptors There are many types of receptors that detect touch, vibration, and related sensations. Many of these receptors are hair cells. When the hairs are bent in one direction, the neuron associated with that hair cell fires. Firing is inhibited when the hairs are bent in the opposite direction. In vertebrates, the semicircular canal system helps in the detection of direction of body movement. The correct orientation of the head with respect to gravity is determined by the utriculus, which corresponds to the statocyst of many invertebrates. Water moving through openings in the lateral line system of fishes enables them to detect other organisms in the water to determine their own direction and velocity. Many kinds of mechano-receptors are found in the skin, responding to pressure, touch, and pain.

Neurons that inform the CNS of the position of limbs and of movements in muscles and joints continuously feed information to the brain. In crustaceans, dendrites of stretch-receptor neurons fire when abdominal muscles are stretched. In each vertebrate muscle bundle there is a muscle spindle. These are modified muscle fibres surrounded by a sensory nerve ending. When the muscle stretches, the neuron fires. Many proprioceptors may adapt to the stretch position and fire only when a change in position occurs. Skeletal movement is detected by informa-tion from receptors in the joints.

Other types of receptors include baroreceptors, which monitor changes in blood pressure by detecting the bulging of the carotid artery. Thermoreceptors monitor temperature and are found in the hypothalamus of birds and mammals and on the heads of pit vipers. Birds and mammals employ information from thermoreceptors to maintain body temperature, and pit vipers use them to locate warm-blooded prey. Electroreceptors allow certain fishes to detect interruptions of the electrical fields that these animals generate, enabling them to navigate and to locate prey.

14. Explain how vertebrates are able to detect the direction of their movement and to orient themselves with respect to gravity.

15. Describe the functioning of at least three of the following: lateral line system, pacinian corpuscles, Merkel's disks, Meissner's corpuscles, muscle spindle, thermoreceptors (in birds and mammals), thermoreceptors (in snakes), electroreceptors, and baroreceptors.

ARTICLE RESOURCES

Information Filtering in Sensory Systems

Types of Sensory Receptors

Jastrow, Robert. Evolution: Selection for perfection. Science Digest, 1981, 89(11), 84-87, 115. Explanation of how the eye and the brain may have developed in prehistoric animals, reaching their perfection in the human. Touches on Darwin's theories. Raises the difficulties inherent in the role of chance in human evolution.

Receptor Cells

Visual Systems

Hellman, Hal. Sight and survival. Science Digest, 1982, 90(2), 54-59, 113-115. Excellent photographic essay on animal eyes, their adaptations, and how these can and do aid in survival.

Vertebrate and Cephalopod Eyes

Photoreceptors of the Vertebrate Eye

Events in the Photoreceptors

Color Vision

Information Processing in the Invertebrate Eye

Scharf, David. Electron eye catchers. Science 81, 1981, 2(7), 62-65. Striking electron micrographs of flower petals and insects. Tells how the scanning electron microscope works. It can give magnifications of 10,000x to 200,000x.

Information Processing in the Vertebrate Retina

Visual Information Processing in the Vertebrate Brain

Auditory Systems

   Maurer, Allan. Rings in their ears. Omni, 1982, 4(10), 39. Otoliths in fish
   ears are being used by commercial fishermen to determine the age of fish with
   remarkable precision. The otoliths are formed each day.

Chemoreception

Olfaction

Chemoreception in Arthropods

Mechanoreception

   Camhi, Jeffrey M. The escape system of the cockroach. Scientific American, 1980,
   243(6), 158-172. The removal of the cerci, two posterior sensory appendages of
   the cockroach, makes it impossible for the insect to run in response to puffs of
   air. These cerci have an array of about 220 hairs that, when stimulated, cause
   the insect to run. Each hair excites a sensory neuron. The sensory neurons
   excite large nerve cells called giant interneurons. There are 14 of these in each
   insect. A series of ingenious experiments demonstrated that it was the wind
   currents striking the cerci that caused the insect to run away from the source of
   the wind currents.

Proprioception

Other Receptors

   Dunkle, Terry. A perfect serpent. Science 81, 1981, 2(8), 30-35. Excellent
   article on a poisonous snake. Includes information on heat sensors, venom, and
   how snakes swallow their prey whole. Well illustrated.

   Newman, Eric A., and Peter H. Hartline. The infrared "vision" of snakes.
   Scientific American, 1982, 246(3), 116-127. A neurophysiological approach to the
   sensitivity of pit organs in rattlesnakes and pythons. Individual axons are very
   sensitive to infrared stimuli. The trigeminal nerve fibers increase their firing
   rate appreciably when a small warm object is introduced as far away as half a
   meter from the pit. The structure of the pit, the nerve pathways to the brain,
   and the structures in the brain are discussed.

ESSAY QUESTIONS

1. List at least five types of sensory receptors and describe their functions.

        Photoreceptors detect light. They may detect only light or darkness, or they
        may detect specific elements of images. Thermoreceptors detect temperature;
        they may assist in maintenance of homeostasis or in prey location. Chemo-
        receptors detect taste, scents, and internal chemical concentrations, thereby
        aiding both in feeding and in the maintenance of homeostasis. Mechanorecep-
        tors detect touch, vibration, and related sensations, alerting the organism
        to movements of its body, pain, and body orientation. Electroreceptors
        detect changes in electrical fields and are used by some organisms to detect
        water movements. Baroreceptors monitor and detect changes in blood pressure
        in the mammalian carotid artery. Proprioceptors provide the cerebellum (in
        vertebrates) with information about the degree of stretch of muscles and
        the position of joints.

2. How does a sensory receptor pass on its information, and how does the central
    nervous system determine the intensity of such information?

        If the receptors are neurons, they pass the information via synapses with neighboring neurons. Neuron-like cells are usually surrounded by neurons, from which information is transmitted to the CNS via synapses with other neurons.

3. Why does a simple patch of pigmented epithelium suffice as an "eye" for an autotroph?

   Autotrophs do not need vision to be able to capture prey. They need only to detect areas of light, enabling them to carry on photosynthesis. Therefore a very rudimentary "eye" will serve this purpose.

4. Explain how animals perceive different colors.

   There are three types of cone cells, each of which perceives a specific wavelength of light because it contains a specific photoreceptive pigment. The composite formed by information from all three of these cone cell types is perceived as a color image by the visual cortex.

5. Give some examples in which the possession of a specific sensory system affords an organism a selective advantage with regard to finding and selecting a mate of the appropriate species and sex.

   Some fish use sight to detect the presence of a likely mate by determining whether shape, size, and coloration are correct. Many birds use sight to determine whether a nearby bird of the same species is of the "correct" sex by using color vision; males are usually brightly colored and females are usually drab. Female frogs respond only to specific auditory cues in the call of the males of their species. Some insect males detect chemicals produced by the females of their own species and follow the chemical "trail" to a prospective mate. Many birds, fish, and mammals (including humans) use visual cues to recognize particular courtship behaviors performed by potential mates.

6. How do animals orient themselves with respect to the earth's gravity?

   Vertebrates monitor the position of an otolith, which lies in a chamber (utriculus) lined with sensory hair cells. When the otolith moves, the hair cells are stimulated. Similarly, invertebrates monitor the position of a statolith, which lies in the statocyst, a chamber filled with hair cells.

TEST SET 27-A

Name _____   Section _____

Choose the best answer to each of the following questions, and write the appropriate letter in the space provided.

Ans: d
p. 680
_____ 1. The function of an ocellus is to sense: (a) movement; (b) temperature changes; (c) the different patterns of objects; (d) light-dark changes.

Ans: a
p. 680
_____ 2. One of the outstanding features of ommatidia is their ability to detect: (a) movement; (b) light-dark changes; (c) color; (d) subtle changes in resolution.

Ans: a
p. 687
_____ 3. Lateral inhibition in the human eye is produced by the activity of the: (a) horizontal cells; (b) amacrine cells; (c) bipolar neurons; (d) ganglion cells.

Ans: b
p. 684
_____ 4. Visual acuity is the primary function of the: (a) blind spot; (b) fovea; (c) pupil; (d) lens.

Ans: c
p. 685
_____ 5. The different colors in vision are the result of the: (a) ability of each cone to absorb all wavelengths of light; (b) ability of each rod to absorb all wavelengths of light; (c) absorption of different wavelengths of light by different classes of cones; (d) absorption of different wavelengths of light by different classes of rods.

Ans: a
p. 684
_____ 6. The pigmented epithelium at the back of the vertebrate eye functions to: (a) absorb extraneous wavelengths of light; (b) reflect extraneous wavelengths of light; (c) change the sensitivity of the cones; (d) protect the retina.

Ans: c
p. 684
_____ 7. The generator potential in photoreceptor cells is initiated by the: (a) splitting of opsin and the fusion of retinaldehyde; (b) fusion of opsin and the splitting of retinaldehyde; (c) splitting of opsin and retinaldehyde; (d) fusion of opsin and retinaldehyde.

Ans: b
p. 684
_____ 8. The initiation of a generator potential in one ommatidium results in: (a) the increased initiation of generator potentials in surrounding ommatidia; (b) the decreased initiation of generator potentials in surrounding ommatidia; (c) a decrease in signals sent to the central nervous system from the ommatidia; (d) a decrease in signals sent from the central nervous system to the ommatidia.

Ans: d
p. 687
_____ 9. _____ do not directly receive information from photoreceptor cells. (a) Horizontal cells; (b) Bipolar neurons; (c) Ganglion cells; (d) Amacrine cells.

Ans: b
p. 690
_____ 10. The "pressure release valve" in the organ of Corti is the: (a) oval window; (b) round window; (c) tectorial membrane; (d) cochlea.

Ans: a
p. 688
_____ 11. Detection of sound requires the conversion of: (a) mechanical energy to electrical energy; (b) electrical energy to mechanical energy; (c) chemical energy to electrical energy; (d) mechanical energy to chemical energy.

Ans: a
p. 691

_____ 12. It is hypothesized that olfactory stimuli are recognized by the interaction between the stimulus and: (a) a specific macromolecule in the membrane of a chemoreceptor; (b) the mucous coating of the chemoreceptor; (c) the electrical potential of the chemoreceptor; (d) the chemical potential of the chemoreceptor.

Ans: b
p. 694

_____ 13. In vertebrates, changes in body position are sensed by the: (a) otolith; (b) semicircular canals; (c) utriculus; (d) pacinian corpuscles.

Ans: c
p. 694

_____ 14. The touch receptors that are located very close to the epidermis are called: (a) Meissner's corpuscles; (b) the pacinian corpuscles; (c) Merkel's disks; (d) otoliths.

Ans: c
p. 690

_____ 15. Chemoreceptors are specialized: (a) lateral geniculate cells; (b) ganglion cells; (c) bipolar neurons; (d) cortical cells.

TEST SET 27-B

Name _____   Section _____

Choose the best answer to each of the following questions, and write the appropriate letter in the space provided.

Ans: c
p. 680

_____   1. The receptors that carry out only a light-dark sensing function are called: (a) compound eyes; (b) ommatidia; (c) ocelli; (d) rhabdomeres.

Ans: c
p. 680

_____   2. The sensory neuron in an ommatidia is a(an): (a) ocellus; (b) rhabdom; (c) retinula cell; (d) rhabdomere.

Ans: d
p. 687

_____   3. Initial processing of visual messages in the brain begins in the: (a) visual cortex; (b) ganglion cells; (c) bipolar neurons; (d) lateral geniculate body.

Ans: a
p. 683

_____   4. The region of the vertebrate eye where the optic nerve passes through the retina is called the: (a) blind spot; (b) fovea; (c) ganglia; (d) pupil.

Ans: c
p. 682

_____   5. As light enters the human eye, the first set of neurons it encounters are the: (a) rods and cones; (b) bipolar neurons; (c) ganglion cells; (d) retinal neurons.

Ans: c
p. 684

_____   6. Vertebrates with eyes that "glow" in the dark have a(an): (a) extra layer of rods; (b) extra layer of cones; (c) tapetum; (d) pigmented epithelium.

Ans: a
p. 685

_____   7. Color vision is the result of different _____ in the photoreceptor cells. (a) opsins; (b) rods; (c) retinaldehydes; (d) windows.

Ans: a
p. 686

_____   8. The ultimate benefit of lateral inhibition in ommatidia is an improvement in: (a) resolution; (b) color patterns; (c) light-dark sensing mechanisms; (d) rhodopsin synthesis.

Ans: d
p. 686

_____   9. One ganglion cell achieves generator potential by the firing of one: (a) rod cell; (b) cone cell; (c) horizontal cell; (d) receptive field.

Ans: b
p. 688

_____   10. The membrane that separates the middle ear from the inner ear is the: (a) round window; (b) oval window; (c) tympanic membrane; (d) tectorial membrane.

Ans: c
p. 690

_____   11. The pitch of sound is determined by the: (a) size of the action potential from the sound-sensitive cells; (b) location of the sound-sensitive cells generating the action potential; (c) displacement of the round window; (d) displacement of the oval window.

Ans: c
p. 691

_____   12. It is hypothesized that there is(are) _____ basic types of olfactory stimuli. (a) one; (b) four; (c) seven; (d) twelve.

Ans: b
p. 694

_____   13. In vertebrates, position in space is determined by the: (a) statolith; (b) otolith; (c) semicircular canals; (d) pacinian corpuscles.

Ans: b
p. 694

_____   14. The touch receptors that are located very deep in the dermis are called: (a) Meissner's corpuscles; (b) the pacinian corpuscles; (c) Merkel's disks; (d) otoliths.

Ans: b
p. 690   _____   15.   Taste buds are composed of: (a) many receptor cells and many bipolar neurons; (b) many receptor cells and one bipolar neuron; (c) one receptor cell and many bipolar neurons; (d) one receptor cell and one bipolar neuron.

CHAPTER TWENTY-EIGHT
EFFECTORS

CHAPTER SUMMARIES AND OBJECTIVES

28-1 Ciliated Cells / Microtubules in Cilia and Flagella / Amoeboid Cells   Effectors are structures that perform tasks, including execution of the commands of the nervous system.   Cilia are hairlike structures used for locomotion and/or filter feeding by many protozoa and other invertebrates; cilia are also used to circulate fluids and other materials, as in the human respiratory system.   The motion of cilia includes an effective stroke, which effects movement of the cell or fluid, and a recovery stroke, which returns the cilium to its original position. The action of cilia appears to be coordinated by a chemical mechanism.   The cilia of all eukaryotes have nine pairs of outer microtubules; in most cases there are also two central microtubules, but there may be one or none.   Flagella are similar to cilia, but they are longer and fewer in number.   Amoeboid movement characterizes not only amoebas, but also various cells of higher animals.   It involves the extension of fluid pseudopods, which are converted to a relatively rigid phase and thus fixed in position.   The chemical mechanism of amoeboid movement is related to that of muscle contraction.

1.   Define effectors.

2.   Explain the structure and function of cilia and flagella.

3.   Describe amoeboid movement and explain its significance to higher animals.

28-2 The Skeleton / The Vertebrate Skeleton / Types of Bone / Other Aspects of the Vertebrate Skeleton   A skeleton is a rigid or semirigid body support structure. Noncompressible liquid in body segments of annelids constitutes a simple hydrostatic skeleton.   In arthropods, muscles attach to a hard outer shell, or exoskeleton, composed of protein and polysaccharide.   The exoskeleton of mollusks also provides protection and/or locomotion.   Vertebrates have an endoskeleton composed of flexible cartilage (collagen fibers in a protein-polysaccharide matrix) and rigid bone (mainly collagen and calcium phosphate, secreted by osteocytes).   Bone may be hard and dense or light and spongy in structure; the latter is called cancellous bone.   Haversian bone is one type of compact bone that is composed of concentric sheets surrounding a blood vessel.   The vertebrate skeleton also includes tendons, which connect muscles and bones, and ligaments, which join bones together; both are composed of connective tissue.

4.   Describe the major types of skeletons.

5.   Identify the components of the vertebrate skeleton.

6.   Differentiate among the major types of bone.

28-3 <u>Muscle / Muscle Function / The Sliding Filament Model / Controlling the Actin-
Myosin Interaction</u> Most animals use contractile units called muscles for locomotion. Muscles are categorized as smooth (composed of elongated cells with
contractile protein), striated (composed of cells fused into muscle fibers), or
cardiac (combining the characteristics of both). Smooth muscle was the earliest
type to evolve, followed by striated or skeletal muscle, of which cardiac muscle
is a specialized form. When muscle cells and fibers contract, in response to
nervous stimulation, they exert force in only one direction. Two sets of muscles
with opposite effects are said to be antagonistic.

A muscle fiber contains 4 to 20 myofibrils, which in turn are composed of contractile units called sarcomeres. Each sarcomere contains thin filaments of
actin and thick filaments of myosin. When these two types of protein filaments
slide past each other, the muscle contracts. The energy for this process is
supplied by ATP.

A single nerve fiber can supply up to 100 muscle fibers; together these are called
a motor unit. When an action potential reaches a (vertebrate) muscle fiber, it
spreads over the plasma membrane and travels along the T-system to the sarcoplasmic reticulum (SR). During the resulting depolarization, Ca^{2+} ions leak from the
SR into the myofibrils and cause contraction by changing the configuration of
troponin such that actin and myosin bind to one another.

7. Trace the evolution of muscle and describe its major types.

8. Explain how muscles contract and how this process is controlled.

28-4 <u>Integrating the Activities of Muscle Fibers / Fast and Slow Muscles / Graded
Contraction in Invertebrate Skeletal Muscle / Cardiac Muscle / Smooth Muscle</u>
A vertebrate muscle <u>fiber</u> is either contracted or relaxed; it cannot give a
graded response. However, an entire muscle is capable of graded response because
of variation in the number of fibers activated by the nervous system. Also,
rapidly repeated twitches build greater tension in the elastic component of
muscle. Most animals have both fast-contracting and slow-contracting types of
skeletal muscle. Some insects have a specialized type of flight muscle called
asynchronous muscle, in which rapid cyclic contractions are regulated by the
fibers themselves. Most invertebrate skeletal muscle fibers, unlike those of
vertebrates, are capable of graded contraction. These fibers are innervated at
several points, and no action potential is formed in the fiber itself. Cardiac
muscle, found only in vertebrate hearts, has branched cells mixed with noncon-
tractile cells; its contraction is myogenic and synchronous. The fibers are in
electrical contact with each other in certain regions, and the SA node acts as a
pacemaker. The central nervous system and endocrine glands also influence the
heart rate. Smooth muscle has two forms: multiunit muscles contract only under
nervous or hormonal stimulation, whereas visceral muscles contract spontaneously.
Catch muscles, found in bivalve mollusks, are specialized to hold the shell shut.

9. Explain the nature of graded muscle responses in vertebrates and in inverte-
    brates.

10. Differentiate between fast and slow muscles.

11. Describe cardiac muscle and smooth muscle.

28-5 <u>Other Effectors / Trichocysts and Nematocysts / Chromatophores / Glands / Bioluminescence / Electric Organs</u> Effectors other than muscle occur in certain
animal groups. Ciliated protozoans use threadlike trichocysts for defense and
for attachment to the substratum. Jellyfish and their relatives use nematocysts,
coiled threads activated by triggerlike devices, to secure and subdue prey.
Several groups, including cephalopod mollusks, fishes, amphibians, and reptiles,
have pigment-bearing cells called chromatophores. These cells effect color
changes by means of nervous and/or hormonal control. There are three major types
of chromatophores: those in which pigment granules are moved by microfilaments
within fixed cell boundaries, those capable of amoeboid movement, and those that
change shape due to the action of muscle fibers (in cephalopods only).

Glands are organs or groups of cells that release substances. Endocrine glands secrete their products into the bloodstream, whereas exocrine glands secrete into the digestive tract or external body surface. Bioluminescence, found in many deep-water and nocturnal animals, is generated when luciferin is oxidized in the presence of the enzyme luciferase. Electric organs, evolutionarily derived from muscle, occur in at least seven families of fishes. These organs consist of long stacks of large disk-shaped cells, which can generate enough current to stun enemies and prey or to use for orientation purposes.

12. Describe trichocysts, nematocysts, chromatophores, bioluminescence, and electric organs. State which organisms possess each and for what purpose(s) they are used.

13. Explain what is meant by a gland, and differentiate between endocrine and exocrine glands.

ARTICLE RESOURCES

Ciliated Cells

Microtubules in Cilia and Flagella

Amoeboid Cells

Lazarides, Elias, and Jean Paul Revel. The molecular basis of cell movement. Scientific American, 1979, 240(5), 100-112. The movement of cells is mediated by a small set of proteins whose functioning is controlled by a larger set of proteins. These cells are found not only in muscles but also in nonmuscle cells. When calcium is present, adenosine triphosphate drives the cyclic formation and dissociation of bridges between actin and myosin filaments, causing the filaments to move past one another. Excellent photomicrographs and drawings. A flowchart of the regulation of cell movement at the molecular level in nonmuscle cells is given.

The Skeleton

The Vertebrate Skeleton

McAuliffe, Kathleen. I. Using the body electric. Omni, 1980, 3(2), 70-73, 98-99. Discusses bone repair by means of pulsating electromagnetic fields which induce tiny currents in the bone. Peripheral nerves can be stimulated to grow by the same means. Electromagnetic fields surrounding the body have long been known by acupuncturists. When the acupuncture sites were charted, they were found to follow the major pathways of the nervous system. Electricity has become the preferred treatment for difficult bone nonunions at many clinics in the United States and abroad. Astro-osteoporosis, in which the bones of astronauts become thin and brittle, can be prevented by use of electromagnetic coils.

Types of Bone

Other Aspects of the Vertebrate Skeleton

Muscle

Matthews, P. B. C. Review lecture: Evolving views on the internal operation and functional role of the muscle spindle. Journal of Physiology, 1981, 320, 1-30. A history of muscle receptors. Reviews the research on muscle spindles from 1860 to 1980. Fascinating reading.

Muscle Function

The Sliding Filament Model

Controlling the Actin-Myosin Interaction

Integrating the Activities of Muscle Fibers

Alexander, R. McN., and A. S. Jayes. Estimates of the bending moments exerted by the lumbar and abdominal muscles of some animals. Journal of Zoology, 1981, 194(3), 291-303. Rat, ferret, dog, and rabbit (hare) abdominal and lumbar muscles were measured and their arrangement noted. These muscles make up a substantial fraction of the body mass. They are obviously important in posture and in activities involving forceful flexion or extension of the back.

Fast and Slow Muscles

Hudlicka, O., K. R. Tyler, T. Srihari, A. Heilig, and D. Pette. The effect of different patterns on contractile properties and myosin light chains in rabbit fast muscles. Pflüger's Archiv European Journal of Physiology, 1982, 393(2), 164-70. The amount of activity in a muscle is important. Even 28 days of stimulation of fast muscles with bursts of higher-frequency activity every 100 seconds with a relatively low total number of stimuli did not affect the contraction time. Fast muscle can be transformed into the slow type by the total amount of activity rather than the frequency of the activity. Useful for the instructor.

Graded Contraction in Invertebrate Skeletal Muscle

Elmer, Robert W., and Alan Campbell. Force, function, and mechanical advantage in the chelae of the American lobster Homerus americanus (Decapoda, Crustacea). Journal of Zoology, 1981, 193(2), 269-286. The American lobster possesses the largest chelae of all crustaceans. It has dimorphic chelae. The larger crusher has blunt processes on the occluding surfaces. The smaller pincer has small needlelike teeth. The strength, mechanics, and function of these appendages were studied. The height of the crusher chelae is greater than that of the pincer. Male lobsters had significantly larger chelae than females. However, there was no significant change in mechanical advantage with chelae height over size.

Cardiac Muscle

Cave, A. J. E. On the cardiac anatomy of the Sumatran rhinoceros. Journal of Zoology, 1981, 193(4), 541-561. This is the first description of the rhinoceros cardiac conducting system. It implies that the internodal transmission mode of the heart contractile impulse may be different from conventional physiology. The rhinoceros heart shows greater papillary musculature as well.

Smooth Muscle

Other Effectors

Trichocysts and Nematocysts

Chromatophores

Glands

Bioluminescence

Electric Organs

ESSAY QUESTIONS

1. Give examples of effectors that are controlled by (a) nerves; (b) hormones; (c) some other mechanism. Explain the function of each, using diagrams if appropriate.

    (a) Examples include skeletal muscles and chromatophores. See textbook pages 712-716.
    (b) Examples include chromatophores and certain glands. See textbook pages 714-717.
    (c) One good example would be nematocysts, which fire on mechanical

stimulation.   See textbook pages 714-715.

2. Explain how bioluminescence is generated, and list its known and hypothesized functions.

See textbook page   717.

3. Describe the differences between cartilage and bone.

See textbook page   703.

4. Explain why a deficiency of calcium in the diet might cause muscle cramps.

The role of calcium in muscle contraction should be explained here. See textbook pages 711-712.

TEST SET 28-A

Name _____   Section _____

Choose the best answer to each of the following questions, and write the appropriate letter in the space provided.

Ans: b
p. 698

_____   1. Structures that carry out the commands of the nervous system are called: (a) muscles; (b) effectors; (c) action potentials; (d) motor units.

Ans: a
p. 699

_____   2. All of the following animals use cilia for locomotion except: (a) mammals; (b) mollusks; (c) flatworms; (d) Paramecium.

Ans: d
p. 714

_____   3. Humans possess all of the following effectors except: (a) flagella; (b) cilia; (c) amoeboid cells; (d) nematocysts.

Ans: d
p. 700

_____   4. How are microtubules arranged in a typical cilium? (a) nine outer, two inner; (b) nine pairs outer, two pairs inner; (c) nine outer, two pairs inner; (d) nine pairs outer, two inner.

Ans: c
p. 702

_____   5. A skeleton is found in: (a) vertebrates only; (b) higher vertebrates only; (c) most animals; (d) all organisms.

Ans: b
p. 704

_____   6. Haversian bone is common in: (a) birds; (b) mammals; (c) sharks; (d) bony fishes.

Ans: c
p. 704

_____   7. Cancellous bone occurs only in: (a) birds; (b) mammals; (c) vertebrates; (d) Haversian systems.

Ans: a
p. 703

_____   8. Osteocytes secrete: (a) proteins and salts; (b) canaliculi; (c) hormones; (d) sarcoplasm.

Ans: b
p. 713

_____   9. Cardiac muscle is a subtype of: (a) smooth muscle; (b) striated muscle; (c) visceral muscle; (d) multiunit muscle.

Ans: a
p. 708

_____   10. The thick filaments of the sarcomere are composed of: (a) myosin; (b) actin; (c) sarcoplasm; (d) ATP.

Ans: c
p. 709

_____   11. Under the sliding filament model, is ATP necessary for actin and myosin to bind together or for them to separate? (a) bind together; (b) separate; (c) both; (d) neither.

Ans: d
p. 711

_____   12. An action potential is: (a) the capacity for action--that is, potential energy; (b) the energy level of action, which is one of the two proteins involved in muscle contraction; (c) all the muscle fibers supplied by one nerve cell; (d) a nerve impulse.

Ans: c
p. 712

_____   13. What is the structure of troponin? (a) double-stranded; (b) globular, two subunits; (c) globular, three subunits; (d) double helix.

Ans: c
p. 713

_____   14. How do fast skeletal muscles differ from slow ones? (a) The molecular basis of contraction is different. (b) Insects have fast molecules, whereas vertebrates have slow ones. (c) Slow muscles have less sarcoplasmic reticulum. (d) Fast muscles are not under the control of Ca^{2+}.

Ans: a
p. 713

_____   15. All of the following statements about cardiac muscle are true except: (a) It is a form of smooth muscle. (b) The cells are branched. (c) Its contraction is myogenic. (d) Its contraction is synchronous.

TEST SET 28-B

Name _____   Section _____

Choose the best answer to each of the following questions, and write the appropriate letter in the space provided.

Ans: b
p. 698
_____ 1. Are all effectors under direct nervous control? (a) Yes. (b) No, some are under hormonal or other control. (c) No; few, if any, effectors are controlled directly by nerves. (d) No; most are under electrical control.

Ans: a
p. 699
_____ 2. Cilia serve all of the following functions except: (a) defending the organism against bacteria; (b) locomotion; (c) clearing of respiratory passages; (d) filter-feeding.

Ans: c
p. 700
_____ 3. What use do sponges make of flagella? (a) none, they have only cilia; (b) locomotion; (c) water circulation; (d) secretion of poison.

Ans: d
p. 700
_____ 4. All of the following groups possess flagella except: (a) Mastigophora; (b) Porifera; (c) Vertebrata; (d) Ciliophora.

Ans: d
p. 703
_____ 5. Chitin is: (a) a protein; (b) an exoskeleton; (c) calcium carbonate; (d) a polysaccharide.

Ans: a
p. 703
_____ 6. The two main types of supportive tissue in the vertebrate skeleton are: (a) cartilage and bone; (b) endoskeleton and exoskeleton; (c) proteins and polysaccharides; (d) ectoplasm and endoplasm.

Ans: d
p. 703
_____ 7. The main function of bone marrow is to: (a) fill empty spaces; (b) lighten the skeleton as an adaptation for flight; (c) regulate body temperature; (d) manufacture blood cells.

Ans: b
p. 713
_____ 8. _____ muscle has highly branched fibers. (a) Smooth; (b) Cardiac; (c) Striated; (d) Flexor.

Ans: a
p. 707
_____ 9. Muscles with opposing actions are called: (a) antagonists; (b) synergists; (c) flexors; (d) extensors.

Ans: b
p. 707
_____ 10. Does a flexor straighten a leg or bend it? (a) straighten; (b) bend; (c) both; (d) either, depending on the animal.

Ans: a
p. 709
_____ 11. What happens if you isolate pure actin and myosin and then mix them? (a) Small fibrils form. (b) They contract. (c) Nothing happens because the H band is absent. (d) They have not yet been isolated in pure form, so the result is unknown.

Ans: c
pp. 710-
711
_____ 12. The text describes rigor mortis in dead animals. Why does such an animal soon become limp again? (a) In the absence of ATP, actin and myosin separate. (b) It doesn't; hence the term "stiff." (c) The muscles and other tissues begin to decompose. (d) Muscle fibers will not contract without nervous stimulation.

Ans: b
pp. 711-
712
_____ 13. Is the movement of Ca^{2+} ions between the sarcoplasmic reticulum (SR) and myofibrils an active or a passive process? (a) active transport in both directions; (b) passive movement out of SR, active transport into SR; (c) active transport into myofibrils, diffusion into SR; (d) no such movement occurs.

Ans: c   _____   14. All of the following are types of smooth muscle <u>except</u>:
p. 713   (a) catch muscle; (b) multiunit muscle; (c) asynchronous muscle;
                          (d) visceral muscle.

Ans: d   _____   15. Paramyosin occurs in: (a) all muscles; (b) all vertebrate
p. 714   muscle fibers; (c) cardiac muscle; (d) catch muscle.

CHAPTER TWENTY-NINE
BEHAVIOR

CHAPTER SUMMARIES AND OBJECTIVES

29-1 <u>Costs and Benefits / Sign Stimuli / Behavioral Genetics</u> Behavior varies widely among species and higher taxa. It has a genetic component, but in general it depends also on individual experience. The costs of behavior are categorized as energetic cost (net energy expended), risk cost (potential danger), and opportunity cost (missed opportunity to engage in other behavior). Offsetting these costs are benefits: improved survival and reproductive success. Trade-off refers to the evolutionary compromises that must be made between one form of risk and another. The study of the evolution of behavior is ethology.

Sign stimuli are those features of a stimulus situation that evoke specific behavior. For example, male robins react aggressively to red feathers, with or without the bird to which they are normally attached. Innate releasing mechanisms in the brain may respond to sign stimuli.

Genes provide the potential for behavior. Instinctual behavior is determined entirely by genes, whereas learned behavior results from experience. In practice, behavior rarely can be classified as one or the other. A self-differentiating behavior develops without experience or practice.

1. Explain the principal costs and benefits of behavior.

2. Explain what is meant by sign stimuli and innate releasing mechanisms.

3. Explain the interaction of genes and experience in shaping behavior.

29-2 <u>Learning / Mechanisms of Learning / Song Learning</u> Learning is the modification of behavior in response to experience. Its principal forms are habituation (cessation of response to neutral stimuli), associative learning (association of two or more stimuli with one experience), operant conditioning (associative learning that results from exploratory behavior), latent learning (stored information used later), and insight learning (adaptation of past experience to solve a new problem). Neural mechanisms of learning are not well understood, but short-term and long-term memory apparently involve separate processes. Memory may be restricted to specific regions of the brain, but not to individual cells. Much research in this area has been performed with invertebrates, such as the mollusk <u>Aplysia</u>.

Song learning in birds is illustrated by the white-crowned sparrow, which exhibits such typical phenomena as a "template" (hypothetical patterned system that learns only appropriate input), a critical period (age range outside of which learning will not occur), and imprinting (learning at a specific age, without reinforcement). The structures involved in bird song are convenient to study, so the neural and muscular control of this behavior is fairly well understood.

4. Define learning and cite its principal forms.

5. Explain what is meant by a template, a critical period, and imprinting.

29-3 <u>Finding and Choosing Food / Human Food Choices / Avoiding Being Eaten</u> Foraging and eating are important, complex, and risky activities. The animal must recognize and ingest appropriate foods; omnivores generally must learn which foods are suitable. Rats, for example, test foods by sampling them. However, they do have genetically determined mineral-selection systems. The role of learning in human food selection is more difficult to study, because subjects cannot be reared in isolation. Some studies have shown that infants tend to select a balanced diet but that adults do not necessarily do the same. Knowledge of nutrition is an attribute of culture, acquired by trial and error over generations.

Bats locate flying insects by sonar. The sounds they emit, however, are detected by certain prey insects, which then perform evasive maneuvers. Young ducks and turkeys crouch and freeze when a hawk (or a hawk-shaped model) passes overhead, but they do not respond similarly to a goose-like model. This response is partly learned, however. Other prey animals maintain a specific "flight distance" from a predator. Some prey appear to initiate pursuit at a distance, perhaps to teach avoidance. Small birds may mob predators.

6. Explain genetic influences on food selection in rats; in humans.

7. Describe some strategies used by prey animals to avoid predators.

29-4 <u>Orientation / Navigation / The Evolution of Behavior</u> Animals have various methods of determining where they are with respect to their nests or home ranges. Taxes are simple orientation movements with respect to light (phototaxis), gravity (geotaxis), or a chemical substance (chemotaxis); they may be positive (toward the stimulus) or negative (away from the stimulus). The light-compass reaction of many invertebrates involves orientation by the angle between the sun or moon and the direction of movement. Many vertebrates show remarkable ability to navigate over long distances. More than one sensory modality may be used; for example, salmon use both a sun compass and their sense of smell for orientation. Birds may use the sun or the stars; in some cases, the direction of orientation to these stimuli depends on physiological conditions, which in turn are influenced by seasonal change in day length. Birds may also orient by the earth's magnetic field.

The capacity for learning varies among species. Human language acquisition involves extensive learning, but it also depends on complex neural structures. Learning ability carries with it both costs and benefits. It is a difficult, but flexible, way to acquire behavior.

8. Describe the principal methods of animal orientation and navigation.

9. Explain the general costs and benefits associated with learning capacity.

ARTICLE RESOURCES

<u>Costs and Benefits</u>

<u>Sign Stimuli</u>

Lloyd, James E. Mimicry in the sexual signals of fireflies. <u>Scientific American</u>, 1981, 245(1), 138-145. Male fireflies signal female fireflies by means of flashes of bioluminescence. The female response is temperature-sensitive and is time-delayed. This male signal pattern combined with the delayed female answer is species-specific. The female of some firefly species can mimic other species and lure the male to her side where he becomes her dinner. The evolution of the firefly signal and its mimicry by predatory species is discussed.

Behavioral Genetics

Learning

Mechanisms of Learning

Song Learning

Finding and Choosing Food

Eberhard, William G. Horned beetles. Scientific American, 1980, 242(3), 166-168, 173-182. Since the time of Darwin, scientists have been interested in the function of the elongated mandibles of staghorn beetles. They are not used as pincers, but rather as tweezers for picking up their rivals and dropping them from trees, or as levers to pry the rival out of a burrow or off a twig. Excellent description of how these horns are used. Photomicrographs of horn designs are given. Comparison of horns, antlers, and tusks with the horn of beetles.

Beckoff, Marc, and Michael C. Wells. The social ecology of coyotes. Scientific American, 1980, 242(4), 130-133, 136-144, 148. Coyotes may pack together to defend large food resources such as carrion of large animals such as deer and elk. If small prey is more readily available, however, solitary living is selected. Only rarely do coyote packs attack and kill large prey. Solitary living is more common from October to May, whereas pack living is more common in summer. There is a much higher mortality rate for ungulates during the winter, hence the greater availability of food. Mating groups, pack individuals, and single individuals were observed.

Human Food Choices

Avoiding Being Eaten

Partridge, Brian L. The structure and function of fish schools. Scientific American, 1982, 246(6), 114-123. Although the schooling of fish has been observed for centuries, little has been understood about it because of the difficulties in observing the small changes in position and velocity that occur under normal conditions. Reasons for schooling are explored, as are the principles of organization. The role of the lateral line in helping the individual fish determine its position in the school is shown. Both the lateral line and vision play a role in position within the school.

Orientation

Blair, John G. Salmon with a southern accent. Science 81, 1981, 2(3), 76. Successful transplantation of coho and chinook salmon eggs from the Pacific northwest to Chile by Domsea Farms has resulted in significant numbers of salmon returning to the transplant site.

Navigation

McCosker, John E. The great white shark. Science 81, 1981, 2(6), 42-51. Everything you ever wanted to know about sharks. Excellent discussion of the great white shark. Includes information on navigation, physiology, and behavior.

Hamer, Blythe. Why do whales run aground? Science 81, 1981, 2(9), 128. Various hypotheses on this phenomenon are explored--dense plankton blooms may prevent radar location of the ocean floor, the leader may have beached himself and others follow, or roundworms in the whale's ear may disrupt its navigation.

The Evolution of Behavior

Wickstein, Mary. Decorator crabs. Scientific American, 1980, 242(2), 146-154. Well-illustrated article on crabs that camouflage themselves with bits of algae and wood chips. The evolution of this behavior pattern is discussed, as is the function of camouflage. Other behavior patterns in crabs are cited.

ESSAY QUESTIONS

1. Discuss neural control of (a) behavior of _Aplysia_; (b) song of male canary.

    This information is taken directly from the text.

2. Distinguish between associative learning and operant conditioning.

    The main point to be clarified is the animal's participation in the learning experience.

3. A young bird is likely to be surrounded by singing males of many different species. What might be the adaptive value of a song-learning template in the nervous system?

    It might help prevent the bird from accidentally learning the wrong song and thus being unable to mate later on.

4. The text states that sodium is an important but scarce element in the diet of most animals. Yet we are constantly warned that our food is overloaded with sodium and that we must reduce our consumption. What accounts for the difference? Has it held true throughout human history, or is it a modern phenomenon?

    Human food is now highly processed and has a lot of sodium added to it. Also, table salt is no longer a scarce commodity, and people tend to use excessive amounts of it. Other animals do not have access to this concentrated source.

TEST SET 29-A

Name _____   Section _____

Choose the <u>best</u> answer to each of the following questions, and write the appropriate letter in the space provided.

Ans: b
p. 720

_____ 1. What determines the potential range of an organism's behavior? (a) experience; (b) DNA; (c) orientation; (d) the endocrine system.

Ans: c
p. 721

_____ 2. When an organism performs one of two mutually exclusive behaviors, each of which involves risks, the phenomenon is known as: (a) risk cost; (b) insight learning; (c) trade-off; (d) habituation.

Ans: c
p. 722

_____ 3. What is a sign stimulus? (a) a hypothetical system in the brain, physical nature unknown; (b) a well-defined physical structure in the central nervous system; (c) those features of a stimulus situation that evoke a specific behavior pattern; (d) instinctive (vs. learned) behavior.

Ans: a
p. 726

_____ 4. Self-differentiating behavior: (a) develops regardless of experience; (b) continuously modifies itself; (c) is used to signal individual identity; (d) is not influenced by genes.

Ans: c
p. 727

_____ 5. Exploratory trial-and-error learning is called: (a) habituation; (b) associative learning; (c) operant conditioning; (d) latent learning.

Ans: d
pp. 726-
727

_____ 6. A baby in a playpen first studies her environment, then stacks all of her stuffed animals in one corner, climbs up the pile and goes over the wall, landing on the floor. What type(s) of learning is(are) probably observed here? (a) imprinting, because every baby does this at a certain age; (b) latent learning, because the behavior was based on information acquired earlier in life; (c) instinct (climbing) and latent learning (stacking objects); (d) insight learning (stacking), possibly followed by operant conditioning (fear of heights).

Ans: d
p. 728

_____ 7. Where is short-term memory stored in humans? (a) midbrain; (b) SCN; (c) cerebral ganglia; (d) this is not known.

Ans: c
p. 729

_____ 8. Does song learning in male white-crowned sparrows depend entirely on hearing the adult song between the ages of 10 and 50 days? (a) Yes. (b) No. The bird must relearn the song in every breeding season, when the SCN enlarges. (c) No. The bird must also hear itself sing. (d) No. Experiences during this time interval are apparently irrelevant.

Ans: c
p. 731

_____ 9. Learning that occurs at a specific age, without reward or punishment, is called: (a) operant conditioning; (b) associative learning; (c) imprinting; (d) habituation.

Ans: d
p. 731

_____ 10. The song of male canaries is controlled or influenced by: (a) SCN only; (b) SCN and midbrain only; (c) SCN, midbrain, and twelfth cranial nerve; (d) SCN, midbrain, twelfth cranial nerve, experience, and hormones.

Ans: a
p. 736

11. What is flight distance? (a) the distance to which a predator can approach a given prey before the prey attempts to escape; (b) the distance a given species of bird can fly before stopping to feed; (c) the distance over which migratory birds can orient using a sun compass; (d) the distance to which prey must approach before a predator will give a pursuit invitation signal.

Ans: b
p. 738

12. Many winged aquatic insects migrate at night and will land when they see moonlight reflected from the surface of a body of water. This orientation method is an example of: (a) positive geotaxis; (b) positive phototaxis; (c) negative chemotaxis; (d) negative geotaxis.

Ans: c
p. 738

13. Are invertebrates with a light-compass reaction able to correct for the sun's movement? (a) Yes. (b) No. (c) Some taxa do, others don't. (d) This has not been determined.

Ans: a
p. 740

14. How do salmon find their way from the ocean to their home river? (a) sun compass and sense of smell; (b) moon compass and the earth's magnetic field; (c) key constellations of stars; (d) topographic landmarks.

Ans: d
p. 740

15. How do small night-migrating birds "know" whether to fly north or south? (a) If it is cold, they fly south. (b) They simply alternate, returning to the place they came from on the previous migration. (c) The choice of direction is random; if they go far enough in either direction, eventually they will be home again. (d) Their physiological condition determines the direction.

TEST SET 29-B

Name _____   Section _____

Choose the best answer to each of the following questions, and write the appropriate letter in the space provided.

Ans: b
p. 721

_____ 1. Which expression refers to the net energy expenditure associated with a behavior? (a) energy flow; (b) energetic cost; (c) benefit; (d) opportunity cost.

Ans: d
p. 720

_____ 2. By what means do most animals optimize their behavior, in terms of relative risks and benefits? (a) insight learning; (b) sign stimuli; (c) imprinting; (d) natural selection.

Ans: d
p. 726

_____ 3. The modification of behavior in response to experience is called: (a) self-differentiation; (b) taxis; (c) evolution; (d) learning.

Ans: d
p. 727

_____ 4. Acquiring information for later use is called: (a) habituation; (b) associative learning; (c) operant conditioning; (d) latent learning.

Ans: a
p. 726

_____ 5. Which is the simplest form of learning? (a) habituation; (b) associative learning (c) latent learning; (d) insight learning.

Ans: c
p. 728

_____ 6. Why is Aplysia convenient for study of the nervous system? (a) It is capable of highly complex learning. (b) Its nervous system is comparable to that of humans. (c) It has only a few distinct ganglia, and their functions are known. (d) It does not habituate to electric shocks.

Ans: d
p. 729

_____ 7. Does song learning in white-crowned sparrows depend entirely on hearing the adult song during the first ten days of life? (a) Yes. (b) No. The young bird must also hear its own song. (c) No. The young bird must also be male. (d) No. Experience before the tenth day of life is apparently irrelevant.

Ans: c
p. 729

_____ 8. If a young male white-crowned sparrow hears only the song of a different species during the critical period, what happens? (a) It learns the other species's song. (b) It develops the song of its own species anyway. (c) It will be unable to sing any normal song. (d) It will sing the songs of both species.

Ans: d
p. 731

_____ 9. Where is a bird's syrinx located? (a) in the larynx, or voice box; (b) in the lungs; (c) in the bill; (d) where the trachea and bronchi join.

Ans: d
p. 731

_____ 10. Why might bird species vary in the number of muscles controlling the syrinx? (a) They don't; all have seven pairs. (b) Only the male species sings; the female species doesn't sing and therefore does not have these muscles. (c) Some birds have more syrinxes than others. (d) Some birds have more complex songs than others, requiring greater control.

Ans: b
p. 735

_____ 11. Why do humans rarely hear bats in flight? (a) The sounds bats make are not loud. (b) The sounds are too high-pitched. (c) Bats fly at night, when people are asleep. (d) Bats rarely produce any sounds.

Ans: a
p. 738

_____ 12. If you place an inverted jar over a fly, it will tend to fly upward. This may be an example of: (a) negative geotaxis; (b) positive phototaxis; (c) positive geotaxis; (d) negative phototaxis.

Ans: d
p. 738
_____ 13. By what primary means do indigo buntings orient themselves during migration? (a) sun; (b) moon; (c) sense of smell; (d) stars.

Ans: d
p. 740
_____ 14. Where is the magnetic receptor of birds located? (a) forebrain; (b) midbrain; (c) SCN; (d) unknown.

Ans: c
p. 740
_____ 15. Human learning ability carries with it very high costs and risks. What is its main compensatory advantage? (a) It enables us to survive, unlike other animal species, which eventually become extinct. (b) It enables us to respond appropriately to every stimulus we encounter. (c) It makes us more adaptable and flexible. (d) It is a necessary test of our willingness to choose good over evil.

CHAPTER THIRTY
SOCIAL BEHAVIOR

CHAPTER SUMMARIES AND OBJECTIVES

30-1 <u>Costs and Benefits of Group Living / Individual Adjustments to Group Living / Roles of the Sexes</u> Social behavior, the interaction of conspecific animals, has both benefits and costs. Associates may aid or interfere with hunting, for example, depending on the strategy employed. A group may be easier than a solitary animal for a predator to find, but harder for it to attack. Disease, parasitism, and competition increase with group living. The benefits and costs of social behavior often differ between the sexes because of the difference in their reproductive investment. Individuals also differ in size, dominance, and competitive ability, factors that influence their optimum social behavior strategies. Birds in flocks interfere and fight with one another, but individuals can spend more time feeding and less time scanning for hawks. Thus flock size varies, depending on which need is greater at any given time. Such flexibility is characteristic of much social behavior.

Sexual roles vary considerably among species. The female's investment in gamete production is always greater, but parental care may be performed by either sex or by both. Except in carnivorous or insectivorous species, males usually do not feed young, but they often guard or defend them. Male parental care is more prevalent in taxa wherein paternity is relatively certain--for example, in those groups with external fertilization.

1. Define social behavior and explain its general costs and benefits.

2. Explain why optimum social strategies differ among individuals.

3. List some major factors that influence the roles of the sexes in parental care.

30-2 <u>Communication: Maintenance of Social Groups / Chemical Communication / Visual and Auditory Communication / Tactile and Electrical Communication</u> Communication involves displays or signals--behavior that has evolved to influence the actions of other conspecifics. It may involve one or more sensory modes. Chemical communication includes taste and smell. Chemicals are easily produced with little energy expenditure; they are the oldest and most widespread form of communication. Those chemicals that influence conspecifics are called pheromones. Chemicals can travel great distances, but they cannot be turned on and off rapidly. Visual signals require appropriate receptors, but they are easily produced, versatile, and rapidly changeable. They also indicate location effectively. Auditory signals are more difficult to locate and can be turned off more readily than visual information. Sounds may be produced vocally or by other means, such as friction or vibration. Tactile communication involves either direct contact or less direct means, such as the vibration of a spider's web. Communication systems may combine several sensory modes; the honeybee's dance combines auditory, chemical, tactile, and visual components. Certain fishes generate electric fields, both for capturing prey and for communication.

4. Define communication.

5. List the principal sensory modes used in communication, and give the main characteristics, limitations and advantages of each.

30-3 Origins of Communication Signals / Evolution of Communication Signals / Interspecific Communication / Choosing Associates Intention movements (preparatory behaviors performed by animals prior to some other behavior) are believed to have been modified and elaborated by natural selection to produce many communication signals. Other signals may have evolved from physiological changes produced by the autonomic nervous system. Still others may have arisen from displacement activities--seemingly irrelevant behaviors performed under stress. Elaboration of such behavior reduces its ambiguity. For signals to evolve, they must benefit the individual giving them as well as some receivers. Interspecific signaling includes many examples of deceit, as in the case of brood parasitism. Choices of conspecific associates include primarily mates and kin. Mating behavior often includes male displays, female selection, and/or male territorial defense. Associates other than mates are often kin, because related animals share genes. By helping a relative to survive and reproduce, an individual promotes the survival of many of its own genes as well.

6. List the major evolutionary origins of communication signals, and explain under what general circumstances such signals evolve.

7. Explain the principle of kin selection.

30-4 The Evolution of Animal Societies / Spiders / Insects / Birds / Mammals Little is known of the behavior of extinct organisms, but the earlier stages of social organization may be inferred from the simpler patterns existing today. Two principal routes to sociality are hypothesized: the familial route (prolonged parental care leads to extended family groups) and the parasocial route (cooperative behavior arises among adults of the same generation). These two routes are observed in certain spiders, although most spiders are solitary. Among insects, bees, wasps, ants, and termites are known for their complex social systems. Most wasps appear to have arrived at sociality via the familial route; a few exceptions probably took the parasocial route. Both routes are well represented among bees. Modern ants and termites are all social, so early stages cannot be observed, but the familial route is inferred. The familial route is prevalent among birds (such as the scrub jay), but there are exceptions (such as the groove-billed ani). In mammals, too, the familial route is believed to predominate; the principal exceptions are bats. All levels of social organization are found among primates. Human societies are believed to have evolved via the familial route.

8. Explain the two principal hypothesized routes to sociality in animals, and give examples of each.

ARTICLE RESOURCES

Costs and Benefits of Group Living

Individual Adjustments to Group Living

Roles of the Sexes

Batten, Mary. Sexual choice: The female's newly discovered role. Science Digest, 1982, 90(3), 84-86, 112-113. A good discussion of evolutionary ecology and the theory of social biology. Females have a much larger parental investment in their offspring than males. Therefore, the female not only has more at stake than the male, but evolution also has more at stake in her.

Walker, William F. Sperm utilization strategies in nonsocial insects. American

Naturalist, 1980, 115(6), 780-799. In social insects there is but one female who mates with one male, but in nonsocial insects all females mate with numerous males. These females can store sperm for long periods of time. Walker theorizes that there is some sort of sperm precedence determined by the female.

Communication: Maintenance of Social Groups

Tyack, Peter. Why do whales sing? The Sciences, 1981, 21(7), 22-25. Apparently whales use their calls to locate others of their species, thus maintaining the pod's cohesiveness. Humpback whales have the most complex songs, but they sing only during the mating season to attract females.

Chemical Communication

Levine, Joseph S. Chemical warfare flourishes among the creatures of the reef. Smithsonian, 1981, 12(6), 120-126. Toxins are a common offensive and defensive adaptation among reef animals both vertebrate and invertebrate. Mechanisms of ingestion are discussed. Ingestion of fish that produce these toxins can be fatal even if the person has not been infected with a toxin. Marine toxins have become a focus of investigation in biomedical research.

Visual and Auditory Communication

Wiersig, Bernd. Dolphins. Scientific American, 1979, 240(3), 136-148. Dolphins have a wide range of vocal signals that enable them to communicate with each other in their herds. Their behavior and social systems are discussed. Echolocation is explained. This is the method by which dolphins can scan the environment to detect potential danger. A well-illustrated article.

Tactile and Electrical Communication

Origins of Communication Signals

Evolution of Communication Signals

Interspecific Communication

Choosing Associates

The Evolution of Animal Societies

Spiders

Insects

Heinrich, Bernd. The regulation of temperature in the honeybee swarm. Scientific American, 1981, 244(6), 146-160. A swarm of bees is different from a hive. Swarming is a late-spring behavior in which a queen bee and several thousand others leave the hive, leaving another queen bee and other bees behind. The primary function of swarming is to establish a new colony by relieving over-crowding. A discussion of how hives are cooled in summer and warmed in winter is given. The necessity for a high temperature is shown. The metabolic rates for swarms of 1,800 to 16,000 bees were comparable.

Fergus, Charles. Lord and master of June. Science 82, 1982, 3(5), 54-59. Beautifully illustrated article on the insects of June, particularly the dragon-fly.

Birds

Mammals

ESSAY QUESTIONS

1. Amplify on the textbook's statement that "male carnivores can make more significant contributions than male herbivores to the care and feeding of the offspring." Why might this be so?

    Some possible responses: Herbivores eat large quantities of low-energy food, which would be hard for the male to transport; hunting with young would be hard for a female carnivore, but grazing with young would be feasible for a female herbivore, so she needs the male's help less.

2. The text states that no mammalian species is known in which the male is primarily responsible for care of the young. Among humans, however, single fathers are relatively common. What might account for this difference?

    Some possible responses: As in so many areas, human behavior is more flexible than that of other species; because of technology and agriculture, men are able to "produce" milk as well as women; biological selection pressures for specific sex roles are largely absent in humans.

3. What conditions might promote (a) the evolution of electrical communication? (b) that of tactile communication?

    For (a), low visibility plus the availability of a suitable medium: water with high electrical conductivity due to a high concentration of dissolved salts. For (b), low visibility; usefulness of a close-range communication system. For both (a) and (b), the availability of suitable receptors might also be a factor. For example, the electric fishes already had the mechanism for capturing prey, and spiders were already able to detect vibrations of their web.

4. A person may clear his throat because it is dry or because he wants to attract someone's attention.

    a. Consider whether these two behaviors are outwardly identical or not.

    b. If there are differences, how are they analogous to the differences between autonomic functions and the displays derived from them in other species?

    c. Does this throat-clearing behavior in humans qualify as a display?

        a. The main differences are that the attention-getting behavior is exaggerated or emphasized and that the person who engages in it usually also looks directly at the person whose attention he wants. Perhaps there are also other differences.

        b. The analogy is that such displays are similarly exaggerated and/or combined with other signals to make them less ambiguous and more noticeable.

        c. The answer depends on one's definition of a display and on one's willingness to apply ethological explanations to human behavior. Throat-clearing as an attention device is probably a cultural pattern and not built into the human genome, but this is true of most human behavior.

TEST SET 30-A

Name _____   Section _____

Choose the best answer to each of the following questions, and write the appropriate letter in the space provided.

Ans: c
p. 745
_____   1. Is cooperative hunting always more effective than solitary hunting? (a) Yes, because it results in more efficient division of labor. (b) No, because only very intelligent species can cooperate effectively. (c) No, because many types of prey must be approached stealthily. (d) Yes; that is why humans have dominion over every living thing, because of our cooperative instinct.

Ans: d
p. 747
_____   2. Why are females generally more selective in their choice of mates than males are? (a) Their innermost drives and needs are not as strong. (b) They are interested in sex only during estrus, whereas males of most species are sexually active all year long. (c) They are simply "playing hard to get" in order to attract more males. (d) They invest much more energy in reproduction than males do, so it is more important for them to select a superior mate.

Ans: a
pp. 745-
    746
_____   3. Why do many small bird species flock, even though flocking promotes intraspecific aggression? (a) This problem is evidently outweighed by the protective value of flocking. (b) Fighting promotes survival of the fittest and hence improves the species. (c) Fighting between males helps females to select superior mates. (d) They are unaware of the problem.

Ans: b
p. 748
_____   4. Adults feed young in all of the following groups except: (a) bees; (b) fishes; (c) mammals; (d) birds.

Ans: d
p. 749
_____   5. What is the relationship between male parental care and mode of fertilization? (a) No correlation is known. (b) Species with internal fertilization can be more sure of paternity; hence males are more likely to care for young. (c) Species with multiple fertilization have more male parental care, because no male can be certain that a given offspring is not his. (d) In vertebrate species with external fertilization, male parental care is more marked.

Ans: b
p. 750
_____   6. All of the following behaviors influence the behavior of conspecifics. Which one is most likely to qualify as a signal? (a) A dog walks down the street. (b) A dog raises its hackles and snarls. (c) A dog sees and chases a rabbit. (d) A dog vomits.

Ans: a
p. 750
_____   7. Pheromones are chemical agents that: (a) influence conspecifics; (b) attract mates; (c) act exclusively on olfactory receptors; (d) are energetically expensive.

Ans: b
p. 751
_____   8. Which of the following communication modes conveys the location of the sender least accurately, under most circumstances? (a) visual; (b) chemical; (c) auditory; (d) electrical.

Ans: d
pp. 750-
    752
_____   9. Is human perspiration odor a pheromone? (a) No, because it can't be turned on and off at will. (b) No, because the response of others is neither consistent nor predictable. (c) Yes, because it influences the behavior of other humans. (d) It may have served as a signal at some point in human evolution, but its status now is not entirely clear.

Ans: b   10. Auditory communication exhibits all of the following characteris-
p. 752   tics, <u>except</u>: (a) being rapidly changeable; (b) pinpointing the
                            location of the source; (c) being easily turned off; (d) a great
                            variety of possible levels and values of stimuli.

Ans: d   11. Are most mute species also deaf? (a) Yes, because they have no
p. 753   need for auditory receptors. (b) Very few animal species are
                            truly mute. (c) Most do not <u>react</u> to sounds, but it is not
                            known whether they can hear. (d) No; they need to hear nonvocal
                            signals as well as other sounds.

Ans: a   12. A bull lowers its head and scrapes the ground with one front
p. 756   foot. What sort of communication is this <u>most</u> likely to be?
                            (a) threat display; (b) sign stimulus; (c) tactile signal;
                            (d) waggle dance.

Ans: c   13. What is the probable origin of the behavior described in Question
p. 755   12? (a) autonomic behavior; (b) familial route; (c) intention
                            movement; (d) kin selection.

Ans: d   14. Do females choose males on the basis of their vigor and other
p. 759   characteristics, or on the basis of the territory they hold?
                            (a) vigor; (b) territory; (c) neither; (d) either, depending on
                            the species.

Ans: b   15. A bird that guards its nest and tries to divert predators en-
p. 761   dangers itself but helps its offspring. Such behavior evolves
                            because of: (a) resource-defense polygamy; (b) kin selection;
                            (c) brood parasitism; (d) displacement activity.

TEST SET 30-B

Name _____   Section _____

Choose the <u>best</u> answer to each of the following questions, and write the appropriate letter in the space provided.

Ans: c
p. 766

_____ 1. Why was cooperative hunting advantageous to early humans? (a) It taught them a spirit of compassion and altruism. (b) It strengthened the pair bond, by excluding females from food gathering and hence making them dependent on the males. (c) It enabled them to kill very large animals, such as mastodons. (d) It made it necessary for humans to become bipedal.

Ans: c
p. 746

_____ 2. Is a social system equally beneficial to all of its members? (a) Yes; otherwise it would never have evolved. (b) Yes, but in different ways. (c) No; small or subordinate individuals may never do as well as dominant ones. (d) No; only a few dominant individuals benefit, while the majority suffer.

Ans: a
p. 747

_____ 3. Is social behavior generally flexible? (a) Yes. (b) It is in vertebrates but not in insects. (c) It is in mammals but not in lower vertebrates or invertebrates. (d) No. In general it is characterized by its inflexibility.

Ans: d
p. 748

_____ 4. Why do males in most mammal species contribute little to feeding the young? (a) In highly evolved taxa, the sexes keep to their proper roles. (b) They are too busy inseminating other females and fighting one another. (c) In the majority of mammals, males <u>do</u> help feed the young. (d) The female produces milk, so it <u>isn't</u> necessary for the male to bring food.

Ans: a
p. 748

_____ 5. Greater care of offspring by males than by females occurs (in some species) in all of the following groups <u>except</u>: (a) mammals; (b) birds; (c) fishes; (d) insects.

Ans: b
p. 750

_____ 6. All of the following behaviors influence the behavior of conspecifics. Which one is <u>most</u> likely to qualify as a signal? (a) A chipmunk runs away as a hawk dives toward it. (b) A chipmunk gives a specific call when a hawk is overhead. (c) A chipmunk struggles and squeals after a hawk captures it. (d) A chipmunk resumes feeding after a hawk has left.

Ans: c
p. 750

_____ 7. Pheromones have all of the following characteristics <u>except</u>: (a) long fade-out time; (b) widespread occurrence; (c) conveying rapid changes in mood; (d) being energetically inexpensive.

Ans: d
p. 750

_____ 8. Is musk after-shave lotion a human pheromone? (a) Yes, because it attracts women. (b) Yes, because it is made partly from the glandular secretions of male mammals and is therefore "natural." (c) Yes, because it affects the behavior of conspecifics, if only to make them open windows. (d) No, because it did not evolve as a human signal.

Ans: a
p. 750

_____ 9. What is the chief difference between taste and smell? (a) Taste involves contact; smell occurs at a distance. (b) Only air-breathing organisms can smell; aquatic animals are limited to taste. (c) There is no significant difference. (d) A chemical that is to be tasted is far more energetically expensive to produce.

Ans: c
p. 752

_____ 10. Visual communication has all of the following characteristics <u>except</u>: (a) being rapidly changeable; (b) being easy to produce; (c) being easily turned off; (d) pinpointing location.

Ans: c
p. 753

_____   11. Vocal communications are: (a) synonymous with auditory communications; (b) the most prevalent method of animal signaling;
(c) sounds produced by forced movement of air across vocal
cords; (d) found only in humans.

Ans: a
pp. 753-
   755

_____   12. Bee communications include all of the following modes <u>except</u>:
(a) electrical; (b) olfactory; (c) auditory; (d) tactile.

Ans: b
p. 756

_____   13. After comic strip character Dagwood Bumstead has a bad day of
being harried by his boss and ridiculed by his family, he goes
into the kitchen and eats a big sandwich. Regardless of what
Freudians would say of such behavior, how might it be characterized in ethological terms? (a) intention movement; (b) displacement activity; (c) territorial defense; (d) chemical communication.

Ans: d
p. 762

_____   14. Which route to sociality typically begins with solitary individuals reproducing without <u>any</u> interaction with conspecifics?
(a) familial; (b) parasocial; (c) both; (d) neither; almost all
animals must mate to reproduce.

Ans: d
pp. 762-
   763

_____   15. Which route to sociality has occurred in the majority of species
of spiders? (a) Spiders followed the familial route. (b) Spiders followed the parasocial route. (c) Most spiders are now
social, so the intermediate stages are difficult to infer.
(d) Neither; most spiders are solitary.

CHAPTER THIRTY-ONE
ORIGINS

CHAPTER SUMMARIES AND OBJECTIVES

31-1 <u>Spontaneous Generation: Old Ideas / Spontaneous Generation Revisited</u> Until about 200 years ago, the formation of living things from nonliving matter was an accepted explanation for the origin of life. Some scientists questioned this doctrine, however, and various experiments, performed by Redi, Spallanzani, and Pasteur, demonstrated that all living organisms arise from preexisting organisms.

A question remained, however, about the origin of the first organisms. Oparin and Haldane independently arrived at a theory of life's origins. Several lines of evidence support the general outline of this theory. This evidence will be examined in detail in later sections of this chapter.

   1. Cite the contributions of each of the following scientists to our understanding of life's origins: Redi, Spallanzani, Pasteur, Oparin, Haldane.

31-2 <u>A Common Metabolic Heritage / What the Common Heritage Tells Us</u> The most important metabolic reactions of all eukaryotes are identical and, where they do differ, they are often only variations on the same theme. Prokaryotes, however, are metabolically diverse. Some bacteria use sulfur or nitrogen compounds or carbon dioxide as electron acceptors for respiration. Photosynthetic bacteria may use sulfur compounds, fatty acids, other organic compounds, or molecular hydrogen instead of water in carrying out photosynthesis. Other bacteria are able to obtain energy for the synthesis of carbohydrates from sulfur, iron, or nitrogen compounds.

All organisms extract energy from organic compounds via the process of glycolysis. There are only a few organisms that do not use glucose as their primary energy source. This suggests that the first living organism was an anaerobic heterotroph, absorbing food from its surroundings and extracting energy in the absence of molecular oxygen. Early life must have originated under conditions of an abundance of organic molecules and a scarcity of free oxygen.

   2. Explain why it is believed that there was little free oxygen in the earth's early atmosphere.

   3. Give several examples of metabolic diversity among the prokaryotes.

31-3 <u>The Fossil Record / Precambrian Rocks / Ancient Prokaryotes / Eukaryotic and Multicellular Organisms</u> A fossil is any recognizable trace of an organic structure preserved from prehistoric times. Fossils may be dated by estimating the age of the rocks in which they occur. This can be accomplished by radioisotope techniques. Rock strata from different locations are correlated by the nature of the fossils they contain. The oldest fossils so far discovered are about 3.5 billion years old. Earth's crust probably became solid about 4 billion years ago. By 600 million years ago, representatives of every modern group of organisms

295

except vertebrates and terrestrial plants, had appeared.

Prior to the Cambrian Period, fossils are scarce. Several explanations have been offered to account for this observation. Modern techniques have enabled us to discover fossils in Precambrian rocks. Stromatolites, for example, are the remains of bacteria and cyanobacteria that lived in shallow aquatic habitats. Several locations have yielded Precambrian fossils. These include the Gunflint chert of the Minnesota-Ontario border and North Pole, Australia.

The first green plants are present in fossil strata that are about 1.4 to 1.2 billion years old. Other eukaryotic microorganisms have been discovered in rocks ranging in age from 900 million to 700 million years old. The first multicellular organisms left fossils in rocks about 670 million years old. The existence of a variety of ancient organisms has led to the conclusion that life appeared during the first few hundred million years of earth's existence.

4. Answer: Why is the period from about 3.5 billion to 800 million years ago referred to as "the Age of Prokaryotes?"

5. Explain how the age of a rock may be determined using radioisotopes.

6. Cite at least two of the important Precambrian fossil locations discussed in this section, and explain the nature and significance of the fossils found in each.

31-4 Meteorites and Extraterrestrial Life Most of the evidence that might suggest the presence of living matter in certain meteorites is rather ambiguous. Though the possibility of the existence of extraterrestrial life remains open, our failure to discover it in meteorites lends support to our interpretations of Precambrian fossils. If "life" were readily demonstrated in meteoritic material, the argument could be presented that we are able to discover "life" wherever we look.

7. Describe carbonaceous chondrites and some of the kinds of organic material found in them.

8. Explain how gleaned information from the study of meteorites tends to support our conclusions about the origin of life on earth.

31-5 Primitive Earth: Atmosphere and Energy Sources / Laboratory Simulation Scientists agree that, regardless of its exact composition, the earth's early atmosphere did not contain free oxygen in significant amounts. Carbon dioxide, water vapor, carbon monoxide, ammonia, and methane were probably present. There were abundant energy sources in the primitive environment. Any of these could have driven endergonic reactions, resulting in the formation of complex organic molecules from simpler substances in the atmosphere.

Laboratory evidence indicates that a great variety of relatively complex organic molecules can be produced simply by exposing mixtures of gases (always in the absence of oxygen) to various forms of energy. The fact that almost any combination of gases, or any energy source, will work provides assurance that such reactions must have taken place on the early earth, leaving the oceans filled with amino acids, lipids, organic acids, simple sugars, and the like. The probable reason for the universal employment of ATP as an energy carrier in living organisms is simply that adenine is the easiest nucleic acid monomer base to obtain in experiments such as these.

9. Describe the probable composition of the primitive atmosphere of earth.

10. List at least three sources of energy that could have driven the synthesis of the first organic molecules.

11. Answer: What is the significance of laboratory experiments such as those carried out by Stanley Miller and others?

31-6 From Monomers to Polymers / Microspheres / Coacervates / Late Stages of Chemical Evolution There are a number of ways in which simple organic molecules can become polymerized into more complex structures. Although the exact mechanism that produced the first components of living systems will probably never be known, it is quite clear that such polymerizations occurred under the postulated early earth conditions.

There are also many ways in which biological polymers and other compounds can be accumulated into isolated droplets in the laboratory. Any of these processes could have led to the formation of the first complex, pre-life systems. Among the types of chemical systems that could have formed the basis for the first cells are microspheres and coacervate drops. Such systems exhibit a number of properties found in living systems.

The precursors of living organisms may have been encapsulated collections of chemical reactants similar to coacervate drops. Coacervates that enclosed reactions, making them more stable, would tend to survive. Any catalysis that speeded up favorable reactions would tend to be retained. In short, a period of chemical evolution occurred, which led to the development of increasingly complex chemical systems, culminating in the first living cells. The development of reproduction led to the first truly living organisms.

12. Differentiate between microspheres and coacervate drops, and indicate at least three properties that the latter share with living cells.

13. Explain how chemical evolution could have led to the first truly living systems. Be sure to include the two most significant events in this process.

ARTICLE RESOURCES

Spontaneous Generation: Old Ideas

Spontaneous Generation Revisited

A Common Metabolic Heritage

King, G. A. M. Evolution of the coenzymes. Bio Systems, 1980, 13(1), 23-45. The author maintains that the earliest biochemicals reproduced by autocatalytic pathways and evolution occurred through a series of symbiotic unions. An elaborate chart of biosynthetic pathways to the coenzymes is given. Five criteria for evolution through symbiosis are discussed.

What the Common Heritage Tells Us

The Fossil Record

Precambrian Rocks

Schopf, J. William. The evolution of the earliest cells. Scientific American, 1978, 239(3), 110-120, 126-138. Microscopic organisms appeared in the Precambrian era. These cells gave rise to the biochemical systems and the oxygen-enriched atmosphere of today. These cells evolved a distinct nucleus and a limiting membrane surrounding the cell. Photographs of stromatolites both fossil and living are given. Metabolic pathways showing how cells extract energy from food are shown. Precambrian microfossils are shown, and a chart on fossil cells shows the evidence for the origin of eukaryotic cells. An excellent summary of the environmental conditions that existed during each of the eras from the Precambrian to the emergence of eukaryotes is included.

Ancient Prokaryotes

Graves, Daniel I., John S. R. Dunlop, and Roger Buick. An early habitat of life. Scientific American, 1981, 245(4), 64-73. Fossil deposits in northwestern Australia indicate probable microfossils and stromatolites--layered structures formed by accretion of fine grains of sediment by matted layers of microorganisms. Through carbon dating it has been established that the early Precambrian environment of these deposits was one in which primitive organisms could have flourished. These fossils go back approximately 3.5 million years. Excellent photographs and clear explanations.

Eukaryotic and Multicellular Organisms

Meteorites and Extraterrestrial Life

Ponnamperuma, Cyril. Our most remote ancestors. Chemistry, 1978, 51(9), 6-12.
Living matter has a biochemical unity founded on two basic molecules: nucleic acid
and proteins. These may be termed the primary building blocks of life. The author
concludes that all life must have a common chemical origin. The history of this
theory is traced. The presence of 53 different amino acids in the Murchison
Meteorite that fell in Australia in 1969 supports the extraterrestrial theory of the
origin of amino acids. A chart showing some of the common molecules found in outer
space is given.

Primitive Earth: Atmosphere and Energy Sources

The earth's evolving atmosphere. Sci Quest, 1979, 52(2), 22-23. An unsigned article detailing the loss of the earth's original atmosphere as a result of intense
heat from the sun. Our present atmosphere developed by the "outgassing" of various
gasses as a result of volcanic activity and meteorite bombardment, thus liberating
carbon dioxide and water vapor. Evaporation and condensation formed oceans, and
large quantities of carbon dioxide were incorporated into rocks and the ocean. The
earliest forms of life apparently formed in the subsurface of the oceans and released oxygen into the atmosphere.

Dickerson, Richard E. Chemical evolution and the origin of life. Scientific
American, 1978, 239(3), 70-86. The history of efforts to show how life on earth
may have arisen. A discussion of the various theories is given. Excellent charts
and drawings are included. It is speculated that life and photosynthetic life
evolved within a billion years of the formation of the earth. The next two billion
years were a period in which the earth's atmosphere went from a reducing atmosphere
with little or no oxygen to an oxidizing one in which one out of five molecules was
oxygen. Consequently, the ozone layer was formed, which sharply reduced the amount
of ultraviolet radiation at the surface of the earth.

Laboratory Simulation

From Monomers to Polymers

Microspheres

Coacervates

Late Stages of Chemical Evolution

ESSAY QUESTIONS

1. Although carbonaceous chondrites have been found to contain a variety of organic
    molecules, including optical isomers and some amino acids not found on earth, none
    has ever been shown conclusively to contain the remains of living organisms. In
    what two ways is this "good news" for scientists studying early life on earth?

    If we were to find many examples of "life" in meteorites, the validity of our
    interpretations of many Precambrian fossils might be called into question.
    It should be argued that we are able to find "life" wherever we look. Secondly,
    the presence of a variety of organic compounds in these meteorites does suggest
    that the processes by which complex organic molecules are formed occurs readily
    throughout the solar system.

2. Summarize our present understanding of the events that led to the formation of the
    first living systems.

    The early atmosphere contained a number of simple molecules but very little
    free oxygen. Under the action of solar radiation, radioactivity in the earth's

crust, vulcanism, and other energy sources, complex organic molecules were
formed from the constituents of the atmosphere. These accumulated in the seas.
The oceans became solutions of organic molecules, which became isolated into
droplets, probably similar to coacervate drops or microspheres. Such complex
chemical systems are capable of a number of activities that we associate with
living cells. The process of chemical evolution led to the formation of increasingly stable systems. Acceleration of chemical reactions by enzymatic
catalysis would have been a significant step in the development of stability
in these systems. With the development of reproduction, stable chemical systems became able to ensure their survival, thus becoming the first living
organisms.

3. One of the most interesting Precambrian prokaryotes so far discovered is Kakabeckia.
    This organism has the distinction of being one of the few organisms first identi-
    fied as a fossil and later discovered alive in modern times. Modern strains of
    Kakabeckia are anaerobic and are found only in environments that have a very high
    ammonia concentration. (Such environments are rare today, because the ammonia is
    oxidized by nitrifying bacteria.) How does the discovery of this organism support
    the theory of life's origins discussed in this chapter?

        The early earth is postulated to have had a high concentration of ammonia and
        no free oxygen. Discovery of such an ammonia-requiring, anaerobic organism
        from the Precambrian suggests that our hypotheses about the early environment
        are correct.

4. What did Stanley Miller, Sidney Fox, and A. I. Oparin, respectively, contribute to
    our understanding of the origin of life?

        Miller conducted the first experiments demonstrating that the hypothesized
        formation of organic molecules form the simple components of the early atmos-
        phere could actually have occurred. Fox discovered that amino acids form
        proteinoids when a dry mixture is heated and that cooling of a hot, concen-
        trated solution of these proteinoids results in the formation of microspheres.
        Oparin is credited with being the first to espouse the theory that life ori-
        ginated from the action of natural energy sources on the constituents of the
        early atmosphere.

5. Do you think it probable that life exists elsewhere in the universe? If so, cite
    some supporting evidence for your conclusion.

        There are vast numbers of stars similar to the sun, any of which might have
        planets suitable for life. Because the reactions leading to the formation of
        complex organic molecules and chemical systems can occur readily, and under a
        wide variety of conditions, the existence of life elsewhere in the universe is
        highly probable.

TEST SET 31-A

Name _____   Section _____

Choose the best answer to each of the following questions, and write the appropriate letter in the space provided.

Ans: c
p. 785

_____   1. Multicellular organisms first appeared on earth about _____ years ago. (a) 3.5 billion; (b) 1.5 billion; (c) 670 million; (d) 350 million.

Ans: d
p. 775

_____   2. The experiments of Francisco Redi led to the conclusion that: (a) maggots appear spontaneously in decaying meat; (b) micro-organisms are abundant and ubiquitous in the environment; (c) all life comes from preexisting life; (d) maggots appear in decaying meat only if flies are allowed to lay their eggs therein.

Ans: a
p. 775

_____   3. Fossil traces of ancient organisms are studied by: (a) paleo-biologists; (b) comparative biochemists; (c) molecular biologists; (d) exobiologists.

Ans: c
p. 778

_____   4. The first living cells on earth were probably: (a) aerobic heterotrophs; (b) anaerobic eukaryotes; (c) anaerobic hetero-trophs; (d) aerobic autotrophs.

Ans: d
p. 787

_____   5. Which of the following was a factor contributing to the appear-ance of complex organic molecules on earth? (a) a primitive atmosphere high in free oxygen; (b) the absence of readily available sources of energy; (c) a planetary mass of just the right amount; (d) a primitive atmosphere containing ammonia, methane, carbon dioxide, and water vapor.

Ans: a
p. 778

_____   6. The real diversity of life is apparent in prokaryotes, because these organisms: (a) display mechanisms for food synthesis and energy storage that are not found in eukaryotes; (b) are mor-phologically more varied than eukaryotes; (c) never use glucose as a source of energy; (d) inhabit many environments that are not suitable for eukaryotes.

Ans: b
p. 777

_____   7. Photosynthetic sulfur bacteria use sulfur compounds as a source of: (a) oxidizing agents; (b) hydrogen atoms; (c) carbon atoms; (d) energy for carbohydrate synthesis.

Ans: b
p. 778

_____   8. Which of the following is not a fossil? (a) the impression of a footprint; (b) traces of radioisotopes in sedimentary rocks; (c) frozen muscle tissue from an extinct mammoth; (d) the calci-fied shell of a mollusk.

Ans: d
p. 779

_____   9. The earliest of the four eons into which earth's history is divided by geologists is the: (a) Phanerozoic; (b) Proterozoic; (c) Archean; (d) Hadean.

Ans: c
p. 781

_____   10. If the entire history of the earth is condensed into one month, representatives of every modern group except vertebrates and terrestrial plants have appeared by the: (a) third day; (b) fifteenth day; (c) twenty-seventh day; (d) thirtieth day.

Ans: b
p. 781

_____   11. Stromatolites are produced by: (a) photosynthetic eukaryotes; (b) photosynthetic prokaryotes; (c) green algae; (d) no modern organisms.

Ans: b
p. 787

_____ 12. The fact that meteorites have been found to contain organic compounds, but no well-developed organisms indicates that: (a) life
has originated only once in the universe; (b) the conditions for
the formation of complex organic molecules exist even in outer
space; (c) most of the fossil evidence for Precambrian life is
suspect; (d) organisms probably abound outside the earth, but
their remains rarely reach us in meteorite fragments.

Ans: a
p. 788

_____ 13. In experiments performed by Stanley Miller and others to study
the origin of life: (a) complex organic compounds are produced,
often in a matter of hours; (b) proteinoids dominate the reaction
products; (c) only electrical discharge is sufficient to supply
energy for the observed reactions; (d) the experiment works only
if oxygen is present in the reaction mixture.

Ans: b
p. 790

_____ 14. Various types of monomers may be polymerized by: (a) exposing
them to extremes of heat and pressure; (b) alternately wetting
and drying them in the presence of clay; (c) biologically
mediated processes only; (d) a variety of means, but none of
those cited here.

Ans: b
p. 792

_____ 15. Chemical selection probably favored: (a) the formation of less
stable coacervates; (b) catalysis of stabilizing reactions;
(c) the formation of microspheres lacking the capacity to absorb
materials from the environment; (d) the development of new forms
of life under modern conditions.

TEST SET 31-B

Name _____   Section _____

Choose the best answer to each of the following questions, and write the appropriate letter in the space provided.

Ans: a
p. 775

_____   1.   The importance of Pasteur's work on spontaneous generation lies in the fact that he: (a) demonstrated conclusively that living organisms are produced only by other living organisms; (b) was unable to account for the presence of microorganisms in his experimental vessels; (c) produced organic molecules by passing a spark through a mixture of gases; (d) was the first who sought to test the idea of spontaneous generation.

Ans: b
p. 776

_____   2.   The study of extraterrestrial forms of life is the province of: (a) paleobiology; (b) exobiology; (c) comparative biochemistry; (d) genetics.

Ans: a
p. 778

_____   3.   The absence of oxygen from the primitive atmosphere of earth suggests that the first organisms were probably: (a) anaerobic heterotrophs; (b) photosynthetic; (c) cyanobacteria; (d) chemo-autotrophs.

Ans: c
p. 778

_____   4.   One compound that provides a common metabolic link among all organisms is: (a) thiosulfate; (b) bacteriochlorophyll; (c) glucose; (d) methane.

Ans: c
p. 778

_____   5.   Rock strata from different locations are correlated by comparison of: (a) the composition of the sediments from which they were formed; (b) the relative amounts of radioisotopes they contain; (c) the nature of the fossils found in them; (d) No method currently available allows us to make such correlations.

Ans: d
p. 779

_____   6.   Abundant fossils of a variety of multicellular organisms are found in rocks of which eon? (a) Hadean; (b) Archean; (c) Proterozoic; (d) Phanerozoic.

Ans: b
p. 781

_____   7.   The oldest signs of life found on earth so far come from: (a) Beck Springs, California; (b) North Pole, Australia; (c) the Ontario-Minnesota border; (d) the Bitter Springs Chert of Australia.

Ans: b
p. 784

_____   8.   The earliest eukaryotes may be: (a) 3.5 billion years old; (b) 1.4 billion years old; (c) 670 million years old; (d) 800 million years old.

Ans: a
p. 781

_____   9.   One reason for the scarcity of Precambrian fossils is the fact that: (a) early organisms lacked easily fossilized hard parts; (b) the conditions for fossilization did not exist on earth prior to about one billion years ago; (c) there was insufficient hydrogen in the primitive atmosphere to support abundant primitive life; (d) unicellular organisms never form fossils.

Ans: d
p. 788

_____   10.   An absolute requirement for the success of experimental attempts to synthesize organic molecules from mixtures of simple gases is: (a) the presence of hydrogen; (b) the presence of thiosulfate ions; (c) the absence of either methane or carbon dioxide, but not both; (d) the absence of molecular oxygen.

Ans: a
p. 788

_____ 11. All living organisms use ATP as their energy currency, probably because: (a) adenine is readily produced from simpler molecules; (b) only ATP contains high-energy phosphate bonds; (c) early organisms used GTP but switched as the level of oxygen in the atmosphere increased; (d) ATP is much more difficult to synthesize than the other nucleotides.

Ans: b
p. 791

_____ 12. The study of coacervate drops is important to our understanding of life's origins because coacervates: (a) are able to reproduce themselves precisely; (b) carry out many processes that are similar to those of living systems; (c) are surrounded by a lipid membrane; (d) are formed only under a very limited set of conditions.

Ans: c
p. 792

_____ 13. The critical step in the development of living organisms from complex chemical systems was the development of: (a) DNA; (b) glucose; (c) reproduction; (d) photosynthesis.

Ans: b
p. 781

_____ 14. The remains of ancient cyanobacteria and other photosynthetic prokaryotes form structures called: (a) carbonaceous chondrites; (b) stromatolites; (c) Gunflint chert; (d) microspheres.

Ans: c
p. 792

_____ 15. The process by which less stable coacervates were gradually replaced by more stable ones, culminating in the production of the first living cells, is called: (a) frequency-dependent selection; (b) non-Darwinian evolution; (c) chemical evolution; (d) speciation.

CHAPTER THIRTY-TWO
TAXONOMY AND PHYLOGENY

CHAPTER SUMMARIES AND OBJECTIVES

32-1 <u>The Goals of Classification / Cladistic, Phenetic, and Evolutionary Systematics</u>
In order to be useful, a classification system must separate the objects under
consideration into categories that serve to guide our behavior toward the objects.
Classification systems can be judged only in terms of their usefulness and their
internal consistency. As a result, there are several types of classification
systems in use by biologists, each emphasizing a different set of criteria.

Phenetic systematics is concerned with the level of similarity among the phenotypes of living organisms. Many characters are studied, and all are weighted
equally.

Cladistic systematics is concerned only with the length of time that has elapsed
since two species shared a common ancestor. The degree of difference is not
considered, only the duration of the separation.

Evolutionary systematics combines both phenetic and cladistic approaches. An
attempt is made to elucidate not only the duration of separation of groups, but
also the nature of evolutionary changes that have occurred since that separation.
The result is a phylogenetic tree depicting the history of the lineage under consideration.

   1. Distinguish among phenetic, cladistic, and evolutionary systematics, and give
      one advantage and one disadvantage of each method.

32-2 <u>Taxonomic Hierarchies</u> The basic unit of all biological classification systems in
use today is the species. Each species is given a unique name, which identifies
the species itself and the genus to which the species belongs. This method of
naming organisms was introduced in the eighteenth century by the Swedish botanist
Linnaeus. Genera are groups of closely related species. Genera are grouped into
families, families into orders, orders into classes, classes into phyla, and phyla
into kingdoms.

   2. Cite the names of the various taxa that are used by biologists and give the
      position of each taxon in the classification hierarchy.

32-3 <u>Taxonomic Characters / Selection of Taxonomic Characters / Homology and Analogy</u>
A variety of characters are employed for the classification of organisms. These
include gross morphology, developmental stages, similarities in behavior, ability
to hybridize with other organisms, amino acid sequences of proteins, antigen-
antibody reactions, electrophoretic analysis of proteins, the nucleotide sequence
of the genes themselves, and the ability of single-stranded DNA from one species
to form double-stranded molecules with the DNA of another species.

The selection of appropriate characters on which to base a classification depends on many factors. A useful trait must be measurable, describable, and not subject to variation due to environmental influences. Not all organisms possess similar features, so different traits are used for classification of different groups. Traits that are useful for classification of certain organisms may not be suitable for use, (may not even be present) in the case of other groups.

Homologous structures are those that owe their similarity in two species to inheritance from a common ancestor. Analogous structures are those that are similar in two species as a result of similar selection pressures. Different species can develop analogous structures as a result of convergent evolution. This concept may be reduced to the simple statement "organisms often respond to similar environments in similar ways." Much of the work of evolutionary systematics involves determining whether an observed similarity represents homology or analogy.

3. List and define at least five types of characters commonly used in the construction of taxonomic classification schemes.

4. Differentiate between homology and analogy. Explain why evolutionary systematics is especially interested in cases of homology but ignores analogies.

5. Cite the considerations that govern the selection of taxonomic characters and explain why the same set of traits cannot be.

32-4 Taxonomic Keys / Phylogenetic Trees of Life   One of the important uses of a classification system is in the identification of unknown organisms. For this purpose, a dichotomous taxonomic key is usually constructed. Such a key uses various traits to separate unknown organisms into smaller and smaller categories, eventually arriving at the species designation. Such keys need not reflect evolutionary relationships. Those that do are called natural keys; those that do not are called artificial keys.

Phylogenetic trees of life, on the other hand, always postulate evolutionary relationships, on the basis of available information. Because of the rapid accumulation of information, such models of life's history are frequently modified.

6. Describe a taxonomic key and then construct a simple key to differentiate among the kingdoms of organisms that will be covered in the next three chapters of the text. (As you read these chapters, you may want to expand your key to include the phyla and classes discussed.)

ARTICLE RESOURCES

The Goals of Classification

Whittaker, R. H., and Lynn Margulis. Protist classification and the kingdoms of organisms. Bio Systems, 1978, 10, 3-18. A history of biological classification is given. Reasons for including protists in a separate phylem are explained. The evidence for evolutionary relationships is the guiding factor here, as well as physiological and biochemical characteristics. There may be as few as 10 phyla of protists and 8 algal and fungal phyla or as many as 30 phyla of protists alone. Extensive classification schemata from earliest times to the present are cited.

Cladistic, Phenetic, and Evolutionary Systematics

Stewart, Kenneth D., and K. and R. Mattox. Structural evolution in the flagellated cells of green algae and land plants. Bio Systems, 1978, 10, 145-152. Scaly flagellates may have given rise to all green algae. Divergences in these early ancestors may account for the groups of advanced green algae. An evolutionary chart illustrating this theory is given.

Valentine, James W. The evolution of multicellular plants and animals. Scientific American, 1978, 239(3), 140-153, 156-158. Excellent photographs of fossils and diagrams of phylogenetic trees of life illustrate this article. Summarizes much of the material given in the chapters on evolution, biogeography, and origins of life. Fish, reptile, and mammal evolution are shown.

Taxonomic Hierarchies

Taxonomic Characters

Selection of Taxonomic Characters

Barr, Donald J. S. Taxonomy and phylogeny of chytrids. Bio Systems, 1978, 10, 153-165. The chytrids are an order of primitive fungi characterized by having zoospores with posterior uniflagella. There is a great deal of morphological variation within the species. The advent of the electron microscope made it possible for chytrid's cytological features to be studied and relationships between the taxa to be established. The author gives five fundamental principles occurring during the evolution of chytrids. Excellent photomicrographs and drawings.

Round, F. E. The evolution of pigmented and unpigmented unicells--A reconsideration of the protista. Bio Systems, 12(1), 61-71. Possible lines of algal phyla are shown. A Precambrian network of interacting heterotrophs is given. Pigmented and nonpigmented chrysophytes, dinoflagellates, and euglenoids seem to have separated at an early evolutionary stage. Histories are not found universally in dinoflagellates, which strengthens this theory.

Homology and Analogy

Taxonomic Keys

Phylogenetic Trees of Life

Woese, Carl R. Archaebacteria. Scientific American, 1981, 244(6), 98-122. The archaebacteria are fundamentally different from the true bacteria, although both are prokaryotic cells. Eukaryotes and prokaryotes are compared and contrasted. Arguments for placing archaebacteria in a separate kingdom are given. These include molecular sequencing and details of function at the molecular level. Molecular traits of archaebacteria, eubacteria, and eukaryotes are compared in a well-organized chart.

ESSAY QUESTIONS

1. From an overall reading of this chapter, indicate at least three criteria that should be met by a good biological classification system.

    Most taxonomists would agree that a classification system should (1) be based on readily identifiable and describable traits that can be measured, (2) organize knowledge about the group being classified in such a way as to reflect as nearly as possible the evolutionary history of the group, and (3) be useful in the assignment of unknown organisms to appropriate taxa. Most taxonomists would also agree that no one system yet devised meets all of these criteria.

2. List, from most inclusive to least inclusive, the taxa employed in the classification of organisms.

    Kingdom, phylum, class, order, family, genus, species.

3. There are two large groups of marine segmented worms (polychaetes). The first, composed of worms that actively move about on the ocean floor in search of food, is characterized by having many similar segments, appendages on each segment, and well-developed vision. In the second group, members spend their time in burrows or tubes, filtering the seawater for microorganisms. These worms are characterized by few segments, few segments bearing appendages, poor vision, and elaborately developed feeding appendages. Explain why the structures of these two groups of

worms can be used as examples of both homology and analogy.

The structures are homologous in that all the segmented worms are derived from a common ancestor; the differences observed are simply variations on the basic body plan: a segmented, wormlike body with numerous appendages. However, the patterns of structure within the two groups reflect a certain level of convergent evolution. Evidently the elongate, many-legged form is more efficient for an errant lifestyle. The sedentary worms, on the other hand, face a different set of needs. For example, any organism that filters seawater for its food must construct a "net" of some kind. Hence all the sessile worms have elaborate feeding appendages.

TEST SET 32-A

Name _____   Section _____

Choose the best answer to each of the following questions, and write the appropriate letter in the space provided.

Ans: b
p. 798
_____ 1. The assignment of a two-part name to a particular species is known as: (a) cladistics; (b) binomial nomenclature; (c) phenetics; (d) systematics.

Ans: a
p. 797
_____ 2. A major element of the definition of a species is that species: (a) maintain genetic isolation from each other; (b) hybridize freely; (c) never change; (d) are artificial constructions of taxonomists.

Ans: d
p. 803
_____ 3. Which of the following is(are) not used in preparing classifications? (a) DNA hybridization; (b) embryology; (c) gross morphology; (d) structural traits that vary with the environment.

Ans: c
p. 794
_____ 4. The study of the pathways along which organisms have evolved is called: (a) phenetics; (b) binomial nomenclature; (c) phylogeny; (d) taxonomy.

Ans: c
p. 795
_____ 5. Phenograms are: (a) graphical representations of evolutionary relationships; (b) estimates of the time elapsed since two species diverged from a common ancestor; (c) graphical presentations of the degree of structural similarity among groups of organisms; (d) always based on common ancestry.

Ans: b
p. 797
_____ 6. The most restrictive classification that two organisms that interbreed freely in nature belong in is the same: (a) genus; (b) species; (c) phylum; (d) class.

Ans: d
p. 797
_____ 7. Only phenetic information can be used to separate species of: (a) plants; (b) animals; (c) sexual organisms; (d) asexual organisms.

Ans: c
p. 797
_____ 8. Binomial nomenclature was the brainchild of: (a) Pasteur; (b) Oparin; (c) Linnaeus; (d) Spallanzani.

Ans: d
p. 799
_____ 9. Which of the following indicates a family of plants? (a) Fagales; (b) Rosa; (c) Tracheophyta; (d) Orchidaceae.

Ans: d
p. 799
_____ 10. Hominidae is a family of: (a) plants; (b) fossil mammals; (c) bacteria; (d) primates.

Ans: a
p. 800
_____ 11. One advantage of using amino acid sequences in taxonomy is: (a) the relative absence of environmental influences on these traits; (b) the simple methods needed for their determination; (c) the marked differences between systems based on these sequences and those based on gross morphology; (d) all of the above.

Ans: c
p. 801
_____ 12. Electrophoretic techniques separate proteins on the basis of differences in their: (a) amino acid sequences; (b) nucleotide sequences; (c) mobility in an electric field; (d) solubility in electrolytes.

Ans: a
p. 803
_____ 13. DNA hybridization studies are based on measurements of: (a) the thermal stability of DNA; (b) the relative mobility of different DNAs in an electric field; (c) the inability of double-stranded DNA to re-form after heating; (d) the molecular weight of DNA from different species.

Ans: c
p. 801

_____ 14. In an immunological reaction, antigen-antibody aggregates form when antibodies are mixed with: (a) only the protein from which they were produced; (b) only proteins other than those from which they were produced; (c) the proteins from which they were originally produced and other, similar proteins; (d) albumin.

Ans: b
p. 806

_____ 15. The wing of a bird and the wing of an insect are: (a) homologous; (b) analogous; (c) evidence for a common ancestry of these two groups; (d) the result of divergent evolution.

TEST SET 32-B

Name _____   Section _____

Choose the best answer to each of the following questions, and write the appropriate letter in the space provided.

Ans: c
p. 797
_____ 1. The correct scientific name for the common anglerfish is:
(a) Histrio Histrio; (b) Histrio histrio; (c) Histrio histrio;
(d) Histrio Histrio.

Ans: d
p. 797
_____ 2. Musca domestica belongs to the genus: (a) Muscidae;
(b) domestica; (c) Domestica; (d) Musca.

Ans: c
p. 797
_____ 3. Genetic isolation is a major criterion for the delimiting of:
(a) kingdoms; (b) phyla; (c) species; (d) genera.

Ans: b
p. 795
_____ 4. The classification of organisms based on the amount of time that
has elapsed since two species shared a common ancestor is called:
(a) phenetics; (b) cladistics; (c) systematics; (d) phylogeny.

Ans: b
p. 807
_____ 5. Lord Kelvin's calculations of the age of the earth turned out to
be incorrect because: (a) he did not account for the heating of
the surface due to friction with the atmosphere; (b) he neglected
to include in his calculations the heat produced by radioactive
decay; (c) his estimate of the original temperature of the earth
was too low; (d) he made a simple mathematical error that was not
detected until after his death.

Ans: d
p. 799
_____ 6. In which of the following taxa would you expect to find the
least degree of similarity among the members? (a) species;
(b) genus; (c) order; (d) phylum.

Ans: c
p. 794
_____ 7. Graphical representations of lines of descent among organisms are
called: (a) cladograms; (b) phenograms; (c) phylogenetic trees;
(d) analogies.

Ans: a
p. 796
_____ 8. The most important criterion for evaluating a given classification scheme is its effectiveness in: (a) serving the purpose for
which it was intended; (b) presenting absolute truth about relationships among organisms; (c) pointing out the similarities
among unrelated organisms; (d) enabling unknown organisms to be
assigned to species.

Ans: c
p. 796
_____ 9. The branching of any phylogenetic tree is based on the assumption
that: (a) most species hybridize frequently; (b) convergent
evolution rarely occurs; (c) similar species descended from a
common ancestor; (d) homology can rarely be distinguished from
analogy.

Ans: c
p. 794
_____ 10. Ideally, the different categories in a classification scheme
should: (a) exclude all organisms that do not fit in; (b) contain at least two species; (c) reflect evolutionary relationships;
(d) be based on developmental stages.

Ans: b
p. 803
_____ 11. DNA strands will not form stable duplexes if their nucleotide
sequences differ by more than: (a) 5 percent; (b) 20 percent;
(c) 1 percent; (d) 10 percent.

Ans: c
p. 803
_____ 12. Which of the following is the best taxonomic character? (a) gross
morphology; (b) amino acid sequences; (c) nucleotide sequences;
(d) embryology.

Ans: d   _____ 13. Taxonomic keys that reflect evolutionary relationships within
p. 806   the groups they cover are called: (a) artificial; (b) phylo-
                          genetic; (c) cladistic; (d) natural.

Ans: d   _____ 14. Analogy is to convergent evolution as homology is to: (a) gross
p. 805   morphology; (b) divergent evolution; (c) hybridization;
                          (d) common ancestor.

Ans: c   _____ 15. Which of the following types of information is most readily
p. 800   obtained from fossils? (a) behavior; (b) amino acid sequences;
                          (c) morphology; (d) embryology.

CHAPTER THIRTY-THREE
MONERA AND PROTISTS

CHAPTER SUMMARIES AND OBJECTIVES

33-1 <u>Viruses / Reproduction of Viruses / Classification of Viruses</u> Viruses are acellular organisms that are obligate parasites of animals, plants, and bacteria. The individual virion is composed of a core of nucleic acid (either DNA or RNA, never both) surrounded by a protein capsid. Once inside the host cell, viruses harness the cellular machinery of the host to cause the production of more virus particles. In animal cells, viruses are usually taken up by endocytosis and released by exocytosis, whereas in plants and bacteria, viruses leave the cell by lysis. Bacterial viruses are usually supplied with specialized structures for injecting their nucleic acid into the host, while plant viruses are introduced on the mouthparts of insects. The interaction between host and virus is very specific and is determined by the characteristics of the viral capsid and the host cell membrane.

Viruses are classified on the basis of the following criteria: (1) nature of the nucleic acid, (2) whether the nucleic acid is single-stranded or double-stranded, (3) structure of the capsid, and (4) presence or absence of envelope. The distribution of viruses in terms of host organisms is scattered, with some groups frequently infected and other groups seldom or never infected.

1. Describe the life cycle of a typical animal virus, and contrast it with that of a plant virus.

2. Explain what arboviruses are and why they are important to humans.

3. List the criteria for classification of viruses.

33-2 <u>The Kingdom Monera (Prokaryotes) / Prokaryotes vs. Eukaryotes / Metabolic Diversity in the Kingdom Monera / Some Other Distinguishing Characteristics of Bacteria</u> The kingdom monera includes all prokaryotic organisms. Modern monera probably represent the result of many independent lines of evolution. They are the most ancient of all living organisms and are extremely diverse. Only about half of the groups of monera recognized by bacteriologists will be presented here.

For a comparison of prokaryotic and eukaryotic cells, refer to Chapter 4. Prokaryotes lack a membrane-bound nucleus, do not divide by mitosis, have a cell wall constructed of peptidoglycan and do not possess membrane-bound organelles. They do possess a variety of internal membrane structures that function in cell division, respiration, and photosynthesis. Bacterial flagella are different in structure from those of eukaryotes. The flagellum is composed of a single strand of the protein flagellin, which rotates about its base.

Facultative anaerobes can obtain energy by either fermentation or respiration. Obligate anaerobes can carry on fermentation only and are poisoned by oxygen. Obligate aerobes cannot survive for very long without oxygen, just like other respiring organisms. There are four nutritional categories within the prokaryotes:

photoautotrophs, photoheterotrophs, chemoautotrophs, and chemoheterotrophs. Photoautotrophs use light as their source of energy and carbon dioxide as their source of carbon. Some carry out photosynthesis in the same manner as plants, whereas others use hydrogen sulfide, rather than water, as an electron donor. Photoheterotrophs obtain energy from light, but they must have carbon in the form of organic compounds synthesized by other organisms.

Chemoautotrophs synthesize organic compounds from carbon dioxide, using chemical energy liberated from the oxidation of inorganic substances, such as ammonia or hydrogen sulfide. Chemoheterotrophs obtain both energy and carbon in the form of organic compounds.

Only monera are capable of carrying out nitrogen fixation.

Gram-positive bacteria contain significant amounts of peptidoglycan in their cell walls and stain blue with the Gram stain. Gram-negative bacteria stain pink with the Gram stain, because their cell walls contain less peptidoglycan. Bacteria may be categorized by the shape and arrangement of the cells and by the presence or absence of various structures.

4. Define obligate anaerobe, facultative anaerobe, obligate aerobe, photoautotroph, photoheterotroph, chemoautotroph, and chemoheterotroph.

5. Discuss the metabolic diversity of the prokaryotes in terms of the theories of life's origins given in Chapter 31.

6. Cite the differences between prokaryotic and eukaryotic cells; contrast eukaryotic and prokaryotic flagella.

7. Answer: What is peptidoglycan, and how is its presence or absence detected in laboratory studies of bacteria?

33-3 Kingdom Monera, Phylum Bacteria, Gliding Bacteria / Spirochetes / Gram-Negative Rods / Gram-Positive Rods / Actinomycetes / Rickettsias / Mycoplasmas / Methanogens / Other Groups Gliding bacteria are short rods that move by a gliding mechanism which is not fully understood. They produce spores in characteristic fruiting bodies. Spirochetes are motile by means of an axial filament, which runs the length of the cell. The cells are spirally coiled rods. The organism that causes syphillis is a spirochete.

Curved and spiral bacteria are all Gram-negative and diverse in properties. Gram-negative rods are an extremely diverse group, including aerobic, facultatively anaerobic, and aerobic forms. Some are nitrogen fixers. Escherichia coli is a Gram-negative rod, as are a number of disease-producing bacteria. Gram-positive bacteria fall into two groups, depending on whether they produce spores. Spore producers include Clostridium and Bacillus, whereas Lactobacillus does not form spores.

Actinomycetes resemble fungi in their mycelial growth habit. Many of these organisms are the source of antibiotics used in medicine. Rickettsias are obligate intracellular parasites, several of which produce serious human diseases, such as Rocky Mountain spotted fever. Mycoplasmas are the smallest cellular organisms. They do not have cell walls. Most are parasitic. Methanogens are unique organisms that have been placed in their own kingdom by some biologists. All are obligate anaerobes, and they produce all of the methane present in the earth's atmosphere today.

Other important groups of bacteria include the photosynthetic green sulfur, purple sulfur, and purple nonsulfur bacteria that were discussed earlier and Gram-positive and Gram-negative cocci.

8. Name one distinguishing characteristic and one example of at least five of the groups of bacteria discussed in this portion of the chapter.

33-4 Bacteria and Disease Many serious diseases are produced by bacteria, but the vast majority are harmless and play numerous beneficial roles in the biosphere.

The causative agent of a particular disease may be identified through the application of Koch's postulates:

> ...The organism must always be found in diseased individuals.

> ...The organism taken from the diseased individual must be grown in pure culture.

> ...Organisms from the culture must produce the disease when introduced into a suitable host.

> ...A new, pure culture of the same organism must be isolated from the new host.

The ability of an organism to produce disease is related to its invasiveness and to its toxigenecity.

9. State Koch's postulates.

10. List three ways in which bacteria perform beneficial roles in the biosphere.

33-5 <u>Kingdom Monera, Phylum Cyanobacteria</u> Cyanobacteria are often called blue-green algae, because they carry out oxygen-liberating photosynthesis, unlike other photosynthetic prokaryotes. Many can fix nitrogen. They are single-celled, colonial, or filamentous, and they exhibit a variety of cell types including vegetative cells, heterocysts, and spores. (Heterocysts are cells specialized for nitrogen fixation.) **None possess true sexuality.**

11. Cite the characteristics of cyanobacteria.

12. Explain why the cyanobacteria are the most nutritionally independent prokaryotes.

33-6 <u>The Kingdom Protista</u> The kingdom Protista is composed of unicellular or colonial eukaryotes. They are extremely diverse in terms of nutrition and mode of life. Most have contractile vacuoles that serve to remove excess water from the cells. Protists are involved in a number of symbiotic associations with other organisms. It is believed that the multicellular kingdoms arose by different means from various lines of protists.

13. Describe the function of the contractile vacuole, and cite experimental evidence in support of your answer.

14. Account for the fact that protists are classified in a variety of ways, depending on the person doing the classification.

15. Explain what distinguishes protists from fungi, plants, and animals.

33-7 <u>Kingdom Protista, Phylum Mastigophora</u> The Mastigophora are characterized by the possession of flagella. They are probably the most primitive protists. This group exhibits a variety of nutritional modes, including organisms like <u>Euglena</u>, which can switch from an autotrophic to a heterotrophic mode of nutrition. Many heterotrophic flagellates live symbiotically in the bodies of other animals--including humans. Some forms (such as <u>Volvox</u>) are strict autotrophs and may be classified with the plants.

Flagellates are said to bridge the gap between multicellular and unicellular organisms, because certain colonial forms, like <u>Volvox</u>, can be seen to be aggregates of individual cells resembling free-living, unicellular forms. Intermediates between these two extremes are also known.

16. List the characteristics of the phylum Mastigophora and cite two organisms that belong in this group.

33-8 <u>Kingdom Protista, Phylum Sarcodina</u> These organisms are motile by means of pseudopodia, which are also used in the capture of prey. The amoeba is the most commonly recognized member of the group. The Sarcodina probably **evolved from flagellate** ancestors. The Sarcodina may be naked or covered with a shell made of sand grains or of secreted material. In foraminiferans, the shell is of calcium carbonate, containing pores through which the pseudopods are extruded. Radiolarians have shells of silica. Radiolarians are exclusively marine; **foraminiferans are** mostly marine, with a few freshwater species. The latter are important fossils, used in the location of oil deposits. Heliozoans are found in fresh water. They are shell-less, but they use their long pseudopods to trap smaller organisms, as do the foraminiferans and radiolarians.

17. Distinguish among the major groups of Sarcodina, and explain how the members of this phylum differ from other protists.

33-9 Kingdom Protista, Phylum Sporozoa The sporozoans are all parasitic protists that generally have an amoeboid form. Some produce spores. Like many other parasites, they have elaborate life cycles, sometimes involving several hosts. The sporozoan Plasmodium vivax causes malaria in humans and is transmitted by mosquitoes.

18. Distinguish the sporozoans from other protists.

19. Outline the life cycle of the malaria parasite.

33-10 Kingdom Protista, Phylum Ciliophora / Cytoplasmic Organization in the Ciliates Ciliates are all animal-like in nutrition and move by means of cilia. They include the most structurally complex of all eukaryotic cells. All ciliates have two types of nuclei. The micronuclei are involved in reproduction and are essential carriers of the genetic information for the reproductive process, whereas the macronucleus is involved in the expression of the genes.

Cell division in Paramecium, a typical ciliate, consists of mitotic division of the micronuclei, with simple fission of the macronucleus. Conjugation, a form of sexual recombination in ciliates, occurs in the following manner. Each micronucleus undergoes mitosis to produce eight haploid nuclei. One of these divides by mitosis to produce two "gametes." The remaining micronuclei break up. The macronucleus also breaks up. Conjugating paramecia reciprocally exchange micronuclei, one remaining in each cell, the others migrating to the opposite "parent." Note that this form of sexual recombination is not associated with the reproductive process! The two "parents" exchange genetic material without accompanying cell division.

Certain ciliates have elaborately developed organelles, including structures analogous to legs, muscles, skeletons, and other structures typical of animals. The most highly specialized of these live as symbionts in the rumens of cattle and other hooved mammals.

20. Distinguish the ciliates from other groups of protists.

21. Explain in detail the process of conjugation in Paramecium.

ARTICLE RESOURCES

Viruses

   Butler, P., et al. The assembly of a virus. Scientific American, 1978,239(5), 62-70. Tobacco mosaic virus has a helical structure consisting of a single strand of RNA packed between the turns of a helical coat of protein made up 2,130 identical subunits. This helix protects the RNA from damage until it infects the host. Then the RNA is released and the viral genes are expressed by enzymes in the host. Well illustrated with diagrams and photomicrographs.

Reproduction of Viruses

Classification of Viruses

The Kingdom Monera (Prokaryotes)

Prokaryotes vs. Eukaryotes

Metabolic Diversity in the Kingdom Monera

Some Other Distinguishing Characteristics of the Bacteria

   Costerton, J. W., G. G. Gaesey, and K.-J Cheng. How bacteria stick. Scientific

American, 1978, 238(1), 86-95. Surface polysaccharides of bacteria taken from natural environments are called the glycocalyx. It is essential to the success of these bacteria in their environment. The most frequently studied bacteria were those from human teeth and lungs, from rocks in fastmoving streams and from the bovine intestine. Pioneer studies of glycocalyx formation were done on Steptococcus mutans, an organism colonizing human teeth, and Streptococcus salivarus, which colonizes the human gum. These are the organisms that free glucans and glucosyl-transferase to build a polysaccharide trap for a mixed population of bacteria to form plaque. Well illustrated.

Kingdom Monera, Phylum Bacteria, Gliding Bacteria

Kingdom Monera, Phylum Bacteria, Spirochetes

Kingdom Monera, Phylum Bacteria, Curved and Spiral Bacteria

Kingdom Monera, Phylum Bacteria, Gram-Negative Rods

Kingdom Monera, Phylum Bacteria, Gram-Positive Rods

Kingdom Monera, Phylum Bacteria, Actinomycetes

Aharonowitz, Yair, and Gerald Cohen. The microbiological production of pharmaceuticals. Scientific American, 1981, 245(3), 140-152. The range of organisms that make antibiotics is narrow, but the molecules involved are diverse both chemically and physiologically. Methods of manufacturing antibiotics are discussed, as are strategies for overcoming the enzyme-based inactivation of the beta-lactam antibiotics. The use of recombinant DNA is discussed. Supplemented with information charts.

Kingdom Monera, Phylum Bacteria, Rickettsias

Kingdom Monera, Phylum Bacteria, Mycoplasmas

Brill, Winston J. Agricultural microbiology. Scientific American, 1981, 245(3), 198-215. There are three main techniques that can be applied to traditional agricultural practices. Microorganisms beneficial to the plants can be grown in fermentation tanks and introduced into the soil. Individual cells can be grown in nutrient solutions to select promising strains. The third method is to engineer directly the genetics of the plant itself. Applications of these methods are discussed and illustrated.

Kingdom Monera, Phylum Bacteria, Methanogens

Kingdom Monera, Phylum Bacteria, Other Groups

Bacteria and Disease

Kingdom Monera, Phylum Cyanobacteria

The Kingdom Protista

Bamforth, Stuart S. Terrestrial protozoa. Journal of Protozoology, 1980, 27(1), 33-36. Litters and soils are specialized ecosystems that contain water in surface films and pore spaces. There are large amounts of organic material present. Extreme temperature and moisture fluctuations occur. Yet these very conditions are exploited by over 300 species of protozoa. Comparisons of the habitats available to protozoa are made.

Kingdom Protista, Phylum Mastigophera

Kingdom Protista, Phylum Sarcodina

Bynum, R. D., and R. D. Allen. Torsional movements in the amoeba. Chaos carolinensis, suggest a helical cytoskeletal organization. Journal of Protozoology, 1980, 27(4), 420-423. Amoeboid movement is discussed. It requires an intact gel cytoskeleton. A selection of amoebae was made of organisms that formed a single pseudopod, which form they retained for periods of time ranging from a few minutes to several hours before resuming their polypodial form. Torsion seems to be a normal event in amoeboid movement. Excellent photographs showing torsional movement.

Kingdom Protista, Phylum Sporozoa

Kingdom Protista, Phylum Ciliophora

Hinrichsen, Robert D. An analysis of temperature blocks in the conjugation sequence of Paramecium tetraurelia. Journal of Protozoology, 1981, 28(4), 417-423.
Temperatures ranging from 19°C-37°C were used to block the conjugation sequence of
this paramecium. Temperatures above 27°C reduced pair formation and nuclear exchange. At 37°C conjugation was completely inhibited. Temperatures below 19°C
inhibited pair formation. Below 19°C the cells were unable to fuse membranes in
the holdfast region.

Cytoplasmic Organization in the Ciliates

ESSAY QUESTIONS

1. Explain why arboviruses are important to humans.

    Many of these viruses are the agents of serious diseases, such as encephalitis.

2. Contrast eukaryotic and prokaryotic cells.

    Prokaryotes lack a membrane-bound nucleus, organelles, and mitosis, and they
    have cell walls constructed of peptidoglycan. The flagella of prokaryotes are
    different in both composition and behavior from those of eukaryotes. The meta-
    bolic diversity of prokaryotes is much greater than that of eukaryotes.

3. Explain why a shot of penicillin is good for a Streptococcus infection but useless
    in the case of a viral infection. (Penicillin exerts its effect by interfering
    with the synthesis of cell wall components.)

    Viruses do not have cell walls and are therefore unaffected by the antibiotic.

4. Explain the process of conjugation. Is it a form of sexual reproduction? Why or
    why not?

    Two conjugating paramecia line up with their oral grooves in contact. The
    macronucleus breaks down and disappears. The micronucleus divides by meiosis
    to produce four haploid nuclei. All but one of these disintegrate, with the
    remaining one dividing mitotically to yield two "gametes." One of these re-
    mains with each cell and the other is exchanged. Conjugation is not a repro-
    ductive process, because it is not associated with cell division. It is, how-
    ever, a sexual process, in that reshuffling of the genetic material between two
    individuals occurs.

5. Certain species of cyanobacteria are often a problem for municipal water-treatment
    facilities, as the cyanobacteria clog pipes, drains, filter intakes, and so on.
    All of these problem organisms possess heterocysts. What is the connection here?

    Heterocysts are characteristic of nitrogen-fixing cyanobacteria. These organ-
    isms are nutritionally independent, requiring only light, water, carbon dioxide,
    atmospheric nitrogen, and a few minerals for growth. They are thus among the
    few organisms that can grow in the nutritionally depleted water that results
    from municipal purification processes. Only the absence of light or the pre-
    sence of chlorine keeps them at bay.

6. Explain why encephalitis, typhus, malaria, and sleeping sickness are a much greater
    problem in underdeveloped nations than they are in the United States.

    Each of these diseases is transmitted by an insect or other arthropod. Control
    of these pests is much better in the United States because of the efforts of
    health departments, spraying of pesticides, better education, and the like.
    Also, the majority of the United States lacks a tropical climate that can
    support the arthropod carrier.

7. Explain why flagellates may be thought of as the link between unicellular and
    multicellular organisms, as well as between prokaryotes and eukaryotes.

In the first case, there exists a series of flagellate forms that progresses from simple, unicellular organisms (similar to Euglena) to complex, almost truly multicellular colonies like Volvox, with appropriate intermediates such as Gonium and Pandorina. In the second case, it appears that the flagellates are the most primitive and fundamental of the eukaryotic phyla. If we accept the hypothesis that eukaryotes evolved from prokaryotes, it follows that the primitive flagellates bridge the gap. Further, the other protist phyla can be thought of as derived from the Mastigophora through the loss of flagella, in the case of Sarcodina and Sporozoa, or the elaboration of the flagella into the locomotor organelles (cilia) of ciliates.

TEST SET 33-A

Name _____   Section _____

Choose the best answer to each of the following questions, and write the appropriate letter in the space provided.

Ans: d
p. 812

_____ 1. Individual viruses are: (a) noninfective; (b) composed of protein surrounded by nucleic acid; (c) taken up by plant cells via endocytosis; (d) found to contain DNA or RNA, with a protein coat.

Ans: d
p. 814

_____ 2. Viruses are classified on the basis of any of the following criteria except: (a) structure of the capsid; (b) nature of the nucleic acid; (c) distribution in the host organisms; (d) presence or absence of an envelope.

Ans: c
p. 811

_____ 3. Viruses are incapable of all of the following except: (a) energy metabolism; (b) protein metabolism; (c) mutation; (d) self-replication outside the host.

Ans: a
p. 814

_____ 4. An important difference between plant and animal viruses is that: (a) plant viruses are introduced into the cell by insects, whereas animal viruses enter by endocytosis; (b) plant viruses usually kill the plant cells, whereas animal viruses do not; (c) viruses leave plant cells via exocytosis and cause lysis of animal cells; (d) plant viruses are frequently surrounded by a membrane derived from the host, whereas animal viruses are not.

Ans: b
p. 811

_____ 5. Viruses are: (a) unicellular; (b) acellular; (c) eukaryotic; (d) prokaryotic.

Ans: a
p. 821

_____ 6. Peptidoglycan: (a) is a cell wall constituent found only in prokaryotes; (b) comprises the bacterial flagellum; (c) is found only in Gram-negative bacteria; (d) is synthesized by a variety of protists.

Ans: b
p. 819

_____ 7. Organisms with flagella that spin around a base would be classified as: (a) cyanobacteria; (b) bacteria; (c) mastigophora; (d) foraminifera.

Ans: d
p. 819

_____ 8. Facultative anaerobes: (a) are incapable of fermentation; (b) are incapable of respiration; (c) are incapable of respiration but capable of fermentation; (d) are capable of both fermentation and respiration.

Ans: d
pp. 819,
    821

_____ 9. An organism that obtains energy from the oxidation of nitrite to nitrate and is a spherical cell would be called a: (a) heterotrophic bacillus; (b) heterotrophic coccus; (c) chemoautotrophic bacillus; (d) chemoautotrophic coccus.

Ans: c
p. 819

_____ 10. Purple sulfur bacteria liberate _____ into the environment. (a) oxygen; (b) carbon dioxide; (c) sulfur; (d) hydrogen sulfide.

Ans: a
p. 820

_____ 11. _____ are often important as a source of antibiotics. (a) Actinomycetes; (b) Methanogens; (c) Rickettsias; (d) Gram-positive rods.

Ans: b
p. 836

_____ 12. The primary function of the macronucleus is: (a) unknown; (b) to carry DNA that is translated into proteins; (c) to participate in conjugation; (d) similar to that of the micronucleus.

Ans: a   _____   13.   Which of the following phyla contains the most primitive
p. 831   eukaryotes?   (a) Mastigophora; (b) Sarcodina; (c) Sporozoa;
                                    (d) Ciliophora.

Ans: a   _____   14.   During conjugation: (a) micronuclei are exchanged by two
p. 838   paramecia; (b) cell division occurs in Paramecium; (c) mitosis
                                    leads to the production of gametes; (d) haploid paramecia become
                                    diploid for a brief period, then resume the haploid state after
                                    an exchange of genetic material.

Ans: c   _____   15.   Exxon would be most likely to be interested in the study of:
p. 834   (a) cyanobacteria; (b) ciliates; (c) foraminiferans; (d) Euglena.

TEST SET 33-B

Name _____   Section _____

Choose the best answer to each of the following questions, and write the appropriate letter in the space provided.

Ans: c
p. 819
_____ 1. A new organism has been discovered that is capable of using sulfate as an electron acceptor for respiratory electron transport. Such an organism is similar in this regard to: (a) Rhizobium; (b) Euglena; (c) denitrifying bacteria; (d) cyanobacteria.

Ans: b
p. 814
_____ 2. The outer protein coat of a virus: (a) never exhibits binal structure; (b) aids in determination of the specificity of the virus for a host; (c) is rarely involved in the invasion of bacterial cells by viral DNA; (d) is one giant protein molecule.

Ans: b
p. 814
_____ 3. The mechanism of entry of viruses into animal cells differs from that of bacterial and plant cells, largely because: (a) animal cells have a protective membrane; (b) bacterial and plant cells have a cell wall; (c) the capsids of animal viruses are not involved in the entry of the virus; (d) plant viruses are typically covered by an envelope of material derived from the host.

Ans: b
p. 813
_____ 4. Viral proteins are synthesized from: (a) instructions already present in the host's genome; (b) instructions from the viral genome; (c) capsid proteins; (d) some instructions from the host's genome and some instructions from the viral genome.

Ans: b
p. 814
_____ 5. The most ancient of the following organisms is a: (a) sporozoan; (b) bacterium; (c) cyanobacterium; (d) flagellate.

Ans: c
p. 820
_____ 6. Based on the information given in Chapter 31, which of the following are likely to have appeared on earth first? (a) chemoautotrophs; (b) photoautotrophs; (c) chemoheterotrophs; (d) photoheterotrophs.

Ans: b
p. 833
_____ 7. Trypanosoma may be controlled by: (a) administration of antibiotics; (b) vaccination of susceptible populations; (c) eradication of mosquitoes; (d) better hygiene.

Ans: c
p. 831
_____ 8. An organism that can readily shift between the autotrophic and heterotrophic modes of nutrition is probably closely related to: (a) Volvox; (b) Bacillus; (c) Euglena; (d) Bdellovibrio.

Ans: b
p. 835
_____ 9. A term that applies equally well to Treponema and Plasmodium is: (a) eukaryote; (b) parasite; (c) protist; (d) flagellate.

Ans: a
p. 832
_____ 10. Highly complex unicellular eukaryotes found in the gut of certain termites belong to the phylum: (a) Mastigophora; (b) Sarcodina; (c) Sporozoa; (d) Ciliophora.

Ans: c
p. 835
_____ 11. Contractile vacuoles are found in all of the following groups except: (a) Mastigophora; (b) Sarcodina; (c) Sporozoa; (d) Ciliophora.

Ans: a
p. 834
_____ 12. A radiolarian is a(an): (a) marine sarcodine; (b) phytoflagellate; (c) colonial autotroph; (d) actinomycete.

Ans: b
p. 828

_____ 13. Which of the following organisms would be likely to have heterocysts? (a) <u>Paramecium</u>; (b) a cyanobacterium; (c) <u>Rhizobium</u>; (d) a mycelial spore-forming prokaryote.

Ans: a
p. 827

_____ 14. Cyanobacteria can be distinguished from other autotrophic prokaryotes by virtue of their: (a) source of hydrogen atoms for photosynthesis; (b) respiratory terminal electron acceptor; (c) ability to use nitrate as an energy source; (d) inability to carry out nitrogen fixation.

Ans: a
p. 840

_____ 15. The <u>most</u> complex unicellular eukaryotes are associated with: (a) hooved mammals; (b) the tsetse fly; (c) mosquitoes; (d) the human intestine.

CHAPTER THIRTY-FOUR
FUNGI AND PLANTS

CHAPTER SUMMARIES AND OBJECTIVES

34-1 <u>The Kingdom Fungi</u> Fungi are heterotrophic eukaryotes with absorptive nutrition, which may be differentiated from protists, from plants or animals, and from each other by their modes of reproduction. There are three phyla of fungi: Myxomycota, the slime molds; Mastigomycota, the water molds; and Eumycota, the "true" molds. The Eumycota are divided into four classes: Zygomycetes (conjugating fungi), Ascomycetes (sac fungi), Basidiomycetes (club fungi) and Deuteromycetes (fungi in which the mode of reproduction is unknown). The classes are distinguished on the basis of reproductive structures. The fungal body is composed of a mass of filaments, the hyphae, which together comprise the mycelium. Some fungi contain cellulose in their cell walls; others have chitin.

   1. Cite the identifying characteristics of the kingdom Fungi.

   2. List the phyla of the kingdom Fungi and the classes of Eumycota.

34-2 <u>Phylum Myxomycota, Class Myxomycetes / Phylum Myxomycota, Class Acrasiomycetes</u>
The myxomycetes may have arisen independently from the other classes of the fungi. There are two classes of Myxomycota: acellular slime molds (Myxomycetes) and cellular slime molds (Acrasiomycetes).

The feeding stage of a myxomycete is a wall-less mass of protoplasm containing many nuclei, which is called a plasmodium. The plasmodium spreads over the substrate, engulfing food. When conditions become harsh, the plasmodium may become a resistant resting place called a sclerotium, or it may form spore-bearing fruiting structures called sporangiophores. The plasmodium is diploid. Nuclei divide by meiosis during spore formation to form haploid spores. Spores germinate to produce flagellated swarmers that can either divide mitotically or function as gametes. The fusion of two swarmers restores diploidy, and the zygote divides mitotically to form a new plasmodium.

The vegetative stage of an acrasiomycete is made up of swarms of independent, isolated cells. When conditions become harsh, the cells aggregate, forming an irregular lump called the pseudoplasmodium. During the pseudoplasmodium stage, nuclei of some cells fuse and meiosis occurs; the result is the formation of a fruiting structure that releases spores.

   3. Delineate the characteristics of the Myxomycota.

   4. Describe the life cycles of a typical myxomycete (such as <u>Physarum</u>) and a typical acrasiomycete (such as <u>Dictyostelium</u>).

34-3 Phylum Mastigomycota The water molds and their terrestrial relatives comprise this phylum. In the Mastigomycota, the hyphae are not separated into individual cells by cross walls, and they are the only fungi that possess flagellated reproductive cells. The life cycle of a typical water mold is typified by Allomyces. Both male and female gametes are motile, the former being attracted to the latter by pheromones. When two gametes come together, the diploid sporophyte is formed; this divides by mitosis to produce numerous zoospores. Alternatively, haploid zoospores may be produced by meiosis in sporangia. Diploid zoospores give rise to new sporophytes. Haploid zoospores develop into gametophytes, which release both male and female gametes, completing the cycle.

5. List the distinguishing characteristics of the Mastigomycota and explain the life cycle of a typical water mold (such as Allomyces).

34-4 Phylum Eumycota, Class Zygomycetes The Zygomycetes, or conjugating fungi, may be distinguished as follows: hyphae are without cross walls, no motile cells are produced, and only one stage in the life cycle--the zygote--is diploid. A common example of this group is Rhizopus stolonifer, the black bread mold. The mycelium spreads over the substrate, giving rise to sporangiophores at intervals. When hyphae of two different mating strains come together, they fuse and form a thick-walled zygote. Pheromones are involved in the attraction of hyphae to each other.

6. List the distinguishing characteristics of the Zygomycetes and describe the life cycle of a typical member of this group (such as Rhizopus).

34-5 Phylum Eumycota, Class Ascomycetes / Subgroups of the Ascomycetes The Ascomycetes, or sac fungi, are distinguished by the formation of a specialized sac, or ascus, in which nuclear fusion and subsequent meiosis take place. Ascomycete hyphae have cross walls that are perforated. Euascomycetes have the asci contained within a specialized fruiting structure, whereas hemiascomycetes do not. Euascomycetes are also distinguished by heterokaryon formation, in which compatible strains fuse as if in conjugation, but the nuclei remain distinct, dividing simultaneously, until ascus formation. The formation of the ascus is a complex process that culminates in the formation of haploid ascospores.

Hemiascomycetes are small, often unicellular. The best known example is baker's or brewer's yeast, Saccharomyces cerevisiae. Individual yeast cells are haploid and may divide by fission or through the formation of buds. Conjugation between compatible cells occasionally occurs and is followed by nuclear fusion, meiosis, and mitosis. The entire structure thus is an ascus, and there is no heterokaryon stage. Euascomycetes may form asexual conidiospores. Common genera include Penicillium, Aspergillus, and Neurospora.

7. Cite the distinguishing characteristics of the Ascomycetes.

8. Distinguish between the sexual cycles of hemiascomycetes and euascomycetes, and give an example of each.

9. Explain heterokaryon formation in euascomycetes.

34-6 Phylum Eumycota, Class Basidiomycetes The Basidiomycetes produce the fruiting structures known as mushrooms and toadstools, and they include the destructive plant parasites called rusts and smuts. They are distinguished by separated hyphae and the formation of the basidium, a club-shaped structure that functions in sexual reproduction in this group as the ascus does in the Ascomycetes. All possess the heterokaryon stage. The basidiomycete fruiting structure consists of a cap, or pileus, elevated above the substrate by a stalk. On the underside of the pileus are numerous gills, each with many basidia along its margins.

10. Know the distinguishing characteristics of the Basidiomycetes.

11. Cite one example of a parasitic basidiomycete.

34-7 Phylum Eumycota, Class Deuteromycetes (Fungi Imperfecti) / Lichens The Deuteromycetes are fungi that do not produce sexual structures which would make it possible to assign them to one of the other groups. Some may have lost these stages in evolution, others may reproduce sexually so rarely that the process has not been observed. When a fungus assigned to this group is found to have a sexual stage, which occasionally happens, it is reclassified in the appropriate group.

Lichens are symbiotic associations of fungi and algae. The fungus is usually an ascomycete; the photosynthetic partner may be either a green alga or a cyanobacterium. Lichens are able to survive in some of earth's harshest environments. The fungus derives photosynthetically fixed carbon for its nourishment from the alga, while the alga obtains nutrients from the fungus, and the meshwork of the mycelium prevents dessication. Lichens reproduce by fragmentation, or the production of specialized fruiting structures, the soredia.

12. Explain why the Deuteromycetes cannot be regarded as a "natural" group.

13. Answer: What are lichens, and how do these organisms exemplify symbiosis?

34-8 <u>The Kingdom Plantae / Algae</u> Plants are eukaryotic photosynthetic organisms. They are the producers of about 90 percent of the world's biomass. They may be roughly separated into the algae and the terrestrial plants. The former are distinguished on the basis of their photosynthetic pigments, their food storage products, and the composition of the cell wall; many may be justifiably classified with the protists. The terrestrial plants are distinguished by their life cycles, the presence or absence of xylem, and the nature of the xylem.

Algae carry out the majority of the photosynthesis on the planet. They may be unicellular, filamentous, or multicellular, with branched filaments or leaflike structures. Their life cycles are variable, and all but the red algae produce motile, flagellated cells at some point in their life cycles.

14. Identify the major features of the plant kingdom and the characteristics by which algae and terrestrial plants are distinguished.

34-9 <u>Phylum Pyrrophyta / Phylum Chrysophyta / Phylum Phaeophyta / Phylum Rhodophyta</u>
The Pyrrophyta are called dinoflagellates, and most are unicellular forms with two flagella. One of the flagella lies in an equatorial groove, and the other lies in a longitudinal groove and extends into the surrounding medium. These organisms are mostly marine. Sometimes they reproduce in huge numbers, producing a "red tide." Often the red tide organisms produce a potent toxin that can kill fish. Many dinoflagellates are bioluminescent.

The most important members of the Chrysophyta are the diatoms. They are yellowish or brownish in color, store oils and chrysolaminarin, deposit silicon in their cell walls, and usually produce a two-part cell wall. The cell walls of diatoms bear elaborate ornamentation, upon which their classification is based. Asexual reproduction would result in the gradual diminution of half the new cells at each generation, because the two halves of the cell wall are of different sizes and each new cell receives either one or the other half. This problem is solved by the production of gametes, which shed the cell wall and fuse to form a zygote, which then constructs a completely new wall.

The brown algae, Phylum Phaeophyta, include some of the largest of algae, the kelps. They are multicellular, form branched or leaflike growths called thalli, and are exclusively marine. Alternation of the gametophyte and sporophyte phases may be isomorphic or heteromorphic. In either case, the sporophyte releases flagellated zoospores, derived by meiosis (and thus haploid), which develop into a gametophyte. The gametophyte releases male and female gametes that fuse to form a zygote, which then develops into a new sporophyte.

The red algae, or Rhodophyta, are found in marine habitats almost exclusively. Almost all are multicellular, and they derive their red color from unique pigments. They exhibit chromatic adaptation, the ability to alter the relative amounts of photosynthetic pigments to take maximum advantage of available light wavelengths. Red algae are unique in their photopigments (phycoerythrin and phycocyanin) and storage product (floridean starch), and they never produce motile, flagellated cells. Male gametes are amoeboid; female gametes are completely immobile. Various patterns of sexual reproduction are known in this group.

15. Distinguish each of the four groups of algae we have discussed in terms of

structure, mode of reproduction, and other characteristics.

34-10 <u>Phylum Chlorophyta</u>   The green algae (Chlorophyta) and certain protists are the only photosynthetic organisms with the same photopigments as terrestrial plants. All possess chlorophyll <u>a</u> and chlorophyll <u>b</u>.   Chlorophyll <u>b</u> is found in no other algae.   The carotenoids of green algae and land plants are also identical, as is the principal storage product, starch.   The green algae are extremely diverse in body form and mode of sexual reproduction.

16. Cite the distinguishing characteristics of the Chlorophyta.

34-11 <u>The Terrestrial Plants / Phylum Bryophyta / Phylum Bryophyta, Class Musci / Phylum Bryophyta, Class Hepaticae / Phylum Bryophyta, Class Anthocerotae</u>   Green multi-cellular land plants belong to two phyla: Bryophyta and Tracheophyta.   Most of the characteristics that distinguish these two groups from the algae are adaptations to a terrestrial habitat.   The two phyla are probably derived from different ancestral groups.

Bryophytes exhibit heteromorphic alternation of generations.   The sporophyte begins development in the archegonium, which is a multicellular structure that protects the female gamete and nourishes the sporophyte.   Male antheridia release motile sperm that swim to the archegonium to fertilize the female gamete.

Mosses are the most familiar bryophytes.   When a typical moss spore germinates, it produces a filamentous structure called the protonema.   Some filaments are photosynthetic;   others, rhizoids, anchor the plant to the substrate and absorb minerals and water.   Some filaments form buds from which the gametophyte develops.   Archegonia and antheridia develop at the tips of the gametophytes.   The sporophyte develops at the tip of the gametophyte stalk after fertilization.   Meiosis within the tip of the sporophyte leads to the formation of haploid spores, which are dispersed when conditions are favorable for their germination.   A few mosses lack this pattern of sporophyte development.

Liverworts do not produce a protonema, have variable gametophyte structures, and produce very simple spore capsules.   Spore dissemination occurs by a variety of mechanisms.   Hornworts resemble liverworts and have very simple gametophytes.   The archegonia are embedded in gametophyte tissue, and the sporophytes are capable of indefinite growth.

17. Cite the distinguishing characteristics of the Phylum Bryophyta and the traits that characterize the classes Musci, Hepaticae and Anthocerotae.

18. Explain the life cycle of a typical moss and relate this pattern to survival on land.

34-12 <u>Phylum Tracheophyta / Phylum Tracheophyta, Subphylum Psilopsida / Leaves of Tracheophytes / Phylum Tracheophyta, Subphyla Lycopsida and Sphenopsida</u>   Vascular plants are a large and diverse group.   They all probably descended from a single common ancestor in which the sporophyte generation possessed a specialized water-conducting cell, the tracheid.

Psilopsids are now extinct, but they dominated the landscape during the Paleozoic Era.   Early psilopsids had a simple vascular system of tracheids and phloem and lacked true leaves and roots.   Fossil sporophytes, but not gametophytes, have been found.

A leaf is a flattened photosynthetic structure emerging from a main axis or stem and possessing vascular tissue.   There are two different types.   Megaphylls are found in Subphylum Pteropsida and are thought to have evolved from the flattening of a branching stem system.   Megaphylls have a vascular system that creates a "leaf gap" where the leaves depart from the main stem.   Microphylls probably arose as the result of development of increasingly complex vascular tissues within scales along the stem.   The vascular system of a microphyll does not create leaf gaps. Microphylls are found in Subphyla Lycopsida and Sphenopsida.

Lycopsida (club mosses) and Sphenopsida (horsetails) have true roots, microphylls, and a life cycle resembling that of the ferns.   The two groups are distinguished on the basis of leaf arrangement, the location of the spore-bearing structures, and the distribution of the meristematic tissue. These groups were much more important components of the vegetation in Carboniferous times than they are today.

19. Differentiate between a megaphyll and a microphyll. In which groups is each type found?

20. Distinguish between Sphenopsida and Lycopsida.

34-13 Phylum Tracheophyta, Subphylum Pteropsida, Class Filicinae / Phylum Tracheophyta, Subphylum Pteropsida, Class Gymnospermae / Fossil Gymnosperms The ferns (Filicinae) undergo heteromorphic alternation of generations, with independent gametophytes and sporophytes. Fern fronds are megaphylls, with spores being produced on the undersurfaces, sometimes within sori. Fossil ferns show characteristics that resemble other tracheophyte subphyla. Ferns rarely produce a cambium.

Gymnosperms include the pines, firs, and their relatives. All produce true seeds. The formation of pollen grains (male gametophytes) frees the plant from dependence on water as a medium for fertilization. The female gametophyte is formed within an ovule, bearing an outer integument which will later form the seed coat. Pollination occurs through an opening in the integument called the micropyle. All gymnosperms have sporophytes in which the sexes are on separate plants. The seed of gymnosperms is not enclosed within a fruit but is borne on the upper surface of the sporophyll. All gymnosperms possess tracheids as the sole water-conducting elements of the xylem.

Certain fossil gymnosperms have woody tissue composed of tracheids, but they share other characteristics with ferns and psilopsids. Several lines of gymnosperms had developed by Carboniferous times, and this group at present is either at peak or still emerging.

21. List the distinguishing characteristics of ferns and gymnosperms and explain the typical life cycle pattern of each group.

34-14 Phylum Tracheophyta, Subphylum Pteropsida, Class Angiospermae / Angiosperm Gametophytes / Double Fertilization / Types of Flowers / Subclasses of the Class Angiospermae The class Angiospermae is composed of the flowering plants. The reproductive structure of angiosperms is the flower. The leaves bearing megasporangia are called carpels; those bearing microsporangia are stamens. There are frequently specialized sterile leaves, the petals and sepals, below the sporophylls. Pollen is produced at the tips of the stamens; the carpel encloses ovules. The ovule has two integuments and is entirely enclosed in carpel tissue.

The female gametophyte of angiosperms is composed of eight haploid nuclei enclosed within one membrane. These include the egg nucleus and two synergids, two polar nuclei, and three antipodal nuclei. The microspore, contained within the pollen grain, is haploid and undergoes a single mitotic division to produce a tube nucleus and a generative nucleus. When the pollen grain contacts the stigma of an appropriate female flower, it germinates and a pollen tube grows toward the ovule. The generative nucleus divides to produce two sperm nuclei, which enter the cytoplasm of the female gametophyte.

One of the sperm nuclei unites with the egg nucleus to produce a diploid zygote, while the other sperm nucleus unites with the two polar nuclei to form a triploid nucleus. The triploid nucleus undergoes mitotic development to yield a specialized nutritive tissue, the endosperm, while the zygote develops into the embryo. The remaining gametophyte nuclei degenerate. The endosperm accumulates food while the embryo differentiates into root, shoot, and cotyledons. The integuments of the ovule develop into the seed coat and fruit wall. The double fertilization process is found in all angiosperms and only in angiosperms. Other characteristics of the group include the presence of vessels in the xylem, the presence of companion cells in the phloem, a carpel that participates in fruit formation, and specialized sterile leaves associated with the sporophylls.

Some flowers bear both male and female organs and are said to be "perfect flowers." Imperfect flowers bear either functional carpels or functional stamens, but not both. In plants with imperfect flowers, a given plant may produce flowers of only one sex (monoecious), or both sexes may be present on the same plant (dioecious). Primitive flowers had multiple parts spirally arranged, whereas more modern types exhibit reduction in the number of parts, differentiation of petals and sepals, stabilization of a fixed number of sterile parts, arrangement in whorls, and a shift from radially to bilaterally symmetrical flowers, often with fusion of floral parts.

The angiosperms are divided into two subclasses. Monocots have a single cotyledon, parallel venation in the leaves, little or no cambium, vascular bundles scattered throughout the stem, roots with rings of vascular bundles surrounding pith, and floral parts in multiples of three. Dicots have two cotyledons, netlike leaf venation, cambium, vascular bundles arranged in a ring in the stem, roots with a central core of xylem surrounded by phloem, and floral parts in multiples of four or five.

22. Answer: What are the distinguishing characteristics of angiosperms?

23. Name all the parts of an "ideal" flower.

24. Explain the process of double fertilization, including the fate of each of the products of nuclear fusion.

25. Differentiate between monocots and dicots.

26. List four trends in the evolution of flowers.

ARTICLE RESOURCES

The Kingdom Fungi

Phylum Myxomycota

Lee, Douglas. Slime mold, the fungus that walks. National Geographic, 1981, 160(1), 131-136. Beautiful colored photographic essay portraying many Myxomycetes. Most are shown in their native habitat. They live in tropical and temperate climates and are widely distributed.

Phylum Myxomycota, Class Myxomycetes

Bonner, J. T., T. A. Davidowski, W. L. Hsu, et al. The role of surface water and light on differentiation in the cellular slime molds. Differentiation, 1982, 21(2), 123-126. Whether or not light is in itself a stimulus for the transition from the migration phase to the final fruiting phase is a question that has puzzled mycologists for nearly forty years. If the tip of the cell mass touches the surface of the water, it will migrate. If the tip rises above the surface, the cell mass will go into final differentiation and will fruit.

Muller-Haeckel, Agnes, and Ludmila Maranova. Freshwater hyphomycetes in brackish and sea water. Botanica Marina, 1979, 22, 421-424. Some freshwater hyphomycetes are able to survive, grow, and sporulate in brackish water and even in sea water. This is unusual. Experiments involving the fate of hyphomycetes living on and in floating decayed leaves were investigated near the Angeran River and at ten places in its estuary and the Bothnian Sea in Sweden. Mycelial growth and spore germination were possible at higher percentages of sea water than was sporulation.

Phylum Myxomycota, Class Acrasiomycetes

Phylum Mastigomycota

Phylum Eumycota, Class Zygomycetes

Phylum Eumycota, Class Ascomycetes

Subgroups of the Ascomycetes

Phylum Eumycota, Class Basidiomycetes

Phylum Eumycota, Class Deuteromycetes (Fungi Imperfecti)

Lichens

The Kingdom Plantae

Algae

Phylum Pyrrophyta

Phylum Chrysophyta

Hoover, Richard B. Those marvelous, myriad diatoms. National Geographic, 1979, 155(6), 871-878. Many species of diatoms are photographed in color to illustrate this article on "living jewels." They are the main components of plankton. Sexual and asexual reproduction are diagrammed. More than 4,000 species are known.

Phylum Phaeophyta

Earle, Sylvia A. Undersea world of a kelp forest. National Geographic, 1980, 158(3), 410-426. A kelp forest off the coast of Santa Catalina Island near Los Angeles is the site of the University of Southern California's Marine Science Center. Elk kelp are tagged and their growth charted: 25 feet per year. A single plant can sustain as many as 500,000 small animals. Excellent color photographs illustrate the article.

Phylum Rhodophyta

Phylum Chlorophyta

The Terrestrial Plants

Marden, Luis. Bamboo, the giant grass. National Geographic, 1980, 158(4), 450-528. There are over 1,000 species of bamboo with about 50 genera. It is a beautiful grass of great versatility. It ranges in size from small grasses to giants 120 feet high and a foot in diameter. It grows from sea level to 13,000-foot mountain slopes. It grows rapidly. Some species can attain a growth of 4 feet in 24 hours! Others can grow 15 inches in 14 hours. The life cycle is shown, as are methods of cultivation. Its many uses are outlined and several species are illustrated.

Phylum Bryophyta

Phylum Bryophyta, Class Musci

Phylum Bryophyta, Class Hepaticae

Phylum Bryophyta, Class Anthocerotae

Phylum Tracheophyta

Phylum Tracheophyta, Subphylum Psilopsida

Leaves of Tracheophytes

Phylum Tracheophyta, Subphyla Lycopsida and Sphenopsida

Phylum Tracheophyta, Subphylum Pteropsida, Class Filicinae

Phylum Tracheophyta, Subphylum Pteropsida, Class Gymnospermae

Fossil Gymnosperms

Phylum Tracheophyta, Subphylum Pteropsida, Class Angiospermae

Angiosperm Gametophytes

Double Fertilization

Types of Flowers

Subclasses of the Class Angiospermae

Starbird, Ethel. The bonanza bean coffee. National Geographic, 1981, 159(3), 388-405. A history of coffee, its uses, and its impact on the economy of several nations. How the shrub grows, how the beans are harvested, and how they are prepared for roasting are shown. A map indicates the major coffee-producing nations and the major coffee-consuming ones.

Rhodes, Robert E. The incredible potato. National Geographic, 1982, 161(5), 668-693. The history of the discovery and introduction of the potato into Europe from Peru is given. The hardiness and range of this tuber--only the deserts of the world are inhospitable to its growth--are remarkable. The nutritional value and versatility of the potato are discussed. Its uses are shown diagrammatically. There are 8 species of potatoes, of which 50 varieties are grown to produce a harvest of 16 million tons in the United States alone.

ESSAY QUESTIONS

1. Some biologists prefer to separate the mastigomycetes and myxomycetes from the other fungi and to assign these two groups to the kingdom Protista. Give one reason that would justify such placement in the case of each group of organisms.

    The mastigomycetes and myxomycetes are the only members of the kingdom Fungi that produce flagellated cells. The myxomycetes do not form a mycelium, as do other fungi. Only the mastigomycetes have cellulose in their cell walls. In both cases, the diploid phase of the life cycle is predominant, where the haploid phase predominates in the Eumycota.

2. Speaking from the standpoint of evolution, what would be considered the primitive state of the fungi with regard to each of the following: hyphal cross walls, duration of heterokaryon stage, structure of the fruiting body?

    The Zygomycetes are probably the most primitive of the Eumycota, with hyphal cross walls absent, very brief heterokaryon stage, and simple fruiting structures. This is in contrast to the advanced basidiomycetes, in which the hyphae are completely divided by unperforated cross walls, the heterokaryon stage persists for a long period, and the fruiting structure is complex, consisting of the pileus, with its stalk and gills, and the basidia, wherein nuclear fusion and meiosis occur.

3. Explain why the biomass of the plant kingdom is estimated to be about ten times the biomass of all other organisms combined.

    Plants are the primary producers of organic matter in almost all food chains. The transfer of chemical energy from producer to consumer is only 2 to 20 percent efficient. Thus roughly ten times the consumer biomass must be present in the producer biomass in order for the producers to support the consumers.

4. Give reasons to support the conclusion that terrestrial plants have evolved from within the phylum Chlorophyta.

    The chlorophytes share with terrestrial plants a number of important characteristics. All chlorophytes, and no other algae, possess chlorophylls a and b. All have starch as the principal storage product, as do all terrestrial plants. The carotenoids of chlorophytes and land plants are identical.

5. Cite three trends in the evolution of the terrestrial plants that represent increasingly efficient adaptations to a terrestrial lifestyle.

These trends include (1) protection of the female gamete and developing sporophytes within specialized organs, (2) decreased dependence on water as a medium for sperm transfer, (3) elaboration of the vascular system, (4) increased efficiency of pollen transfer owing to the evolution of flowers to attract pollinators, and (5) reduction in the gametophyte stage.

6. Describe in detail the process of double fertilization, giving the source of all nuclei involved, the fates of each of these nuclei, and the products of each fertilization event.

The male gametophyte, or pollen grain, is produced in the anthers and consists of one tube nucleus and one generative nucleus. The female gametophyte is contained in the ovule and consists of the embryo sac, a single membrane in which are one egg nucleus, two synergid nuclei, two polar nuclei, and three antipodal nuclei. When the pollen grain touches the receptive surface of the stigma, it germinates and a pollen tube grows toward the ovule, under the direction of the tube nucleus. During this time, the generative nucleus divides mitotically to yield two sperm nuclei. Upon reaching the ovule, the two sperm nuclei enter the embryo sac. One sperm nucleus unites with the egg nucleus to produce the zygote, which divides by mitosis to form the embryo. The other sperm nucleus unites with the two polar nuclei, forming a triploid nucleus, which divides by mitosis to yield the endosperm. Endosperm tissue will nourish the embryo.

TEST SET 34-A

Name _____   Section _____

Choose the best answer to each of the following questions, and write the appropriate letter in the space provided.

Ans: c
p. 849

_____ 1. You have discovered an organism with heterotrophic nutrition, heterokaryon formation, and club-shaped cells that bear sexual spores. This organism would be classified in: (a) phylum Eumycota, class Myxomycetes; (b) phylum Mastigomycota; (c) phylum Eumycota, class Basidiomycetes; (d) phylum Tracheo-phyta, subphylum Pteropsida.

Ans: d
p. 849

_____ 2. Which of the following is a parasitic basidiomycete? (a) Rhizo-pus; (b) Allomyces; (c) Dictyostelium; (d) Puccinia.

Ans: d
p. 873

_____ 3. Each of the following is the site of meiosis and nuclear fusion in a fungal group except: (a) ascus; (b) basidium; (c) mastigo-mycete sporangium; (d) carpel.

Ans: a
p. 849

_____ 4. Asexual spores produced by many ascomycetes are called: (a) conidiospores; (b) ascospores; (c) zoospores; (d) auxospores.

Ans: c
p. 855

_____ 5. Diatoms are classified on the basis of the structure of their: (a) storage polysaccharides; (b) motile reproductive cells; (c) cell walls; (d) chloroplasts.

Ans: d
p. 853

_____ 6. Each of the following is a criterion for the classification of algae except: (a) nature of storage product; (b) photopigments; (c) composition of cell wall; (d) presence or absence of xylem.

Ans: c
p. 848

_____ 7. Which of the following is a hemiascomycete? (a) Rhizopus; (b) Puccinia; (c) Saccharomyces; (d) Allomyces.

Ans: a
p. 858

_____ 8. Red algae are unique in the plant kingdom in that they: (a) never produce flagellated cells; (b) possess chlorophyll a; (c) form branching filaments, or thin sheets of cells; (d) have nonmotile female gametophytes.

Ans: b
p. 852

_____ 9. Soredia are: (a) specialized filaments found in some algae; (b) reproductive structures of lichens; (c) unique to the dinoflagellates; (d) found on the underside of the pileus.

Ans: a
p. 852

_____ 10. In the symbiotic association between alga and fungus that forms a lichen, the alga: (a) is protected from desiccation by the fungus; (b) derives all of its minerals from the surrounding medium; (c) cannot live in the absence of the fungus under any circumstances; (d) has lost the ability to be photosynthetic.

Ans: b
p. 874

_____ 11. Double fertilization: (a) results in the formation of the zygote and the synergids; (b) produces the endosperm and the zygote; (c) occurs in angiosperms and gymnosperms; (d) is restricted to monocots.

Ans: b
p. 871

_____ 12. The fact that the pollen grain of some gymnosperms releases a motile sperm: (a) means that these plants must rely on free water for fertilization; (b) suggests evolutionary affinities with ferns, horsetails, and club mosses; (c) demonstrates the importance of chemicals released by the carpel; (d) suggests that they descended from a common ancestor with the bryophytes.

Ans: d
p. 871

_____ 13. Each of the following is found <u>only</u> in angiosperms <u>except</u>:
              (a) double fertilization; (b) vessels; (c) companion cells;
              (d) seeds.

Ans: a
p. 855

_____ 14. Auxospore formation in diatoms: (a) ensures that the line will
              not die out due to diminution of the cells; (b) is controlled by
              the availability of silicon; (c) is an asexual process;
              (d) causes a change in the architecture of the cell wall,
              resulting in several types of walls in a particular species of
              diatom.

Ans: c
p. 875

_____ 15. A flower in which the petals, stigma, sepals, and anthers are
              elevated above the ovary is probably a: (a) monocot; (b) monoe-
              cious flower; (c) relatively modern flower; (d) fossil specimen.

TEST SET 34-B

Name _____   Section _____

Choose the <u>best</u> answer to each of the following questions, and write the appropriate letter in the space provided.

Ans: c
p. 848

_____ 1. A fungus with perforated cross walls between the cells of the hyphae and a special sac in which meiosis occurs would be classified in: (a) phylum Mastigomycota; (b) phylum Myxomycota, class Acrasiomycetes; (c) phylum Eumycota, class Ascomycetes; (d) phylum Eumycota, class Basidiomycetes.

Ans: b
p. 856

_____ 2. Which of the following is a member of the phylum Phaeophyta? (a) <u>Ulva</u>; (b) <u>Sargassum</u>; (c) <u>Dictyostelium</u>; (d) <u>Mnium</u>.

Ans: c
p. 860

_____ 3. Each of the following is a criterion for the classification of terrestrial plants <u>except</u>: (a) nature of the life cycle; (b) presence or absence of xylem; (c) primary photosynthetic storage product; (d) nature of the xylem.

Ans: c
p. 854

_____ 4. Diatoms differ from dinoflagellates in that the <u>latter</u>: (a) have cell walls made of silica; (b) have cell walls made of chitin; (c) have cell walls made of cellulose; (d) do not have cell walls.

Ans: a
p. 846

_____ 5. The only fungi that produce flagellated cells are: (a) water molds; (b) slime molds; (c) sac fungi; (d) club fungi.

Ans: c
p. 849

_____ 6. Corn smut is caused by a parasitic: (a) zygomycete; (b) ascomycete; (c) basidiomycete; (d) deuteromycete.

Ans: d
p. 854

_____ 7. A severe toxin that can accumulate in shellfish and make them unfit for human consumption is sometimes produced by certain: (a) water molds; (b) diatoms; (c) red algae; (d) dinoflagellates.

Ans: b
p. 858

_____ 8. <u>Chondrus crispus</u> may appear bright green when growing in a tide-pool and deep red when growing at extreme depths. This is: (a) one reason why these algae are so difficult to classify; (b) the phenomenon of chromatic adaptation; (c) a color change that occurs in these algae when they produce floridean starch; (d) a result of the influence of xanthophylls on leaf coloration.

Ans: d
p. 874

_____ 9. The products of double fertilization are the _____ and a _____ nucleus. (a) endosperm, haploid; (b) seed, tube; (c) egg, generative; (d) zygote, triploid.

Ans: a
p. 867

_____ 10. Pteropsids with strobili, microphylls, and motile gametes are either: (a) sphenopsids or lycopsids; (b) gymnosperms or ferns; (c) angiosperms or ferns; (d) ferns or lycopsids.

Ans: d
p. 863

_____ 11. Which of the following is(are) characteristic of <u>all</u> vascular plants? (a) pollen; (b) endosperm; (c) vessels; (d) tracheids.

Ans: a
p. 873

_____ 12. The angiosperms: (a) have a double-walled ovule; (b) produce pollen grains that contain a single nucleus; (c) evolved during Carboniferous times; (d) share with gymnosperms the process of double fertilization.

Ans: c
p. 875

_____ 13. Each of the following is a trend in the evolution of flowers <u>except</u>: (a) fusion of floral parts; (b) reduction in the number of floral parts; (c) elevation of the ovary above the petals; (d) stabilization of the number of floral parts.

Ans: d
p. 850

_____ 14. Each of the classes of Eumycota is probably a natural group
except: (a) Zygomycetes; (b) Ascomycetes; (c) Basidiomycetes;
(d) Deuteromycetes.

Ans: b
p. 846

_____ 15. Which of the following structures is(are) produced by acrasiomycetes? (a) soredia; (b) pseudoplasmodium; (c) conidiospores;
(d) auxospores.

CHAPTER THIRTY-FIVE
ANIMALS

CHAPTER SUMMARIES AND OBJECTIVES

35-1 <u>Criteria for Classifying Animals: Symmetry / Embryology / Body Cavities</u>   Animals are heterotrophic organisms with ingestive nutrition. The change from protists to animals entailed a division of labor among cells. Animals either are sessile, extracting food from the medium, or are active hunters. The three main criteria by which animals are classified are symmetry, embryology, and body cavities.

Animals with radial symmetry are usually sessile. They have no definite front or rear. Bilaterally symmetrical animals have a definite head and tail and left and right sides. These animals usually hunt their food. These two groups constitute the Radiata and Bilateria, respectively.

Bilateria may be subdivided on the basis of embryological development. In one group, protostomes, the mouth arises near the blastopore, and cleavage is spiral and determinate. In the other group, deuterostomes, the mouth develops at a location distant from the blastopore, and cleavage is radial and indeterminate. Protostomes and deuterostomes may be subdivided on the basis of body cavities.

Some protostomes are acoelomate, without an internal cavity. Psuedocoelomate protostomes have a body cavity derived from the embryonic blastocoel. Coelomate animals have a true coelom, derived from mesoderm and lined with peritoneum. In coelomate deuterostomes, the coelom arises as an outpocketing of the gut--enterocoelous development. In coelomate protostomes, the coelom develops by splitting of masses of mesoderm--schizocoelous development.

   1. List the major distinguishing features of the animal kingdom.

   2. Explain the meaning of each of the following terms used in the classification of animals: radial symmetry, bilateral symmetry, protostome, deuterostome, determinate cleavage, indeterminate cleavage, spiral cleavage, radial cleavage, acoelomate, coelomate, pseudocoelomate, enterocoelous, schizocoelous, and coelom.

35-2 <u>Phylum Porifera / Phylum Cnidaria / Alternation of Generations in Cnidarians / Corals / Phylum Ctenophora</u>   Sponges (Porifera) probably arose independently of the rest of the animal kingdom. Sponges feed by drawing water through their porous bodies, filtering out nutrients. Flagellated cells, (choanocytes) line the body cavities and produce the water movement. Sponges have no mouth, muscles, digestive cavity, or nervous system. Specialized cells (porocytes) contain the pores through which water enters. Water passes through a layer of mesenchyme containing cells which carry out the essential functions of channeling water and capturing food.

Phylum Cnidaria consists of the jellyfish, sea anemones, and corals. The name comes from their unique cnidoblast cells, which discharge stinging structures called nematocysts that are used in capturing prey. These organisms are radially symmetrical, have a mouth and a gastrovascular cavity, and exhibit a netlike nervous system.

Cnidarians exhibit one of two body plans. The polyp is a sessile stage in which the body is a cylinder anchored at one end, with the mouth at the other end surrounded by a crown of tentacles. The medusa may be thought of as an upside-down polyp and is shaped like an umbrella with the tentacles pendant. Medusae are usually free-swimming. Some cnidarians exhibit only one of these stages during their lives, others alternate between the two stages. In the latter group, the polyp produces medusae by asexual budding, and the medusae produce gametes that are released into the water. Fertilization results in the development of a planula larva, which settles to the bottom and becomes a polyp.

In corals, the polyps are housed in a calcium carbonate cup. Coral colonies grow by budding of polyps, with new polyps forming atop the skeletons of old ones. In time, the corals produce a reef. All corals possess symbiotic dinoflagellates in their tissues, which contribute to the coral's nutrition.

The comb jellies (Ctenophora) are similar to cnidarians but lack nematocysts. They have rows of cilia on the body surface that aid in their locomotion.

3. Cite the two phyla of Radiata and their characteristics.

4. Distinguish between a polyp and a medusa, and explain how each plays a role in the life cycle of certain cnidarians.

35-3 Phylum Platyhelminthes / Flatworms and the Origins of the Animal Kingdom   The flatworms comprise the phylum Platyhelminthes. Best known are the tapeworms and flukes, but there are a number of nonparasitic species. A typical free-living flatworm is the planarian. This organism has a head, with the sensory structures concentrated at the front of the body. The intestinal tract is a blind sac with a mouth but no anus, there is no circulatory system, and the excretory system is rudimentary. Respiration occurs across the epidermis, the flattened body thus being adaptive.

Certain flatworms, the acoels, strongly resemble the planula larva of cnidarians. It has been suggested that the Bilateria have their origins in a planula larva that did not continue its development, evolving instead into an acoel. Another hypothesis suggests that the first animal was a two-layered plakula, originating independently among the protists and giving rise to both cnidarians and flatworms.

5. List the distinguishing characteristics of the flatworms.

6. Explain two hypotheses concerning the origins of the multicellular animals.

35-4 Phylum Rotifera / Phylum Nematoda   The two most important phyla of pseudocoelomate animals are the rotifers and the roundworms. Rotifers possess a corona of cilia that sets up currents to draw food into the mouth. Most are microscopic though multicellular. The digestive tract has both a mouth and an anus. Most rotifers are freshwater organisms with several unique features: the number of body cells is fixed for a given species, each species is found over the entire planet, and many are able to live on the surfaces of plants that become dried out and grow active only when rain provides a film of moisture.

Roundworms (Nematoda) are adapted for life within their food. They are abundant and found everywhere. Many are parasitic, such as Trichinella spiralis, which can be transmitted to humans through the consumption of infected, poorly cooked pork. Nematodes are distinguished from rotifers by their possession of a cuticle that covers the body surface and provides protection. The cuticle also prevents the ingestion of large meals, so nematodes must feed continuously.

7. List the distinguishing characteristics of the Rotifera and the Nematoda.

35-5 Phylum Annelida   The segmented worms (Annelida) are related to Mollusca and Arthropoda. Each is coelomate, all are protostomes, and annelids and mollusks both possess the trochophore larva. Arthropods do not have a trochophore, but other evidence suggests they were derived from the annelids. Most of the distinctive features of annelids are related to improvements in locomotory ability. The fluid-filled coelom is separated into chambers by the septa between segments. Each chamber can contract and expand independently, owing to the presence of separate sets of muscles and nerves in each segment. This enables the organism to move efficiently through soil or mud. Most annelids are marine, with lateral appendages on each segment. These forms are quite diverse. Some annelids are parasitic.

8. Explain why annelids, mollusks, and arthropods are thought to be closely related.

9. Answer: How does the architecture of the annelid body allow for improved loco-motion, compared with the pseudocoelomates and flatworms?

35-6 Phylum Mollusca / Classes Amphineura and Monoplacophora / Classes Bivalvia, Gastro-poda, and Cephalopoda Mollusks include such animals as snails, clams, and squid. They are characterized by a muscular foot, mantle, ctenidia, radula, and shell. Not all mollusks exhibit all these features. The various classes of mollusks are differentiated on the basis of their body plans, which are modifications for their particular lifestyles.

Chitons (class Amphineura) are adapted to life on rocky surfaces. They are the most primitive mollusks. The gills and shell plates of chitons are repeated down the length of their bodies. In Neopilina (class Monoplacophora), until recently known only as a fossil form, muscles, excretory organs and other structures are re-peated down the length of the body. The relationship between mollusks and annelids is still being debated.

Clams, oysters, and scallops constitute the class Bivalvia, and are characterized by reduction in the head and a two-part shell.

Gastropods, commonly called snails, exhibit torsion. The gastropod larva is bi-laterally symmetrical, but it undergoes a twisting of the body which brings the internal organs into position above the head. From this position, the vital parts can be quickly retracted into the shell when danger threatens. The advantages this confers can readily be appreciated when one considers that the gastropods are the most diverse and successful mollusks.

Cephalopods, the octopus and squid, have the foot modified into manipulative tenta-cles. They have a well-developed head and a large brain and are probably the most skillful and intelligent invertebrates.

10. List the features of a typical mollusk, and be able to differentiate among the molluscan classes.

11. Discuss the adaptive value of each of the basic molluscan body plans.

35-7 Phylum Arthropoda / Diversification of Arthropods / Ancient Arthropods and Modern Descendants / Phylum Arthropoda: Arachnids, Crustaceans, and Myriapods / Phylum Arthropoda, Class Insecta The arthropods, including spiders, insects, crustaceans, and related types, make up one of the most successful phyla. They are character-ized by a protective exoskeleton. The exoskeleton covers the entire body, inclu-ding the numerous jointed appendages, which are often specialized for various functions. One disadvantage to the exoskeleton is the necessity of periodically replacing it.

The exoskeleton confers numerous advantages on the arthropods. It functions equally well in water or on land, and it prevents desiccation. Thus arthropods have numerous terrestrial representatives, most notably the insects. The insects constitute the majority of arthropods, and indeed of all animal species. The ancient origin of insects gave them an early start in exploiting the newly evolved land vegetation. This, coupled with the evolution of flight, has contributed to their spectacular success.

That arthropods evolved from annelids is supported by considerable evidence. The onychophorans, wormlike forms with segmented legs but internal structures resembling annelids, suggest what forms intermediate between ancient annelids and primitive arthropods might have looked like. Trilobites and horseshoe crabs are ancient arthropod groups; the latter are still alive today.

Arachnids are terrestrial arthropods with six pairs of appendages, the first two of which are adapted for feeding. The body is divided into the prosoma and abdomen. Examples are spiders, scorpions, mites, and ticks. Crustaceans are primarily marine forms with two pairs of antennae. They include copepods, shrimp, crayfish, lobsters, crabs, barnacles, and sowbugs. Myriapods include four groups of terrestrial arthropods with a distinct head, a wormlike, segmented body, and many legs. Included are centipedes, millipedes, pauropods, and symphylans, the latter being the probable ancestors of insects.

Insects (class Insecta) have three basic body parts: the head, thorax, and abdomen. All have one pair of antennae on the head, three pairs of legs on the thorax, and breathing structures called tracheae. Some adult insects can fly. There are almost one million species of insects, which may be divided into four groups (evolutionary grades). The apterogytes are wingless and primitive, and there are three grades of pterygotes (winged) forms. Paleopterous insects cannot fold the wings over the back; neopterous insects, in which the wings can be so folded, are of two types: those with incomplete metamorphosis and those with complete metamorphosis.

12. Describe each of the arthropod classes and state the characteristics shared by all members of the phylum Arthropoda.

13. Explain why insects have been so spectacularly successful.

14. Describe the four evolutionary grades of insects, giving an example of each type.

35-8 The Lophophorate Animals / Phylum Echinodermata / Classes of the Echinoderms / Phylum Hemichordata   Lophophorate animals include three phyla of coelomate organisms that feed by means of a lophophore. This structure is a U-shaped organ bearing many ciliated tentacles. Examples include brachiopods, or lamp shells, which superficially resemble mollusks, and bryozoans, or moss animals, which superficially resemble corals. The phyla so far considered are all protostomes. Deuterostomes include echinoderms, chordates, and a few other phyla. Many produce a characteristic larva, the tornaria.

Phylum Echinodermata includes sea stars, sea urchins, sea cucumbers, crinoids, and related forms. They are very different from other animals. All have a water-vascular system that is used primarily in locomotion and food capture. The tube feet of echinoderms operate like suction cups to grasp prey, or they may serve as organs of touch or movement or as sites of gas exchange. Echinoderms have an internal skeleton of limy plates, often with spiny projections affording protection. All are secondarily radially symmetrical, beginning as bilaterally symmetrical larvae.

There are five major living classes of echinoderms. These are Asteroidea (sea stars), Ophiuroidea (brittle stars), Echinoidea (sea urchins), Holothuroidea (sea cucumbers), and Crinoidea (sea lilies). Each group shows adaptations of the basic echinoderm body plan to meet the demands of different lifestyles.

Acorn worms (phylum Hemichordata) are burrowing marine creatures that were once thought to be related to the chordates.

15. Explain why echinoderms and chordates (including ourselves) are thought to be closely related.

16. List the distinctive features of the echinoderms.

35-9 Phylum Chordata / Subphylum Vertebrata / Class Agnatha / Class Placodermi / Class Chondrichthyes / Class Osteichthyes   The chordates (phylum Chordata) include the vertebrates and two subphyla of invertebrates. All share several features: dorsal, hollow nerve cord; gill slits; a notochord. All are enterocoelous deuterostomes and are bilaterally symmetrical. The invertebrate chordates include tunicates (subphylum Urochordata) and lancets (subphylum Cephalochordata). Vertebrates are the most important chordates.

In vertebrates, the notochord and nerve cord are enclosed in a series of bony vertebrae. All vertebrates have a closed circulatory system, a well-developed nervous system, and a complex sensory apparatus, and all probably evolved from a common ancestor similar to the tadpole larva of tunicates. The mechanism by which vertebrates are believed to have developed is through neoteny, the overtaking of reproductive functions by a larval organism that displaces the original adult form.

Perhaps the most primitive vertebrates are the jawless fishes, (class Agnatha), of which only a few types are still living. The placoderms (class Placodermi) were a group of fishes in which jaws evolved from some of the gill arches. Some of these may have given rise to the modern fishes. All are now extinct. Cartilaginous fishes (class Condrichthyes) include the skates, sharks, and rays. In these, the internal skeleton is composed of cartilage. Most forms are marine.

The bony fishes (class Osteichthyes) include most of the familiar modern fishes. These have a bony skeleton and gills protected by a bony flap, the operculum. Bony fishes originated in fresh water and possess several unique features. Osmotic balance is maintained in bony fishes in a different fashion than in sharks. Many bony fishes developed lungs to supplement the gills in respiration. Only a few species retain lungs, most having modified them into swim bladders. One group of ancient lunged fishes, the lobe-fin fishes, were probably ancestral to the terrestrial vertebrates.

17. Cite the distinguishing characteristics of chordates, vertebrates, and the four classes of fishes.

18. Describe how each of the following structures has played a role in vertebrate evolution: notochord, jaws, lungs, and lobe-fins.

35-10 Phylum Chordata, Subphylum Vertebrata, Class Amphibia / Class Reptilia / Class Aves / Class Mammalia   The first amphibians (Amphibia) are thought to have evolved from crossopterygian fishes, through the evolution of limbs as an improvement on the lobe fins. Crossopterygians probably moved from one body of water to another across short stretches of land. Amphibians still remain close to the water; their larvae are aquatic, and their eggs must be laid in the water.

Reptiles (Reptilia) have a number of adaptations that permit them to survive under conditions where water is too scarce for the needs of amphibians. The major adaptation is in the structure of the reptilian egg, which is surrounded by a protective shell, and in which the embryo is surrounded by membranes providing not only protection but also assistance in excretion and respiration. The egg contains enough yolk to permit the embryo to achieve a relatively advanced level of development before hatching. Other adaptations of reptiles include internal fertilization, which permits mating on land; more efficient respiratory and circulatory systems; and a well-developed brain. During the Mesozoic era, reptiles were the dominant land organisms.

Three groups of reptiles achieved the ability to fly. Two of these are now extinct; the other gave rise to the birds (Aves). The major differences between reptiles and birds reflect adaptations to flight. Feathers are modified scales, which probably served originally to insulate the warm-blooded birds and later became employed in flight. Modifications of the skeletal and muscular systems are evident. Warm-bloodedness also demands an efficient circulatory system. Birds have evolved a completely divided ventricle, which serves to separate oxygenated from deoxygenated blood, increasing efficiency. The lungs of birds are more efficient than those of mammals, and the visual and coordination centers of the brain are highly developed.

With the decline of the dinosaurs at the end of the Mesozoic, mammals, which had evolved from the reptiles much earlier, began to diversify and increase in numbers. All mammals suckle their young with milk produced by the mother's body. With a few exceptions, the young of mammals develop within the body of the mother. The heart is similar to that of birds, and respiratory efficiency is improved by means of the diaphragm. Because they are warm-blooded, mammals can maintain a high level of activity despite varying external temperatures. All are covered with hair, affording both protection and insulation. Mammals are the most intelligent organisms. There are 18 modern mammalian orders. Humans belong to the family Hominidae, of the order Primates. Our nearest relatives are the gorrilla, chimpanzee, and orangutan.

19. Identify the distinguishing characteristics of amphibians, reptiles, birds, and mammals.

20. Relate the evolutionary developments that gave rise to each of the major vertebrate classes.

ARTICLE RESOURCES

Criteria For Classifying Animals: Symmetry

Criteria For Classifying Animals: Embryology

Criteria For Classifying Animals: Body Cavities

Phylum Porifera

Phylum Cnidaria

Alternation of Generations in Cnidarians

Corals

   Kohl, Larry. British Columbia's cold emerald sea. National Geographic, 1980, 157(4), 526-551. A photographic essay depicting the inhabitants of the Straits of Georgia separating Vancouver Island from the rest of British Columbia. Shown are metridia, sea urchins, strawberry anemones, soft corals, scallops, eels, nudibranches, jelly fish, sea pens, octopi, clown shrimp, starfish, and hermit crabs.

Phylum Ctenophora

Phylum Platyhelminthes

Flatworms and the Origins of the Animal Kingdom

Phylum Rotifera

Phylum Nematoda

Phylum Annilida

Phylum Mollusca

Phylum Mollusca, Classes Amphineura and Monoplacophora

Phylum Mollusca, Classes Bivalvia, Gastropoda, and Cephalopoda

   Roper, Clyde F. E., and Kenneth J. Boss. The giant squid. Scientific American, 246(4), 96-105. The first description of the giant squid was given by Herman Melville, author of Moby Dick. In 1861 a partial specimen was hauled in by the crew of a French warship. A squid of medium size--10 meters long--became available for study by zoologists in 1980. It was stranded off Plum Island, Massachusetts. The anatomy is thoroughly discussed. A map shows strandings, sightings, and capture of the giant squid. Numerous diagrams and photographs illustrate this interesting article.

   Ward, Peter, Lewis Greenwald, and Olive E. Greenwald. The bouyancy of the chambered nautilus. Scientific American, 1980, 243(4), 190-203. Chambered cephalopods were important inhabitants of the sea during the Cretaceous period and reached their greatest diversity during the Triassic, Jurassic, and Cretaceous periods before diminishing in numbers. The genus Nautilus is the only remaining chambered cephalopod. Their bouyancy system is remarkable in its precision. They inhabit the sea outside the coral reefs of the islands of the tropical Western Pacific. Excellent discussion of their anatomy and reproduction.

Phylum Arthropoda

Diversification of Arthropods

Ancient Arthropods and Modern Descendants

Rudloe, Anne, and Jack Rudloe. The changeless horseshoe crab. National Geographic, 1981, 159(4), 562-576. Horseshoe crabs have remained unchanged in appearance since Devonian times. They are often called living fossils. These animals are arthropods-arachnids. Their anatomy is shown. Their breeding habits are discussed and their embryological development is illustrated. The uses of the horseshoe crab range from fertilizers and baits to biomedical research.

Phylum Arthropoda: Arachnids, Crustaceans, and Myriapods

Phylum Arthropoda: Class Insecta

Bowser, Hal. Revealed: The secrets of flight. Science Digest, 1982, 90(6), 46-53. Photographs of insects in flight with an explanation of how they are able to fly. Bowser discusses forward flight, hovering, and upside-down landings.

Boraiko, Allan A. The indomitable cockroach. National Geographic, 1981, 159(1), 130-142. Cockroaches have been on the earth since the Upper Carboniferous era. Little change has taken place in them since these early times. They seem to be ubiquitous, living in every region of the earth except the polar regions. There are 3,500 known species of cockroaches, of which only a dozen or so are pests in the United States (only 5 are truly common). They are uniquely adapted to their environments, as is illustrated.

The Lophophorate Animals

Phylum Echinodermata

Classes of the Echinoderms

Phylum Hemichordata

Phylum Chordata

Phylum Chordata, Subphylum Vertebrata

Phylum Chordata, Subphylum Vertebrata, Class Agnatha

Phylum Chordata, Subphylum Vertebrata, Class Placodermi

Phylum Chordata, Subphylum Vertebrata, Class Chondrichthyes

Phylum Chordata, Subphylum Vertebrata, Class Osteichthyes

Phylum Chordata, Subphylum Vertebrata, Class Amphibia

Phylum Chordata, Subphylum Vertebrata, Class Reptilia

Phylum Chordata, Subphylum Vertebrata, Class Aves

Calder, William A., III. The kiwi. Scientific American, 1978, 239(1), 132-142. The kiwi is a flightless bird native to New Zealand and found nowhere else in the world. Except for two species of bats, there were no mammals to be found in New Zealand until the arrival of humans about the ninth century A.D. The evolution, anatomy, and immunological relations between the egg white proteins of the ratite families are discussed. Despite the fact that the kiwi is a bird, it exhibits some mammalian characteristics.

Phylum Chordata, Subphylum Vertebrata, Class Mammalia

Ray, G. Carlton. Learning the ways of the walrus. National Geographic, 1979, 156(4), 565-580. Graphically illustrated article detailing the behavior of this large mammal. A herd was followed for 12 days in the Chukchi Sea off the Bering Straits. Its known natural enemies are few. Only the polar bear, an occasional killer whale, and humans prey on it. Walruses spend the majority of their time in the sea, hauling themselves out to rest and bear their young. Their food consists primarily of mollusks. Walruses have the ability to vasodilate and vasoconstrict, which provides them with an efficient means of thermoregulation.

ESSAY QUESTIONS

1. Discuss two contrasting hypotheses that attempt to explain how multicellular animals originated.

    One view supposes that the simplest flatworms (Acoela) were derived neotenously from the planula larva of some cnidarian, the cnidarians having arisen from the protists. The other hypothesis derives all animals directly from the protists through an intermediate stage similar to the modern Trichoplax adhaerans.

2. Explain how the animal kingdom is divided on the basis of symmetry, body cavities, and embryology, and list the phyla included within each major subdivision.

    Symmetry--Radial symmetry (Radiata): Cnidaria, Ctenophora. Bilateral symmetry: all other animals. Body cavities--Acoelomate: Platyhelminthes. Pseudo-coelomate: Rotifera, Nematoda. Coelomate: all other animals. Embryology--Protostomes, spiral and determinate cleavage, schizocoelous development: Annelida, Mollusca, Arthropoda, Onychophora, Brachiopoda, Bryozoa. Deuterostomes, radial and indeterminate cleavage, enterocoelus development: Echinodermata, Hemichordata, Chordata.

3. Trace the evolutionary history of the chordates, beginning with the tunicates and concluding with the mammals, indicating at each transition the major development that made the transition possible.

    Tunicates gave rise to the vertebrates, it is believed, through the neotenous development of jawless fishes from the tunicate tadpole larva, followed by development of the endoskeleton. Jawless fishes produced bony fishes as a result of the development of movable jaws from certain of the hyoid arches. Of the jawed fishes, some developed lungs to aid the gills in respiration, and some of these also developed lobed fins. These, the crossopterygians, evolved into the amphibians, probably as a result of spending more and more time on land as they moved from pond to pond. Amphibians are not fully terrestrial; they must return to the water to lay their eggs. The major development leading to the evolution of reptiles from amphibians was the shelled egg with membranes to surround and protect the embryo. Birds and mammals evolved from separate lines of reptiles. In the case of birds, the major adaptations are related to flight; in the mammals, major adaptations are directed toward an increase in speed, alertness, and intelligence, and better protection of offspring. Birds and mammals are able to maintain these high levels of activity because of their warm-bloodedness and efficient circulatory systems.

4. Based on your knowledge of their structural adaptations, match each of the following lifestyle descriptions with one or more of the classes of mollusks: (1) filter-feeding organisms buried in the sand of a beach; (2) grazers on algal films growing on rocks in the intertidal zone; (3) active predators of shrimp, crabs, and fishes; (4) herbivorous grazers, specialized predators on various sessile invertebrates, scavengers, and omnivorous predators are all found within this group; (5) deep-water herbivores or scavengers.

    (1) Bivalvia, (2) Amphineura, (3) Cephalopoda, (4) Gastropoda, (5) Monoplacophora.

5. Explain why the two major classes of terrestrial arthropods are believed to have had separate evolutionary origins.

    The distinctions between the arachnids and the insects are based not only on the fossil record but also on the distinct anatomical differences between the two groups. Arachnids have two body segments, no antennae, and four pairs of walking legs, with the first two pairs of appendages modified for feeding. Insects have three body divisions, one pair of antennae, and three pairs of walking legs. Many insects have wings, while no arachnids do. Arachnids were probably derived from primitive marine forms related to the horseshoe crab, whereas the insects are thought to be descended from wormlike terrestrial arthropods, probably very similar to symphylans (Myriapoda).

6. Describe the structures and uses of the two types of appendages possessed by echinoderms.

Tube feet are like tiny suction cups operated by hydraulic action of the water vascular system. When many tube feet are attached to a single object, they can exert great force. Tube feet are used in feeding, locomotion, touch, and (because of their thin surface) in respiration. Pedicellariae are shaped like small clamps or pincers and are found on the body surface of some sea stars and sea urchins. They function in cleaning the surface, in protection, and sometimes in food capture.

7. Why are annelids, arthropods, and mollusks thought to be closely related?

Annelids develop from trochophore larvae. Not all mollusks have a trochophore stage, but a sufficient number do to provide strong evidence for an evolutionary affinity with the annelids. Arthropods do not possess a trochophore stage, but they are believed to have been derived from the annelids through an intermediate stage similar to the modern onychophorans, which have both annelid and arthropod characteristics. In addition, mollusks, annelids, and arthropods are all coelomate protostomes, bilaterally symmetrical with spiral, determinate cleavage and schizocoelous development.

TEST SET 35-A

Name _____   Section _____

Choose the <u>best</u> answer to each of the following questions, and write the appropriate letter in the space provided.

Ans: c
pp. 918,
    920

_____ 1. Birds and mammals evolved separately from the: (a) crossopterygians; (b) amphibians; (c) reptiles; (d) placoderms.

Ans: d
p. 884

_____ 2. Radial, indeterminate cleavage would be expected in: (a) cnidarians; (b) annelids; (c) lophophorates; (d) humans.

Ans: a
p. 885

_____ 3. Which of the following has a coelom derived from an outpouching of the embryonic gut? (a) sea urchin; (b) clam; (c) spider; (d) rotifer.

Ans: b
p. 919

_____ 4. Most of the adaptations that distinguish birds from reptiles have to do with: (a) the circulatory system; (b) flight; (c) insulation; (d) the structure of the egg.

Ans: a
p. 892

_____ 5. The acoelomate Bilateria include: (a) Platyhelminthes; (b) Mollusca; (c) Bryozoa; (d) Rotifera.

Ans: d
p. 894

_____ 6. Nematodes: (a) are coelomate; (b) probably evolved from annelids; (c) are rarely parasitic; (d) live within their food.

Ans: b
p. 905

_____ 7. The lophophore is used primarily for: (a) attracting a mate; (b) feeding; (c) protection; (d) respiration.

Ans: a
p. 899

_____ 8. An organism with a radula and ctenidia would be closely related to a(an): (a) octopus; (b) sea star; (c) hemichordate; (d) sponge.

Ans: d
p. 884

_____ 9. Which of the following statements is <u>not</u> true? (a) All animals can be arranged into a phylogenetic tree that is a straight line with no side branches. (b) All of the animals with radial symmetry are protostomes. (c) Enterocoelous development is always associated with spiral cleavage. (d) All vertebrates have a notochord, but not all animals with a notochord are vertebrates.

Ans: d
p. 905

_____ 10. Paleopterous insects: (a) are unable to fold their wings over their backs; (b) represent the most primitive evolutionary grade; (c) always exhibit complete metamorphosis; (d) include the grasshoppers, crickets, and their relatives.

Ans: a
p. 915

_____ 11. The major evolutionary development that enabled the first vertebrates to venture from the water was: (a) limbs; (b) jaws; (c) a scaly skin; (d) an exoskeleton.

Ans: c
p. 902

_____ 12. Crayfish, shrimp, and lobsters all belong to the class: (a) Arachnida; (b) Insecta; (c) Crustacea; (d) Xiphosaura.

Ans: b
p. 901

_____ 13. Symphylans are related to insects and centipedes in the same way that onychophorans are related to: (a) fishes and amphibians; (b) arthropods and annelids; (c) sponges and protists; (d) tunicates and vertebrates.

Ans: a
p. 923

_____ 14. Humans are classified in: (a) phylum Chordata, class Mammalia,
order Primates; (b) phylum Chordata, class Mammalia, order
Hominidae; (c) phylum Chordata, subphylum Mammalia, order
Primates; (d) phylum Hemichordata, subphylum Vertebrata, class
Mammalia.

Ans: b
p. 905

_____ 15. A bivalved organism with a lophophore is not: (a) a brachiopod;
(b) a mollusk; (c) a protostome; (d) schizocoelous.

TEST SET 35-B

Name _____   Section _____

Choose the best answer to each of the following questions, and write the appropriate letter in the space provided.

Ans: b
p. 886

_____   1. Sponges have bodies that are adapted to: (a) life in open water; (b) filtering seawater for food; (c) capturing prey with nematocysts; (d) acute sensory discrimination.

Ans: c
p. 891

_____   2. An organism with radial symmetry, a mouth but no anus, and rows of cilia providing locomotion would be classified as a member of the phylum: (a) cnidaria; (b) echinodermata; (c) ctenophora; (d) annelida.

Ans: a
p. 888

_____   3. Which of the following is characteristic only of organisms with a coelenteron? (a) nematocysts; (b) lophophore; (c) radula; (d) notochord.

Ans: c
p. 892

_____   4. Tapeworms are parasitic: (a) annelids; (b) nematodes; (c) platyhelminthes; (d) rotifers.

Ans: d
p. 893

_____   5. An unusual characteristic of rotifers, shared by no other animal, is the: (a) lophophore; (b) cuticle; (c) pseudocoelom; (d) corona.

Ans: b
p. 893

_____   6. Trichinella spiralis parasitizes: (a) the intestines of birds and fishes; (b) the muscle tissue of mammals; (c) the skin of fishes and human swimmers; (d) Trichoplax adhaerans.

Ans: c
p. 891

_____   7. Corals grow only in shallow water because: (a) they must have abundant algae on which to feed; (b) only here are they safe from predation; (c) symbiotic dinoflagellates live within their tissues; (d) they must periodically be exposed by receding tides.

Ans: a
p. 895

_____   8. One of the main advantages of segmentation in annelids was: (a) the ability to use the chambers to exert pressure by hydraulic action; (b) an increase in the efficiency of the excretory system; (c) improved reproductive capacity owing to the duplication of gonads in each segment; (d) protection from predators--if a few segments are lost, the animal can still survive.

Ans: b
p. 899

_____   9. The evolution of torsion: (a) occurred in all classes of mollusks; (b) enabled the animals to draw the vital organs into the shell more efficiently; (c) is a primitive condition in mollusks; (d) provided for more efficient locomotion in gastropods.

Ans: b
p. 900

_____   10. The main disadvantage of the arthropod skeleton is that it: (a) offers little protection from desiccation; (b) must be periodically shed, leaving the animal defenseless temporarily; (c) inhibits muscular movement; (d) makes internal fertilization of the female a necessity.

Ans: d
p. 903

_____   11. An arthropod with a well-formed head, numerous segments bearing appendages, and an elongate body is probably a(an): (a) arachnid; (b) insect; (c) annelid; (d) myriapod.

Ans: b   _____   12. Amphibians never became truly terrestrial because: (a) their
p. 916   primitive limbs do not allow efficient locomotion; (b) they must
                              lay their eggs in water; (c) their lungs are not fully function-
                              al; (d) they could not compete with the reptiles, which evolved
                              at about the same time they did.

Ans: b   _____   13. The brains of birds and mammals developed along different lines.
p. 919   Those of the birds: (a) show an increase in the size of the
                              cerebral hemispheres; (b) exhibited an increase in the size of
                              the areas devoted to vision and coordination; (c) demonstrated
                              a marked increase in intelligence and visual acuity; (d) showed
                              relatively little change from the reptilian condition.

Ans: c   _____   14. Which of the following is not a characteristic feature of the
p. 884   chordates? (a) a dorsal, hollow nerve cord; (b) gill slits;
                              (c) spiral cleavage; (d) bilateral symmetry.

Ans: c   _____   15. Gorillas, chimpanzees, and orangutans are: (a) ancestral to
p. 923   humans; (b) unrelated to humans; (c) closely related to humans;
                              (d) members of Hominidae.

CHAPTER THIRTY-SIX
THE PROCESSES OF EVOLUTION

CHAPTER SUMMARIES AND OBJECTIVES

36-1 <u>The Structure of Populations / The Hardy-Weinberg Law</u> The total genetic information present in a population at any given time is the gene pool of that population. Consider a gene locus with two alleles, <u>A</u> and <u>a</u>. The total number of <u>A</u> alleles in the population is the gene frequency of that allele, and is designated <u>p</u>. There are only two alleles at this particular locus, so the frequency of the other allele, <u>a</u>, is equal to 1 - <u>p</u> and may be designated <u>q</u>.

The Hardy-Weinberg law states that the processes of sexual reproduction do not themselves change allele frequencies. The frequencies of different genotypes in the population remain constant and are related to the allele frequencies by the formula $p^2 + 2pq + q^2 = 1$. This formula is a null hypothesis. When observed allele frequencies do not correspond to the frequencies predicted by the Hardy-Weinberg formula, either nonrandom mating is occurring or some agent of evolution is in operation.

1. Derive the Hardy-Weinberg law and explain what each of the terms in the equation $p^2 + 2pq + q^2 = 1$ represents.

2. Describe the usefulness of the Hardy-Weinberg equation to the study of evolutionary genetics in real populations.

36-2 <u>Variability within Populations</u> The process of evolution converts the variability within a group (population) into variation between groups (species). Some of the variation observed within populations is the result of environmental influences and is not passed on to future generations. Genetic variation is heritable; it results from differences in the genotypes of individuals within the population. It is via genetic variation that evolution operates.

Genetic variation cannot be measured directly, but it can be estimated by a variety of techniques. Such studies have revealed that genetic variation is extensive. Different species (or different populations of the same species) do not, however, show the same degree of variation. Also, some groups may be highly variable at certain gene loci and show little variation at other gene loci. There is still much to be learned about the extent and nature of variability.

3. Answer: What is variation and why is it fundamental to the evolutionary process?

4. Give an example of a way in which genetic variability in a population may be estimated.

36-3 <u>Agents of Evolution / Sexual Recombination / Mutation Pressure / Genetic Drift /</u>

<u>Gene Flow</u> Any factor that causes a change in allele frequencies is an evolutionary agent. Agents of evolution include sexual recombination, mutation pressure, genetic drift, gene flow, and natural selection.

As stated in the previous section, sexual reproduction does not change the frequency of alleles. It does create new combinations of alleles on which other agents of evolution can act. The primary advantage of sexual reproduction is that it creates variability in the offspring, increasing the chances that some will survive the changing environment.

Genetic variation begins with mutation, any change in the genetic material. Mutation can involve changes in individual bases in the DNA molecule, as well as major changes in the chromosomes.

Genetic drift is any change in allele frequencies that occurs by chance. Genetic drift exerts its effects most strongly on small populations. An important type of genetic drift is the founder effect, the starting of a new population by a small number of pioneers. The allele frequencies in the pioneer population will differ from those of the parent population as a whole.

Gene flow occurs as the result of immigration of individuals from one deme to another within a population. The effects of gene flow depend on how many immigrants there are and on how much they differ genetically from the members of their new deme. The extreme case of gene flow is hybridization, interbreeding between individuals considered to belong to different species.

5. Define each of the agents of evolution discussed in this section.

6. Describe the relative contributions of sexual recombination, mutation pressure, genetic drift, and gene flow to the process of evolution.

36-4 <u>Natural Selection / Frequency-Dependent Selection / Fitness / Levels of Natural Selection</u> Natural selection adapts populations to their environments. As a result of the differential reproductive success of individuals, certain alleles increase in frequency at the expense of others. There are several possible results of this process. Stabilizing selection tends to eliminate individuals from the extremes of the population. In some cases, balanced polymorphisms may result, in which extreme individuals are present in fairly constant proportions, even though they are inferior to intermediate individuals.

A certain phenotype may be more successful when it is rare than when it is common. This phenomenon is called frequency-dependent selection.

Fitness is defined as the contribution of a given phenotype or genotype to the composition of future generations, relative to other genotypes or phenotypes. This concept is central to the idea of natural selection. <u>Relative</u> reproductive success of a given type determines its contribution to the next generation's gene pool. A given individual can affect the gene pool of future generations in two ways: leaving offspring (individual selection) and/or aiding the survival of other individuals with which it shares alleles (kin selection). These two components together determine the fitness of an individual.

Natural selection can operate on levels higher than individuals. Local demes may be "born" and "die" just as individuals do. Species may be "born" and exist for millions of years before becoming extinct. Though these events do not direct evolution within populations, they do influence the pattern of evolutionary change over long periods of time.

7. Explain how natural selection operates and distinguish among the different types of natural selection.

8. Define fitness. Explain the meaning of individual selection and kin selection as components of inclusive fitness.

9. Define interdemic selection, using the example of rabbits and myxomytosis virus described in the text.

36-5 How Much of Genetic Variability is Adaptive? Small variations in proteins that do not affect the functioning of the protein are produced by translation of neutral alleles. Because the functioning of the gene product is unchanged, neutral alleles are not acted on directly by natural selection. For proteins that have been studied enough, such as cytochrome c, it is possible to recognize areas of the molecule in which the amino acid sequence does not vary. This suggests that these areas are crucial to the functioning of the molecule. However, there are areas where extensive variation does occur. Is there some significance to these variations, or are they functionally neutral? This question has not been answered yet.

10. Answer: What are neutral alleles?

36-6 Industrial Melanism / The Sickle-Cell Trait / House Sparrows in North America / Camelina in Flax Fields What follows is a discussion of several examples of microevolution that have been studied well. Industrial melanism refers to the spread of darkly colored moths and butterflies near polluted industrial centers. The classic example is that of moths living on lichen-covered tree trunks in Great Britain. Prior to the industrialization of Britain, light-colored moths predominated in the population. As increased air pollution killed off the lichens, light-colored moths began to decrease in favor of dark-colored ones. The selective force operating in this instance is birds, which capture a larger percentage of moths that do not match their backgrounds.

Persons homozygous for the sickle-cell trait usually die in childhood. Nevertheless, the gene is abundant in certain regions of Africa and Asia because heterozygotes are more resistant to malaria than are persons who lack the sickle-cell allele.

The initial variability of the population of English sparrows released in North America was probably much less than in the European parent population. Nevertheless, after 120 years, the sparrows have evolved as much variability in the North American population as is present in the European population.

The seeds of Camelina linicola are so similar in appearance to those of flax that they are planted with the flax by farmers. Camelina seeds ripen at the same time as the seeds of the flax plants. Natural selection has thus enabled Camelina to reproduce successfully, despite the attempts of farmers to eradicate it from flax fields.

11. For any two of the examples of microevolution described in this section, identify the nature of the selective force and explain how natural selection produced the observed results.

ARTICLE RESOURCES

The Structure of Populations

Valentine, James W. The evolution of multicellular plants and animals. Scientific American, 1978, 239(3), 140-158. The fossil record is rich in multicellular organisms, and these fossils have made it possible to chart the major events in their evolution. The evolution of kingdoms and of species such as fish, reptiles, mammals, and vascular plants is charted. Well illustrated. Supplements textbook materials.

The Hardy-Weinberg Law

Variability within Populations

Ayala, Francisco J. The mechanisms of evolution. Scientific American, 1978, 239(3), 56-69. Genetic variation within a species is illustrated by the wing covers of the Asiatic lady beetle. Point mutations occur during replication of DNA molecules and they occur randomly. Genetic variation is necessary for evolutionary changes: Large numbers of genetic variations in 125 animal and eight plant species were studied by means of gel electrophoresis. Well illustrated by charts and diagrams.

Agents of Evolution

Eigin, Manfred, William Gardner, Peter Schuster, and Ruthild Winkler-Oswatitsch. The origin of genetic information. Scientific American, 1981, 244(4), 86-92, 96, 99-118. The discovery of de nova RNA synthesis is explained. Clear diagrams and informative boxes clarifying the mathematical investigations of the "quasispecies model" and the "hyper cycle model." Expands on the textbook material.

Sexual Recombination

Mutation Pressure

Genetic Drift

Gene Flow

Natural Selection

The bird is back. Science 82, 1982, 3(1), 6. Unsigned article telling of the return of the bower bird, long thought to be extinct. Plain birds build more elaborate bowers than spectacularly feathered birds to attract females. However, they are successful in mating only once in 200 tries.

Frequency-Dependent Selection

Fitness

Buffetant, Eric. The evolution of the crocodilians. Scientific American, 1979, 241(4), 130-144. Crocodiles are well adapted to life in a semiaquatic environment in warm climates. These were the leading form of life in the Triassic era. Chart traces their evolution to the present. The distribution of crocodiles on major continents is shown.

Levels of Natural Selection

Lewontin, Richard C. Adaptation. Scientific American, 1978, 239(3), 212-230. The evolutionary change whereby organisms are enabled to cope with the difficulties in their environment is adaptation. The "Red Queen" hypothesis is discussed. This hypothesis states that, because the environment is constantly decaying with respect to existing organisms, natural selection operates only to maintain their state of adaptation. Adaptation is accomplished through natural selection. Excellent charts, graphs, and diagrams.

How Much of Genetic Variability is Adaptive?

Sisson, Robert F. Deception: Formula for survival. National Geographic, 1980, 157(3), 395-415. Excellent illustrations of the peppered moth and industrial melanism. Shows the return of the pale moth to areas where progress has been made against air pollution.

Industrial Melanism

The Sickle-Cell Trait

House Sparrows in North America

Camelina in Flax Fields

ESSAY QUESTIONS

1. Among the males of a certain species of birds, allele P, which is dominant, results in bright red coloration of the throat pouch, whereas the recessive allele, p,

produces a white pouch. When a population of these birds was surveyed in 1945, approximately 4 percent of the males had white pouches. A second survey, conducted in 1975, found 64 percent of the males to have red pouches. Explain, in terms of the Hardy-Weinberg law, whether evolution with respect to pouch coloration is occurring within this population.

White-pouched males must be homozygous (pp), because this allele is recessive. The percentage of white-pouched males in 1945 was 0.04, so the frequency (q) of the recessive allele is the square root of 0.04, or 0.2. Since p = 1 - q, the frequency of the dominant allele is 0.8. By the Hardy-Weinberg formula, the percentage of red-pouched birds should be .96 $[0.8^2 + 2(0.8 \times 0.2)]$. This percentage should remain constant so long as no evolution is occurring in the population. The observation that the fraction of red-pouched birds has decreased to 0.64 indicates that some form of selection favoring the white-pouched individuals is occurring.

2. What is meant by genetic variation? Must all variation within a population be adaptive?   Give an example of variation resulting from a neutral allele.

Genetic variation is variability within the genes themselves, as opposed to variability that results from environmental influences. Not all variation is necessarily adaptive. An example of variation produced by a neutral allele is a change in the amino acid sequence of a protein which does not affect the functioning of that protein.

3. For each of the following situations, indicate the specific nature of the evolutionary agent operating.

   a. The planting of trees on the Great Plains by human settlers resulted in an increase in the frequency of hybridization between Baltimore and Bullock's orioles.

   Gene flow.

   b. An unusually cold winter kills the seeds of all but a few individuals in a population of annual plants.

   Genetic drift.

   c. As pollution killed the lichen on British trees, dark-colored moths came to dominate the population.

   Natural selection.

   d. As a result of failure of the chromosomes to separate during meiosis, a polyploid individual developed in a population of orchids. Self-fertilization of this individual produced several thousand seeds, which have germinated to produce healthy new plants.

   Mutation pressure.

4. What is meant by the "individual selection" and "kin selection" components of inclusive fitness? Give an example in which each of these components predominates.

Individual selection refers to the success of the individual in leaving offspring, whereas kin selection refers to the contribution of an individual to the survival of other individuals with whom it shares alleles. Individual selection predominates in species wherein the members are largely solitary; kin selection is likely to be important in highly social groups.

5. Explain in detail how the gene for the sickle-cell trait is maintained in human populations in Africa and Asia. Why do Central Africans suffer so severely from the effects of the sickle-cell trait, while Asians do not?

The sickle-cell trait is controlled by a single allele which, when present, results in the production of abnormal hemoglobin. Persons homozygous for the sickle-cell trait usually die in childhood. Nevertheless, the presence of the allele in the gene pool is maintained. The reason for this apparent anomaly is that persons heterozygous for the sickle-cell trait are more likely to survive an attack of malaria than are persons who do not carry the allele. Malaria has been a problem in Central Africa for a much shorter period of time than in Asia. As a result, Africans have not developed the additional genetic changes that serve to counter the effects of the sickle-cell allele in Asians.

TEST SET 36-A

Name _____   Section _____

Choose the <u>best</u> answer to each of the following questions, and write the appropriate letter in the space provided.

Ans: b
p. 932

_____ 1. A Mendelian population is a deme in which all individuals:
(a) are identical genetically; (b) interbreed; (c) are asexual;
(d) live in the same locality.

Ans: a
p. 934

_____ 2. The Hardy-Weinberg formula predicts that the frequencies of
alleles: (a) remain constant throughout subsequent generations;
(b) will be altered by sexual recombination; (c) fluctuate as
long as no agent of evolution is operating; (d) are invariant
under all conditions.

Ans: b
p. 934

_____ 3. Nonrandom mating will result in: (a) readily predictable gene
frequencies; (b) gene frequencies that do not conform to the
Hardy-Weinberg formula; (c) little significant change in a
population; (d) the stabilizing of gene frequencies in compliance
with the Hardy-Weinberg law.

Ans: a
p. 938

_____ 4. Genetic variation results from: (a) changes in the structure of
genes and chromosomes; (b) environmental induction of variable
traits; (c) point mutations only; (d) changes in gene frequencies.

Ans: c
p. 939

_____ 5. One hundred persons are selected to become the first passengers
on a "space ark." This ship is designed to voyage to another
planetary system, on a journey that will take many hundreds of
years to complete. The primary evolutionary agent that will
operate on this population is: (a) genetic variation; (b) frequency-dependent selection; (c) the founder effect; (d) mutation
pressure.

Ans: d
p. 933

_____ 6. Four percent of the snails in a study sample were found to
possess a trait resulting from a recessive allele. The frequency
of the corresponding dominant allele in these snails is:
(a) 0.4; (b) 0.96; (c) 0.16; (d) 0.8.

Ans: a
p. 941

_____ 7. Disruptive selection is operating when: (a) extreme individuals
are favored at the expense of intermediates; (b) intermediate
individuals are favored at the expense of extreme individuals;
(c) a phenotype is more successful when it is rare; (d) the
natural reproductive cycle of a group of organisms is disrupted.

Ans: c
p. 942

_____ 8. One way in which an organism can influence the gene frequencies
in future generations is by ensuring the survival of other
individuals with which it shares a common ancestor. This is
called: (a) individual selection; (b) interdemic selection;
(c) kin selection; (d) stabilizing selection.

Ans: d
p. 943

_____ 9. That a decrease in the virulence of the myxomytosis virus had
occurred several generations after the introduction of the virus
into the Australian rabbit population was demonstrated by:
(a) the increased survival rate of the rabbits; (b) the rapidity
with which the disease spread through the population; (c) an increase in the toxigenicity of the virus; (d) experimental
injection of the virus into rabbits with no prior exposure to
myxomytosis.

Ans: c
p. 944

_____ 10. _____ are <u>not</u> acted on by natural selection. (a) Gene frequencies; (b) Demes; (c) Neutral alleles; (d) Individuals.

Ans: a
p. 945

_____ 11. In the case of the cytochrome c molecule, it is clear that:
(a) certain regions of the molecule have not been altered throughout almost the entire history of life; (b) variant positions
within the molecule have no evolutionary significance; (c) a
great deal of variation must occur to render the molecule non-
functional; (d) the tertiary structure of this molecule in horses
is strikingly different from that in fish.

Ans: c
p. 948

_____ 12. Predation by birds is to industrial melanism as sickle-cell
trait is to: (a) house sparrows; (b) Central Africans;
(c) malaria; (d) hemoglobin.

Ans: b
p. 950

_____ 13. That founder effects cannot be assumed to be important just
because colonizing populations are small is demonstrated in the
case of: (a) Camelina linicola in flax fields; (b) house
sparrows in North America; (c) sickle-cell trait in Asia;
(d) Darwin's finches in the Galapagos.

Ans: c
p. 951

_____ 14. The selective force operating in the case of Camelina linicola
is: (a) predation; (b) competition with flax for water and
minerals; (c) human farmers; (d) overcultivation of flax fields.

Ans: a
p. 938

_____ 15. Changes in gene frequencies that occur by chance are termed:
(a) genetic drift; (b) gene flow; (c) natural selection;
(d) mutations.

TEST SET 36-B

Name _____   Section _____

Choose the <u>best</u> answer to each of the following questions, and write the appropriate letter in the space provided.

Ans: c
p. 932

_____   1.   The fundamental unit for studying the genetics of evolution is the: (a) panmictic deme; (b) Hardy-Weinberg unit; (c) population; (d) genus.

Ans: c
p. 934

_____   2.   The Hardy-Weinberg law is a null hypothesis in that it: (a) cannot be used to study real populations; (b) is only a statistical estimate; (c) supplies a standard by which the results of experimental observations of real populations may be checked; (d) applies only to situations in which evolutionary agents are operating.

Ans: d
p. 933

_____   3.   If the frequency of an allele is q, the percentage of individuals in a population that possess this allele is given by: (a) $p^2 + 2pq + q^2$; (b) $2pq$; (c) q^2; (d) $2pq + q^2$

Ans: b
p. 941

_____   4.   When the heterozygote is superior to either of the homozygotes, stabilizing selection often results in: (a) gene flow; (b) balanced polymorphism; (c) no effect; (d) hybridization.

Ans: a
p. 939

_____   5.   The founder effect is a type of: (a) genetic drift; (b) frequency-dependent selection; (c) gene flow; (d) kin selection.

Ans: d
p. 933

_____   6.   Nine percent of the finches on an island in the Galapagos Archipelago were found to possess a trait resulting from a recessive allele. The frequency of the corresponding dominant allele in these birds is: (a) 0.49; (b) 0.9; (c) 0.18; (d) 0.7.

Ans: a
p. 942

_____   7.   One way in which an organism can influence the gene pool of future generations is to leave offspring. This is called: (a) individual selection; (b) interdemic selection; (c) kin selection; (d) stablizing selection.

Ans: b
p. 941

_____   8.   Stabilizing selection is operating when: (a) environmental conditions remain relatively stable with time; (b) intermediate individuals are favored at the expense of extreme individuals; (c) a phenotype is more successful when it is rare; (d) extreme individuals are favored at the expense of intermediates.

Ans: c
p. 943

_____   9.   The percentage of rabbits killed by myxomytosis declined from over 99 percent (during the first year after the introduction of the disease into the Australian rabbit population) to about 50 percent (by the third year). This decline is due to: (a) an increase in the virulence of the virus; (b) the operation of natural selection on individual viruses; (c) the operation of interdemic selection on the viruses and natural selection on the rabbits; (d) the operation of interdemic selection on the rabbits.

Ans: b
p. 944

_____   10.   Certain genetic changes may have no influence on the functioning of the proteins coded by the genes in which the changes occur. In this case, we are dealing with: (a) hybrid alleles; (b) neutral alleles; (c) recessive alleles; (d) dominant alleles.

Ans: a
p. 940

_____   11.   The only evolutionary agent that adapts organisms to their environment is: (a) natural selection; (b) gene flow; (c) genetic drift; (d) mutation pressure.

Ans: b _____ 12. The evolutionary agent that results from the immigration of
p. 940   individuals from one population to another is: (a) natural
            selection; (b) gene flow; (c) genetic drift; (d) mutation
            pressure.

Ans: b _____ 13. Human farmers are to <u>Camelina linicola</u> as environmental varia-
p. 950   bility is to: (a) <u>sickle-cell trait;</u> (b) house sparrows;
            (c) certain British moths; (d) flax plants.

Ans: c _____ 14. In highly social species, the inclusive fitness of an individual
p. 942   is dominated by: (a) frequency-dependent selection; (b) the
            individual selection component; (c) the kin selection component;
            (d) neither the kin selection component nor the individual
            selection component.

Ans: a _____ 15. The key selective force operating in cases of industrial melanism
p. 947   is: (a) predation by birds; (b) the destruction of lichens by
            pollution; (c) the growth of industrial centers; (d) the mutation
            rate of the allele for wing coloration.

CHAPTER THIRTY-SEVEN
THE MULTIPLICATION OF SPECIES

CHAPTER SUMMARIES AND OBJECTIVES

37-1 <u>The Species Concept / Weaknesses of the Species Concept</u> A species is a population or series of populations within which a significant amount of gene flow occurs under natural conditions, but which is genetically isolated from other populations. Closely related species may interbreed in the laboratory, but they do not do so under natural conditions.

The species concept is difficult to apply to populations that are similar but widely separated geographically. Fossil organisms are also difficult to assign to species, because the fossils provide no information about interbreeding. Similarly, asexual organisms (those that always self-fertilize) cannot be assigned to species on the basis of interbreeding.

1. Define the term species.

2. Explain how organisms to which the species concept does not readily apply are assigned to species.

37-2 <u>Barriers to Exchange / Geographical Speciation / Parapatric Speciation / Sympatric Speciation</u> The first step in the formation of a species is the development of some barrier to gene flow. The most common way for a gene pool to become split is the development of a physical barrier between two portions of a population. The isolated daughter populations can then diverge from one another in response to local conditions. If enough divergence occurs, the two populations may not interbreed if they later come together again; thus they are new species. The nature of such a barrier between populations depends on the nature of the organisms. In general, smaller, less mobile organisms can speciate in smaller areas than larger, highly mobile organisms.

Parapatric speciation can occur in response to strong local selective pressures, and it results in the development of reproductive isolation within a continuous population in the absence of an actual geographical barrier.

Sympatric speciation occurs when a gene pool becomes divided even though the ranges of the daughter species overlap during the process. Among flowering plants, sympatric speciation commonly occurs via polyploidy. Polyploidy individuals usually cannot mate with their parent types, but only with each other. Disruptive selection can also result in sympatric speciation, at least in the laboratory; no evidence of this type of speciation has as yet been found in nature, however.

3. Distinguish among allopatric, parapatric, and sympatric speciation.

4. Answer: What is polyploidy and how can it occur? What is the role of

polyploidy in the formation of new species?

37-3 <u>Maintaining Separate Gene Pools / Genetic Changes Accompanying Speciation / Where Do Species Divide</u> There are a number of ways in which genetic isolation may be maintained between species, even when closely related species come into contact with each other. Such barriers to genetic exchange are called isolating mechanisms.

Laboratory studies of closely related species indicate that very little genetic change may accompany speciation. However, small genetic differences can result in large morphological differences, especially if the genetic changes affect regulatory genes, such as those that influence the timing and rate of growth.

Barriers to genetic exchange do not occur with equal probability throughout the entire range of a species. Near the periphery of its range, a species tends to become fragmented. Peripheral populations are thus often isolated and may evolve into new species.

5. List several types of isolating mechanisms.

6. Explain why changes in allometric growth patterns are often favored by natural selection.

7. Explain why speciation is less likely to occur among members of a population living near the center of the range than among those living on the periphery of the range.

37-4 <u>Ecological Adjustments</u> When speciation occurs, the daughter species become genetically, but not ecologically, isolated from one another. An alteration in the traits of a species as a result of interactions with closely related species is called character displacement. Many examples of character displacement are known. Sometimes extensive opportunities open up to a lineage as a result of speciation, and many species evolve, each exploiting its environment in a slightly different way. This phenomenon is called adaptive radiation.

8. Define, by means of examples, character displacement and adaptive radiation. You may create hypothetical examples or use those from the text.

ARTICLE RESOURCES

<u>The Species Concept</u>

<u>Weaknesses of the Species Concept</u>

<u>Barriers to Exchange</u>

<u>Geographical Speciation</u>

J. S. Jones. St. Patrick and the bacteria. <u>Nature</u>, 1982, 296(5853), 113-114. Discussion of the McArthur and Wilson equilibrium theory of insular biogeography as applied to bacterial and viral infections in humans. The theory is clarified. Measles occurs only in humans. The virus can continue to exist only in a population of over 500,000, which will, because of its size, have a continuous supply of susceptible individuals. Diseases have been studied in such isolated areas as Iceland and Arctic villages.

<u>Parapatric Speciation</u>

Felman, Moshe, and Ernest R. Sears. The wild gene resources of wheat. <u>Scientific American</u>, 1981, 244(1), 102-112. Wheat plants have a substantial gene pool in wild wheat varieties that may be exploited to improve the productivity of wheat. There are four species of cultivated wheat. All four contain 7 chromosomes in the haploid state. One species has 14 somatic chromosomes, two have 24 somatic chromosomes, and one has 42 somatic chromosomes. This last variety has more than 20,000 cultivated varieties that are adapted to a wide range of habitats. Viable hybrids can be formed within different genera, thus making possible genetic variation that can be exploited for the improvement of wheat.

Puff, C., and D. E. Mantell. Revision and affinities of Galium (Rubiaceae) in Madagascar. Plant Systematics and Evolution, 1982, 140(1), 57-73. The origins of the Galium species in Madagascar and their relationships with other African taxa are discussed. Some species exhibit polyploidy. A key to the taxa and drawings of the flower parts of two species are given.

Sympatric Speciation

Maintaining Separate Gene Pools

Genetic Changes Accompanying Speciation

P. R. Grant. Patterns of growth in Darwin's finches. Proceedings of the Royal Society, London Series B, 1981, 212(1189), 403-432. Beak and body size proportions among six species of Darwin's finches were studied to determine how these were brought about in ontogeny. Prehatching growth appears to differ among the species. The author concludes that differentiation of Darwin's finches has been brought about by evolutionary changes in all states of growth characteristics. Drawings and charts illustrate the article.

Where do Species Divide?

Ecological Adjustments

Bowser, Hal. Vanished! Science 82, 1982, 3(6), 48-55. Well-illustrated article on extinct animals and some possible explanations for their extinction.

ESSAY QUESTIONS

1. Define the term species, and list three cases in which your definition is difficult to apply.

   Species are populations within which a significant amount of gene flow occurs under natural conditions but which are genetically isolated from other populations. Because of the emphasis on gene flow, this definition is difficult to apply to populations that are widely separated from each other in space or time, or to asexual or self-fertilizing populations.

2. A major seed company has recently introduced a new genus of flowers, Pardancanda. The new plants are the result of crossing Pardanthopsis (formerly in the iris family) with Belamcanda, the blackberry lily. What phenomenon is demonstrated by this type of plant breeding? Why do you think botanists have now reclassified Pardanthopsis as a lily?

   This is a good example of allopolyploidy. Pardanthopsis has been moved to the lily family on the basis of its ability to hybridize with Belamcanda. Hybridization among plants is usually most successful between members of the same family.

3. Give at least three mechanisms by which closely related species may maintain genetic isolation.

   Individuals of one population may not find members of the other attractive as mating partners; breeding seasons may differ; structural changes may prevent successful mating; hybrid matings may not produce viable offspring; hybrid offspring may be sterile.

4. Isolated small bodies of water in southeastern California and southwestern Nevada are home to several species of pupfish of the genus Cyprinodon. Each small pool has its own characteristic pupfish species. Geologic evidence indicates that these small "oases" were once part of a larger inland sea that covered what is now the Death Valley region. Based on your knowledge of speciation mechanisms,

explain how these various species of pupfish probably came into being.

This is an example of allopatric speciation. When the inland sea dried up, individuals of the parent pupfish species became isolated in the small ponds and pools. With time, differences in these populations have accumulated.

TEST SET 37-A

Name _____   Section _____

Choose the best answer to each of the following questions, and write the appropriate letter in the space provided.

Ans: c
p. 954

_____ 1. Species are populations: (a) or series of populations within which a significant amount of gene flow does not occur; (b) that experience gene flows with other, differing populations; (c) or series of populations within which there is a significant amount of gene flow but which are isolated genetically from other populations; (d) that experience gene flow within the population and with other, differing populations.

Ans: b
p. 955

_____ 2. The concept of genetic isolation has no meaning in terms of: (a) interbreeding populations; (b) asexual species; (c) allopatric speciation; (d) species whose ranges overlap.

Ans: b
p. 956

_____ 3. The interposition of a physical barrier is required for: (a) any speciation event; (b) allopatric speciation; (c) parapatric speciation; (d) sympatric speciation.

Ans: c
p. 956

_____ 4. Mexican platyfish provide an example of: (a) speciation in the absence of geographical isolation; (b) parapatric speciation; (c) different levels of genetic isolation in geographically isolated populations; (d) three separate species that are interfertile.

Ans: d
p. 957

_____ 5. Which of the following organisms might be expected to undergo speciation in the smallest area? (a) arctic tern; (b) field mouse; (c) grasshopper; (d) ciliated protozoan.

Ans: c
p. 957

_____ 6. The best explanation for the diversity among Darwin's finches is that: (a) they are descended from separate ancestral stocks; (b) they were formed by the splitting of populations on a single island and subsequently spread to other areas of the archipelago; (c) a single ancestral finch species was subjected to different selective regimes on different islands; (d) the environment is uniform throughout the archipelago.

Ans: a
p. 958

_____ 7. The existence of an important environmental discontinuity may result in: (a) parapatric speciation; (b) geographical isolation; (c) allopatric speciation; (d) speciation only in the presence of a physical barrier to gene flow.

Ans: b
p. 958

_____ 8. One of the factors that have influenced gene frequencies in Agrostis tenuis is: (a) availability of moisture; (b) lead tolerance; (c) the occurrence of polyploidy; (d) both (a) and (b).

Ans: c
p. 959

_____ 9. Many species of tubular-flowered gilias in western North America have arisen as a result of: (a) parapatric speciation; (b) allopatric speciation; (c) sympatric speciation; (d) disruptive selection.

Ans: c
p. 960

_____ 10. Members of different host races of insects: (a) feed and mate on the same plant species; (b) feed on, but do not mate on, the same plant species; (c) feed and mate on separate plant species; (d) mate, but do not feed on, the same plant species.

Ans: b   _____   11.   An example of the rapid evolution of isolating mechanisms can be
p. 961   seen in: (a) <u>Camelina linicola</u>; (b) Baltimore and Bullock's
                              orioles; (c) Darwin's finches; (d) <u>Partula</u> snails.

Ans: a   _____   12.   Species so similar that they are difficult to distinguish are
p. 961   called: (a) sibling species; (b) host races; (c) allometric
                              species; (d) subspecies.

Ans: c   _____   13.   What percentage of modern angiosperm species has probably arisen
p. 959   as a result of polyploidy?   (a) less than 10%; (b) between 10%
                              and 25%; (c) between 25% and 60%; (d) over 60%.

Ans: b   _____   14.   When extensive ecological opportunities open up in a given area,
p. 967   some groups of organisms living in the area may experience:
                              (a) character displacement; (b) adaptive radiation; (c) exten-
                              sive hybridization; (d) rapid evolution of neutral alleles.

Ans: b   _____   15.   Character displacement in Darwin's finches is probably the re-
p. 965   sult of: (a) an unusual rate of mutation among these birds;
                              (b) competition for food; (c) natural selection favoring larger
                              bill size; (d) undetermined factors.

TEST SET 37-B

Name _____   Section _____

Choose the best answer to each of the following questions, and write the appropriate letter in the space provided.

Ans: a
p. 956

_____   1. Allopatric speciation implies: (a) the interposition of a physical barrier between two populations; (b) the formation of species over a continuous geographical area; (c) the absence of genetic isolation between the speciating populations; (d) the occurrence of speciation despite hybridization.

Ans: b
p. 955

_____   2. Asexual species: (a) are rarely classified on the basis of phenotypic differences; (b) do not fit into the biological species concept; (c) usually do not occur together in the same locality; (d) are absent from the fossil record.

Ans: c
p. 954

_____   3. Fundamental to the biological species concept is the notion of: (a) character displacement; (b) gene flow between different species; (c) genetic isolation; (d) the frequent isolation of peripheral populations by geographical barriers.

Ans: a
p. 956

_____   4. True speciation occurs when: (a) one lineage divides into two; (b) gradual changes occur in a lineage over time; (c) mutations arise in large populations; (d) disruptive selection is operating only.

Ans: b
p. 956

_____   5. Different levels of genetic isolation are most likely to be observed among Mexican platyfish species because: (a) the populations are not completely geographically isolated from each other; (b) the populations have not been apart long enough for complete genetic isolation to occur; (c) genetic isolation apparently develops very rapidly in these fishes; (d) they live in different streams.

Ans: a
p. 954

_____   6. Ernst Mayr's association with the Arfak people of New Guinea led to the conclusion that: (a) biological species are not merely abstract concepts; (b) the Arfak recognized none of the species that Mayr identified; (c) phenotypic characters are not sufficient for the delineation of species descriptions; (d) biological species are artifacts of the methods used for recognizing them.

Ans: b
p. 957

_____   7. Biologists are interested in the size of the area in which a group of organisms can undergo speciation, because this may have important implications for the ability of new species to arise: (a) in the laboratories of the future; (b) within the confines of national parks; (c) among the larger mammals only; (d) in the absence of barriers to gene flow.

Ans: b
p. 958

_____   8. In parapatric speciation, a species boundary usually forms: (a) in areas of environmental uniformity; (b) where an important environmental discontinuity exists; (c) in the absence of any environmental influences; (d) at random.

Ans: c
p. 960

_____   9. When all matings of a particular insect population occur on the same plant species, _____ may develop. (a) geographical barriers; (b) allopolyploidy; (c) host races; (d) autopolyploidy.

Ans: d   _____   10.   A reduction in the frequency of hybrid mating is usually accom-
p. 960   plished by the development of: (a) panmictic demes; (b) inviable
                            offspring; (c) character displacement; (d) isolating mechanisms.

Ans: c   _____   11.   Which of the following human activities augmented the frequency
p. 960   of hybridization between Bullock's and Baltimore orioles?
                            (a) clearing of cropland; (b) introduction of new food plants;
                            (c) planting of trees; (d) hunting.

Ans: a   _____   12.   Allometric growth refers to: (a) different rates of growth of
p. 963   different body parts; (b) differential increase in the size of
                            favored and unfavored populations in a given environment;
                            (c) rapid exploitation of a variety of new habitats by a lineage;
                            (d) the development of the head of human infants.

Ans: b   _____   13.   In the wasp Eumenes flavopictus, the greatest amount of differen-
p. 964   tiation occurs within: (a) mainland populations; (b) island
                            populations; (c) hybrid populations; (d) populations that have
                            overlapping ranges.

Ans: b   _____   14.   Substantial morphological and physiological differences can
p. 962   result from: (a) hybridization; (b) mutations in regulatory
                            genes; (c) no known biological processes; (d) sexual recombina-
                            tion alone.

Ans: b   _____   15.   The alteration of a trait of a species as a result of inter-
p. 965   actions with associated species is known as: (a) genetic drift;
                            (b) character displacement; (c) parapatric speciation; (d) gene
                            flow.

CHAPTER THIRTY-EIGHT
MACROEVOLUTION

CHAPTER SUMMARIES AND OBJECTIVES

38-1 <u>The Fossil Record / The Formation of Fossils</u> Past life forms are studied via their fossilized remains. About 300,000 fossil species are known, an estimated 6 percent of all species alive today. Six major extinctions have occurred, each followed by extensive speciation. Certain groups of organisms have left more fossils than others because of differences in environment and body structure. Higher taxonomic categories of fossil organisms are better known than lower categories. In general, newer sedimentary deposits yield more fossils than older ones.

    1.   Explain how we can estimate the number of species of organisms that have lived in the past.

38-2 <u>Patterns in the Fossil Record</u> Early organisms fossilized poorly because they lacked hard parts. The oldest metazoan fossils are about 700 million years old. A few modern phyla had appeared by the early Cambrian and most others by the mid-Cambrian. In the Ordovician, life forms were still restricted to the ocean but were evolving rapidly. Some land plants appeared in the Silurian. Fishes diversified during the Devonian; some amphibians and arthropods invaded the land. In the Carboniferous, forests expanded and the first reptiles appeared. Reptiles diversified during the Permian and became dominant in the Triassic, while many early invertebrates and amphibians disappeared. During the Jurassic, reptiles gave rise to birds, mammals, and dinosaurs; the latter became extinct at the end of the Mesozoic, while small mammals and flowering plants became common.

    2.   Explain why there are so few Precambrian fossils.

    3.   Describe the major evolutionary events of each geologic period.

38-3 <u>The Cenozoic / Human Evolution</u> Vertebrates radiated extensively during the Cenozoic. Most modern birds had appeared by the Miocene. Mammals evolved rapidly during the Eocene; by the late Miocene, primates were abundant in Africa and South America. Two groups of African primates returned to the ground and grew larger: the ancestors of baboons and those of humans. The latter group evolved bipedal locomotion, which in turn facilitated the use of tools. Increase in cranial capacity occurred later, probably in the Pleistocene, as cooperative hunting became advantageous. Language and culture are among the consequences of this development.

    4.   Describe vertebrate and primate evolution during the Cenozoic.

    5.   Name the two most distinctive features of humans, as compared with other primates.

38-4 <u>The Changing Pace of Evolution / Rates of Evolution of Nonfossilized Traits</u> The rate of evolution has varied considerably over geologic time. Many groups radiated most rapidly soon after they appeared, then more slowly. Mass extinctions have occurred six times: at the end of the Cambrian, Devonian, Permian, Triassic, and Cretaceous and at the present time. Present extinctions are caused primarily by human activities, but the causes of the other major extinctions are widely debated.

Important information about evolution is obtained by the study of amino acid substitutions in proteins and nucleotide substitutions in DNA. These molecular changes can be checked against the fossil record to determine their accuracy.

6.   Describe the six major extinctions that have occurred, and explain how the most recent one is unique.

7.   Explain how amino acid substitutions can be used to determine rates of evolution.

38-5 <u>Macroevolutionary Mechanisms / South American Mammals: Isolation and Competition / Continental Drift and Macroevolution / Speciation and Adaptive Radiation / How Predictable is Evolution?</u> Evolution has resulted from natural selection combined with unique events in geologic history. New physical and/or behavioral characteristics offer the potential for adaptive radiation; the success or failure of a given characteristic depends largely on the changing physical and biotic environment. Examples of successful major adaptations are the primate's hand, the elephant's trunk, and the snail's operculum.

An example of sudden change in the biotic environment is the interaction of North American and South American mammals after the formation of the Isthmus of Panama. Continental drift has produced major changes in the physical environment, as when shallow seas were eliminated in the late Permian. Asteroid bombardment may also have caused some major extinctions. The "adaptive landscape" model may explain the uneven pace of evolution: Genetic drift, expected to occur in populations decimated by environment change, can be more effective than selection alone in producing optimum adaptation.

8.   Describe the major rearrangements of continents following the breakup of Pangaea.

9.   Explain the adaptive landscape model.

10. Give examples of successful major evolutionary adaptations.

ARTICLE RESOURCES

<u>The Fossil Record</u>

Langston, Wann, Jr. Pterosaurs. <u>Scientific American</u>, 1981, 244(2), 122-136. Pterosaurs evolved during the early part of the Mesozoic age, preceding birds by 50 million years. Their fossils have been found on every continent except Antarctica. They were the largest flying creatures for which fossils have been discovered. Their wing span was 11 to 12 meters. Apparently they lived on fish, but how they caught their prey is not clear. Excellent diagrams and an evolutionary chart are included, as well as a drawing showing homology between the arm of a human, the wing of a bat, a bird's wing, and the wing of a pterosaur.

<u>The Formation of Fossils</u>

Graves, Daniel I., John S. R. Dunlop, and Roger Buick. An early habitat of life. <u>Scientific American</u>, 1981, 245(4), 64-73. Fossil deposits in northwestern Australia indicate probable microfossils and stromatolites--layered structures formed by accretion of fine grains of sediment by layers of microorganisms. Through carbon dating it has been established that the early Precambrian environment of these deposits was one in which primitive organisms could have flourished. These fossils go back approximately 3.5 billion years. Excellent photographs and clear explanations.

Patterns in the Fossil Record

Voorhies, Michael R. Ancient ashfall creates a Pompeii of prehistoric animals. National Geographic, 1981, 159(1). Northeastern Nebraska is the site of one of the richest finds of prehistoric animals in the United States. About ten million years ago, a volcanic eruption blanketed hundreds of square miles with volcanic ash. Rhinoceros, horses, camels, birds, grains, turtles, and diatoms of the late Miocene age were found. More than 200 skeletons were retrieved from the shallow water hole. Well illustrated.

Hoffman, Paul. Asteroid on trail. Science Digest, 1982, 90(6), 58-63. An attempt to show how the extinction of the dinosaurs may have occurred by means of an extraterrestrial agent--an asteroid. Evidence is cited in a cleverly written article.

The Cenozoic

Human Evolution

Johanson, Donald C., and Maitland A. Edey. Lucy, the inside story. Science 81, 1981, 2(2), 44-55. Illustrated account of the discovery of Lucy, the earliest ancestor of humans, in the Rift Valley of Africa.

Johanson, Donald C., and Maitland A. Edey. How ape became man: Is it a matter of sex? Science 81, 1981, 2(3), 44-49. Continuation of the discovery of Lucy suggests a new explanation for erect hominids premised on sexual and social reasons.

Washburn, Sherwood L. The evolution of man. Scientific American, 1978, 239(3), 194-208. The anatomy, brain size, locomotion, and use of objects made of stone and other materials by man over a possible span of 10 million years are traced. Primate divergence is shown. Molecular anthropology is explored from the perspective of immunology.

The Changing Pace of Evolution

Rates of Evolution of Nonfossilized Traits

Macroevolution Mechanisms

South American Mammals: Isolation and Competition

Continental Drift and Macroevolution

Speciation and Adaptive Radiation

How Predictable is Evolution?

ESSAY QUESTIONS

1. Why is natural selection generally slower than artificial selection?

> Artificial selection is based on a specific goal in the mind of an investigator, who seeks to attain that goal as rapidly as possible. The products of such selection need not be viable under natural conditions, so it is possible to select intensively for one desirable trait while sacrificing others. For example, a high-yield crop plant might be developed without regard to its resistance to disease or its ability to survive adverse weather, because pesticides, irrigation and the like can compensate for any deficiencies. Natural selection, by contrast, tends to favor compromises: An individual that is barely adequate in all respects usually has a better chance than one that is superior in a few traits but deficient in many others. Thus natural selection for any given trait proceeds by smaller steps.

2. Discuss the relevance of amino acid and nucleotide substitutions to the study of evolution.

See textbook page 982-983

3. Outline the major theories that have been proposed to account for Cretaceous extinctions.

See textbook page 988

4. Explain the "adaptive landscape" model of natural selection, using the mollusks of Lake Turkana as an example.

See textbook page 989-991

TEST SET 38-A

Name _____   Section _____

Choose the <u>best</u> answer to each of the following questions, and write the appropriate letter in the space provided.

Ans: d
p. 971

_____   1.   Which animal group contains the most species? (a) vertebrates; (b) mollusks and other hard-shelled invertebrates; (c) bacteria; (c) insects.

Ans: b
p. 971

_____   2.   How did the earliest fishes differ from later ones? (a) They were generally smaller and more flexible. (b) They had heavy armor and no jaws. (c) They had short limbs instead of fins. (d) They had lungs, not gills.

Ans: b
p. 974

_____   3.   During which period did the first birds appear? (a) Permian; (b) Jurassic; (c) Cretaceous; (d) Pleistocene.

Ans: b
p. 974

_____   4.   During which period did the first mammals appear? (a) Triassic; (b) Jurassic; (c) Eocene; (d) Pleistocene.

Ans: a
p. 975

_____   5.   Which of the following statements <u>best</u> reflects our present knowledge of the fossil record? (a) The ancestors of humans and of baboons had some very similar adaptations. (b) Humans have evolved from baboons or baboon-like primates. (c) The ancestors of humans lost their prehensile tails when they began living on the ground. (d) The evolution of bipedalism appears to have followed an increase in brain size.

Ans: b
p. 975

_____   6.   All of the following are presently under consideration as possible causes of extinction <u>except</u>: (a) continental drift; (b) racial senescence; (c) asteroids; (d) climatic changes.

Ans: d
p. 975

_____   7.   How is silent substitution (of nucleotides) different from re-placement substitution? (a) The first is harmful; the second is not. (b) The second is neutral; the first is beneficial. (c) The first can be detected; the second cannot. (d) The second causes changes in amino acid sequence; the first does not.

Ans: d
p. 976

_____   8.   How does phyletic evolution differ from speciation? (a) The first is relatively sudden; the second is gradual. (b) The first involves groups of organisms; the second involves individuals. (c) The first refers to an important new adaptation that gives rise to a new phylum; the second is another word for the adaptive radiation that follows. (d) The first means gradual change in one lineage; the second means the splitting of one lineage to form two.

Ans: d
p. 973

_____   9.   Why do most fossil assemblages <u>not</u> represent past ecological communities? (a) Different organisms fossilize at different rates, so an entire community was not preserved at once. (b) Ecological succession occurs too rapidly. (c) The organisms in such an assemblage did not live in one place, but their remains were carried to the fossil site from various places. (d) Complex ecological communities did not evolve until late in the history of life.

376   Chapter 38

Ans: a
p. 977
_____ 10. Why is the fossil record better for higher taxonomic categories (such as phyla) than for lower ones (such as species)? (a) A higher level includes many more individuals than a lower one. (b) Phyla have existed far longer; it takes time for them to begin to differentiate into species. (c) There are many more phyla than there are species. (d) Phyla tend to fossilize more readily than species.

Ans: d
pp. 987-988
_____ 11. Is coal still being formed? (a) No, because the carboniferous forests are gone. (b) No, because it was formed largely from the remains of dinosaurs, which are now extinct. (c) Yes, because all dead organic matter ultimately becomes coal under suitable conditions. (d) No, because swamps are being drained.

Ans: b
pp. 979-980
_____ 12. Since the beginning of the Cenozoic, mammals, as a group: (a) have increased in numbers but decreased in size; (b) have increased in both numbers and size; (c) have increased in size but decreased in number of species, because of extinctions caused by human activities; (d) have remained fairly similar because all major taxa had appeared by then.

Ans: c
p. 983
_____ 13. If a child were born with a single eye in the center of its forehead, would this constitute evidence that humans evolved from cyclopean primate ancestors? (a) Yes, because primitive ancestral characteristics tend to reappear sooner or later. (b) Yes, because folklore and mythology demonstrate the past existence of cyclops. (c) No, because no fossil cyclops has been found and because errors of development do occur from time to time. (d) No, because early primates needed binocular vision to leap through trees.

Ans: c
p. 972
_____ 14. The most likely evolutionary factor influencing enlargement of the human brain was: (a) the need to make weapons; (b) the need to escape from predators; (c) the need for cooperation; (d) the need for improved vision.

Ans: a
p. 969
_____ 15. The fibrinogen molecule: (a) contains two short peptides that hold the chain open; (b) has changed little in the course of evolution; (c) is manufactured largely by pseudogenes; (d) is formed from fibrin by enzyme action.

TEST SET 38-B

Name _____ Section _____

Choose the best answer to each of the following questions, and write the appropriate letter in the space provided.

Ans: b
p. 971

_____ 1. Which animal groups tend to have the best fossil record?
(a) those with the most species; (b) those with hard parts;
(c) flying and swimming animals; (d) large animals.

Ans: b
pp. 970;
980-982

_____ 2. During which period did the first amphibians appear?
(a) Ordovician; (b) Devonian; (c) Carboniferous; (d) Permian.

Ans: b
p. 972

_____ 3. During which period did the first reptiles appear? (a) Devonian;
(b) Carboniferous; (c) Triassic; (d) Cretaceous.

Ans: c
p. 972

_____ 4. All of the following groups have undergone rapid speciation
during the Cenozoic except: (a) snakes; (b) birds; (c) brachiopods; (d) bats.

Ans: d
p. 970

_____ 5. Bipedal locomotion: (a) is used by no animals other than humans;
(b) is faster than quadrupedal locomotion; (c) appeared late in
human evolution, possibly to allow hunters to see farther;
(d) frees the hands to use tools.

Ans: d
p. 974

_____ 6. Which of the following is a unique characteristic of humans?
(a) binocular vision; (b) symbolic communication; (c) use of
tools; (d) culture, as defined in the text.

Ans: c
p. 970

_____ 7. The quality of the fossil record has been significantly affected
by all of the following except: (a) body structure; (b) sediment
survival; (c) asteroid bombardment; (d) taxonomic level.

Ans: b
p. 976

_____ 8. The text states that most species of insects living today have
not yet been described. How, do you suppose, have scientists
determined that? (a) Insects have left a good fossil record, so
it is possible to estimate the number of species that have lived
at other times in the earth's history. (b) Museums are full of
unclassified insect specimens that no one has had time to
describe, and collecting trips yield a high percentage of un-
knowns. (c) It is possible to predict the number of species in
a group on the basis of the length of time it has existed and
the constancy of mutation and speciation rates. (d) Analysis of
an ecosystem reveals accurately the number of species it can
support.

Ans: b
pp. 976-
977

_____ 9. Why has sediment survival increased over time? (a) Rates of
formation of sedimentary rocks have increased. (b) More recent
sediments are less likely to have been buried. (c) Since the
Cretaceous, the oceans have withdrawn from shallow areas.
(d) Volcanic activity was more frequent in the past.

Ans: d
chapter

_____ 10. Which group of animals has not declined appreciably since the
Triassic? (a) amphibians; (b) brachiopods; (c) reptiles;
(d) fish.

Ans: b
p. 970

_____ 11. Which groups of organisms became extinct at the end of the
Cretaceous? (a) dinosaurs only; (b) dinosaurs and ammonites
only; (c) dinosaurs, ammonites, and gymnosperms only; (d) dinosaurs, ammonites, gymnosperms, and brachiopods.

Ans: d
pp. 977-
     978

_____ 12. In 1982, a medical journal reported that a child had been born with a two-inch tail. The event was cited as evidence that humans evolved from apes. Was this a valid inference? (a) Yes, because primitive ancestral characteristics tend to reappear sooner or later. (b) Yes, because early prehumans used prehensile tails for living in trees. (c) No, because our ancestors lost their prehensile tails when bipedalism freed their hands. (d) No, because the Old World primate ancestors of humans did not have tails.

Ans: d
p. 980

_____ 13. The giant ground sloth (a Pleistocene mammal, officially extinct) is claimed by some still to be alive in South America. There are also reports of live dinosaurs in Africa. Which is the more likely claim? (a) The dinosaur, because much of Africa is unexplored. (b) The dinosaur, because a cold-blooded animal could remain dormant and hidden for long periods. (c) The sloth, because it is smaller and harder to spot from an airplane. (d) The sloth, because there are fairly recent fossils of it (aged about 10,000 years rather than 100,000,000).

Ans: b
p. 983

_____ 14. What is meant by the cladistic component of systematics? (a) Species that look similar are grouped together. (b) Species with recent common ancestors are grouped together. (c) Species that occupy the same geographical area are grouped together. (d) Species with identical hemoglobin structure are grouped together.

Ans: b
p. 984

_____ 15. Compared with modern elephants, Miocene elephants: (a) had longer trunks; (b) had more tusks; (c) had a shorter lower jaw; (d) were essentially the same.

CHAPTER THIRTY-NINE
ECOSYSTEMS

CHAPTER SUMMARIES AND OBJECTIVES

39-1 <u>Climates on Earth / Global Atmospheric Circulation / Global Oceanic Circulation / Biogeochemical Cycles</u> The sun provides the earth with heat, the amount and distribution of which vary with latitude, season, elevation, and topography. Global air circulation results from adiabatic heating and cooling, solar energy input, and wind. Objects (or winds) moving toward or away from the equator experience the Coriolis force: Momentum is conserved while relative velocity increases or decreases, so the object is deflected to one side. Oceanic circulation is, in turn, driven by global air circulation and modified by the presence of the continents. Because ocean temperature is relatively stable, coastal land areas have less seasonal temperature fluctuation than inland areas. Seasonal rainfall distribution is also influenced by ocean temperatures. All organisms are composed largely of a few chemical elements, primarily carbon, nitrogen, phosphorus, calcium, sodium, sulfur, hydrogen, and oxygen. These circulate through ecosystems as will be discussed in later sections.

   1. Explain the major factors determining climate patterns.

   2. Describe global patterns of wind and oceanic circulation.

39-2 <u>The Oceanic System / Fresh Waters / The Atmospheric System / The Land / Plants and Soils</u> The global ecosystem includes the oceans, fresh waters, the atmosphere, and the land. The oceans contain much of the earth's carbon, phosphorus, and sulfur, and they respond slowly to outside disturbances; their chief photosynthetic plants are algae. Fresh waters originate as rainfall and carry minerals to the oceans. Mineral circulation in lakes results from wind and biannual overturn. The atmosphere includes the troposphere and stratosphere; it consists largely of nitrogen and oxygen and helps regulate surface temperatures. Airborne compounds are oxidized in the atmosphere and return nutrient elements to the land via precipitation. The land (about one-fourth of earth's surface) is covered by soil, which stores nutrient elements and loses them to the ocean via groundwater. Soil is formed by weathering of rocks and is subdivided into A, B, and C horizons according to its composition. Decomposition of plant litter produces humus.

   3. Name and give the characteristics of the four major ecosystem compartments.

39-3 <u>Carbon and Nitrogen Cycles / Human Activity and Aquatic Cycles / Acid Precipitation / Alterations of the Carbon Cycle</u> All of earth's minerals are continually recycled; some enter deep sea elements and are temporarily removed from circulation. Human activities have altered biogeochemical cycles at local, regional, and global levels. Sewage and other wastes add phosphorus to lakes and rivers, causing algal and bacterial growth which depletes oxygen and alters fish and invertebrate faunas, as illustrated by the example of Lake Erie. Acid precipitation results from fossil fuel combustion, which yields oxides of nitrogen and sulfur; it damages plants and fish and lowers the cation exchange capability of soil. Increased atmospheric CO_2, resulting from fossil fuel combustion and the burning of forests, may cause significant warming over the entire earth.

4. Outline the principal biogeochemical cycles.

5. Describe the major effects of human activities on biogeochemical cycles.

39-4 <u>Energy Flow through Ecosystems / Body Maintenance and Energy Flow / Agriculture and Ecosystem Productivity / Ecosystem Stability</u> The sun is the ultimate source of nearly all energy used by organisms. Net primary production by plants--photosynthesis minus respiration--varies by geographic region because it is correlated with transpired water, which in turn is influenced by soil moisture and air temperature. Organisms are grouped into trophic levels according to their principal energy source. Energy and biomass pyramids represent the relative energy and biomass distributions among trophic levels of an ecosystem; the shapes of these pyramids depend on the nature of the ecosystem (for instance, the digestibility of the dominant plants). Food chains are sequences of species related to each other as prey and predator. The mean production efficiency of a given group of organisms is the energy apportioned to production divided by the total energy used (production plus respiration). Endotherms are less efficient than ecotherms; herbivores are less efficient than carnivores.

Agriculture serves to maximize those species valuable to humans and it does so at the expense of other species. It requires removal of competitors and pests, application of fertilizers, selective breeding, and energy for cultivation and harvesting. Less nutrient cycling occurs in agricultural than in natural systems. Ecosystem stability includes both resistance to disturbance and ability to return to normal after a disturbance. When perturbation exceeds a limit, ecosystems may achieve alternative stable states.

6. Explain the significance of (a) net primary production, (b) trophic levels, (c) energy/biomass pyramids, (d) food chains, and (e) mean production efficiency.

7. Explain how agricultural ecosystems differ from natural ones.

ARTICLE RESOURCES

Climates on Earth

Gore, Rick. An age-old challenge grows. <u>National Geographic</u>, 1979, 156(5), 594-639. Well-illustrated piece depicting the desert regions of Egypt and the United States. A map shows the major deserts of the world and the prevailing winds that may be responsible for their location. Discusses the climate and vegetation of deserts. Sounds an alarm about overpopulation and overgrazing. Shows unique methods applied in Iran for reclaiming the desert.

Global Atmospheric Circulation

Global Oceanic Circulation

Canby, Thomas Y. Our most precious resources--water. <u>National Geographic</u>, 1980, 158(2), 144-179. A long article detailing the hydrologic cycle, pollution, acid rain, aquifers, eutrophication, toxic waste, and irrigation. Covers the United States from ocean to ocean and from border to border.

Matthews, Samuel W. New world of the ocean. <u>National Geographic</u>, 1981, 160(6), 792-833. A history of ocean exploration from 1950 to the present. Satellite photographs of phytoplankton concentrations, undersea vessels for exploration, a model of an ocean showing ocean-to-air boundaries, surface currents, and a wealth of other details are included. Discussion of the Coriolis effect is illustrated. A chart of the ocean layer circulation is shown. Some of the products (other than fish) that are derived from the sea are cited.

Biogeochemical Cycles

The Oceanic System

   Wiehe, Peter H. Rings of the Gulf Stream. Scientific American, 1982, 246(3), 60-70. The Gulf Stream and other oceans traversed by currents produce eddies that affect the hydrography and the biology of the sea. The rings formed can be as large as 300 kilometers in diameter. The salinity, temperature, distribution of life, and oxygen content differ from one side of the eddy to another. They extend to a depth of 4,000 to 5,000 meters. Several rings may be formed at once and may be either warm-cored or cold-cored, depending on where they are formed. Zooplankton is concentrated near the center of cold-core currents. Clear diagrams and graphs are given.

   Goreau, Thomas F., Nora I. Goreau, and Thomas J. Goreau. Corals and coral reefs. Scientific American, 1979, 241(2), 124-138. Excellent descriptions of fringing reefs, barrier reefs, and atolls. The available nutrients in tropical oceans are sparse, yet coral reef environments have the highest rates of nitrogen fixation, photosynthetic carbon fixation, and limestone deposition of any ecosystem. The biology of corals is discussed. Good photographs and diagrams. A map shows coral reefs of the world.

Fresh Waters

The Atmospheric System

The Land

Plants and Soils

Carbon and Nitrogen Cycles

Human Activity and Aquatic Cycles

Acid Precipitation

Alterations of the Carbon Cycle

   Revelle, Roger. Carbon dioxide and world climate. Scientific American, 1982, 247(2), 35-43. The burning of fossil fuels and the clearing of forests have increased atmospheric carbon dioxide about 15 percent in the past 100 years. Carbon dioxide absorbs and radiates some of the infrared radiation that would otherwise be transmitted back into space, so we have what is known as the "greenhouse effect." This is accompanied by increases in the average global temperature. The absorption of carbon dioxide by the oceans is discussed. A diagram of the carbon dioxide cycles is given. The causes and effects of the increase in carbon dioxide in the atmosphere is shown, and steps to control it are suggested.

Energy Flow through Ecosystems

Body Maintenance and Energy Flow

   Vogel, Stephen. Organisms that capture currents. Scientific American, 1978, 239(2), 128-135, 139. Exploration of the physical effects of air and water flow, which are exploited by a variety of animals and plants. Sponges, prairie dogs, termites, stomata of plants, limpets, and birds all utilize velocity differences to extract energy. Well illustrated.

Agriculture and Ecosystem Productivity

Ecosystem Stability

ESSAY QUESTIONS

1. Explain what is meant by "closed" vs. "open" mineral cycles, and give an example of each, using diagrams.

    See Figures 11 and 12. (pages 1005-1006)

2. Draw and label pyramids of biomass to represent (a) a marine ecosystem and (b) a forest ecosystem. Explain the reason for the differences in proportions.

    See Figure 19. (page 1014)

3. Why do you think domesticated plants tend not to do well in competition with wild plants unless farmers invest energy in the form of cultivation, herbicides, pesticides, fertilizers, irrigation, and so on? How do these requirements exacerbate the effects of agriculture on nutrient cycles?

    Domesticated plants have been selected for specific traits desirable to humans, not for general viability. Also, many have been introduced from other parts of the world and have therefore evolved in response to very different conditions. Fertilizers add phosphorus to aquatic systems, causing eutrophication; irrigation may leach minerals from soil and/or divert water necessary to other ecosystems; pesticides and herbicides may be absorbed by organisms other than those for which they were intended, with harmful long-term effects.

4. How do lush tropical forests manage to grow, even though their soils are too impoverished to support agricultural systems?

    See textbook page (pages 1005, 1016)

TEST SET 39-A

Name _____   Section _____

Choose the best answer to each of the following questions, and write the appropriate letter in the space provided.

Ans: b
p. 995

_____   1. Seasonal change in temperature increases at higher latitudes because: (a) the earth rotates more slowly; (b) day lengths vary more; (c) the Coriolis force is greater; (d) the angle of sunlight is greater.

Ans: d
p. 996

_____   2. Why does air descend at the equator? (a) The Coriolis force deflects it. (b) There are fewer mountains to cause it to rise. (c) Winds blow away from the equator, so air from above descends to replace it. (d) It doesn't descend; it rises.

Ans: c
p. 996

_____   3. When an object in either hemisphere moves toward the equator, it is deflected to the: (a) right; (b) left; (c) west; (d) east.

Ans: c
p. 999

_____   4. A gram of water and a gram of soil each cool by one degree centigrade. The amount of heat given off: (a) is greater for the water; (b) is greater for the soil; (c) is the same for both; (d) cannot be determined from the information given.

Ans: b
p. 999

_____   5. Fish are abundant in zones of upwelling because: (a) the water is warm; (b) nutrients are brought up from deeper water; (c) fish are carried to these areas by ocean currents; (d) eutrophication by sewage from the nearby shore increases nutrients in the water.

Ans: c
p. 1000

_____   6. Which of the following elements is least abundant in living organisms? (a) hydrogen; (b) nitrogen; (c) phosphorus; (d) oxygen.

Ans: d
p. 1002

_____   7. Which of the following elements is least abundant in the atmosphere? (a) argon; (b) oxygen; (c) nitrogen; (d) carbon dioxide.

Ans: d
p. 1003

_____   8. What is the main source of the OH radical in the atmosphere? (a) smog from internal combustion engines; (b) fluorocarbons from aerosol sprays; (c) photosynthesis; (d) ozone bombardment by ultraviolet light.

Ans: c
p. 1004

_____   9. Why are desert soils potentially very fertile? (a) Many animals die in the desert, and their decomposition enriches the soil. (b) Heat causes chemical reactions that provide bases necessary to plant growth. (c) Evaporation exceeds precipitation, so nutrients are not leached from the soil. (d) There is little plant life to remove nutrients from the soil.

Ans: d
p. 1004

_____   10. Why are tropical soils potentially very fertile? (a) Vegetation is abundant and decomposes rapidly. (b) Rainfall is taken up by plant roots before it can penetrate far enough to leach minerals out of the soil. (c) Nitrogen-fixing bacteria are abundant. (d) They are not very fertile; bases have been leached out.

Ans: a
p. 1008

_____   11. How can reversal of eutrophication best be achieved? (a) Reduce the amount of sewage entering a lake. (b) Treat sewage before it is dumped into the lake. (c) Introduce fish to eat the algae. (d) Introduce chemicals to kill the bacteria.

Ans: a
p. 1009

_____ 12. How can the effects of acid rain <u>best</u> be reversed? (a) Remove oxides of nitrogen and sulfur from stack gases. (b) Avoid using any nitrogen-containing fuels. (c) Apply alkaline solutions to plants and freshwater systems to neutralize the acidity. (d) Allow natural selection to produce strains of plants and fish that can tolerate low pH.

Ans: d
p. 1010

_____ 13. About what fraction of solar energy received by the earth is utilized in photosynthesis? (a) 1/10; (b) 1/100; (c) 1/1,000; (d) 1/10,000.

Ans: d
p. 1012

_____ 14. An acquaintance of the author claims that sun-dried fruit contains more energy than fresh fruit. As evidence, she cites a biology teacher who once told her that "plants make energy from sunlight." Is the author's acquaintance in error? (a) No; any health-food dealer will tell you the same thing. (b) Yes; the energy produced in the fruit by photosynthesis will be canceled by the energy lost in the evaporation of moisture. (c) No; photosynthesis occurs only in leaves and not in any other plant part. (d) Yes; photosynthesis occurs only in green parts of living plants, not in fruit that has been picked.

Ans: d
p. 1012

_____ 15. The same acquaintance (see Question 14) cites as further evidence a USDA bulletin which shows that a pound of dried pears contains ten times as many calories of energy as a pound of fresh ones. Based on this information, would you like to change your answer to the previous question? (a) Yes; you can't argue with statistics. (b) No; these statistics are probably bogus. (c) Yes; it is true that green pears have chlorophyll in their chloroplasts, so they could generate food energy while drying. (d) No; a pound of dried pears simply contains more <u>pears</u> than a pound of fresh ones.

TEST SET 39-B

Name _____   Section _____

Choose the best answer to each of the following questions, and write the appropriate letter in the space provided.

Ans: b
p. 995

_____   1.   What is the approximate straight-line distance represented by every degree of latitude on earth? (a) 10 km; (b) 110 km; (c) 110 miles; (d) It is too variable to given an approximation.

Ans: a
p. 996

_____   2.   Why does air rise at 60° North latitude? (a) The reason is not known. (b) Direction is conserved, so air moving in a straight line leaves the earth tangentially. (c) Solar energy input is maximum at that latitude, so the heated air rises. (d) It does not rise there; it descends.

Ans: c
p. 998

_____   3.   Oceanic circulation is driven by the: (a) continents; (b) moon; (c) winds; (d) force of gravity.

Ans: a
p. 997

_____   4.   Solar energy input is maximum at: (a) the intertropical convergence zone; (b) the equator; (c) the Tropic of Cancer; (d) the Tropic of Capricorn.

Ans: a
p. 999

_____   5.   A continental climate: (a) shows little seasonal variation; (b) shows marked seasonal variation; (c) has two rainy seasons, unlike a maritime climate, which has one; (d) is characteristic of Europe, particularly the French Riviera.

Ans: b
pp. 999-
       1000

_____   6.   Annual rainfall is greatest in coastal areas; (a) in summer; (b) at high latitudes; (c) when the land is warmer than the ocean; (d) when surface water is warmed by upwelling.

Ans: b
p. 1000

_____   7.   Photosynthesis (vs. respiration) is the dominant biological process in all of the following environments except: (a) oceans; (b) rivers; (c) lakes; (d) land.

Ans: a
p. 1007

_____   8.   Soil, when formed, initially lacks nitrogen. How does nitrogen, in forms usable by plants and animals, get from the atmosphere into the soil? (a) fixation by microorganisms; (b) simple diffusion; (c) photosynthesis; (d) oxidation.

Ans: c
p. 1004

_____   9.   Which cations are most important for plant nutrition? (a) sodium, aluminum, and silicon; (b) hydrogen, oxygen, carbon, and nitrogen; (c) calcium, potassium, and magnesium; (d) iron, sulfur, and copper.

Ans: a
p. 1005

_____   10.   Which of the following statements about humus is not true? (a) It is formed by decomposition of clay. (b) It has two varieties called mull and mor. (c) Its particles are small. (d) It adds nutrients to soil and modifies its consistency.

Ans: b
p. 1007

_____   11.   Cycling of substances is slowest through: (a) living organisms; (b) the oceans; (c) the atmosphere; (d) the terrestrial ecosystem.

Ans: d
p. 1008

_____   12.   The element that most often limits photosynthesis in fresh water is: (a) carbon; (b) oxygen; (c) nitrogen; (d) phosphorus.

Ans: d
p. 1009

_____ 13. What is the <u>most</u> likely effect of increased atmospheric CO_2, resulting from fossil fuel combustion? (a) No effect; it quickly enters marine sediments as calcium carbonate. (b) Plant life will increase, because photosynthesis requires CO_2; atmospheric oxygen and ozone will also increase as by-products of photosynthesis. (c) The excess CO_2 will be converted to carbonic acid in the atmosphere, causing acid precipitation. (d) More infrared radiation will be retained by the atmosphere, thus causing warmer climates.

Ans: c
p. 1012

_____ 14. Organisms are categorized in trophic levels according to: (a) taxonomic group; (b) geographical area; (c) principal energy source; (d) oceanic, atmospheric, freshwater, or terrestrial ecosystem.

Ans: a
p. 1012

_____ 15. The coyote regularly eats berries and other vegetable matter, yet it is invariably described as a carnivore. Is this classification valid? (a) Yes, because the bulk of its diet is meat. (b) No, because true carnivores are strictly hunters. (c) Yes, because "carnivore" is a taxonomic rather than an ecological term. (d) No, because the term "carnivore" alone means nothing; an animal is either a primary carnivore or a secondary carnivore.

CHAPTER FORTY
GROWTH AND DYNAMICS OF POPULATIONS

CHAPTER SUMMARIES AND OBJECTIVES

40-1 <u>Life History as a Concept / Births and Deaths</u> The study of population dynamics encompasses all the activities, life stages, births, and deaths of the members of a deme. The primary life stages--which often overlap extensively--are (1) the growth stage, from newborn (spore, hatchling, and so on) to adult; (2) the dispersal stage, when organisms leave their birthplace; (3) the reproductive stage; and (4) the energy-gathering stage, which may or may not span the entire life history. Energy is allocated among these activities according to a time and energy budget that has resulted from natural selection.

Births and deaths are studied by means of life tables, which plot the proportion of a given cohort still alive at various time intervals. A survivorship curve represents these data in graphical form; the shape of this curve varies widely among taxonomic groups and often between sexes. The relative magnitudes of the birth rate and death rate determine whether the population is increasing or decreasing.

1. Explain the significance of population dynamics.

2. Name the major life stages of an organism, and explain the concept of a time and energy budget.

3. Explain the significance of life tables and survivorship curves.

40-2 <u>Exponential Growth / Logistic Growth / Evolution of Birth and Death Rates</u> The intrinsic rate of increase, <u>r</u>, of a population is equal to the difference between the average birth rate <u>b</u> and the average death rate <u>d</u>. Exponential population growth is expressed by the equation R = (b - d)N, where N is the number of individuals in the population at a given time and R (which is equal to dN/dT, for those who have studied calculus) is the <u>instantaneous</u> rate of population growth. (In order to calculate the actual population size at a given time, the above expression must be integrated, because the population size N will continuously change. The resulting equation contains an exponent, which explains why this form of increase is called "exponential.")

Under real-life conditions, population growth is limited by environmental factors and cannot be described adequately by the exponential model. Thus the concept of carrying capacity, <u>K</u>, is introduced:

$$R = (b - d) \times (\frac{K - N}{K})N$$

where K is the maximum number of individuals that an environment will support. Note that when N = K, R = 0. Several oversimplifications are inherent in this model, but it is often a useful approximation.

Birth and death rates are correlated with each other and with the environment. As population size approaches carrying capacity, the energy available for reproduction may decline. Organisms may be semelparous (reproducing once in a lifetime) or iteroparous (reproducing more than once). The timing of reproduction has been selected to optimize survival of offspring.

4. Explain the exponential and logistic models of population growth.

5. Explain, and give examples of, the selection pressures that have influenced birth and death rates and the timing of reproduction.

40-3 Age Distribution / Populations in Patchy Environments / Habitat Selection / Population Density and Habitat Selection A theoretical population, in which death is independent of age, can be shown to be dominated by young individuals. If a death-causing gene is introduced into this population, its effect on the age distribution depends on the time of expression of this gene: If the phenotypic effects appear relatively late in life, few individuals will have survived long enough to be affected by it, assuming that accidental deaths occur. Thus selection will tend to favor late expression of deleterious genes and early expression of favorable ones.

Environments vary in homogeneity. Most are "patchy" both in time and in space, and a given species favors certain patches within its habitat. Habitat selection involves two choices: where to settle and when to change locations. These choices involve four levels of decision-making: (1) habitat; (2) patch; (3) behavioral mode; and (4) responses to objects encountered. The relative risks and advantages of leaving a suboptimal habitat in search of a better one also influence habitat selection, as does the density of conspecific individuals already present at a given location and their behavior toward new arrivals.

6. Explain the principal factors influencing mortality rates, and cite their effects on age distribution.

7. Explain the concept of a patchy environment.

8. Summarize the major choices an organism must make in the process of habitat selection.

40-4 Dispersal / Population Regulation / Population Regulation and Management Many animals leave habitats when conditions become unfavorable, often on a regular seasonal basis. Other animals may enter a resistant phase or physiological state (such as hibernation). Similarly, plants may die back, dry up, or produce seeds or spores in order to survive such conditions. Regular migrations and less regular irruptions are observed primarily in birds, mammals, and insects.

Most populations fluctuate relatively little in size. Generally speaking, populations become smaller at higher levels on the food chain. Population size is regulated by both density-dependent and density-independent factors. The first category includes those factors that are related to population density, such as food supply, predation, and disease. The second category includes such events as bad weather, but few environmental factors are entirely density-independent. Normally, populations are regulated by several interacting factors influencing birth and death rates.

Human beings consider certain species desirable and certain others undesirable, and they attempt to regulate population sizes accordingly. Such attempts have revealed (1) that population growth is highest at levels well below K, and (2) that logistic population growth is maximum--in numbers, not rate--at about one-third K to one-half K. Optimum harvesting strategies vary among groups of organisms, depending partly on their modes of reproduction. When such strategies are not employed, overharvesting may result.

9. Describe the major ways in which plants and animals respond to or avoid unfavorable environmental conditions.

10. Explain the major factors regulating population size, distinguishing between density dependence and density independence.

11. Explain how strategies are determined for harvesting "desirable" species and eliminating "undesirable" ones.

ARTICLE RESOURCES

Life History as a Concept

Births and Deaths

Exponential Growth

Logistic Growth

Dafni, Amots, and David Heller. Adventive flora of Israel--Phytogeographical, ecological, and agricultural aspects. Plant Systematics and Evolution, 1982, 140(1), 1-18. Adventive plants as colonizing species are usually weeds characterized by extension of seed germination over a long period, wide range of plasticity and adaptability, rapid reproduction under adverse conditions, wide seed dispersal, and large numbers of progeny. Their seeds are generally light and are capable of being carried long distances. Seedlings grow rapidly and several generations per year can be grown. Seventy-three adventive species in Israel were studied. Fifty-three percent were of tropical origin.

Evolution of Birth and Death Rates

Age Distribution

Populations in Patchy Environments

Hamner, William M. Strange world of Palau's salt lakes. National Geographic, 1982, 161(2), 264-282. The Palau Islands are part of the Trust Territory of the Pacific southwest of the Philippines. They were formed twenty million years ago by volcanic forces. They are essentially reef-enclosed marine lakes. Each lake is an isolated marine ecosystem. They present a unique natural experiment in the organization of marine environment food webs. Some support large sharks and crocodiles. Others have minute crustaceans as the principal inhabitants. Well illustrated.

Vietmeyer, Noel D. Rediscovering America's forgotten crops. National Geographic, 1981, 159(5), 702-712. Tepary beans, ground nuts, jojoba, guayule, and amaranth are native American plants growing, except for ground nuts, in isolated patchy habitats in the southwestern United States. These plants may prove valuable substitutes for cultivated beans, potatoes, oil, rubber, and wheat. They are drought-resistant and produce seeds that are extremely versatile.

Habitat Selection

Population Density and Habitat Selection

Dispersal

Fisher, Allan C. Jr. Mysteries of bird migration. National Geographic, 1979, 156(2), 154-194. Methods of navigation by birds are delineated. Navigation routes of birds are shown on a global map. Excellent photographs showing bird capturing and banding. Recovery of banded birds or even just the bands have revealed where and when birds migrate, how long they live, and whether they return to the same area.

Population Regulation

Drewien, Roderick G., and Ernie Kuyt. Teamwork helps the whooping crane. National Geographic, 1979, 155(5), 680-693. Efforts to have sandhill cranes hatch and rear whooping cranes have been successful. The history of the near extinction of the whooping crane is told, and an account is given of the discovery of their nesting grounds. A flock has been established at the Patuxent Wildlife Research Center in Maryland, and a second flock has been established in the U.S. Rockies.

Population Regulation and Management

Wildcat of the North: Canada lynx. National Geographic World, 1981, 76, 3-7. The lynx is a solitary hunter of the wilderness areas of Canada and sections of the United States. Its chief food is the snowshoe hare as well as squirrels, mice, rabbits, and birds. The fur of the lynx has been prized for clothing, but the lynx has been hunted so much that limited hunting seasons are now enforced. Excellent photographs.

Bohlen, Janet. Saving wild species through habitat protection. The Science Teacher, 1980, 47(4), 20-23. The wild ecosystems are being destroyed at record rates. Globally, we are losing a species a day. Therefore, it is imperative that we learn to manage ecosystems. Most endangered species are threatened by habitat destruction. A list of areas being lost is provided, as well as the species that are endangered by this loss. Several techniques for studying ecosystems are given.

ESSAY QUESTIONS

1. Describe the significance of birth, death, and the four conventional life stages with reference to a bacterium. Is it necessary to redefine some terms?

    The point here is that general designations of life history stages do not apply neatly to all organisms. In a bacterium, does the individual cease to exist at the reproductive stage? What is the significance of its dispersal stage? And so on.

2. Discuss the assumptions, limitations, and shortcomings inherent in the logistic model of population growth.

    See textbook page 1026-1027.

3. Define and compare the ideal free distribution and the ideal dominance distribution.

    See textbook pages 1033-1034.

4. The text states that "every one of us will die." Yet many alternatives to death are being seriously considered nowadays: cryogenic freezing of people with terminal illnesses, in the hope of reviving them when cures have been found; improved medicine; elucidation of the aging process; colonization of space, if the earth becomes full; and so on. Discuss the probable effectiveness of each of these measures (and any others that come to mind) in ecological terms.

    Basic points that should be covered: If death by natural causes is overcome, then problems of food and space will be even more critical than they are at present. Besides, since when has advanced medical technology been available to more than a tiny percentage of the human race? It would be impossible to build and launch spaceships full of people at a rate fast enough to compensate for births, even if we knew of a good place to send them. And so on.

TEST SET 40-A

Name _____   Section _____

Choose the best answer to each of the following questions, and write the appropriate letter in the space provided.

Ans: d
p. 1025

_____ 1. The text states that a single bacterium, growing and reproducing without limit, would in one month form a colony outweighing the universe and expanding at the speed of light. What is the fundamental reason why this will not occur? (a) No object can attain the speed of light. (b) The universe is infinite. (c) The whole idea is preposterous. (d) Density-dependent phenomena would restrict growth.

Ans: c
pp. 1024-
   1025

_____ 2. In a hypothetical population, N = 1000, the birth rate b = .10, and the death rate d = .05. Using the formula R = (b - d)N, determine the size of the population after five years. (a) 1250; (b) 250; (c) 1276; (d) The answer cannot be determined using this formula; an exponential expression is required.

Ans: c
p. 1026

_____ 3. In a hypothetical logistically growing population, N = 850, K = 1000, b = .20, and d = .15. R for this population is: (a) 42.5; (b) 425; (c) 6.375; (d) 63.75.

Ans: b
p. 1026

_____ 4. For the same logistically growing population (Question 3), would R be higher or lower if N = 900? (a) higher; (b) lower; (c) the same; (d) The answer cannot be determined from the information given.

Ans: d
p. 1026

_____ 5. For the same population (Questions 3 and 4), at what level of N will the greatest numbers of new individuals be added to the population? (a) At N = 0, because at that point logistic growth is equivalent to exponential growth. (b) At very low, but non-zero, N. (c) At N = 1000, when all environmental resources are being used for maximum growth. (d) At N = 350 to 500.

Ans: d
p. 1026

_____ 6. Under the logistic growth model, what happens when N exceeds K and d exceeds b? (a) N can never exceed K. (b) Because deaths outnumber births, R must be negative. (c) Because the carrying capacity is exceeded, R must be 0. (d) The model breaks down under these conditions. (Hint: Try it on paper.)

Ans: b
p. 1026

_____ 7. All of the following statements about K (carrying capacity) are true except: (a) It varies from one species to another. (b) It is constant for any given species. (c) It may temporarily be exceeded by population size. (d) It limits population growth.

Ans: a
p. 1027

_____ 8. The birth rate for a typical population is usually: (a) similar to the death rate; (b) lower than the death rate; (c) higher than the death rate; (d) independent of the death rate.

Ans: b
p. 1028

_____ 9. Production of smaller, more numerous offspring appears to be a good strategy in: (a) stable environments; (b) fluctuating environments; (c) aquatic environments; (d) highly competitive environments.

Ans: d
p. 1028

_____ 10. Production of larger, fewer offspring appears to be a good strategy in: (a) stable environments; (b) fluctuating environments; (c) aquatic environments; (d) highly competitive environments.

Ans: d
p. 1029

11. In a population with a stable age distribution: (a) individuals are immortal; both the birth rate and the death rate are assumed zero; (b) population size is constant over time; (c) carrying capacity has been reached; (d) birth and death rates are constant.

Ans: a
p. 1029

12. What is meant by the following statement? "The consequences of a deleterious allele are reduced when its phenotypic expression occurs later in the life of the organism." (a) An older individual is more likely to have reproduced already, so its death has little effect on the population. (b) An older individual is less likely to be seriously affected by disease than younger one; this is why infant mortality rates are so high in many parts of the world. (c) Deleterious mutations are less likely to occur later in life. (d) The statement refers to the lifespan of a species, not an individual; late in its history, a species is less likely to be undergoing adaptive radiation, so a harmful allele is less likely to be selected against.

Ans: a
p. 1027

13. In human population A, an average couple has 4 children, born every two years beginning when the wife is 30. In population B, a typical family also has 4 children, but they are born every 18 months beginning when the wife is 20. All other things being equal, which population will increase in size more rapidly? (a) B. (b) A. (c) Neither; the birth rate per capita is the same, so the populations will grow at the same rate. (d) It cannot be determined from the information given.

Ans: c
p. 1030

14. Why do more people die of cancer today than 10,000 years ago? (a) Cancer did not exist until preservatives were added to food. (b) There were no people 10,000 years ago. (c) People now live long enough to get cancer, because many infectious diseases can be cured. (d) There is no evidence that cancer is more common today.

Ans: a
p. 1030

15. In the past 30 years, the average age at death in the United States: (a) has changed slightly; (b) has not changed at all; (c) has changed a great deal; (d) has continued to decline.

TEST SET 40-B

Name _____   Section _____

Choose the best answer to each of the following questions, and write the appropriate letter in the space provided.

Ans: b
p. 1031

_____   1. A "patch," as defined in the text: (a) contains several habitats; (b) is a part of a habitat; (c) is internally heterogeneous; (d) is a piece of the environment large enough to support an individual for a lifetime or a breeding season.

Ans: a
p. 1031

_____   2. The "grain" of an environment is best defined as: (a) the "scale of patchiness" with respect to a given organism; (b) the average size of a patch, expressed in units of area or volume; (c) the ratio of patch size to habitat size; (d) the number of species that a given habitat supports.

Ans: a
p. 1031

_____   3. Suppose we have five species of moth larvae, each feeding on leaves of only one species of tree. Suppose that Sherwood Forest contains eight species of trees, including the five needed by these larvae. With reference to these larvae, how many patches are there in Sherwood Forest? (a) 8; (b) 5; (c) 13; (d) 1.

Ans: d
p. 1031

_____   4. Now suppose we have one species of bird, which feeds on all insects within a given size range in the canopies of trees. With reference to this bird, and to Question 3, how many patches are there in Sherwood Forest? (a) 8; (b) 5; (c) 13; (d) 1.

Ans: b
p. 1032

_____   5. Which of the following sequences of levels represents increasing frequency of choices an organism must make? (a) patch, habitat, responses to objects, behavioral mode; (b) habitat, patch, behavioral mode, responses to objects; (c) habitat, behavioral mode, patch, responses to objects; (d) responses to objects, behavioral mode, patch, habitat.

Ans: b
p. 1033

_____   6. Which of the following is the best definition of a territory? (a) an area occupied by an individual, pair, or group; (b) an area occupied and defended by an individual, pair, or group; (c) a feeding area occupied and defended by an individual, pair, or group; (d) a habitat occupied by one species, and from which other species are excluded.

Ans: a
p. 1033

_____   7. Can territories overlap? (a) No, by the usual definition this is impossible. (b) Yes, because many species occupy the same habitat. (c) Sometimes; it depends on population density. (d) Sometimes; it depends on the species.

Ans: b
p. 1035

_____   8. In which animal group is migration most widespread? (a) mammals; (b) birds; (c) insects; (d) fish.

Ans: d
p. 1035

_____   9. Irruptions occur in all of the following animal groups except: (a) mammals; (b) birds; (c) insects; (d) corals.

Ans: c
p. 1037

_____   10. Is disease a density-dependent factor in the regulation of population size? (a) No, because microorganism population density is generally independent of host population density. (b) No, because the percentage of a population killed by disease is generally constant even though the absolute number varies with population size. (c) Yes, because disease spreads more readily when populations are dense and is more likely to kill animals already weakened by food shortage. (d) Yes, because disease rarely occurs under natural conditions; it appears only when populations are abnormally high--due, for example, to human mismanagement of game species.

Ans: d
p. 1037

_____ 11. Suppose that a power plant releases hot water into a lake. This does not kill fish directly, but it slows their reaction time. Is this a density-dependent factor, a density-independent factor, or neither? (a) Neither, because no fish are killed by it. (b) Density-independent, because the amount of water released is independent of population density. (c) Density-dependent, because the more fish present, the more will be affected. (d) Density-dependent, because it makes the fish vulnerable to predation; the higher the population density, the more predators are likely to be nearby.

Ans: c
p. 1037

_____ 12. Suppose that a new weapon of biological warfare kills all human beings with free (vs. attached) ear lobes. Is this a density-dependent or a density-independent factor, or neither? (a) Neither; everyone has free ear lobes. (b) Density-dependent; the higher the human population, the more people will be killed. (c) Density-independent; the percentage killed will be the same regardless of population density. (d) Neither; an event that occurs only once is not categorized in this manner.

Ans: b
p. 1026

_____ 13. In a logistically growing population, the birth rate: (a) increases initially and then decreases; (b) decreases as population density increases; (c) remains constant; (d) increases as K decreases.

Ans: a
p. 1026

_____ 14. In a logistically growing population, the total number of births: (a) increases initially and then decreases; (b) decreases as population density increases; (c) remains constant; (d) increases as K decreases.

Ans: d
p. 1039

_____ 15. What is generally the best way to increase an endangered animal population? (a) Get rid of predator species. (b) Trap specimens to raise in zoos. (c) Provide feeding stations. (d) Increase its habitat.

CHAPTER FORTY-ONE
INTERACTIONS AMONG POPULATIONS

CHAPTER SUMMARIES AND OBJECTIVES

41-1 <u>Predator-Prey Interactions / Types of Predators and Prey / Foraging Theory</u> The major types of interactions between populations are competition (two species compete for the same limited resource), predator-prey interactions (one species eats the other), mutualism (both species benefit), commensalism (one species benefits, one is unaffected), and amensalism (one species is harmed, the other is unaffected). Predator-prey interactions are further subdivided according to the relative sizes of predator and prey and the mode of predation. Parasites feed on larger or same-size prey but generally do not totally consume them. Parasitoids consume prey of similar size. Prey vary in mode of defense and rate of population renewal. Filter feeders nonselectively filter much smaller prey from water or air. Detritivores eat dead organic matter; herbivores eat the tissues of plants. Typical predators (carnivores) pursue and capture individual prey.

Foraging strategy involves the maximization of energy and/or nutrient intake per unit time and the minimization of risk and of pursuit and handling time. Theory predicts that a prey item will be taken only if other prey types, yielding more energy per unit time, do not exceed a critical level of abundance.

1. Define and give examples of the major types of interactions between populations.

2. Differentiate among the various types of predator-prey interactions.

3. Explain the factors that determine whether or not a given prey item will be taken by a predator, according to foraging theory.

41-2 <u>Short-Term Consequences of Predator-Prey Interactions / Mimicry / Stability of Predator-Prey Interactions</u> Many predator and prey populations tend to oscillate in size, because there is a time lag between a change in the level of one population and a response by the other. Such oscillations have been confirmed both in the field and in laboratory studies. Prey are more vulnerable to predation in certain patches of their habitat than in others. Prey defenses take many forms, one of which is mimicry: A palatable species resembles an unpalatable one (Batesian mimicry), or two or more unpalatable species resemble each other (Müllerian mimicry).

Predator-prey interactions are stabilized, at least in part, by prey refuges and minimum-encounter thresholds; that is, places are available where prey can go to escape, and there is a minimum prey-population level below which predators do not find them. Other stabilizing factors include stronger selective pressures on prey than on predators; limitations on the efficiency of predators, resulting from the evolutionary compromises that enable body structures to serve multiple functions; energetic considerations, which require that a predator move on or starve

396   Chapter 41

when a prey population falls below a critical level; and prohibitive costs of overcoming prey defenses.

4. Explain why predator and prey populations tend to oscillate.

5. Define the two major forms of mimicry and outline the constraints on each.

6. Explain the most probable reasons why predators do not normally overeat their prey.

41-3 Theory of Competition / Competition in Nature When two species overlap in their use of resources, there are several possible outcomes. If the resource is in large enough supply, the two species may not compete. If the resource is limiting, competition may take the form of either exploitation or interference. In the first case, many organisms (such as plants) compete simply by taking nutrients or other resources needed by another species, thereby reducing the amount of that resource available to it. In the second case, animals tend to compete by direct behavioral interactions--for example, by chasing the other species away from a food source. Either form of competition has one of two possible outcomes: (1) The two species may coexist. (2) One species may exclude the other. The outcome depends on the relative strength of interspecific vs. intraspecific competition and on environmental heterogeneity.

Under natural conditions, competition is demonstrated by an increase in one species following removal of another. Plants exhibit the self-thinning rule: Under competition, individuals decrease in numbers but increase in weight. Many examples of habitat restriction in birds and mammals are attributed to competition. Coexistence is favored by stable, abundant resources and by patchy environments. In some cases, predation on a dominant species may create opportunities for its competitors, or a dominant species may force competitors into habitats where they are more vulnerable to predation.

7. Describe and give examples of the major forms of competition.

8. Describe the possible results of competition, and cite some factors favoring coexistence.

41-4 Commensalism and Mutualism / Plant-Animal Mutualisms / Ecological Niches Examples of probable commensalism include the relationship between cattle egrets and cows and that between epiphytes and trees. Mutualisms include nitrogen-fixing bacteria and their plant hosts, corals and tunicates with algae, cleaner fishes with larger parasite-laden fish, fungi and algae that coexist as lichens, ants and aphids, and a variety of plant-animal mutualisms. In the latter, certain animals protect or pollinate plants and receive shelter or food in return. Plants have evolved many complex adaptations to attract and reward pollinating insects, birds, bats, and other animals, and to maximize the probability of successful pollen transfer. However, there is often a basic conflict of interest between plants and pollinators, so that selection on one does not maximally benefit the other.

The role of a species in relation to its physical and biotic environment is called its ecological niche. The fundamental niche of a species refers to the range of conditions that the species would occupy in the absence of competitors and predators. The realized niche is that part of the fundamental niche that the species actually occupies in the presence of other species. Not all niche dimensions are likely to be in short supply; those that are critical are likely to be utilized differently by coexisting species.

9. Define commensalism and mutualism.

10. Discuss the interrelationships between plants and pollinators.

11. Explain the ecological niche concept and differentiate between fundamental and realized niches.

ARTICLE RESOURCES

Predator-Prey Interactions

Types of Predators and Prey

Merritt, Richard W., and J. Bruce Wallace. Filter-feeding insects. Scientific American, 1981, 244(4), 132-136, 141-144. Three orders of aquatic insects hatch under water and gather food with brushes, nets, and other fine-mesh filters. The insects are mayflies, two genera of caddisflies, and blackflies. The blackflies have head fans--retractable mouth parts between the antennae and mandibles. The caddisflies build nets. The mayflies have modified legs for capturing food.

Nielsen, Lewis T. Mosquitos, the mighty killers. National Geographic, 1979, 156(3), 426-440. The most deadly insect is the mosquito, yet it also collects nectar, and some species produce larvae that prey on blood-sucking mosquito larvae. However, the devastation caused by mosquitos is not to be taken lightly. Hundreds of millions of people have died of malaria, yellow fever, dengue fever, and filariasis. The life cycle of the mosquito is photographed. New weapons for combatting the insects are cited.

Moehlman, Patricia D. Jackals of the Serengeti. National Geographic, 1980, 158(6), 840-850. Jackals are monogamous hunters eating small rodents and fruit and scavenging on prey brought down by larger animals. A description of the life cycle and use of helpers to raise the pups is given. These helpers (older siblings from last year's litter) help to increase the survival rate of the younger off-spring. Two types of jackals, golden jackals and silver backed jackals, were studied as examples of social carnivores.

Foraging Theory

Short-Term Consequences of Predator-Prey Interactions

Mimicry

Sisson, Robert F. Deception: Formula for survival. National Geographic, 1980, 157(3), 394-415. Photographic essay on camouflage and mimicry. Graphically illustrates material covered in the text.

Stability of Predator-Prey Interactions

Tuttle, Merlin D. The amazing frog-eating bat. National Geographic, 1982, 161(1), 78-80. Bats prey on frogs on Barro Colorado Island, Panama. They can distinguish edible from poisonous species by depending on frog calls. Calls of edible and of poisonous frogs were recorded. When a bat was captured and placed in a cage with recordings of both species, it immediately flew to the recorder emitting the call of an edible species. When poisonous toad calls were used with calls of previously unheard edible frogs, the bat chose the edible frog. Frogs give off both simple and complex calls, yet the bats responded more often to the more complex call. Excellent photographs.

Theory of Competition

Competition in Nature

Franklin, William L. Living with guanacos--wild camels of South America. National Geographic, 1981, 160(1), 63-76. Guanacos are one of three members of the camel family that live in South America. It is a large herbivore that ranges from the western Peruvian Andes to Tierra del Fuego and Patagonia. Competition of the guanacos with livestock has made it an endangered species. Well illustrated.

Clark, Tim W. The hard life of the prairie dogs. National Geographic, 1979, 156(2), 270-281. Five species of prairie dogs live in twelve states on either side of the Rocky Mountains. Their social behavior is shown. Benefits of living with the bison are outlined. Whether or not they are a pest is open to discussion. Competition with livestock for forage is slight but, where alfalfa is a main crop, prairie dogs eat large amounts of it.

Commensalism and Mutualism

Plant-Animal Mutualisms

Ecological Niches

ESSAY QUESTIONS

1. The following graph shows population levels of a predator and a prey species, plotted against time.

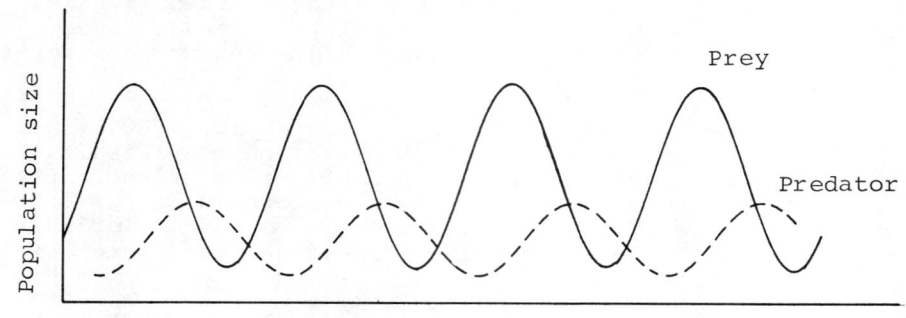

   Explain the following features of the graph: (a) The prey species becomes more abundant than the predator species. (b) Both curves oscillate. (c) Oscillations of the two species are out of synch.

   See textbook page 1048.

2. Discuss the advantages and disadvantages of wind and of specialized insects as pollen-transmitting agents.

   See textbook page 1058-1061.

3. Why might predator-prey oscillations occur more commonly at high latitudes? (Hint: How might prey-species diversity change with latitude?)

   Species diversity is generally lower at high latitudes, so there might be fewer alternative prey available.

4. Discuss factors that are believed to prevent predators from overeating their prey.

   See textbook page 1053.

5. The snapping turtle has a worm-shaped appendage on its tongue, with which it attracts fish into its mouth. (a) In this example of mimicry, which species is likely to be more abundant--the model (worm) or the mimic (turtle)? (b) The turtle obviously benefits from this mimicry. Does the worm benefit, or not?

   (a) The worm, the model in this case, is far more abundant. (b) The idea here is that worms might benefit if fish learned not to attack them, due to narrow escapes from snapping turtles. However, real worms are so numerous relative to snapping turtles, and the percentage of fish that would learn avoidance is so small, that this mimicry can have no effect on the safety of worms.

TEST SET 41-A

Name _____   Section _____

Choose the best answer to each of the following questions, and write the appropriate letter in the space provided.

Ans: a
p. 1043

_____ 1. If one species is harmed by an interaction with another that is not affected, the interaction is called: (a) amensalism; (b) commensalism; (c) competition; (d) predation.

Ans: a
p. 1043

_____ 2. Which of the following is an example of commensalism? (a) Harmless mites live in the follicle of human eyelashes. (b) Beneficial microorganisms live in the human large intestine. (c) An elephant crushes numerous insects every day as it walks through the forest. (d) A lamprey weakens a freshwater fish by drawing blood from it.

Ans: d
p. 1044

_____ 3. Sessile prey are those that: (a) defend themselves; (b) are attacked by parasites rather than by typical predators; (c) have stable populations; (d) are relatively immobile.

Ans: b
p. 1044

_____ 4. All of the following typically consume many individual prey items during their lives except: (a) typical predators; (b) parasites; (c) filter feeders; (d) omnivores.

Ans: c
p. 1047

_____ 5. Whether or not a prey item is included in a predator's diet depends primarily on: (a) its abundance; (b) its size; (c) the abundance of more energy-rich prey; (d) the availability of a foraging strategy.

Ans: a
p. 1053

_____ 6. In Müllerian mimicry, which outcome is typical? (a) All species involved benefit, so the phenotypes converge. (b) All species are harmed except the model, so the phenotypes diverge. (c) Only the predator benefits; the prey phenotypes remain constant. (d) The mimic must evolve toward the model faster than the model evolves away from the mimic.

Ans: a
pp. 1054-
    1055

_____ 7. If two species are in competition, removal of one species causes: (a) an increase in the other; (b) a decrease in the other; (c) no change in the other; (d) predator-prey oscillation.

Ans: c
p. 1054

_____ 8. The text refers to two competing species of Paramecium, one of which has symbiotic algae that enable it to live at the bottom of a container of algae. What do the algae have to do with where their host can live? (a) They provide ballast; a protozoan without them would tend to float to the top. (b) The host eats them; there is relatively little food at the bottom. (c) They provide oxygen by photosynthesis. (d) The algae are unrelated to the host's ability to live in deep water; their presence simply identifies that species.

Ans: d
p. 1057

_____ 9. A lichen is a mutualistic association of fungi, blue-green algae, and green algae. What is the role of the fungus? (a) photosynthesis; (b) defense; (c) anaerobic respiration; (d) absorption of water and nutrients.

Ans: a
p. 1057

_____ 10. In Question 9, what is the role of the blue-green algae? (a) photosynthesis; (b) defense; (c) anaerobic respiration; (d) absorption of water and nutrients.

Ans: d
p. 1058

11. What is an evolutionary advantage of wind (vs. animal) pollination of plants? (a) The direction and timing of wind are not
influenced by natural selection. (b) Wind transmits pollen
farther than most animal pollinators do. (c) The wind blows at
all times of the year, whereas pollinators may be active only in
particular seasons. (d) The plant need not invest any energy to
encourage this form of pollination.

Ans: b
p. 1059

12. In many instances, red flowers have evolved because their color
attracts: (a) ants; (b) birds; (c) bats; (d) pollen.

Ans: c
p. 1063

13. A fundamental niche differs from a niche in that: (a) the latter
is a subdivision of the former; (b) every species has a niche
whereas only competing species are said to share a fundamental
niche; (c) the fundamental niche represents the extent of the
niche in the absence of competitors; (d) the fundamental niche
represents the extent of the niche in the presence of competitors.

Ans: a
p. 1064

14. For which group of animals is temperature most often an important
niche variable? (a) insects; (b) birds; (c) large mammals;
(d) bats.

Ans: d
p. 1064

15. Oxygen is generally an important niche variable for all of the
following except: (a) soil microorganisms; (b) freshwater fish;
(c) marine fish; (d) birds.

TEST SET 41-B

Name _____   Section _____

Choose the <u>best</u> answer to each of the following questions, and write the appropriate letter in the space provided.

Ans: c
p. 1043

_____   1. If two species interact so that both benefit, the relationship is known as: (a) amensalism; (b) commensalism; (c) mutualism; (d) competition.

Ans: b
p. 1043

_____   2. Which of the following is an example of mutualism? (a) Harmless mites live in the follicles of human eyelashes. (b) Beneficial microorganisms live in the human large intestine. (c) An elephant crushes numerous insects every day as it walks through the forest. (d) A lamprey weakens a freshwater fish by drawing blood from it.

Ans: b
p. 1045

_____   3. Sessile predators are those that: (a) renew rapidly; (b) are relatively immobile; (c) use camouflage; (d) are small invertebrates.

Ans: d
p. 1047

_____   4. How does an organism evolve an optimal foraging strategy? (a) by trial and error learning; (b) by insight learning; (c) by reading <u>The American Naturalist</u>; (d) by natural selection.

Ans: b
p. 1048

_____   5. When predator-prey populations oscillate, what is the usual time relationship between peaks for the two species? (a) The prey population peaks after the predator population. (b) The prey population peaks before the predator population. (c) The two peak at the same time. (d) Only the predator population shows distinct peaks.

Ans: d
p. 1053

_____   6. Two species can use the same resource: (a) under no circumstances; (b) only if the two populations are in different geographical areas; (c) only if the two species are not closely related; (d) only if the resource is plentiful.

Ans: c
pp. 1043,
1045

_____   7. A man weeding a garden is a: (a) predator; (b) mutualist; (c) remover of competition; (d) detritivore.

Ans: a
p. 1055

_____   8. On more fertile plots of land, self-thinning occurs: (a) more rapidly; (b) less rapidly; (c) at the same rate; (d) not at all.

Ans: d
p. 1056

_____   9. Under which conditions is the coexistence of two competing species <u>least</u> likely? (a) patchy environment, fluctuating resources; (b) homogeneous environment, stable resources; (c) patchy environment, stable resources; (d) homogeneous environment, fluctuating resources.

Ans: c
p. 1058

_____ 10. Why do aphids process much more sugar than they need? (a) They have increased their rate of carbohydrate processing because ants protect them in order to obtain these secretions. (b) They must store energy for the winter, when leaves are dead. (c) The sap they feed on is low in amino acids, so a great deal must be ingested. (d) The reason is not known; apparently they are just wasteful feeders.

Ans: a
p. 1058

_____ 11. What is an evolutionary disadvantage of wind (vs. animal) pollination of plants? (a) The direction and timing of wind are not influenced by natural selection. (b) Wind transmits pollen farther than most animal pollinators do. (c) The wind blows at all times of the year, whereas pollinators may be active only in particular seasons. (d) The plant need not invest any energy to encourage this form of pollination.

Ans: c
p. 1060

_____   12. Some flowers have ultraviolet markings, which may have evolved
              because they attract: (a) birds; (b) bats; (c) insects;
              (d) pollen.

Ans: a
p. 1062

_____   13. An ecological niche is: (a) a species's role in relation to its
              environment; (b) a subdivision of a habitat; (c) a habitat;
              (d) a subdivision of an ecosystem, but larger than a habitat.

Ans: d
p. 1063

_____   14. A realized niche differs from a niche in that: (a) the latter
              is a subdivision of the former; (b) every species has a niche,
              whereas only competing species are said to share a realized
              niche; (c) the realized niche represents the extent of the niche
              in the absence of competitors; (d) the realized niche represents
              the extent of the niche in the presence of competitors.

Ans: a
p. 1055

_____   15. Coexisting species tend to use: (a) critical resources in
              different ways; (b) critical resources in the same way; (c) non-
              critical resources in different ways; (d) different resources in
              the same way.

CHAPTER FORTY-TWO
ECOLOGICAL COMMUNITIES

CHAPTER SUMMARIES AND OBJECTIVES

42-1 <u>Macrodescriptors</u> An ecological community consists of all organisms living together in a given area. These may have dynamic relationships (the species influence each other's distribution and abundance) and/or coevolved relationships (the species influenced each other's evolution). Macrodescriptors help us understand complex communities. One useful macrodescriptor is the guild--a group of interacting organisms that share a common food resource. Another macrodescriptor is the food subweb--a group of organisms, extending upward through one or more trophic levels, that share common predators. Species richness and species diversity are two additional macrodescriptors. Species richness is the number of species in a given area; species diversity also considers size and abundance.

1. Define a community.

2. Describe two types of interrelationships among the members of a community.

3. Define the following macrodescriptors: guild, food subweb, species richness, and species diversity.

42-2 <u>Plant Growth Forms / Plant Height</u> In studying the plants of any community, it is useful to subdivide these into groups on the basis of growth (or life) forms. One such classification system groups plants by the position of overwintering buds (see Table I). This classification provides a rough estimate of the relative energy cost of existing through the unfavorable season. For example, phanerophytes (on which buds are exposed on upright shoots, and which include the largest of all plants) must expend energy throughout the winter to support considerable living tissue. At the other end of the scale, therophytes (which survive only as seeds) have very low energy costs during the winter.

Phanerophytes, with overwintering buds supported above the ground, are also usually much more susceptible to fire and grazing (or brousing) than are plants with below-ground growing points. On the other hand, several survival advantages go along with tallness. These include (1) avoidance of high microclimatic temperature extremes, (2) if the plant is sufficiently tall, less susceptability to brousing, and (3) greater success in competing for light.

The greater success enjoyed by tall plants in competing for light results because they can position more leaves above a given ground-surface area, shade out their competitors, and intercept more sunlight. The value of tallness in gathering sunlight is increased in tall plants that utilize a multilayer growth form (rather than a monolayer, which works best in the shade).

4. Describe the basis of Raunkaier's life-form classification of terrestrial plants.

5. Explain the reason for the differences in the annual energy costs of phanerophytes compared to the plants of other life-form groups, and indicate other disadvantages of tallness.

6. Cite the advantages of tallness in plants.

42-3 Plant Dependencies and Mutualisms The abundance of one species of plants in a community is commonly influenced by the presence of other species of plants. The species with such relationships are said to be members of a dependency guild, and the relationships involved may be based on either physical or energetic support. For example, parasites and saprophytes obtain their energy from other plants, whereas epiphytes, epiphylls, and vines utilize other plants for physical support (see Table II).

All plants are eaten by animals, but some gain mutualistic survival advantages from their association with animals and compete with other plants for these benefits. Plants benefit most commonly by the dispersal of pollen and/or seeds (see Table III).

Plants that interact with one another in their relationships with animal predators are said to be members of a defense guild. The interacting defenses often involve mimicry, which may be either chemical or visual.

7. Define a dependency guild, and give two examples (one involving physical support and one involving energetic support).

8. Describe a mutualistic relationship between plants and animals.

9. Define a defense guild, and describe the defensive mechanisms that are utilized.

42-4 Animal Utilization of Plants: Folivores / Flower Visitors / Frugivores / Granivores / Wood Eaters Groups of animals that utilize certain specific types of plant tissues form guilds, which are named by adding the suffix -ivore to the word describing what is eaten. Thus a folivore is an eater of foliage.

Folivores are faced with two problems: (1) Leaves are a poor source of food (high in cellulose, low in protein). (2) Leaves are often defended by the presence of two types of chemical substances. Acute toxins interfere with metabolic functions by inhibiting nerve impulses to muscles, by inhibiting insect metamorphosis, or by suppressing protein metabolism. Digestibility-reducing substances (toxins) inhibit extracellular digestion of proteins. The amount of these chemicals in the leaves may increase following periods of heavy usage by folivores.

Among the folivorous (grazing) mammals in an East African grassland community, the larger forms (which arrive first) are less selective and clear out the less nutritious forage. This allows the smaller forms to be progressively more selective and to utilize only the more nutritious portions. In this manner, several species of folivores are able to make use of a single food resource.

A flower-visitor guild is composed of animals that are attracted to flowers to obtain pollen and/or nectar (or other flower visitors). These animals (especially insects) have strong coevolved relationships with the plants whose flowers they visit because of the important mutualistic advantages gained by both: pollen transportation for the plants and a food resource for the animals.

Frugivores are eaters of fleshy fruit. These animals also have coevolved relationships with plants (fruits contain seeds that are transported by the animals). Birds are the dominant frugivores in most communities, but many other types of animals (such as bats and ants) also utilize fruits as a food resource. Fleshy fruits are nutritious but some contain toxins. For example, some unripe fruits contain toxins that discourage consumption by animals until the seeds are mature.

Granivores eat seeds that are not contained in fleshy ovaries. All such animals are predators on plants, but many serve as seed dispersers. In hot deserts, the seed-eating guild is dominated by ants, birds, and rodents. Both ants and rodents transport and store seeds (some of which may later germinate), but they also consume large quantities of seeds and consequently influence each other's populations as well as those of the plants involved. Seeds are very nutritious: they rarely contain toxins because these would hinder seed embryo development.

Examples of members of the wood-eating guild include fungi and termites. Most animals that eat wood cannot digest cellulose directly but utilize cellulase-producing microorganisms in their guts for this purpose.

10. Define the following terms: folivore, frugivore, and granivore.

11. Describe two problems that face a foliage eater.

12. Explain how several species of grazing mammals (ranging from large to small) can exist on a single grassland foliage food resource.

13. Explain the mutualistic relationships that occur between plants and (a) flower visitors, (b) frugivores, and (c) granivores.

14. Explain why toxins can occur in fruits but not in seeds; explain the advantages to a plant for toxins to occur in unripe fruit but not in ripe fruit.

42-5 Food Subwebs Predators of plants also have predators. A single, specialized species of predator may be involved when all the members of one guild are similar. When the members of a guild are quite distinct (like rodents, ants, and birds), the food subweb may include several types of predators.

Many such food subwebs may be quite complex, with various interactions between the component species. The food subweb in which tropical American butterflies of the genus Heliconius are members is an example of such complexity (see Figure 42-18).

15. Explain why a food subweb involving hot desert granivores (ants, rodents, and birds) includes several types of predators.

16. Describe some of the complex interrelationships among the various members of the Heliconius food subweb (Figure 42-18).

42-6 Biomes: Tundra / Boreal Forest / Temperature Deciduous Forest / Grassland / Cold Deserts / Hot Deserts / Mediterranean Biome / Thorn Forest / Tropical Deciduous Forest / Tropical Evergreen Forest / Tropical Montane Forests Biomes are strikingly different assemblages of species that occupy the major climatic regions of the earth. These regions differ in their seasonal patterns of temperature and rainfall (as can be shown graphically in climographs).

The tundra biome occurs above the limits of tree growth. Arctic tundra occurs above the latitudinal limits, and alpine tundra occurs above the altitudinal limits. Tropical alpine tundra and arctic tundra exist under very different climatic regimes (arctic tundra has a short summer with 24 hours of sunlight, while tropical alpine tundra has little seasonal changes and continuous 12-hour days and 12-hour nights).

The boreal forest biome occurs (both latitudinally and altitudinally) just below the tundra, or along the west coast of continents at mid to high latitudes. This biome is dominated by coniferous trees, most of which are evergreen, wind-pollinated, and wind-dispersed.

The temperate deciduous forest biome occurs on the east side of continents at mid-latitudes. The dominant trees are winter-deciduous, and species richness is greater than in boreal forests. Many of these winter-deciduous trees have animal-dispersed pollen and seeds. Seasonal cycles of activity are striking.

The grassland biome may occur wherever potential evaporation exceeds precipitation. Seasonal variation in temperature and/or precipitation may be extreme, but the primary influencing factors are fire and grazing (without which many grasslands will become shrub communities). Grasslands are dominated by nonwoody plants that die back to the ground during unfavorable periods.

The cold desert biome occurs at mid to high latitudes in dry regions located either in continental interiors or on the inland side of coastal mountain ranges. Precipitation is usually rather evenly distributed, but cold deserts experience great seasonal changes in temperature. These deserts are dominated by a few species of low-growing shrubs.

The hot desert biome is found in two belts that occur at about 30° N and 30° S latitude. Seasonal extremes are less than in cold deserts, and summer rain is common. Hot deserts support a rich flora and fauna. The flora may include leafless succulents (cactus) that store water, and annuals may be abundant when conditions are favorable.

The Mediterranean biome occurs on the west side of continents at moderate latitudes where cold ocean waters circulate offshore. The climate involved is mild and wet in the winter and hot and dry in the summer. This biome is dominated by low-growing shrubs with tough evergreen leaves, producing bird-dispersed fruits (the olive is an example).

The thorn forest biome occurs at latitudes just below the hot deserts. The rainy season is brief, and no rain may fall for 8 or 9 months each year. Many plants of the thorn forest are similar to those found in the deserts.

The tropical deciduous forest biome is closer to the equator (at a lower latitude) than the thorn forest. Consequently it has a longer rainy season, more species, and fewer succulents. Most trees are drought-deciduous. The soil is excellent for agricultural purposes.

The tropical evergreen forest biome occurs at or near the equator in regions that receive more than 100 inches of annual rain and where the dry season lasts no more than a few months. This biome has the greatest species richness of both plants and animals and the highest productivity. Food webs are extremely complex. Soil is poor for agriculture.

The tropical montane forest biome occurs at the same latitude, but at higher elevations, than the tropical evergreen forest. As elevations increase, forest height decreases, vines and epiphytes become more important, leaf sizes decrease, and productivity is less.

17. Insofar as possible, describe the dominant vegetation (and/or other biotic characteristics) of each of the terrestrial biomes.

18. Describe the characteristic climatic conditions (and/or geographical location) of each of the terrestrial biomes.

42-7 Aquatic Communities: Rivers and Lakes / Marine Communities Basic interrelationships exist in all communities, but the physical environment of aquatic communities differs from that of terrestrial communities in several respects: (1) Water provides more support but creates greater forces than air. (2) Movement through water is more difficult than through air. (3) Sunlight is rapidly attenuated by passage through water.

Consequently, plants don't use animals to disperse pollen or seeds; plants are flexible and mostly small, and all tissues are readily eaten by animals; herbivores often determine the structure of the communities; animals are usually the largest and longest lived; and animals often compete for space with each other and with plants.

Most organisms in rivers live on food that is carried downstream, but considerable photosynthesis does occur in lakes (especially in the spring when semi-annual turnover brings nutrient-rich waters to the surface).

Organisms suspended in the water are called plankton, and these may be either plants (phytoplankton such as green or blue-green algae) or animals (zooplankton such as small crustaceans). Animals that swim (such as fish) are called nekton. Animal communities in lakes are dominated by insects and fishes, with insects in particular occupying a wide variety of niches (the larvae of some terrestrial forms provide a link between aquatic and terrestrial communities).

The physical environment of marine communities can be subdivided into several zones. As illustrated in Figure 42-40, that portion of the ocean situated over the continental shelf is called the neritic zone, while the rest is called the oceanic zone. At all depths, the bottom is the benthic zone and the water above is the pelagic zone. That part of the benthic zone that is below the level penetrated by light is called the abyssal zone.

Marine communities are based on tiny floating plants (phytoplankton), especially diatoms and dinoflagellates. Pelagic communities in open water are simple, but complex communities occur in the littoral zone at the ocean margins where many plants and animals compete for nutrients and space. The most complex of all marine communities are coral reefs, where the coral animals (like terrestrial plants) create a complex structure within which many lifestyles are possible.

19. Describe three major differences between the physical environments of aquatic and terrestrial communities.

20. Describe the basic difference between the plants and animals that occur in aquatic communities and those that occur in terrestrial communities.

21. Distinguish between plankton (phytoplankton and zooplankton) and nekton.

22. Explain how some insects provide a link between aquatic and terrestrial communities.

23. Define and distinguish between litteral, neritic, oceanic, benthic, pelagic, and abyssal zones in the marine environment.

24. Explain why coral reefs are the most complex and pelagic communities are the most simple of all marine communities.

ARTICLE RESOURCES

Macrodescriptors

Plant Growth Forms

Plant Height

Plant Dependencies and Mutualisms

Folivores

> Degabriele, Robert. The physiology of the koala. Scientific American, 1980, 243(1), 110-117. The koala, a native of Australia, lives on eucalyptus leaves from 35 of the 600 species. These leaves provide nutrients as well as water. Eucalyptus leaves are toxic to other animals. The cecum of the koala takes up 20 percent of the postgastric intestine. It is the site of microorganisms that digest the cellulose of the leaves. Feeding habits, water balance, and protection against the environment are discussed. Clear explanations are given.

Flower Visitors

Frugivores

Granivores

Wood Eaters

Food Subwebs

Biomes

Tundra

Chadwick, Douglas H. "So empty, yet so full." National Geographic, 1979, 156(6), 737-770. Photographic exploration of the tundra with beautiful pictures of the flora and fauna taken in the Arctic National Wildlife Refuge in Alaska. Here an unusual diversity of arctic and subarctic habitats exists.

Boreal Forest

Temperate Deciduous Forest

Grasslands

Farney, Dennis. The tallgrass prairie: Can it be saved? National Geographic, 1980, 157(1). The tallgrass prairie once covered the Midwest from Ohio to Montana. The Nature Conservancy has preserved 1.5 million acres to date. Photographs of virgin prairie, types of grasses, birds, and mammals illustrate this article, depicting a landscape that once covered much of the Midwest.

Cold Deserts

Hot Deserts

Mediterranean Biome

Thorn Forest

Tropical Deciduous Forest

Tropical Evergreen Forest

Tropical Montane Forests

Aquatic Communities

Rivers and Lakes

Marine Communities

Fisher, Allan C., Jr. My Chesapeake--Queen of bays. National Geographic, 1980, 158(4), 428-468. Chesapeake Bay is the largest estuary on the East Coast. It is nearly 200 miles long with 150 rivers and creeks emptying into it. Its shoreline is more than 8,000 miles. It abounds in wildlife, historic sites, villages, and large cities. It is a complicated ecological system. Salinity ranges from 0 to 30 parts per thousand. Over 2,700 species inhabit it. The life cycle of the oyster and threats to its survival at each stage are detailed. Excellent photographs.

Fleming, B. Carrol. Life and flotsam. Science 81, 1981, 2(6), 98-99. Contents of driftage at the shore line are examined. Beach animals reflect their habitat. Hence organisms of the Atlantic Coast, Gulf Coast, and Pacific Coast are different, as are those from cold waters, warmer waters, sandy beaches, and rocky beaches.

Clark, Eugenie. Secrets of the Red Sea. Science Digest, 1982, 90(4), 46-53. Well-illustrated essay on the Red Sea reefs along the Sinai Peninsula. Defines a reef and describes a number of species common to reefs.

Ballard, Robert D., and J. Frederick Grassle. Return to oases of the deep. National Geographic, 1979, 156(5), 388-405. Study of the Galapagos Rift 2.5 kilometers below the surface, where active vents spew water at 13°C into the normally 2°C sea. Hordes of animals congregate around these vents. Pressures of 250 atmospheres exist here. Unusual invertebrates feed off bacteria at these depths. Well illustrated.

ESSAY QUESTIONS

1. In discussing communities, your text occasionally compares complex communities with simple communities (such as a coral reef with a pelagic community).

    a. What is a community?

        A community consists of all the organisms living together in a given area.

    b. From a biotic standpoint only, how do complex communities differ from simple communities?

        A complex community contains more species, more guilds, and more food subwebs, and there are many more relationships. The guilds that occur contain more members, and the food chains are longer. The number of individuals, however, could be less.

    c. Given only the characteristics of a physical environment, would it be possible to predict whether or not the community occupying that area would be simple or complex?

        Such predictions might not be correct, but in general a complex community exists in a larger geographical area, where the microclimatic conditions are heterogeneous (varied) and where the physical conditions are such that many species have evolved the necessary adaptations for survival.

    d. Knowing only that one species was present in abundance in a community, would it be possible to predict whether or not that community would be complex or simple?

        Only when the single species is a dominant one that exerts a very strong influence on the community. If the single species creates an abundance of niches--like coral--the community will be complex. If the single species is a plant that creates few niches and on which there are few predators, the community will be more simple.

2. a. List several advantages associated with tallness in plants.

        More leaves above a given surface area shade out competitors and intercept more light.

    b. Since there are such advantages, why aren't all plants tall?

        There is always "more than one way to skin a cat" and more than a single way to make a living. Shortness is one of these. Short plants, existing in the shade of tall plants, utilize a monolayer growth form, and in temperate deciduous forests they flower early in the spring before the tall plants leaf out. If there are no tall plants in the community, as in grasslands, shortness is no disadvantage. In all communities, with few exceptions, every possible means of making a living is somehow exploited by some type of organism, and the existence of a community consisting of only tall plants seems unlikely (even in a tropical evergreen forest).

    c. Tall plants represent a food resource, and it is reasonable to assume that adaptations would have evolved among animals to allow them to exploit this resource. List some of these adaptations and the animals that belong to such a guild.

        Climbing ability (squirrels, monkeys, and insects), flight (birds, bats, and insects), long necks (giraffes), large body size (elephants), and so on.

    d. Are there any disadvantages associated with tallness in plants?

        The energy costs, especially during the unfavorable season.

3. It can be an advantage for a plant to have its seeds transported away from their source. This is accomplished in some species by animals.

a. Why is this transportation of seeds an advantage?

It allows dispersal into unoccupied but suitable habitats and germination away from the parent plant to avoid competition and shading.

b. What adaptations have been evolved by plants to encourage animals to accomplish this transportation?

Nutritious fruit (which contain seeds).

c. What means is sometimes used to prevent seeds from being removed from a plant before the seeds are mature? How does a plant prevent overuse by animals?

Toxins that occur only in unripe fruit; toxins in fruit may force a frugivore to vary its diet and to move on to other food sources.

d. What possible benefit to the plant could be provided by a granivore?

Many granivores gather and store seeds, some of which germinate when conditions are suitable.

4. Complete the following chart, which compares two biomes: a temperate deciduous forest and a tropical deciduous forest.

	Temperate Deciduous Forest	Tropical Deciduous Forest
Geographical Location	(east side of continents at mid-latitudes)	(low latitudes)
Precipitation Patterns	(evenly distributed throughout the year; both rain and snow)	(rain occurs during the "rainy season" followed by a long dry season; no snow)
Seasonal Temperature Extremes	(strong seasonal fluctuations)	(relatively uniform temperatures throughout the year; seasons based on rain)
Type of Deciduousness	(winter-deciduous)	(drought-deciduous)
Means of Pollen and Seed Distribution	(by animals)	(by animals)

TEST SET 42-A

Name _____   Section _____

Choose the best answer to each of the following questions, and write the
appropriate letter in the space provided.

Ans: d
p. 1066

_____   1.   A community could be all of the following except: (a) an assemblage of species occupying one of the major climatic regions of
the earth; (b) the 3,800 species in 5 square kilometers near
Oxford, England; (c) the species in a temporary pond with a
surface area of only 1 square meter; (d) 180 species of breeding
birds in 3 square miles of tropical forest in Brazil.

Ans: a
p. 1067

_____   2.   A food subweb is: (a) a guild extended upward through one or
more trophic levels; (b) a measure of species richness, except
that it involves food resources and gives weight to size and
abundance; (c) a group of organisms that compete for the same
space and share a common food resource; (d) the ecological equivalent of the "gross national product."

Ans: d
pp. 1068,
   1069

_____   3.   The ability of tall plants to compete more successfully for light
is due to all of the following except: (a) They can position
more leaves above a given ground surface. (b) They can shade out
other plants. (c) They can intercept more light. (d) The intensity of light is considerably greater just a short distance above
the ground, and soil surface temperatures are correspondingly
lower.

Ans: c
p. 1071

_____   4.   Which of the following statements is not true of an epiphyte?
(a) It obtains physical support from another plant. (b) It is
a member of a dependency guild. (c) It obtains energetic support
from another plant. (d) It grows on the trunks and branches of
another plant.

Ans: b
p. 1077

_____   5.   Unripe fruit sometimes contains toxins that do not exist in ripe
fruit. This adaptation serves to: (a) keep folivores away from
the fruit during times when the fruit is most susceptible to
predation; (b) keep frugivores away from the fruit until the
seeds have had time to mature; (c) prevent fruit predators from
digesting the seeds; (d) provide a mutualistic mimicry relationship based on the presence of chemical toxins.

Ans: d
p. 1078

_____   6.   _____ obtain plant food that is highly nutritious and usually
lacks toxins. (a) Folivores; (b) Frugivores; (c) Epiphylls;
(d) Granivores.

Ans: a
pp. 1079-
   1080

_____   7.   The passion-flower/Heliconius food subweb is a: (a) complex food
subweb involving butterflies and vines in tropical American
forests; (b) mutualistic defense guild involving lizards that
mimic leaves; (c) dependency guild in which passion-flower
pollination is accomplished by flower-visiting ants; (d) food
subweb occurring is East African grasslands and including large
grazing mammals.

Ans: a
p. 1081

_____   8.   Which of the following statements about biomes is not true?
(a) A biome is unlike a community because it is an assemblage
of unique climatic conditions. (b) A biome is characterized
by its seasonal patterns of precipitation. (c) A biome is
characterized by its seasonal patterns of temperature. (d) A
biome is characterized by its geographical location, especially
latitude.

Ans: a   _____   9. Boreal forests: (a) are dominated by a few species of wind-
pp. 1082-   pollinated and wind-dispersed trees; (b) occur on the east sides
    1083   of continents at mid-latitudes; (c) are characterized worldwide
                          by winter-deciduous trees that produce leaves well adapted for
                          rapid photosynthesis during the warm moist summers; (d) are sub-
                          jected to mild, dry winters and summers that are relatively long,
                          hot, and wet.

Ans: a   _____   10. A tropical deciduous forest: (a) gets somewhat less than 250 cm
p. 1087   (100 in.) of rainfall each year, and the dry season is long and
                          hot; (b) has soil that is deeply weathered and generally unsuited
                          for modern agriculture; (c) has the greatest species richness of
                          both plants and animals and the highest productivity; (d) is
                          dominated by a few species of evergreen trees that are winter
                          deciduous.

Ans: b   _____   11. A biome occurring at about 30 degrees North latitude (where
pp. 1085-   descending air picks up moisture rather than releasing it), where
    1086   rainfall occurs both in the winter and in the summer, where vege-
                          tation is rich and diverse (and includes stem-succulents), and
                          where populations of rodents and ants are often remarkably high
                          is most likely to be: (a) cold desert; (b) hot desert;
                          (c) grassland; (d) tropical thorn forest.

Ans: d   _____   12. A community in which the plants are mostly very small, provide
p. 1089   little physical structure, and have high reproductive rates, and
                          in which the animals are the largest and longest-lived members
                          and often determine the structure of the community, is most
                          likely to be a: (a) community in the tundra biome; (b) community
                          in the tropical montane forest biome; (c) community in the grass-
                          land biome; (d) marine community.

Ans: b   _____   13. The important plankton of lakes includes all of the following
pp. 1090-   except: (a) green algae; (b) fish; (c) filamentous blue-green
    1091   algae; (d) small crustaceans.

Ans: b   _____   14. The most important grazers of the phytoplankton in the sea are:
p. 1091   (a) diatomes and dinoflagellates; (b) crustaceans; (c) insect
                          larvae; (d) fish of the abyssal zone.

Ans: d   _____   15. Pelagic communities are less complex (more simple) than coral
p. 1092   reef communities for all of the following reasons except:
                          (a) Pelagic communities have fewer species. (b) Pelagic
                          communities include fewer guilds and have shorter food chains.
                          (c) Pelagic communities exhibit fewer interrelationships and
                          interactions. (d) Pelagic communities occur on the ocean bottom,
                          at depths where little light penetrates, causing lowered pro-
                          ductivity.

TEST SET 42-B

Name _____   Section _____

Choose the best answer to each of the following questions, and write the appropriate letter in the space provided.

Ans: b
p. 1066

_____ 1. A community: (a) consists only of plant species, all of which have dynamic relationships; (b) consists of species, each one of which always interacts in some way with other species in the community; (c) consists of animal and plant species, all of which have both dynamic and coevolved relationships; (d) consists of species all of which are members of the same food subweb.

Ans: c
pp. 1067,
    1079

_____ 2. Seed eaters (like rodents, ants, and some birds) feeding on the same plant are all members of each of the following groups except: (a) food subweb; (b) community; (c) multiple-prey/single-predator subunit; (d) guild.

Ans: a
p. 1071

_____ 3. All of the following members of a dependency guild obtain physical support from another plant except: (a) saprophytes; (b) epiphytes; (c) epiphylls; (d) vines.

Ans: d
pp. 1067-
    1070

_____ 4. Which of the following is a disadvantage of tallness in plants? (a) In tall plants, overwintering buds occur at or near the ground surface and consequently are subject to winter freezing. (b) In tall plants, the leaves must be retained throughout the unfavorable season, even though the supporting structures die back to the ground. (c) Tall plants are subjected to higher microclimatic temperatures, because summer above-ground temperatures are higher than surface temperatures. (d) Tall plants have higher energy costs in order to survive unfavorable periods.

Ans: c
p. 1070

_____ 5. Tall plants are able to capture more light per unit of ground area for all of the following reasons except: (a) The sun is not a point source of light, and leaf shadows disappear at a distance of 108 times the leaf diameter. (b) The rate of photosynthesis does not increase linearly with light intensity but reaches maximum values at one-quarter to one-fifth full sunlight. (c) The leaves on a tall plant can be arranged into a vertical monolayer, and in this way they obtain more radiant energy in direct sunlight per unit time. (d) A tall plant can position more leaves above a given piece of ground surface.

Ans: c
pp. 1074-
    1076

_____ 6. The most common mutualistic relationship between plants and animals involves: (a) mutualistic sharing of a common food resource; (b) animal mimicry of edible plant parts; (c) the dispersal of pollen and/or seeds; (d) physical or energetic support.

Ans: b
p. 1072

_____ 7. Folivores: (a) play an important role as seed dispersers; (b) obtain food that has a low nutritional content and usually contains toxins; (c) consume seeds and seed components; (d) are rewarded by the plant by being provided with thick, fleshy fruits which contain seeds rich in acute toxins.

Ans: c
pp. 1072-
    1074

_____ 8. Acute toxins that occur in leaves may do all of the following except: (a) interfere with nerve impulse transmission; (b) interfere with insect metamorphosis; (c) interfere with protein digestion; (d) interfere with protein metabolism and function.

Ans: c
p. 1080

_____ 9. Passion-flower vines utilize all of the following defenses against <u>Heliconius</u> butterflies <u>except</u>: (a) poisonous hairs; (b) toxic chemicals; (c) leaves that mimic butterflies, which fool the butterflies into "thinking" that the vine is fully occupied; (d) extrafloral nectaries that attract <u>Heliconius</u> predators.

Ans: c
p. 1082

_____ 10. The tropical alpine tundra (a) has permafrost; (b) has several species of coniferous trees, all of which are adapted to freezing conditions; (c) has 12-hour days and nights, and a steady rate of photosynthesis all year; (d) has a short, brief period of intense biological activity, following which many animal species leave.

Ans: a
p. 1085

_____ 11. Cold deserts: (a) are dominated by a few species of low-growing shrubs and species richness is low; (b) occur in two belts at about 30 degrees North and 30 degrees South latitude; (c) get most of their rain in the summer; (d) are characterized by moderate seasonal changes in temperature, so plant growth is concentrated in late summer.

Ans: d
p. 1086

_____ 12. The Mediterranean biome is: (a) dominated by low-growing, wind-dispersed shrubs with deciduous leaves; (b) characterized by a climate that is mild and wet in the summer and cold and dry in the winter; (c) a biome in which most growth occurs in late summer when most insects and animals are reproductively active; (d) a biome that occurs on the west coast of continents at mid-latitudes where cold ocean waters circulate offshore.

Ans: c
p. 1088

_____ 13. In tropical latitudes, the closer to the equator (at sea level): (a) the less the productivity; (b) the less the rainfall; (c) the greater the species richness; (d) the better the soil for agriculture.

Ans: b
p. 1089

_____ 14. Aquatic plants have fewer mutualistic interactions with animals than terrestrial plants. This is true because aquatic plants: (a) are not eaten by animals; (b) do not produce animal-transported pollen, fruit, or seeds; (c) are typically the largest and longest lived members of the community in which they occur; (d) usually determine the structure of the community, but are able to support only a small biomass of herbivores.

Ans: b
pp. 1090-
  1094

_____ 15. Most photosynthesis in aquatic communities occurs: (a) in rivers, in the fall; (b) in lakes, near the surface in the spring during the semi-annual overturn; (c) in oceans, in the abyssal zone; (d) in lakes, at mid-depths in autumn.

CHAPTER FORTY-THREE
BIOGEOGRAPHY

CHAPTER SUMMARIES AND OBJECTIVES

43-1 <u>Shaping of the Modern World / Historical Biogeography</u> Biogeography seeks to describe and explain geographical distributions of organisms. Historical biogeography focuses primarily on the evolutionary origins of groups, whereas ecological biogeography is concerned with ecological interactions as they influence distribution. Continental drift, originally proposed by Wegener, has been substantiated by geologic evidence; it has significantly affected the distributions of many groups of organisms. Repeated glaciations have also affected world climates and biogeographical patterns. Endemic species are particularly numerous on land masses that have been isolated for a long time, such as Australia. Biogeographical barriers other than oceans include deserts, mountains, grasslands, and others; such barriers have produced distinctive biotas in each of the four major biogeographical regions of the earth. In addition, many remote oceanic islands have unique biotas.

    1.   Differentiate between historical and ecological biogeography.

    2.   Describe some major physical and biotic factors that influence the distribution of organisms.

    3.   Explain why oceanic islands often have unusual biotas.

43-2 <u>Ecological Biogeography / Tests of Biogeographical Theory / Habitat Islands / Patterns of Species Richness</u> Boundaries of biotic regions are designated where the geographical limits of many species coincide. The number of species, species turnover, and geographical turnover are expressed in terms of alpha-, beta-, and gamma-richness, respectively. Richness is the net result of species arrivals and extinctions. As the number of species on an island increases, the rate of arrival of new species decreases and the extinction rate rises. Species equilibrium occurs when the immigration rate equals the extinction rate. Smaller islands have lower immigration rates and higher extinction rates. More isolated islands have lower immigration rates. Saturation refers to the theoretical maximum number of species that an island or habitat can hold. Tests of biogeographical theory are provided by natural disasters (such as Krakatoa), by experiments (such as the removal of island faunas), and by statistical analysis of island biotas. Habitat islands on the mainland display biogeographical phenomena similar to those of oceanic islands, but their immigration rates are usually higher. Continental islands, which have recently been connected to the mainland, provide information on local extinction rates. In most taxa, more species occur in tropical than in temperate regions. Exceptions include river communities, which are similar at all latitudes; certain groups of birds for which food and other resources are abundant at high latitudes; salamanders, which require cool climates; and several groups of plants.

4. Cite the factors used in differentiating between biotic regions.

5. Explain the concept of species equilibrium, and distinguish between species equilibrium and saturation.

6. Explain the principal factors that determine the number of species present on an island.

7. Compare oceanic islands, terrestrial habitat islands, and continental islands from a biogeographical standpoint.

43-3 <u>Community Convergence / Convergence in Mediterranean-Type Vegetation / Species Introductions</u> Convergent evolution is the evolution of similarities among organisms that originally were more distinct from one another. Parallel evolution, by contrast, involves organisms that started out similar and remained that way. Convergence takes millions of years and has resulted in outwardly similar plant and animal communities in many parts of the world. For example, Mediterranean climates worldwide are associated with specific types of plant communities that are able to optimize their carbon budgets under these environmental conditions. However, other regions of the world with similar climates support very different ecological communities. The main reasons appear to be historical; that is, the colonization of a region often depends on which species arrived there first.

Human activities have transported many organisms from one continent to another. Most introduced species are successful only in manmade environments; natural communities typically resist invasion. Impoverished island biotas, however, are less resistant, and many have been destroyed by introduced predators and competitors.

8. Differentiate between convergent and parallel evolution.

9. Give some possible explanations for (a) similar communities in certain regions of the world with similar climates and (b) dissimilar communities in certain other regions of the world with similar climates.

10. Describe conditions under which introduced species are most likely to be successful.

43-4 <u>Outbreaks of Pests / The Future of Species Introductions / The Preservation of Genetic Diversity</u> Some species introductions have resulted in serious pest outbreaks. Where natural enemies are lacking, either harmful or beneficial species can be highly successful. However, their success is usually temporary, because predators and parasites eventually will adapt or be introduced. Species introduction is now restricted by law, but in practice, the movement of smaller organisms cannot be prevented. Thus a continued effort to combat introduced pests will be necessary. Many larger species that are becoming rare may be reintroduced to their native ranges by governments and conservation organizations. Raising animals in captivity is expensive and often unsuccessful.

The human species exerts a disproportionate effect on other organisms. Agriculture tends to decrease genetic diversity drastically, thus creating opportunities for new pest strains to arise and destroy a significant percentage of a crop. Such disasters can be prevented only by preserving diverse strains with various resistances. Similarly, in the case of wild organisms, it is not possible to predict which species may ultimately prove useful to humans. Thus it is essential that species be preserved, for utilitarian as well as aesthetic reasons.

11. Explain why introduced species can be so harmful.

12. Explain why preservation of genetic diversity is important.

ARTICLE RESOURCES

Shaping of the Modern World

Rowe, Findley. Mountain with a death wish. In the path of destruction. The day the sky fell. National Geographic, 1981, 159(1), 3-66. A series of three articles on the eruption of Mt. St. Helens in 1980. The birth of a volcano, major volcanic areas of the world and uncertain plate boundaries are diagrammed. The total devastation of the area is depicted and the initial return of life to the area is shown.

Johnston, Arch C. A major earthquake zone on the Mississippi. Scientific American, 1982, 246(4), 60-68. Three of the greatest earthquakes in the United States occurred December 1811 to February 1812 on the Mississippi River near where Kentucky and Missouri meet. This area remains seismatically active, with tremors happening every 48 hours. Tectonic influences have shifted the course of the Mississippi River in several places. These are detailed. The river terrain is discussed. An explanation of faults, stresses, and crators is given. A map showing the North American crator is included.

Historical Biogeography

Ecological Biogeography

Tests of Biogeographical Theory

Habitat Islands

Brower, Kenneth. Life by night in a desert sea. National Geographic, 1981, 160(6), 834-847. Photographic essay on the marine desert of the North Pacific Ocean off the coast of Hawaii. Spectacular photographs of larval forms, invertebrates, and fish.

Patterns of Species Richness

May, Robert M. The evolution of ecological systems. Scientific American, 1978, 239(3), 160-178. Explores the coevolution of species within ecological systems. Expands on the textbook discussion of historical biogeography. A dynamic-equilibrium model of island biogeography is discussed. Competition is discussed, together with the total number of species in an ecological system and the relative abundance of such species. Food chains and their importance are explored.

Page, Jake. A dog's worst friend. Science 82, 1982, 3(1), 90-92. Dogs exhibit a number of variations that more or less determine whether or not they will be acceptable as pets. The richness of species is well illustrated.

Community Convergence

Convergence in Mediterranean-Type Vegetation

Species Introduction

Outbreaks of Pests

Boraiko, Allen A. The pesticide dilemma. National Geographic, 1980, 157(2), 144-184. Uses and the results of uses of pesticides. Common pests of agricultural products are discussed, such as gypsy moths, pink bollworms, grasshoppers, and corn pests. Uses of beneficial insects and viruses are discussed. A thought-provoking article, well illustrated, that points up the problems of pesticide use.

The Future of Species Introduction

Ogburn, Charlton. Island, prairie, marsh, and shore. National Geographic, 1979, 155(3), 350-381. Explores Merritt Island, Florida, where more endangered species than anywhere else in the United States outside of Hawaii are located. Bald eagles soar. Manatees scour the coves for food. Lagoons and marshes abound. Farallon Islands off the coast of San Francisco harbors the largest concentration of marine bird nesting sites south of Alaska. Marine mammals, predominantly seals, abound.

Lostwood Wildlife Refuge in North Dakota is a prairie refuge. It is a top water-fowl breeding refuge located in virgin prairie. The delta formed by the Bear River as it flows into Great Salt Lake in Utah is marshland, which is an oasis on the central flyway of one of the four areas of bird migration. Beautiful photographs of the animals, birds, and plants. Wildlife refuges of the United States, Alaska, Hawaii, and Puerto Rico are located on a chart showing major animals present, best seasons for observation, and areas where endangered animals may be seen.

The Preservation of Genetic Diversity

ESSAY QUESTIONS

1. Why do marine biotas tend to be less distinct than terrestrial ones?

    The main factors are continuous contact of seas in past and the early evolution of marine phyla.

2. Any introduced pest--such as the proverbial rabbit in Australia--eventually will be controlled by some native predators or parasites, even if human measures fail. Are the disturbed biotas likely to return to their original state after this occurs? Is there any single "original state" that can be regarded as the ideal, in any case?

    The answer to the first question is no, they probably won't return to "normal." The main point here is that species have moved between continents throughout the history of life, so there isn't any "right" distribution. The only difference humans have made is to accelerate the process enormously.

3. Discuss the ethics involved in destroying island biotas in order to study biogeographical theory.

    This is a pretty open question, but answers should include such points as: What is the value of the study? What new information will be gained? What species will be killed--are they common elsewhere? Will the agent used to kill them have any long-term effects?

4. Discuss some examples of human attempts to (a) combat introduced pests and prevent the introduction of additional ones; (b) preserve large endangered species.

    Answers should include customs searches, the efforts of wildlife organizations, the Medfly, and so on.

TEST SET 43-A

Name _____   Section _____

Choose the best answer to each of the following questions, and write the
appropriate letter in the space provided.

Ans: d
p. 1098

_____ 1. Paleontology is the study of: (a) ancient civilizations; (b) geo-
              graphical distributions of organisms; (c) continental movements;
              (d) past life forms.

Ans: c
p. 1098
(& Ch. 38)

_____ 2. In the early Cambrian, how did terrestrial organisms get from one
              continent to another? (a) The continents were joined together.
              (b) They crossed on land bridges. (c) There were no terrestrial
              organisms in the early Cambrian. (d) There was little exchange
              of organisms between continents.

Ans: d
p. 1099

_____ 3. An endemic species is: (a) one that originated on a given con-
              tinent; (b) one that has been introduced; (c) one that preys on
              or competes with another species; (d) one that is restricted to
              a given continent.

Ans: c
p. 1100

_____ 4. Should eastern and western North America be regarded as separate
              continents? (a) Yes, because they are isolated by grasslands
              and have distinctive biotas. (b) No, because they are not
              separate land masses. (c) No, because definitions of continents
              are arbitrary; these two areas traditionally have been considered
              one continent. (d) Yes, because they are on different tectonic
              plates.

Ans: a
p. 1100

_____ 5. _____ marine biotas are the most distinct. (a) Littoral;
              (b) Pelagic; (c) Cold water; (d) Warm water.

Ans: d
p. 1100

_____ 6. Why are biogeographical regions in oceans better defined than
              those on the land? (a) There are no barriers to the dispersal
              of marine organisms. (b) Marine biotas have received less study.
              (c) Marine phyla evolved before the separation of continents.
              (d) They are not better defined; it's the other way around.

Ans: d
p. 1104

_____ 7. The evolution of honeycreepers in Hawaii is an example of:
              (a) alpha diversity; (b) competitive exclusion; (c) saturation;
              (d) convergent evolution.

Ans: c
p. 1105

_____ 8. The difference between alpha-richness and alpha-diversity is
              that: (a) the first refers to species, the second to habitats;
              (b) the first is an ecological term, the second a biogeographic
              term; (c) the second is weighted by a measure of species size or
              abundance; (d) the two terms are interchangeable.

Ans: b
p. 1106

_____ 9. How does an increase in the number of species on an island affect
              population size? (a) It will be larger, on average, because
              predator species will have more prey to eat. (b) It will be
              smaller, on average, because resources must be shared. (c) It
              will remain the same, because different species use different
              resources. (d) There is no general rule on this.

Ans: a
p. 1106

_____ 10. The term species equilibrium signifies that: (a) the extinction
              rate is equal to the immigration rate; (b) the area in question
              has all the species it can hold; (c) species exist in peace and
              harmony with one another; (d) population levels are stable.

Ans: a
p. 1106

_____ 11. On more isolated islands (compared to the mainland), the equili-
              brium number of species is generally: (a) smaller; (b) larger;
              (c) the same; (d) There is no general rule on this.

Ans: d   _____   12.   Imagine a hypothetical mainland with only one animal species; a
pp. 1100-   yellow bird that lives high up in banana trees. There is also an
1106   offshore island initially without life. Every 5,000 years, a
                        strong wind blows a pair of yellow birds to the island. At the
                        end of 50,000 years, how many species of organisms could be on
                        this island? (a) One--the yellow bird. (b) Ten--between coloni-
                        zations, the island bird populations could become reproductively
                        isolated from each other. (c) None; what would they eat?
                        (d) Any number from zero to a million or more; extinction cannot
                        be predicted, nor can colonization by arthropods and spores.

Ans: d   _____   13.   Mediterranean vegetation occurs where: (a) winters are cold and
p. 1112   dry, summers hot and wet; (b) winters are short, summers long;
                        (c) carbon budgets are essential; (d) summers are hot and dry,
                        winters wet and mild.

Ans: a   _____   14.   The North American deserts have several species of lizards known
pp. 1110-   as horned toads. They are flattened, spiny, and sedentary, and
1111   feed mainly on ants. In Australia, there is a desert lizard with
                        similar characteristics and habits, but it belongs to a different
                        family. This resemblance is an example of: (a) convergent
                        evolution; (b) coincidence; (c) parallel evolution; (d) special
                        creation.

Ans: d   _____   15.   Why do you think large endangered species, particularly mammals
pp. 1115-   and birds, receive more attention from wildlife protection
1116   groups than smaller, less attractive forms? In other words, why
                        would so few people rally to save an endangered slug? (a) Larger
                        species, being higher on the food chain, are more important to
                        our environment. (b) Small organisms are generally difficult to
                        rear in captivity. (c) There are few, if any, small endangered
                        species. (d) Large species have greater economic value (to
                        hunters, birders, and so on) and more sentimental appeal.

TEST SET 43-B

Name _____   Section _____

Choose the best answer to each of the following questions, and write the appropriate letter in the space provided.

Ans: b
p. 1098

_____   1.   What was Alfred Wegener's significant contribution to biogeo-graphy? (a) development of island biogeographical theory; (b) proposal of continental drift; (c) control of cotton blight and other agricultural pests; (d) discovery of the economic value of guayule.

Ans: a
p. 1098

_____   2.   High mountains and volcanoes are located primarily: (a) where plates collide; (b) where plates separate; (c) at low latitudes; (d) in the southern hemisphere.

Ans: c
p. 1099

_____   3.   Which continent has the most distinctive biota? (a) South America; (b) North America; (c) Australia; (d) Eurasia.

Ans: d
p. 1098

_____   4.   Once the Panama Canal had been completed, why didn't North and South America drift apart, since they were no longer joined by land? (a) They did drift apart. (b) The process of separation has probably started, but not enough time has yet passed for it to be detectable. (c) The locks hold them in place. (d) Digging a ditch cannot affect the movements of the plates on which the continents rest.

Ans: c
p. 1100

_____   5.   Wild camels and their relatives survive today in: (a) Africa only; (b) Africa and Asia only; (c) Africa, Asia and South America only; (d) all continents except Australia.

Ans: a
p. 1104

_____   6.   The sunflower forests of St. Helena are an example of: (a) adaptive radiation; (b) parallel evolution; (c) saturation; (d) alpha-richness.

Ans: a
p. 1105

_____   7.   The number of species in a small, local, homogeneous area is called its: (a) alpha-richness; (b) beta-richness; (c) gamma-richness; (d) biota.

Ans: d
p. 1105

_____   8.   When the number of individual birds arriving on an oceanic island equals the number that die there, what happens to species richness? (a) It remains constant. (b) It increases. (c) It decreases. (d) The answer cannot be determined from the information given.

Ans: b
pp. 1105-
    1106

_____   9.   As more species arrive on an island, what generally happens to the extinction rate? (a) It decreases. (b) It increases. (c) It remains the same. (d) There is no general rule on this.

Ans: a
p. 1106

_____   10.   On more isolated islands (compared to those less distant from the mainland), the immigration rate is generally: (a) lower; (b) higher; (c) the same; (d) There is no general rule on this.

Ans: a
p. 1106

_____   11.   _____ is(are) substantiated by field evidence. (a) Species equilibrium only; (b) Saturation only; (c) Neither species saturation nor equilibrium; (d) Both species equilibrium and saturation.

Ans: a   12. A coastal area has a high immigration rate and a high extinction
p. 1106   rate. If the area becomes a continental island due to a rise in
                      sea level, what is likely to happen to the number of species
                      there? (a) It will decrease, unless the extinction rate drops.
                      (b) It will increase, because predator species will be isolated
                      on the mainland. (c) It will generally remain the same, unless
                      the saturation level rises. (d) It will either increase or re-
                      main the same.

Ans: d   13. All of the following groups contain more tropical-zone than
p. 1101   temperate-zone species except: (a) lizards; (b) birds;
                      (c) insects; (d) salamanders.

Ans: d   14. For two outwardly similar organisms to be products of convergent
p. 1111   evolution, do they have to exist at the same time? (a) Yes,
                      because both must be descended from a recent common ancestor.
                      (b) No; but if they coexist in time, the phenomenon is called
                      parallel evolution instead. (c) Yes; otherwise, how would anyone
                      know about them? (d) No; many large extinct reptiles were eco-
                      logically and morphologically similar to modern mammals.

Ans: d   15. What is the principal cause of the movement of species between
p. 1113   continents in the present century? (a) continental drift;
                      (b) warm climates; (c) habitat reduction; (d) humans.

LIST OF TRANSPARENCY MASTERS

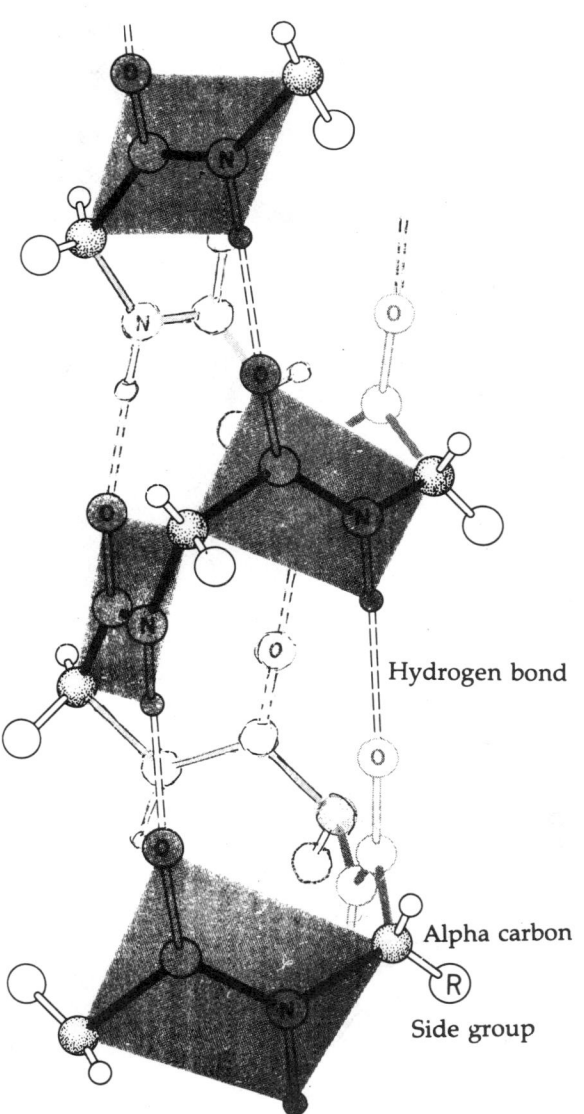

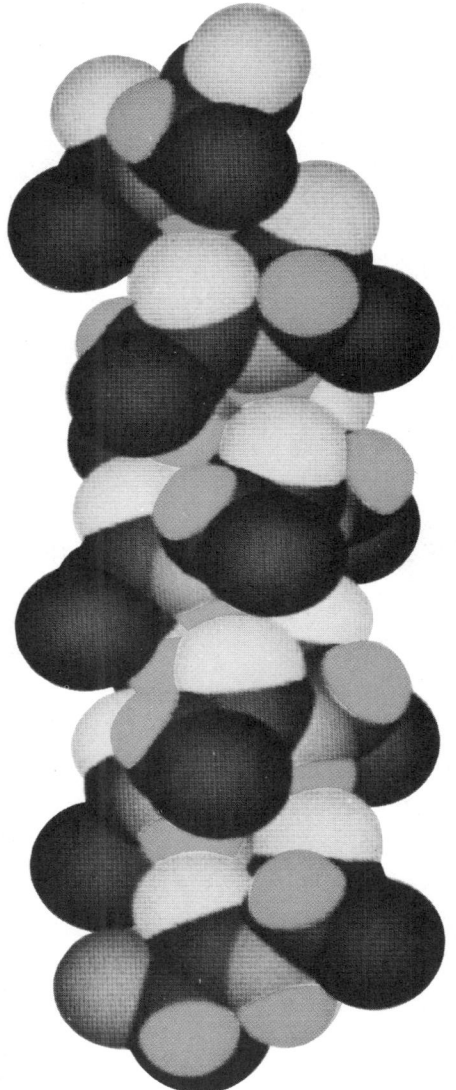

Hydrogen bond

Alpha carbon

Side group

FIGURE 3.17   THE α- HELIX   (text page 61)

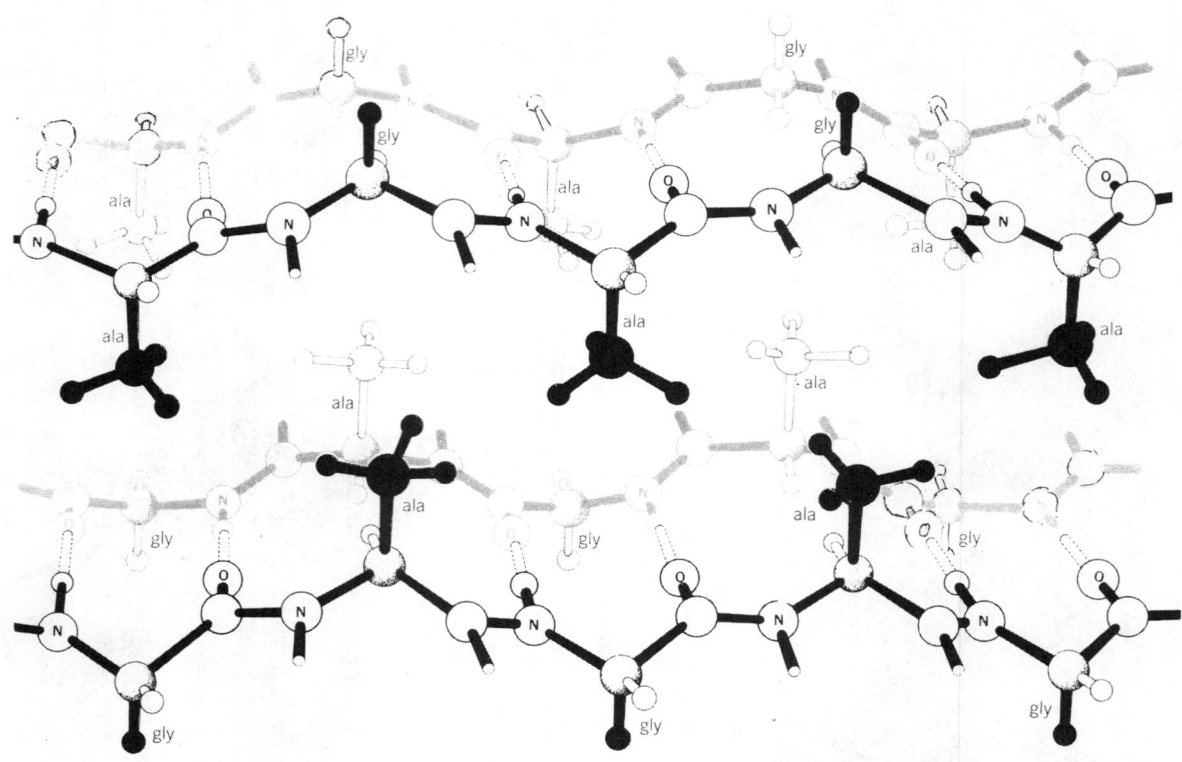

FIGURE 3.18   *β*-PLEATED SHEET   (text page 62)

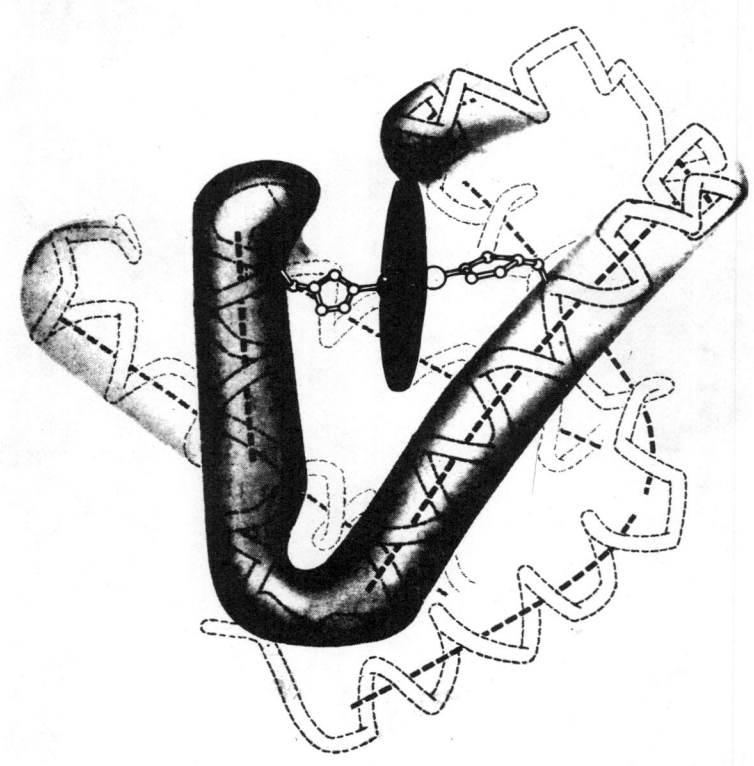

FIGURE 3.20   TERTIARY STRUCTURE OF A PROTEIN   (text page 63)

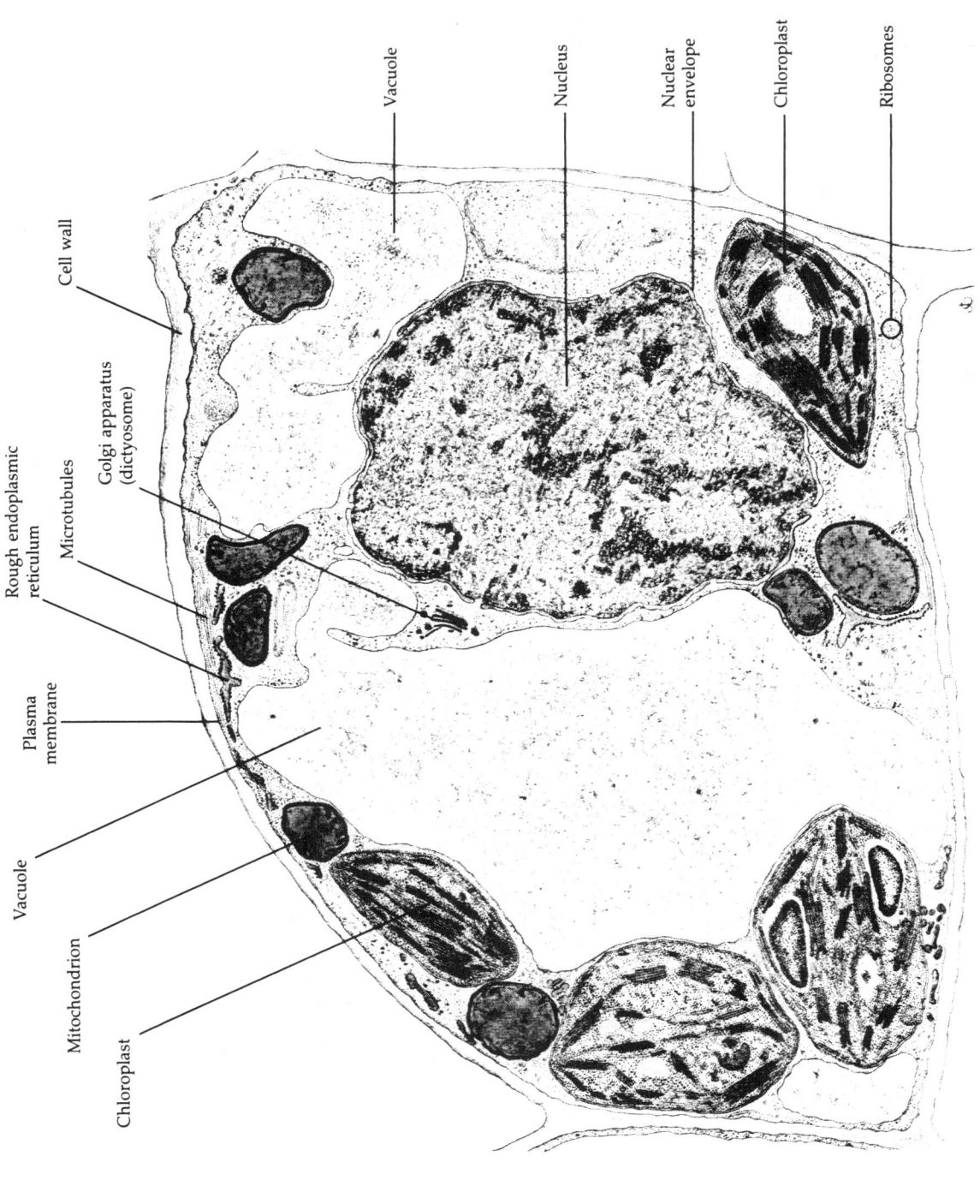

FIGURE 4.7   A PLANT CELL   (text page 81)

Cell wall

Vacuole

Nucleus

Nuclear
envelope

Chloroplast

Ribosomes

Rough endoplasmic
reticulum

Microtubules

Golgi apparatus
(dictyosome)

Plasma
membrane

Vacuole

Mitochondrion

Chloroplast

From Life: The Science of Biology, Copyright © 1983 by William K. Purves and Gordon H. Orians.

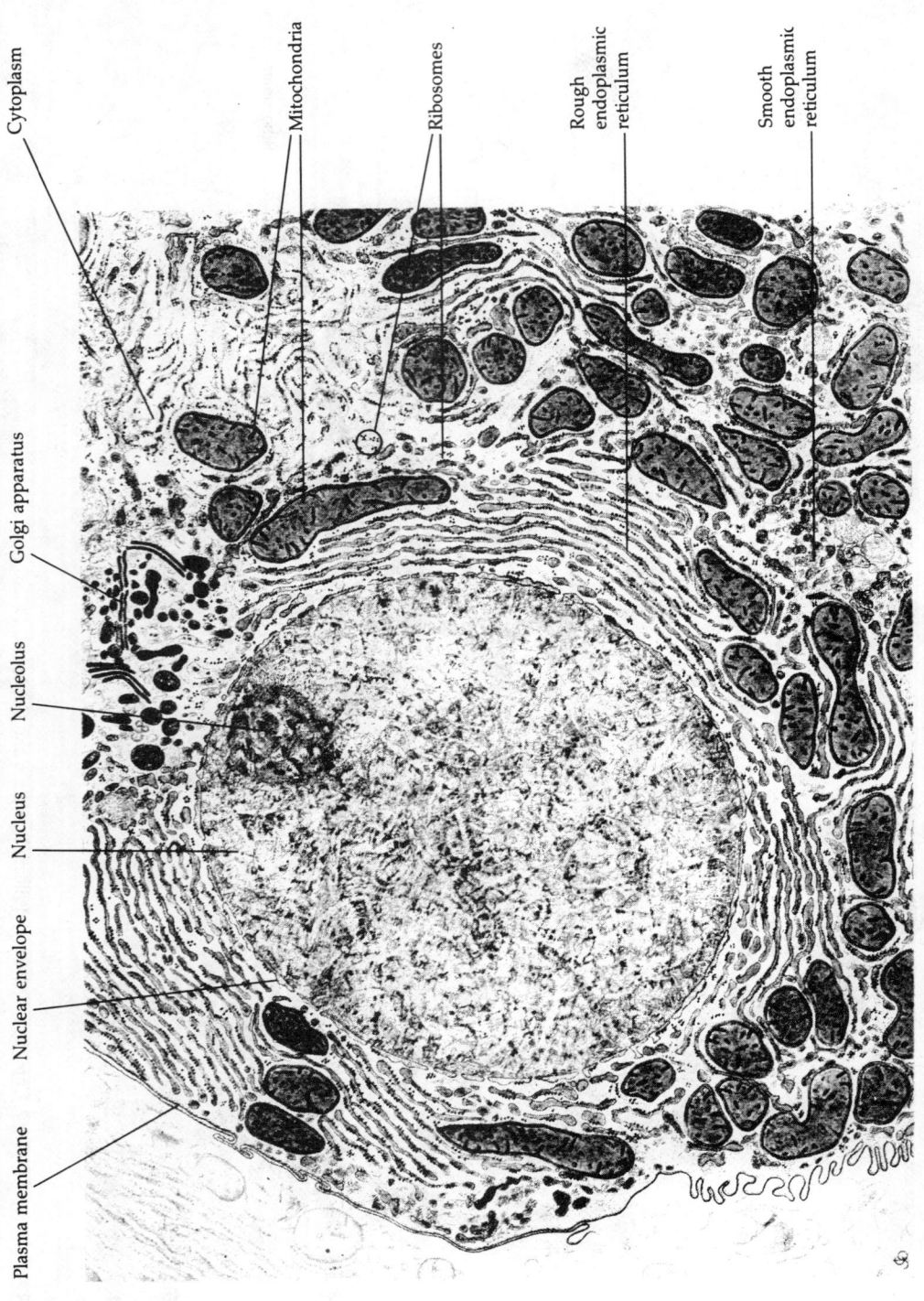

Cytoplasm

Mitochondria

Ribosomes

Rough
endoplasmic
reticulum

Smooth
endoplasmic
reticulum

Golgi apparatus

Nucleolus

Nucleus

Nuclear envelope

Plasma membrane

FIGURE 4.8   AN ANIMAL CELL   (text page 83)

From *Life: The Science of Biology.* Copyright © 1983 by William K. Purves and Gordon H. Orians.

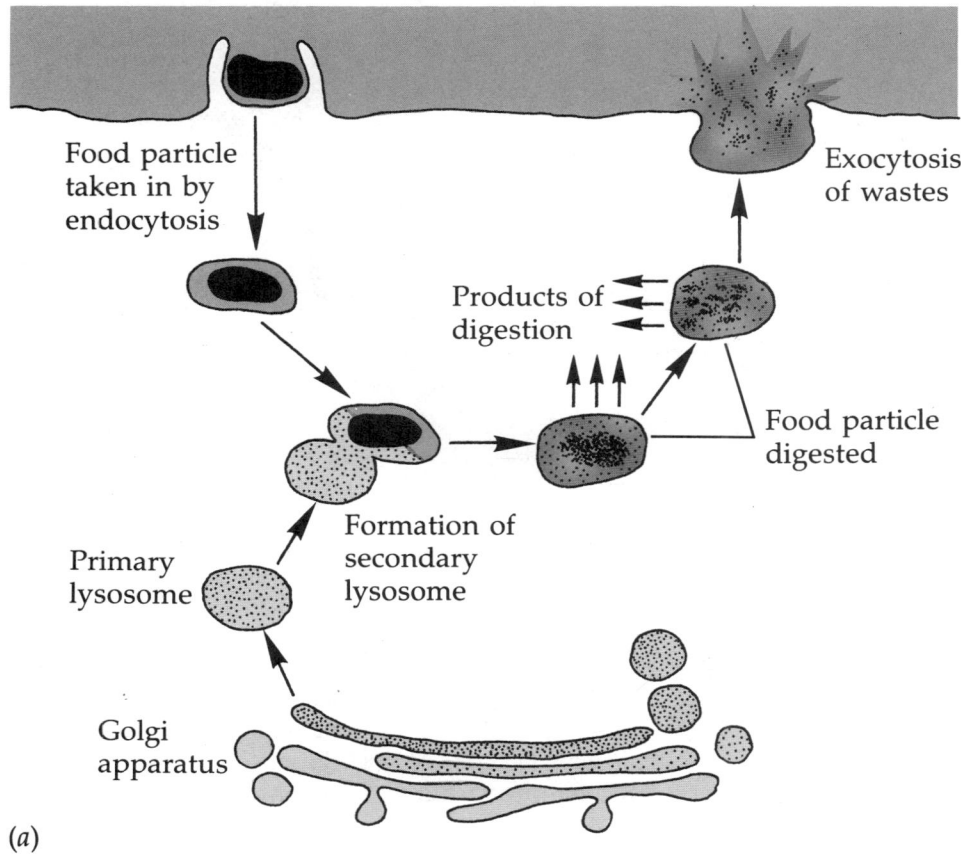

Food particle taken in by endocytosis

Exocytosis of wastes

Products of digestion

Food particle digested

Formation of secondary lysosome

Primary lysosome

Golgi apparatus

(a)

FIGURE 4.26   LYSOSOMES   (text page 98)

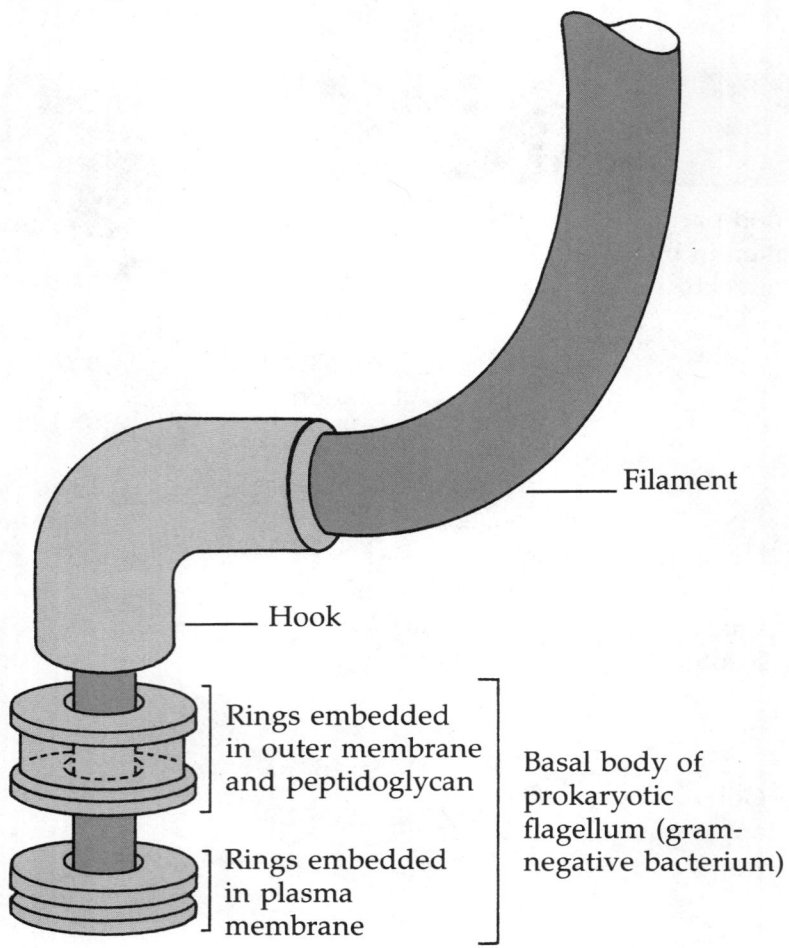

Filament

Hook

Rings embedded
in outer membrane
and peptidoglycan

Rings embedded
in plasma
membrane

Basal body of
prokaryotic
flagellum (gram-
negative bacterium)

FIGURE 4.31   THE PROKARYOTIC FLAGELLUM   (text page 102)
From M.L. Depamphilis and J. Adler, *J. Bacteriol.* 105: 384–395, 1971.

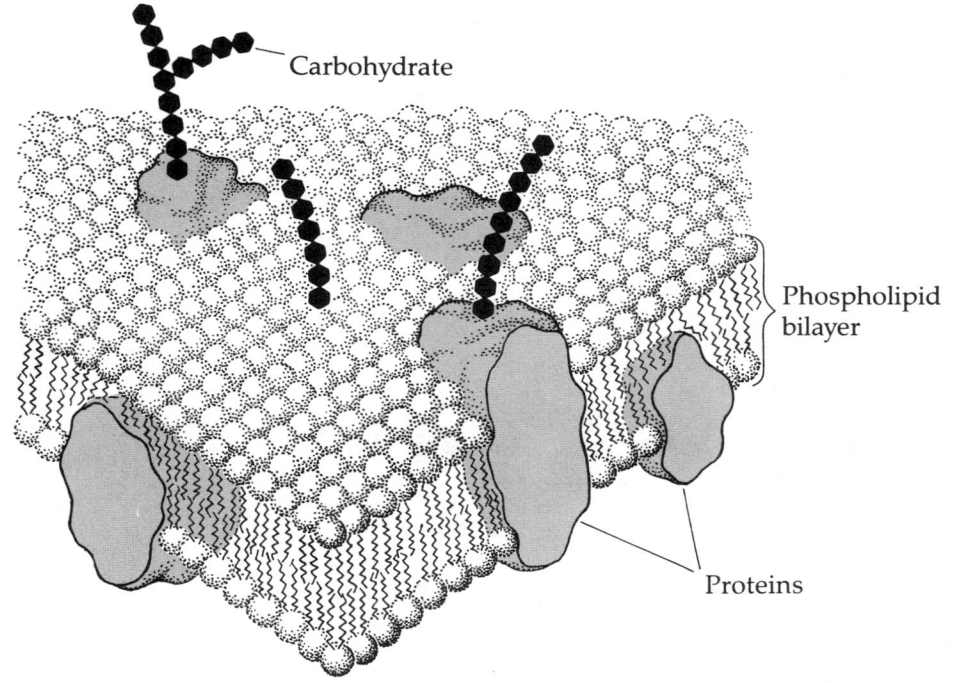

Carbohydrate

Phospholipid
bilayer

Proteins

FIGURE 5.1   BIOLOGICAL MEMBRANES   (text page 111)

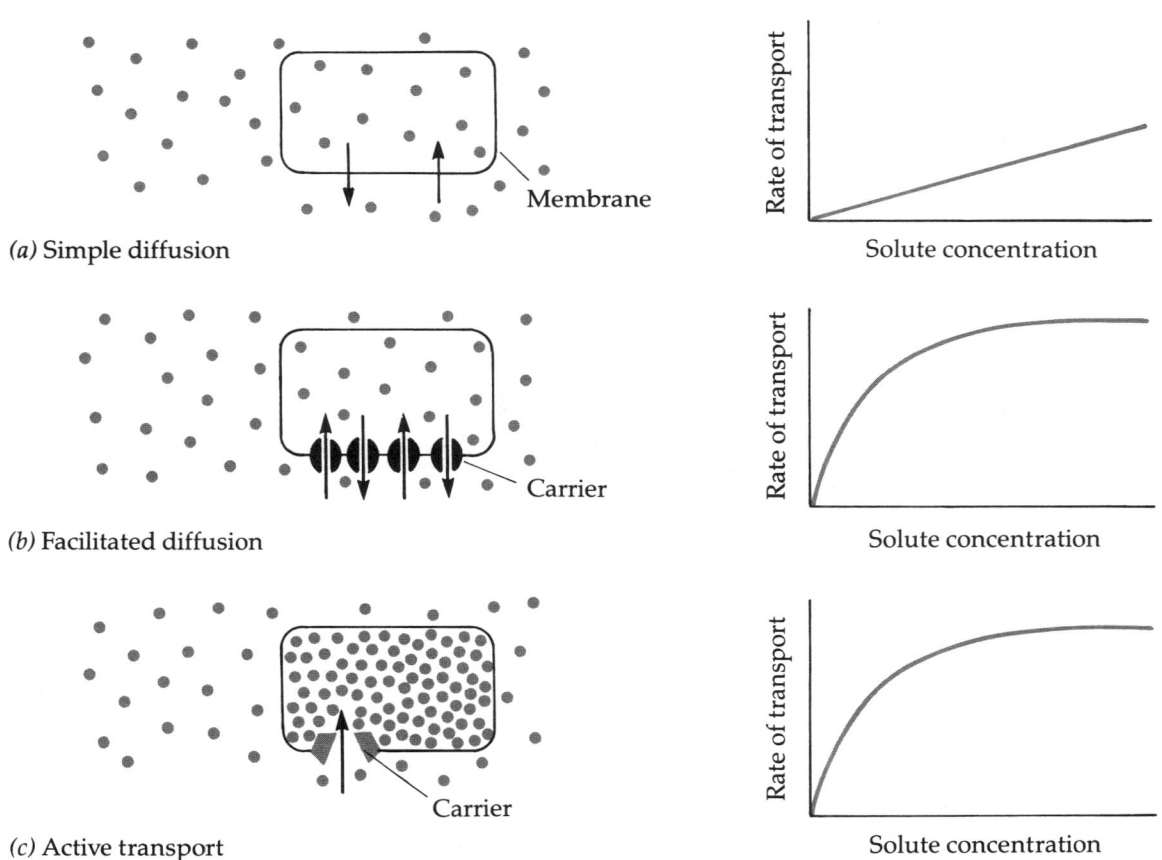

Membrane

(a) Simple diffusion

Carrier

(b) Facilitated diffusion

Carrier

(c) Active transport

Rate of transport

Solute concentration

Rate of transport

Solute concentration

Rate of transport

Solute concentration

FIGURE 5.11   CROSSING BIOLOGICAL MEMBRANES   (text page 120)

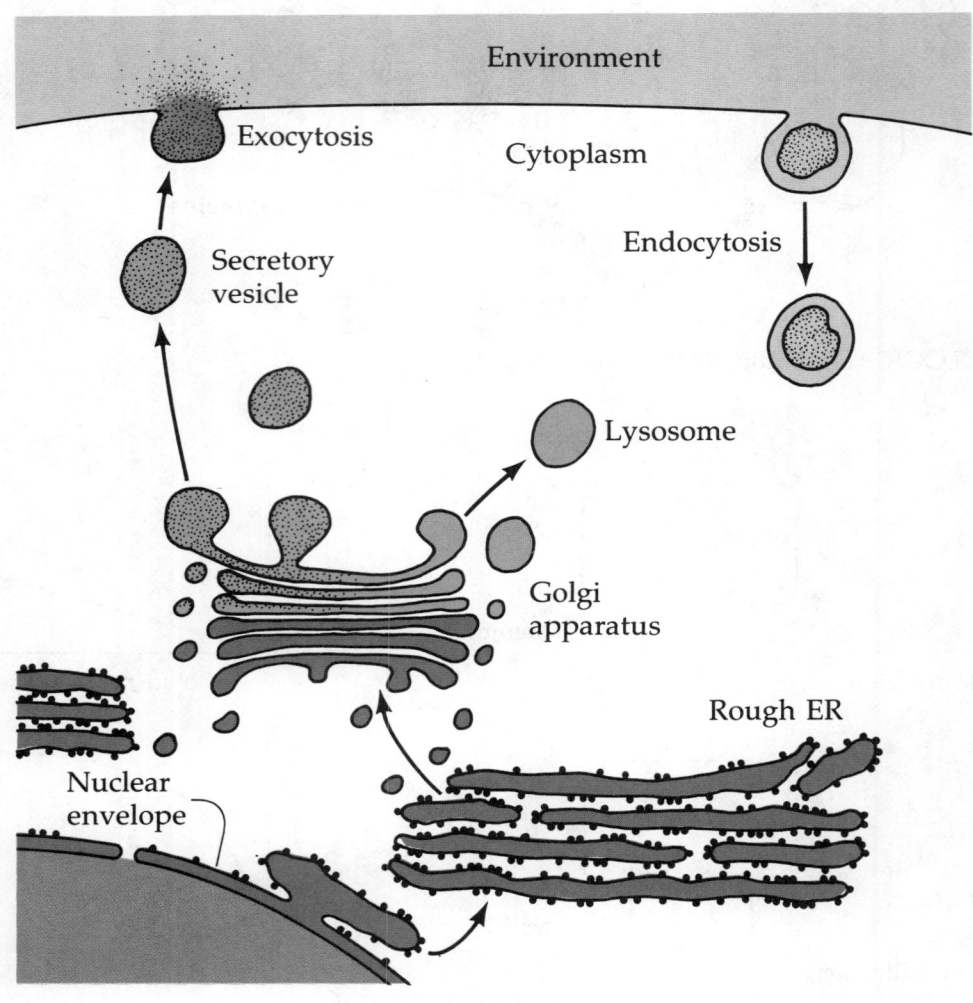

Environment

Exocytosis

Cytoplasm

Endocytosis

Secretory
vesicle

Lysosome

Golgi
apparatus

Rough ER

Nuclear
envelope

FIGURE 5.21   MEMBRANE CONTINUITY IN CELLS   (text page 128)

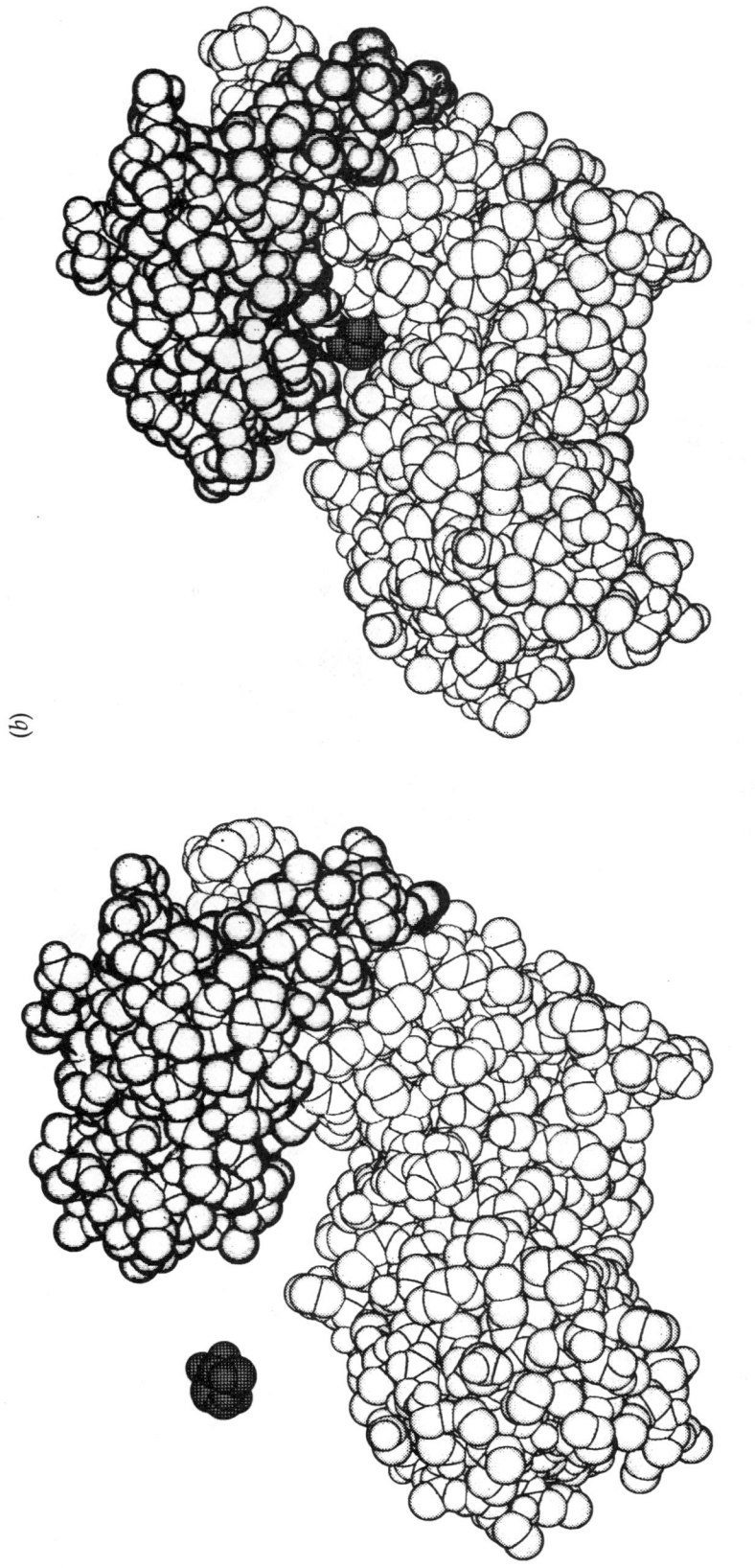

(b)

FIGURE 6.11 INDUCED FIT (text page 141)

From W.S. Bennett and T.A. Steitz, *J. Mol. Biol.* 140: 211–230, 1980.

From *Life: The Science of Biology,* Copyright © 1983 by William K. Purves and Gordon H. Orians.

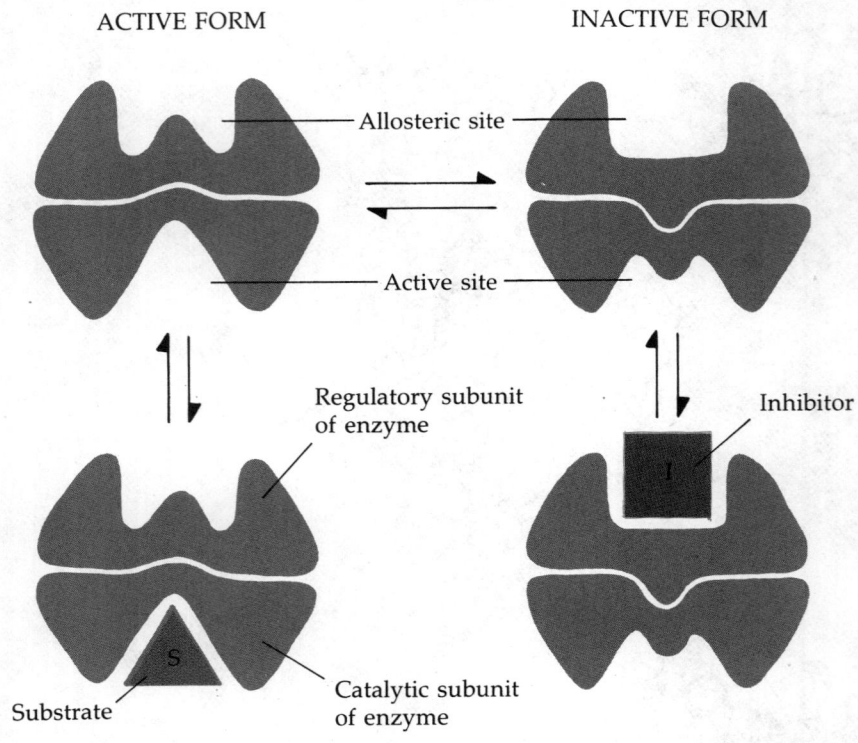

ACTIVE FORM   INACTIVE FORM

Allosteric site

Active site

Regulatory subunit
of enzyme

Inhibitor

Substrate

Catalytic subunit
of enzyme

FIGURE 6.22   ALLOSTERIC REGULATION OF ENZYMES   (text page 149)

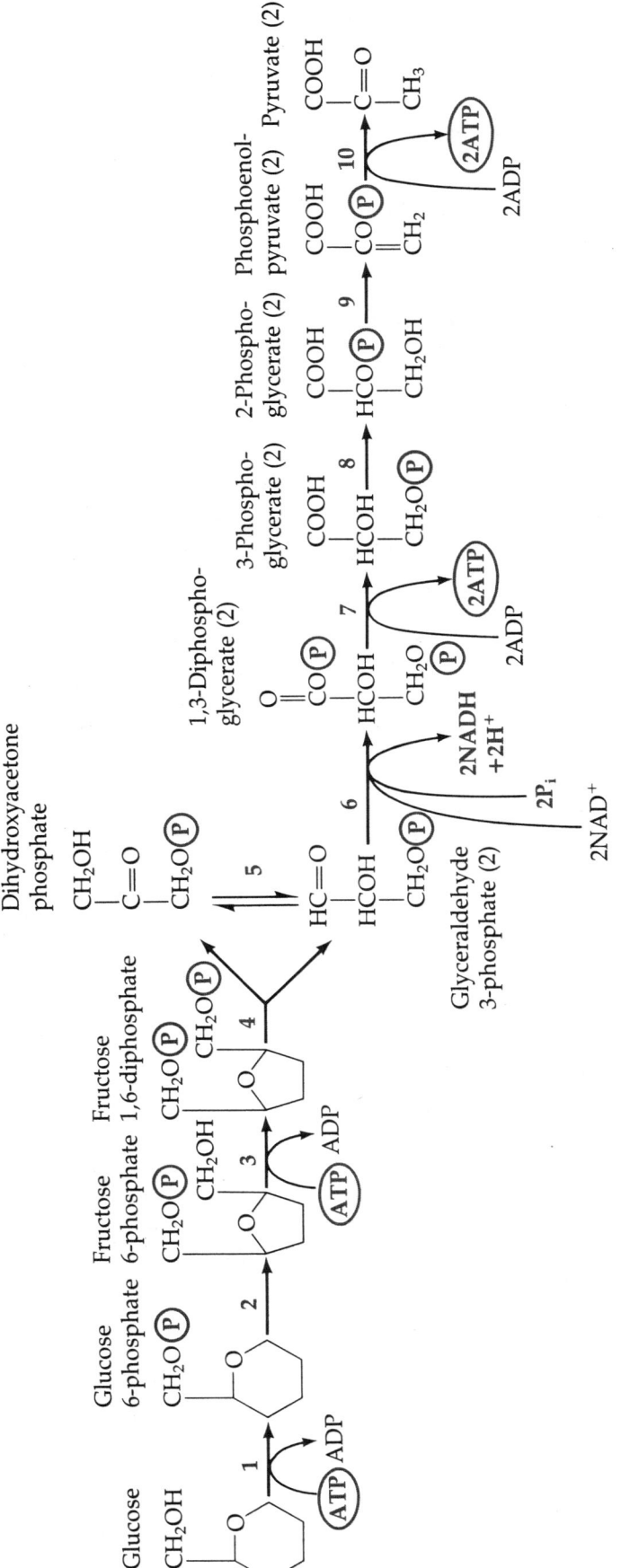

FIGURE 7.12 GLYCOLYSIS (text page 166)

From *Life: The Science of Biology.* Copyright © 1983 by William K. Purves and Gordon H. Orians.

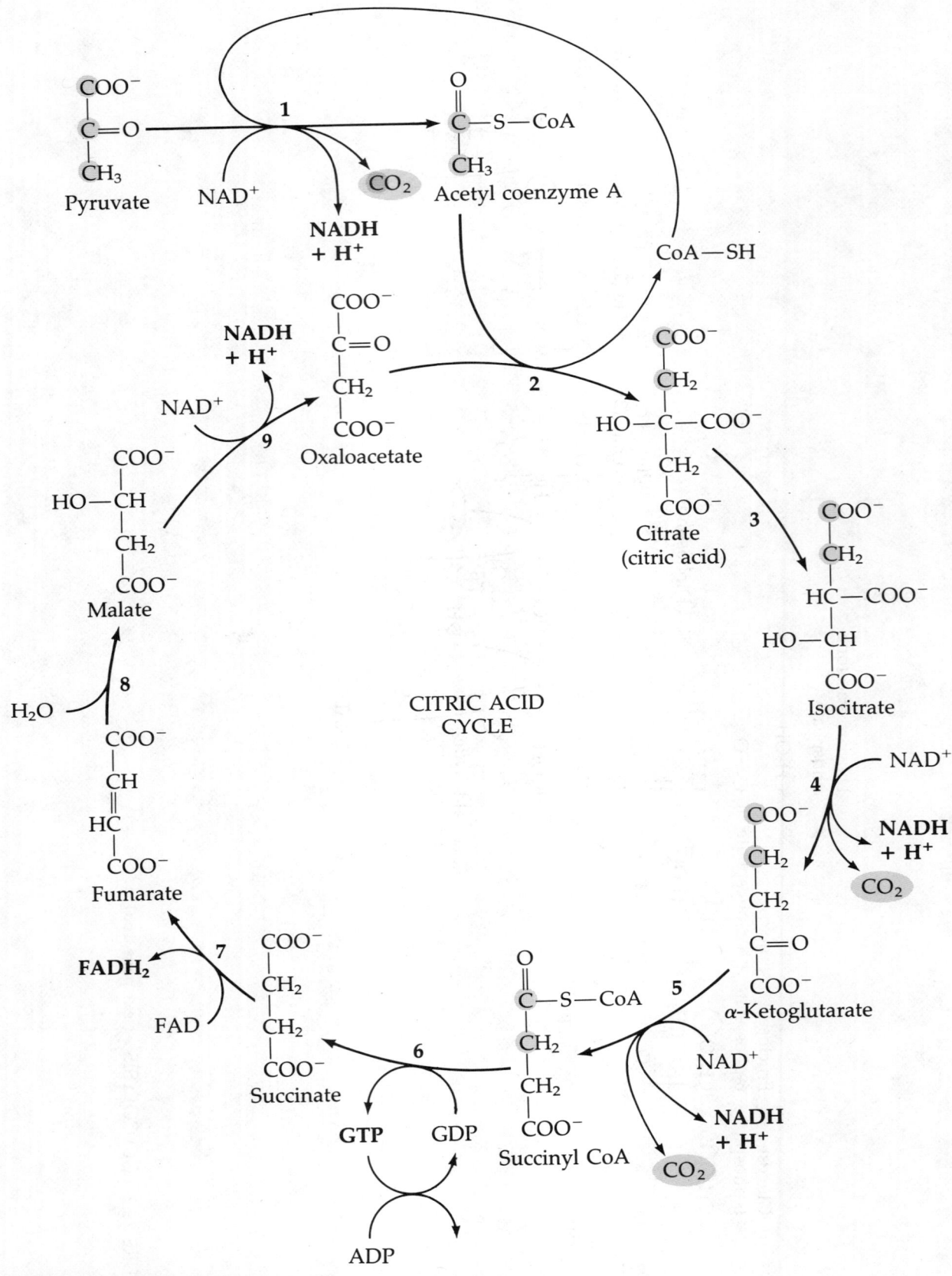

FIGURE 7.14   THE CITRIC ACID CYCLE   (text page 169)

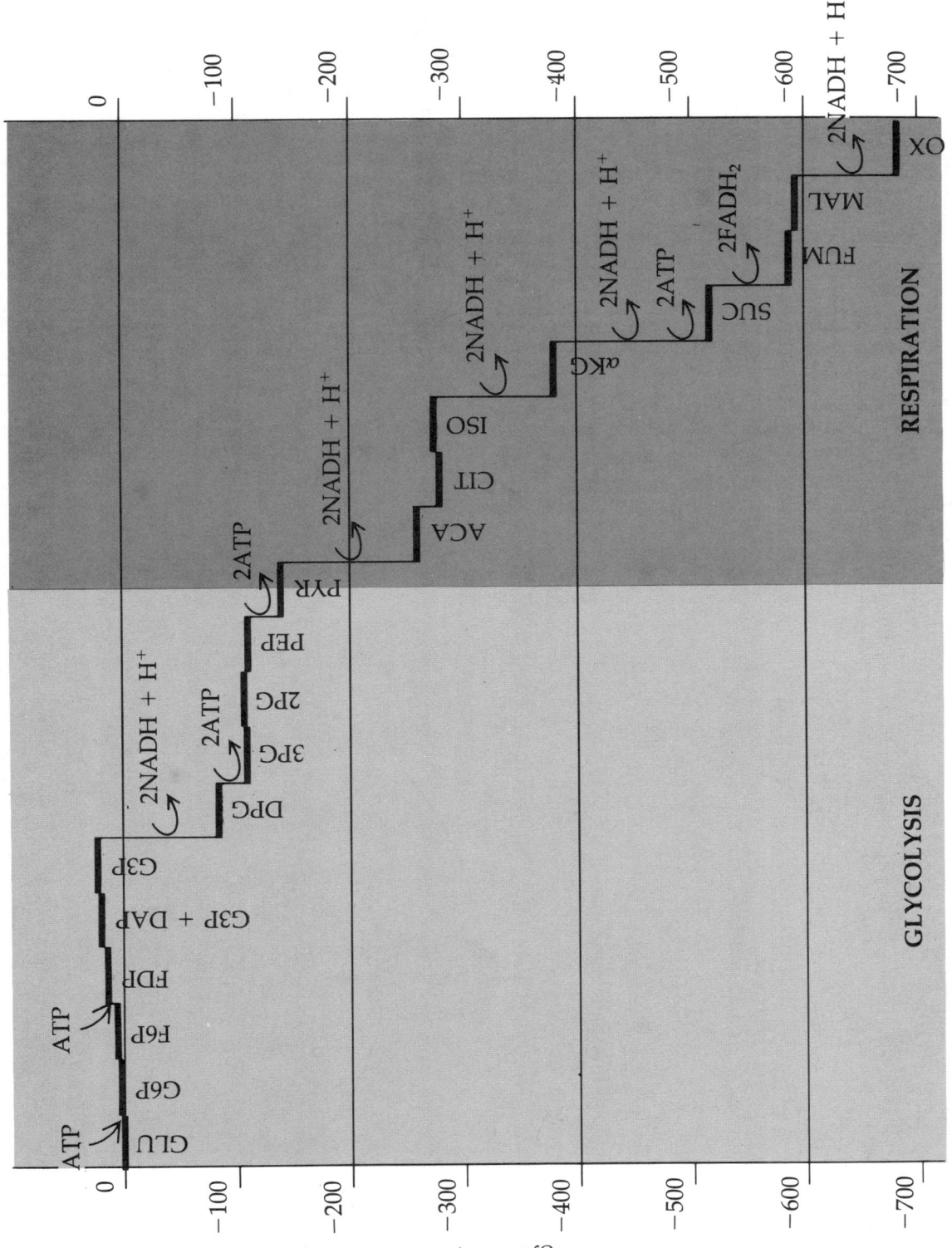

FIGURE 7.15   FREE ENERGY CHANGE DURING GLYCOLYSIS AND CELLULAR RESPIRATION   (text page 170)

From *Life: The Science of Biology*, Copyright © 1983 by William K. Purves and Gordon H. Orians.

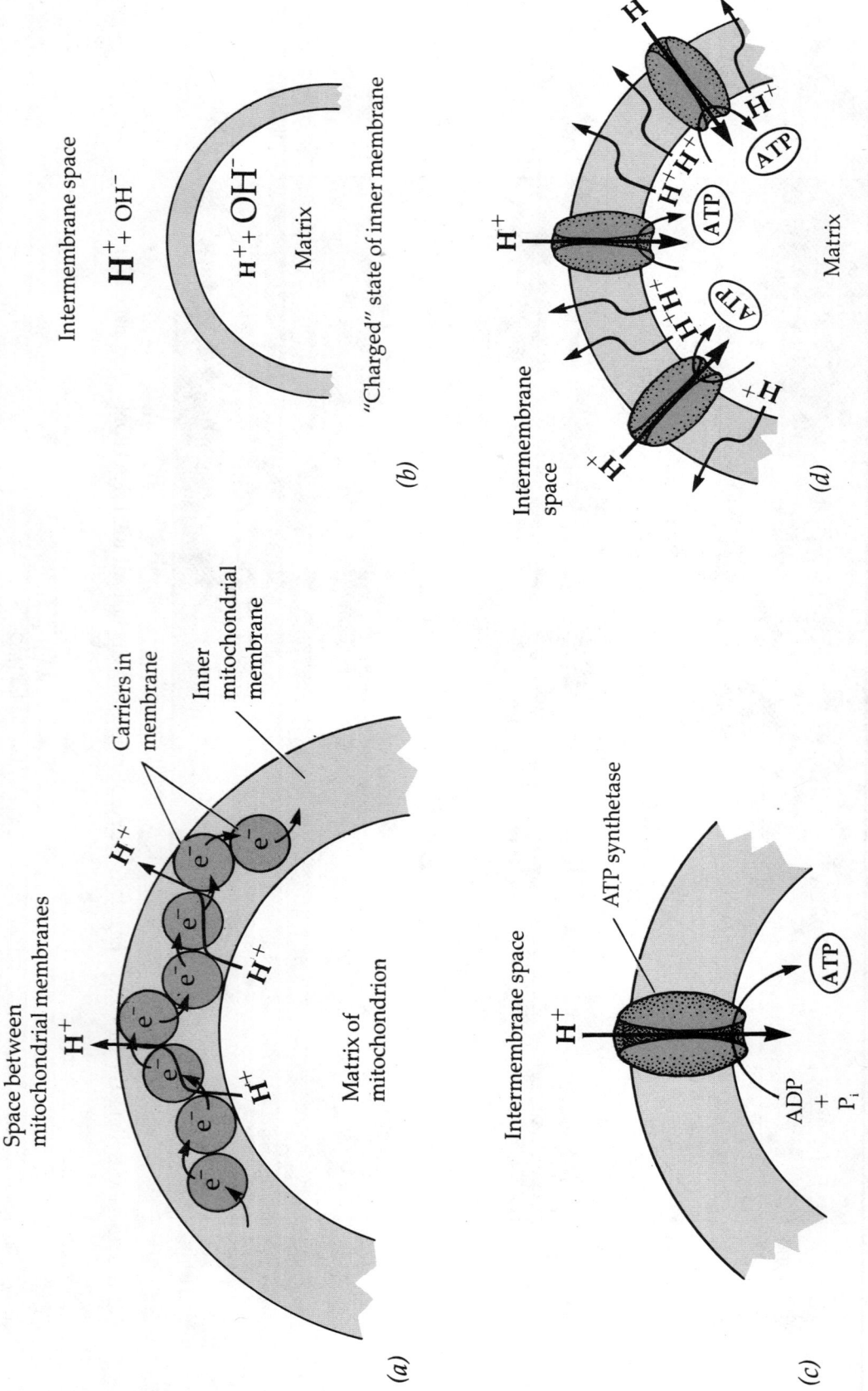

FIGURE 7.19   THE CHEMIOSMOTIC HYPOTHESIS   (text page 174)

From *Life: The Science of Biology*, Copyright © 1983 by William K. Purves and Gordon H. Orians.

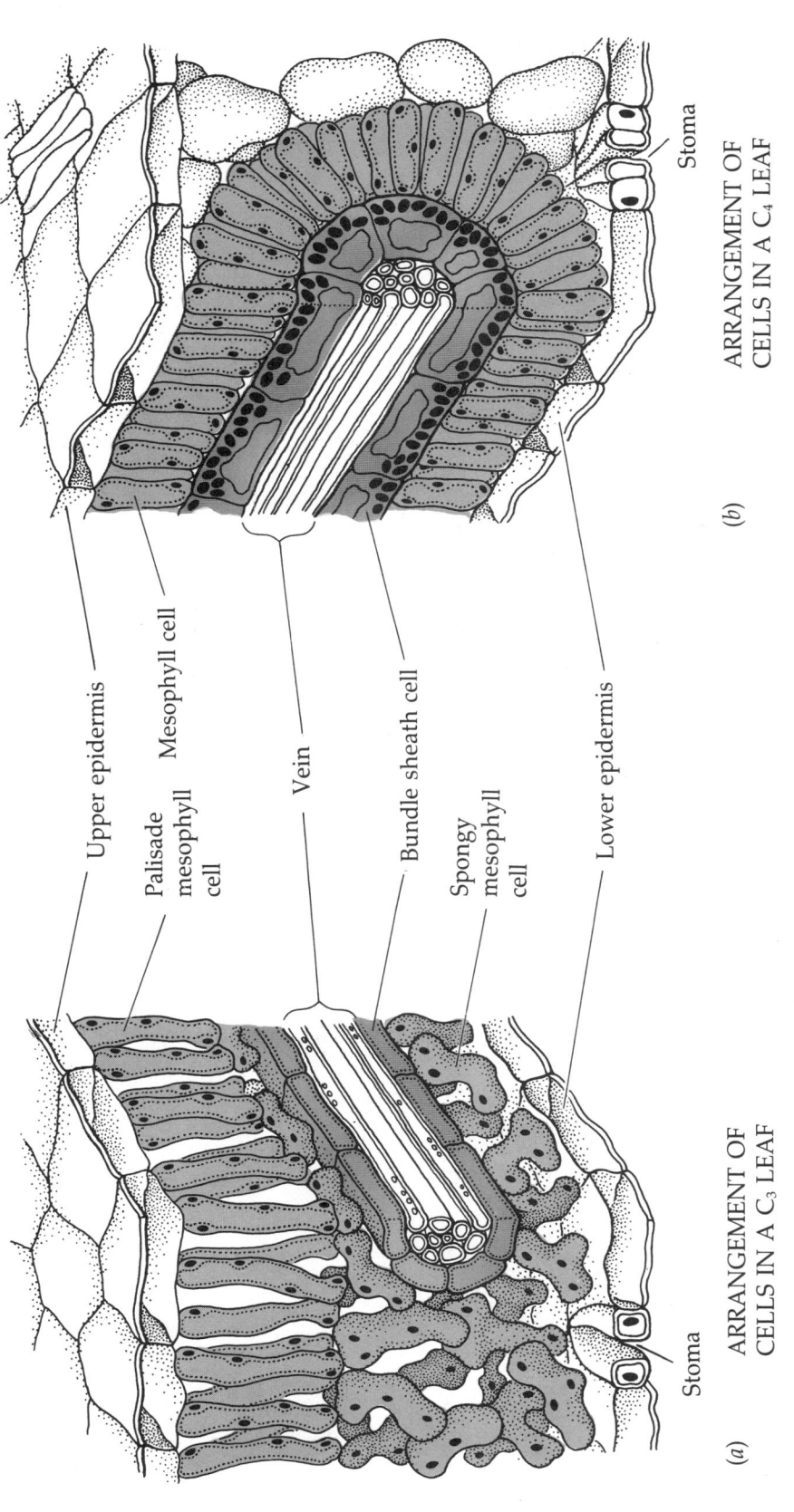

(a)

ARRANGEMENT OF
CELLS IN A C₃ LEAF

Upper epidermis

Mesophyll cell

Palisade
mesophyll
cell

Vein

Bundle sheath cell

Spongy
mesophyll
cell

Lower epidermis

Stoma

(b)

ARRANGEMENT OF
CELLS IN A C₄ LEAF

Stoma

FIGURE 8.24   LEAF ANATOMY OF C₃ AND C₄ PLANTS   (text page 201)

From *Life: The Science of Biology.* Copyright © 1983 by William K. Purves and Gordon H. Orians.

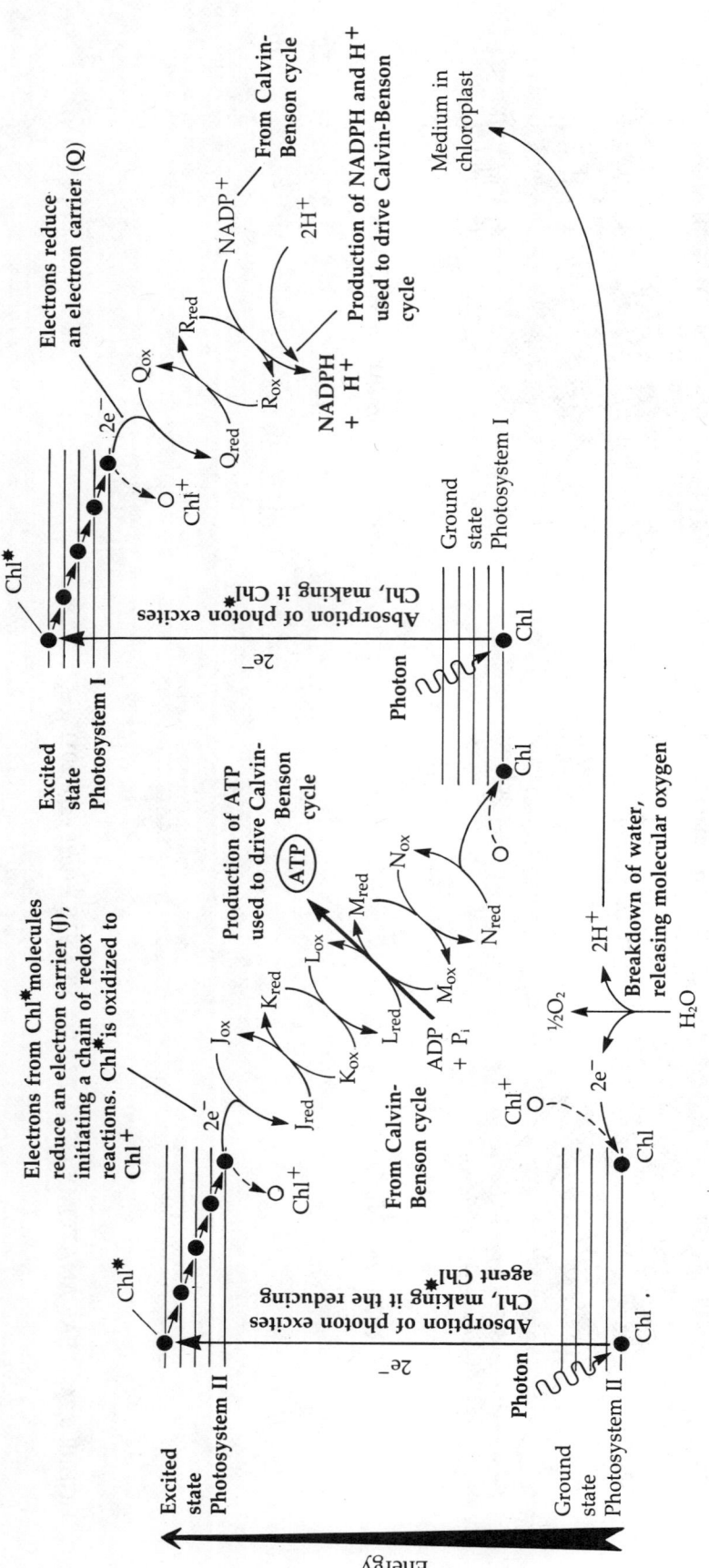

FIGURE 8.31 NONCYCLIC PHOTOPHOSPHORYLATION (text page 207)

From Life: *The Science of Biology.* Copyright © 1983 by William K. Purves and Gordon H. Orians.

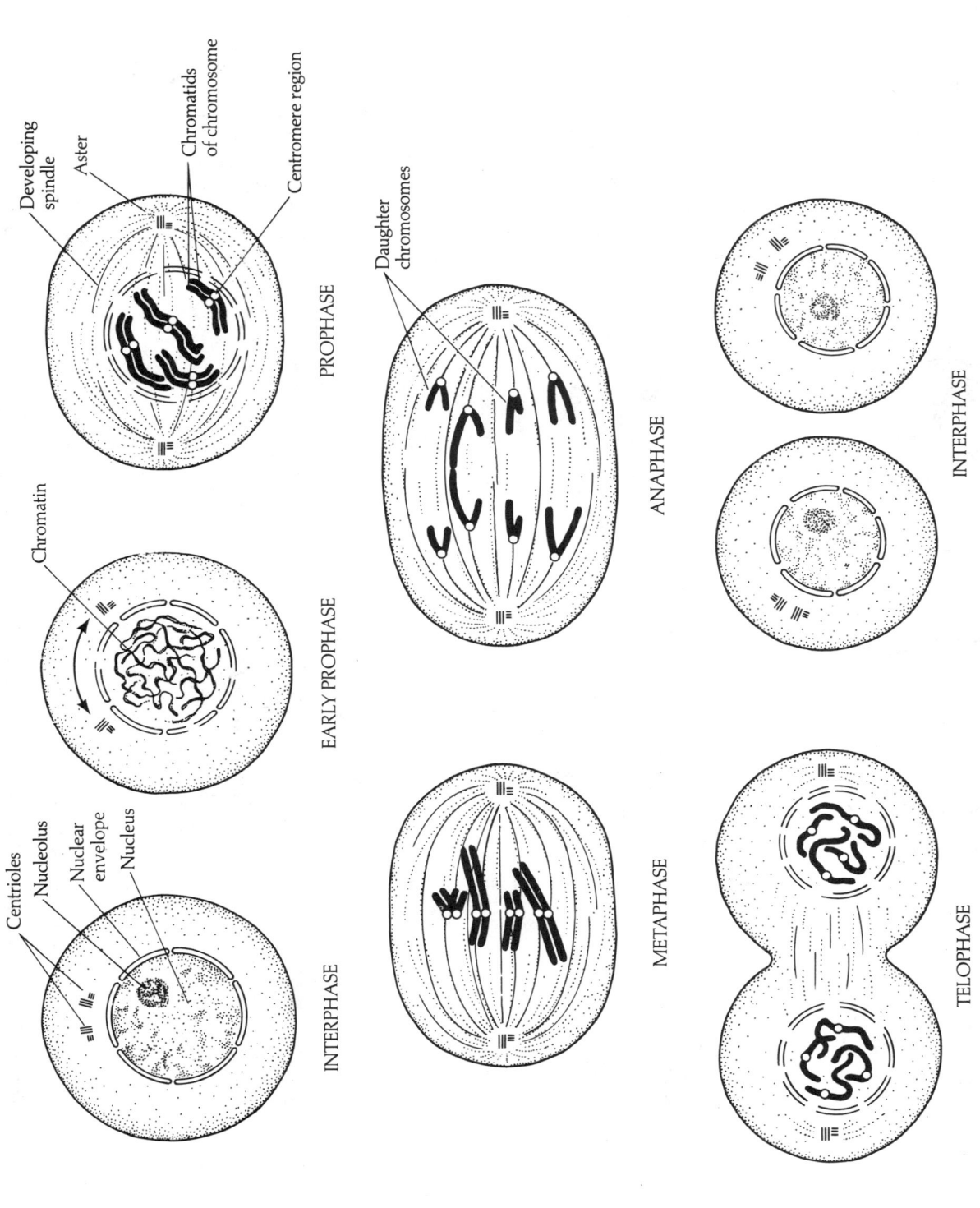

Centrioles
Nucleolus
Nuclear envelope
Nucleus

INTERPHASE

Chromatin

EARLY PROPHASE

Developing spindle
Aster
Chromatids of chromosome
Centromere region

PROPHASE

METAPHASE

Daughter chromosomes

ANAPHASE

TELOPHASE

INTERPHASE

FIGURE 9.2   STAGES OF MITOSIS   (text page 220)

From *Life: The Science of Biology.* Copyright © 1983 by William K. Purves and Gordon H. Orians.

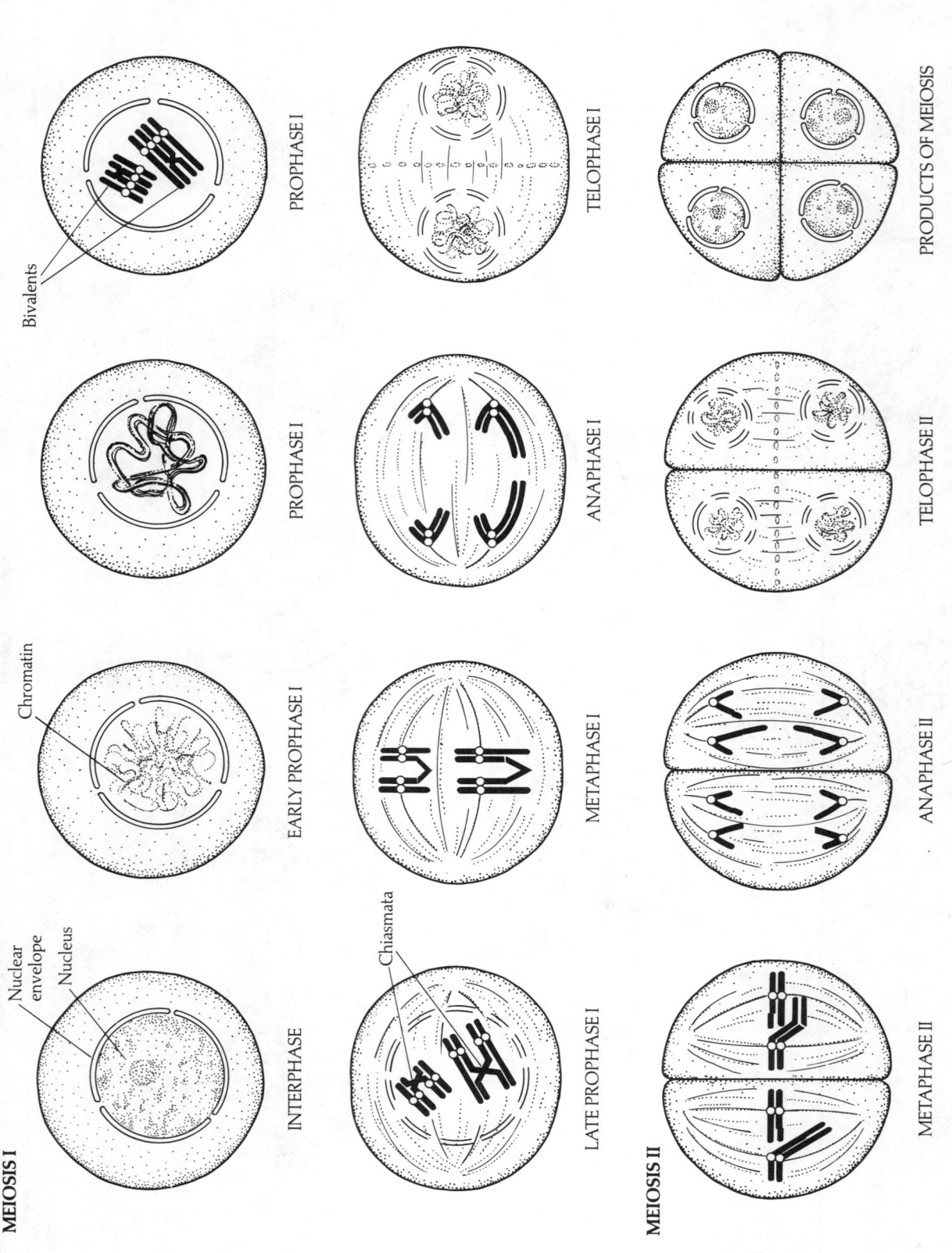

MEIOSIS I

INTERPHASE — Nuclear envelope, Nucleus

EARLY PROPHASE I — Chromatin

PROPHASE I

PROPHASE I — Bivalents

LATE PROPHASE I — Chiasmata

METAPHASE I

ANAPHASE I

TELOPHASE I

MEIOSIS II

METAPHASE II

ANAPHASE II

TELOPHASE II

PRODUCTS OF MEIOSIS

FIGURE 9.15 STAGES OF MEIOSIS (text page 233)

From *Life: The Science of Biology*, Copyright © 1983 by William K. Purves and Gordon H. Orians.

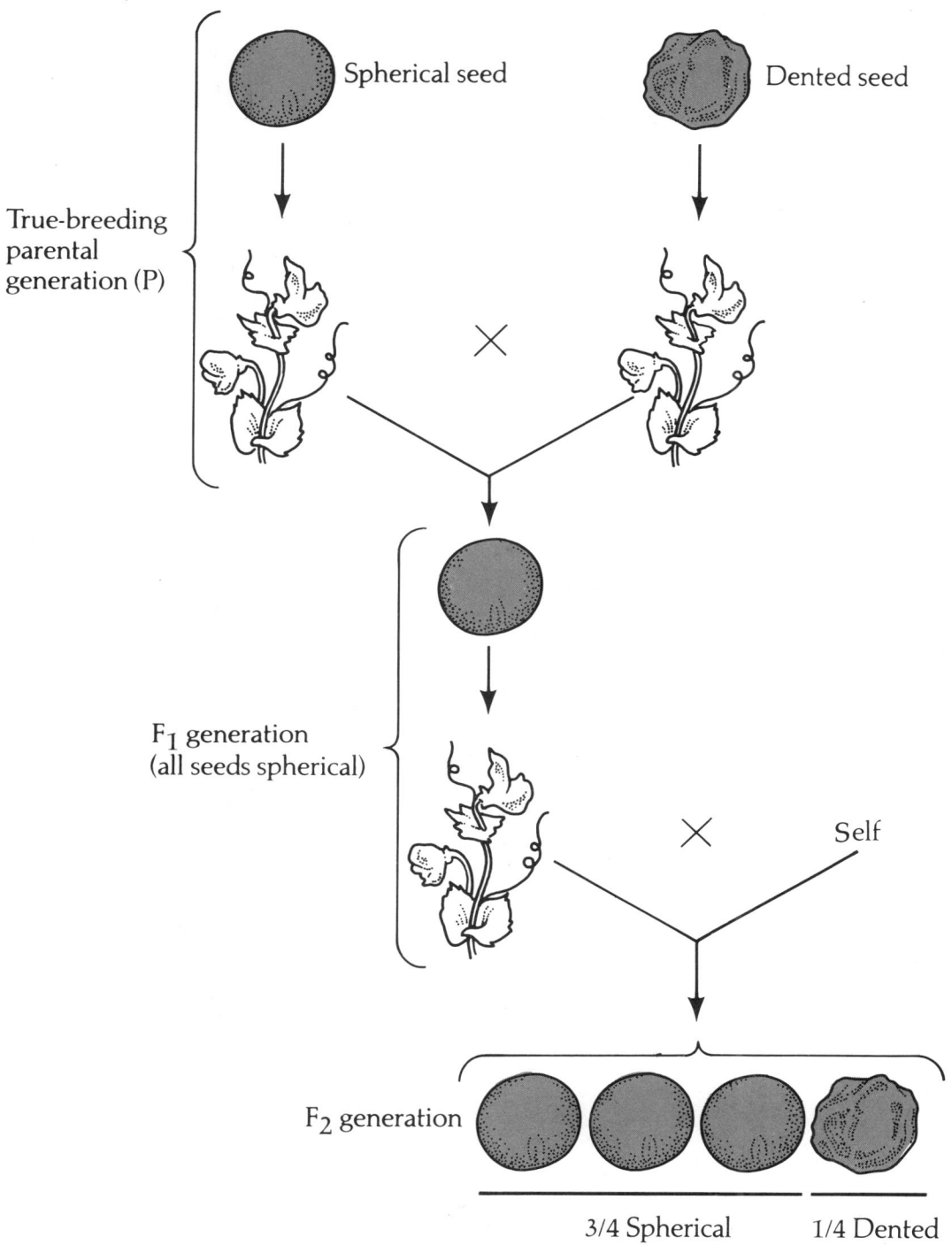

True-breeding
parental
generation (P)

Spherical seed

Dented seed

×

F₁ generation
(all seeds spherical)

×   Self

F₂ generation

3/4 Spherical   1/4 Dented

FIGURE 10.2   MENDEL'S EXPERIMENT 1   (text page 245)

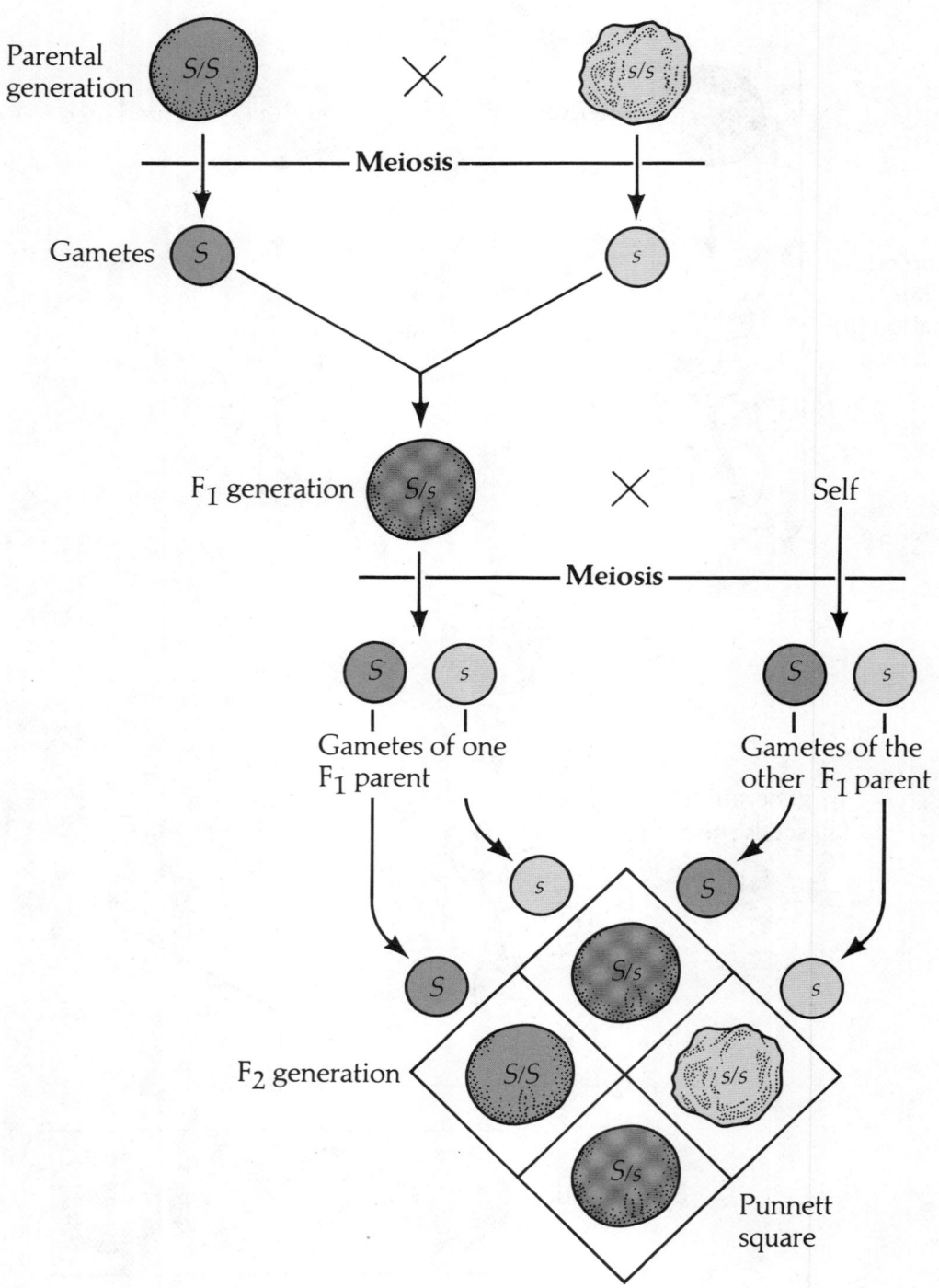

FIGURE 10.3   MENDEL'S EXPLANATION OF EXPERIMENT 1   (text page 248)

Test cross

Spherical pea
of undetermined
genotype

$S/-$ $\times$ s/s

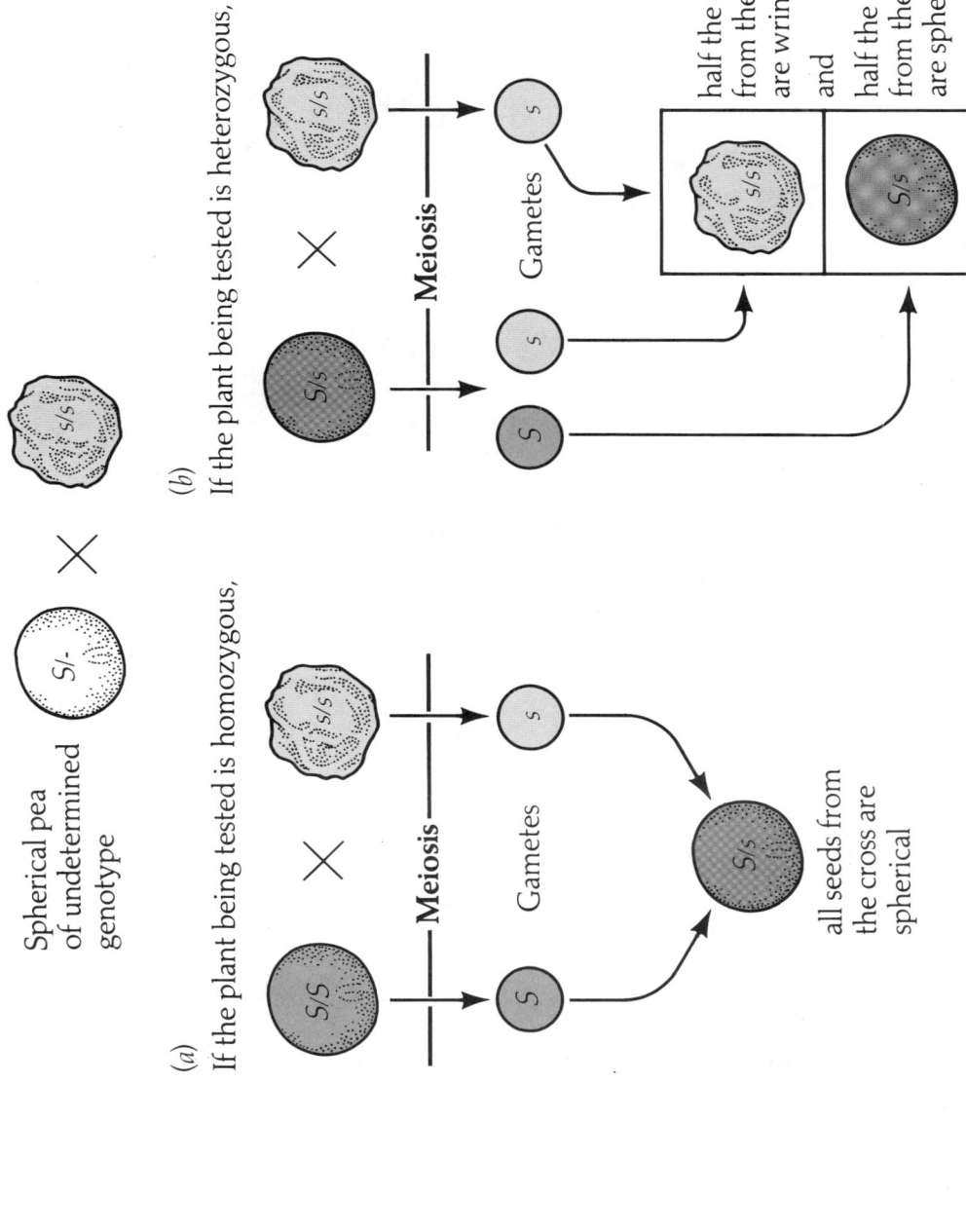

(a)
If the plant being tested is homozygous,

S/S $\times$ s/s

—— **Meiosis** ——

Gametes

S s

S/s

all seeds from
the cross are
spherical

(b)
If the plant being tested is heterozygous,

S/s $\times$ s/s

—— **Meiosis** ——

Gametes

S s s

| S/s | s/s |

half the seeds
from the cross
are wrinkled

and

half the seeds
from the cross
are spherical

FIGURE 10.5   A TEST CROSS   (text page 249)

From *Life: The Science of Biology*, Copyright © 1983 by William K. Purves and Gordon H. Orians.

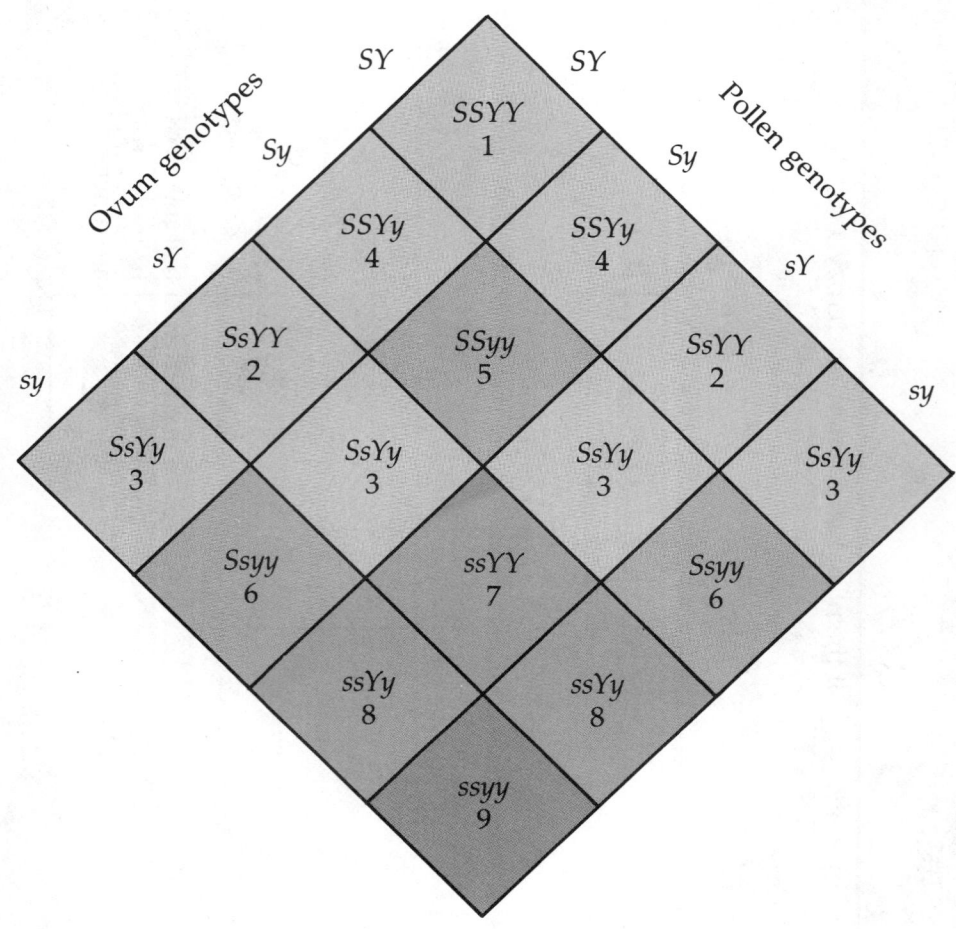

FIGURE 10.7   INDEPENDENT ASSORTMENT   (text page 252)

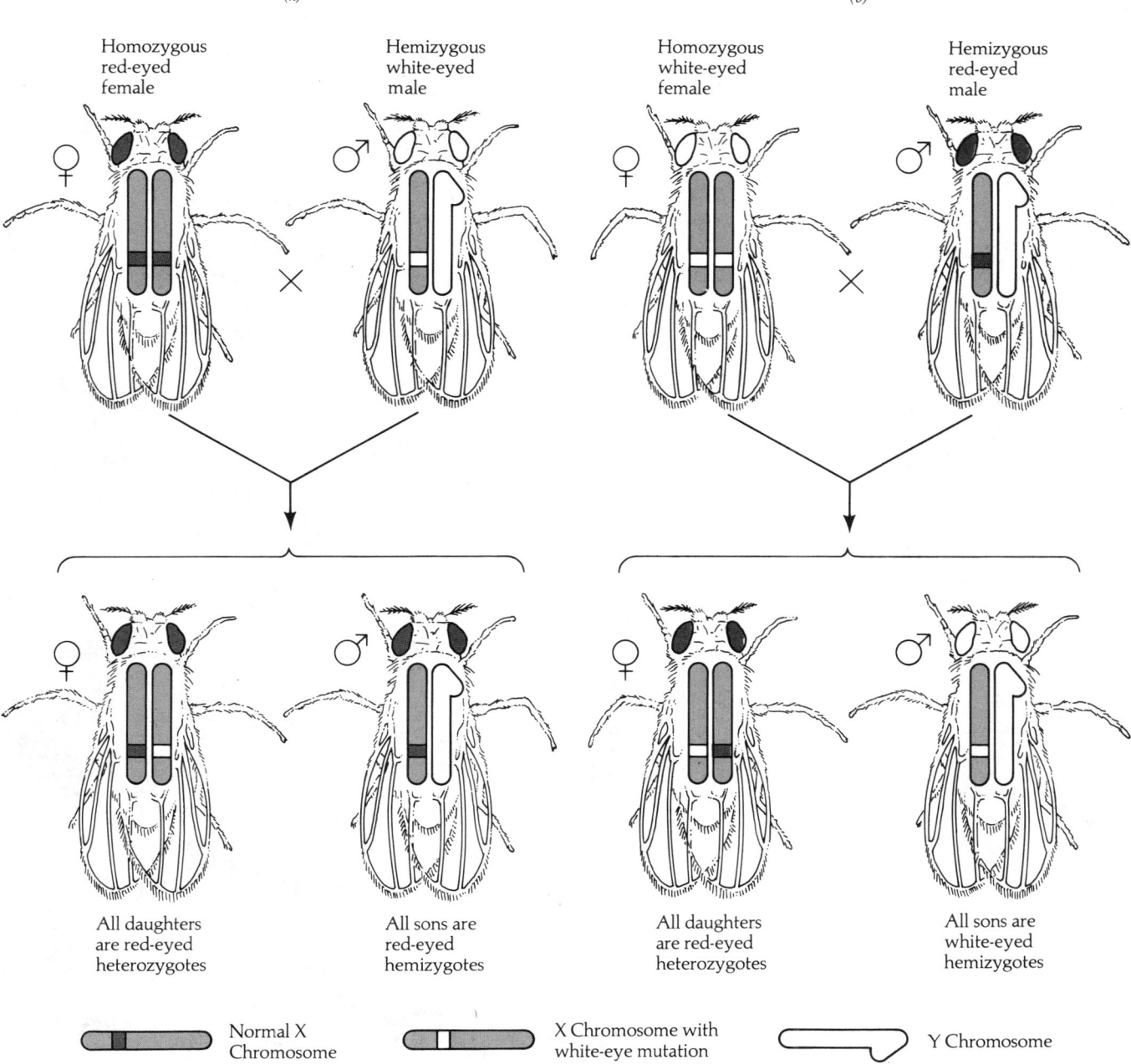

	(a)		(b)	
	Homozygous red-eyed female	Hemizygous white-eyed male	Homozygous white-eyed female	Hemizygous red-eyed male

All daughters are red-eyed heterozygotes   All sons are red-eyed hemizygotes   All daughters are red-eyed heterozygotes   All sons are white-eyed hemizygotes

Normal X Chromosome

X Chromosome with white-eye mutation

Y Chromosome

FIGURE 10.10 SEX-LINKED INHERITANCE (text page 256)

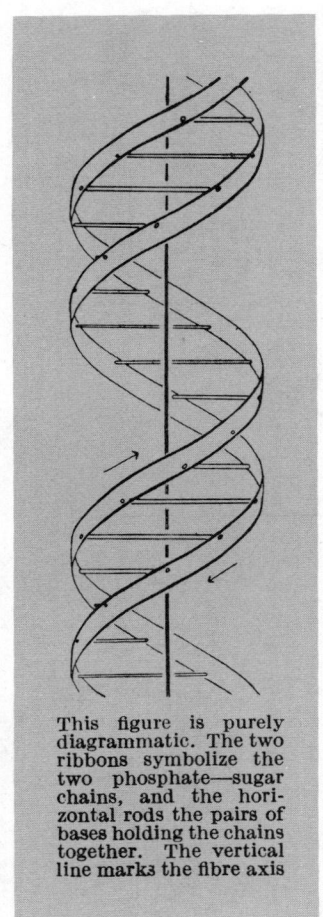

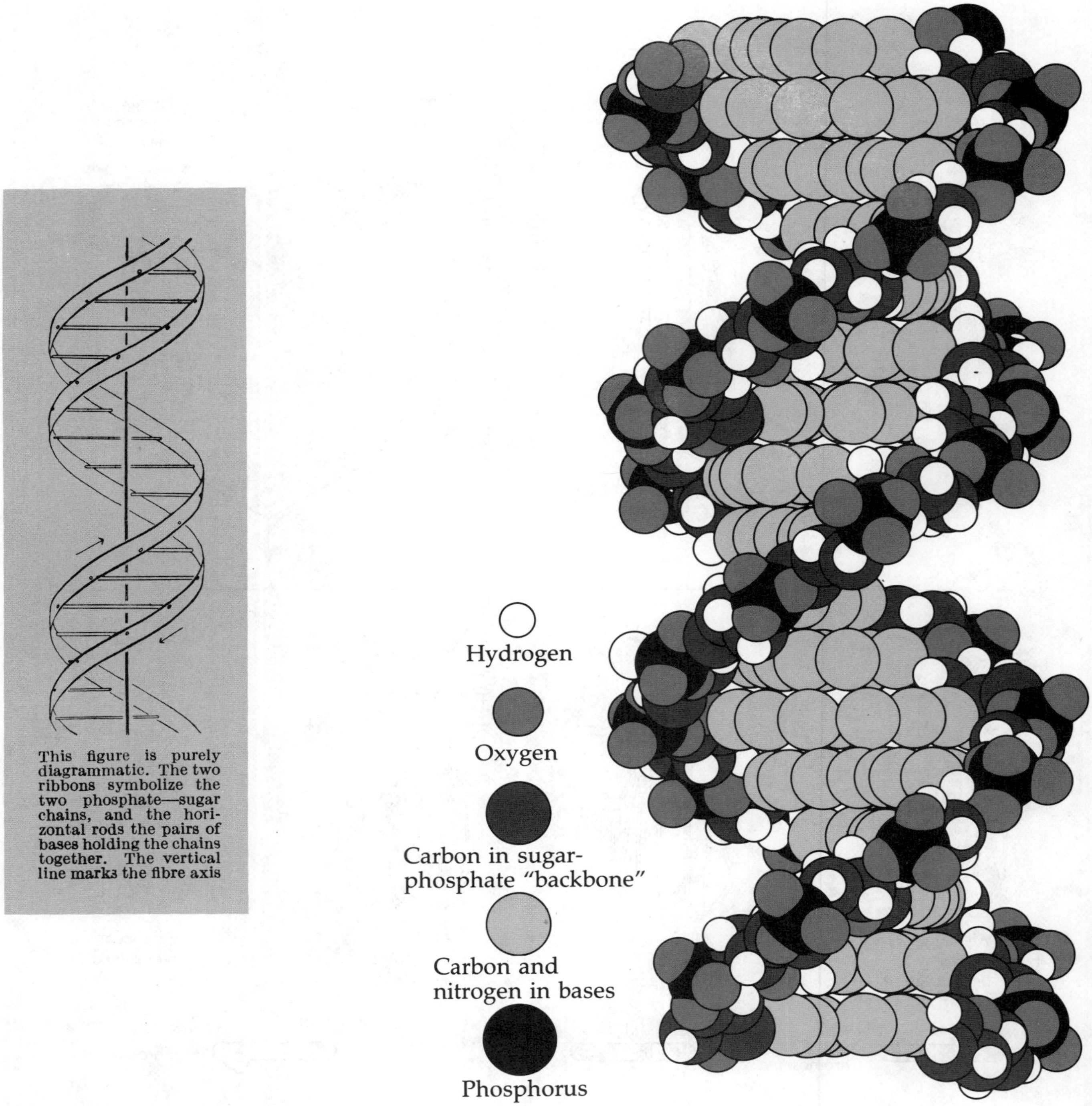

This figure is purely diagrammatic. The two ribbons symbolize the two phosphate—sugar chains, and the horizontal rods the pairs of bases holding the chains together. The vertical line marks the fibre axis

Hydrogen

Oxygen

Carbon in sugar-phosphate "backbone"

Carbon and nitrogen in bases

Phosphorus

FIGURE 11.7   THE WATSON-CRICK MODEL OF DNA   (text page 280)

Part (a) from J.D. Watson and F.H.C. Crick, *Nature* 171: 737–738, 1953; part (b) after F.J. Ayala and J.A. Kiger, Jr., *Modern Genetics*, Benjamin/Cummings, 1980.

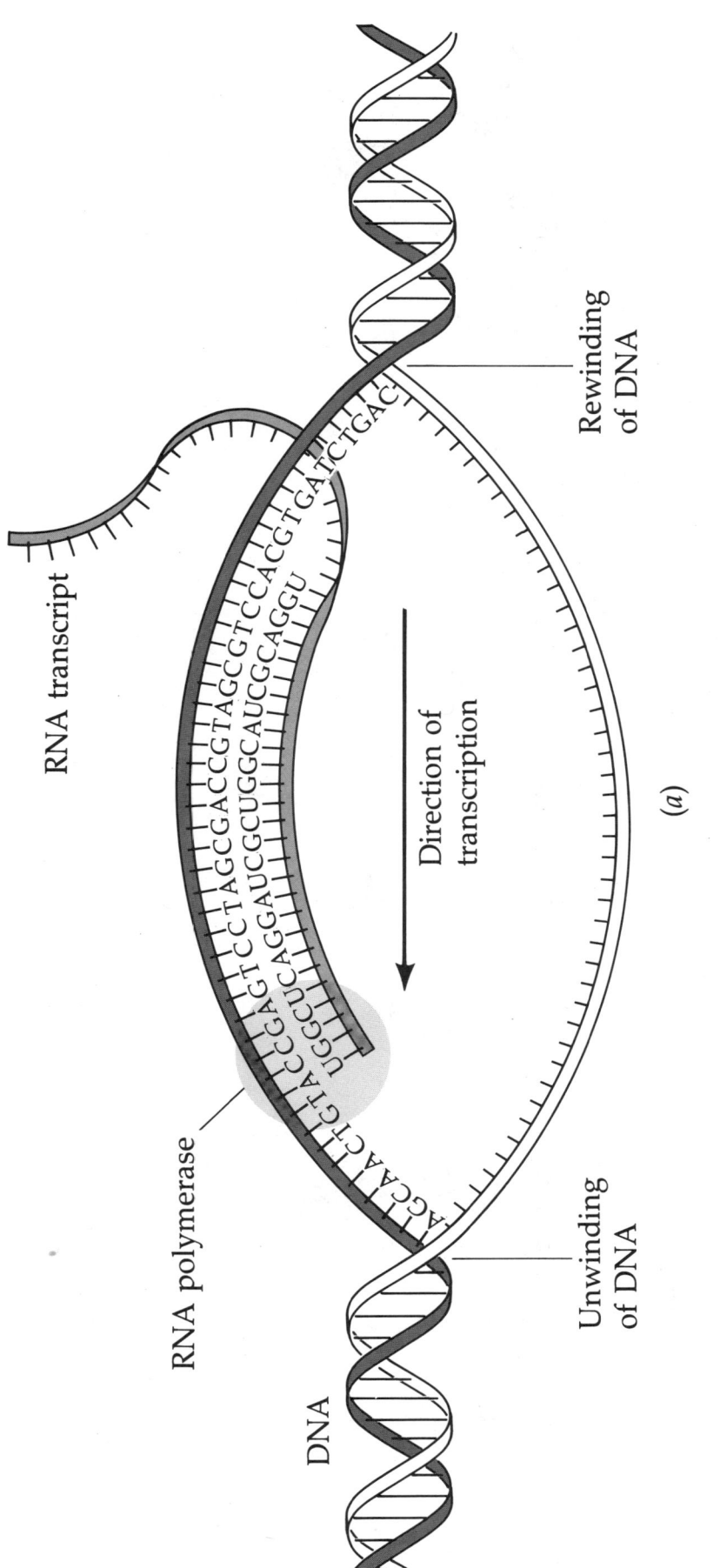

RNA transcript

RNA polymerase

DNA

Rewinding
of DNA

Direction of
transcription

Unwinding
of DNA

(a)

FIGURE 11.17a   DNA IS TRANSCRIBED INTO RNA   (text page 289)

From *Life: The Science of Biology*, Copyright © 1983 by William K. Purves and Gordon H. Orians.

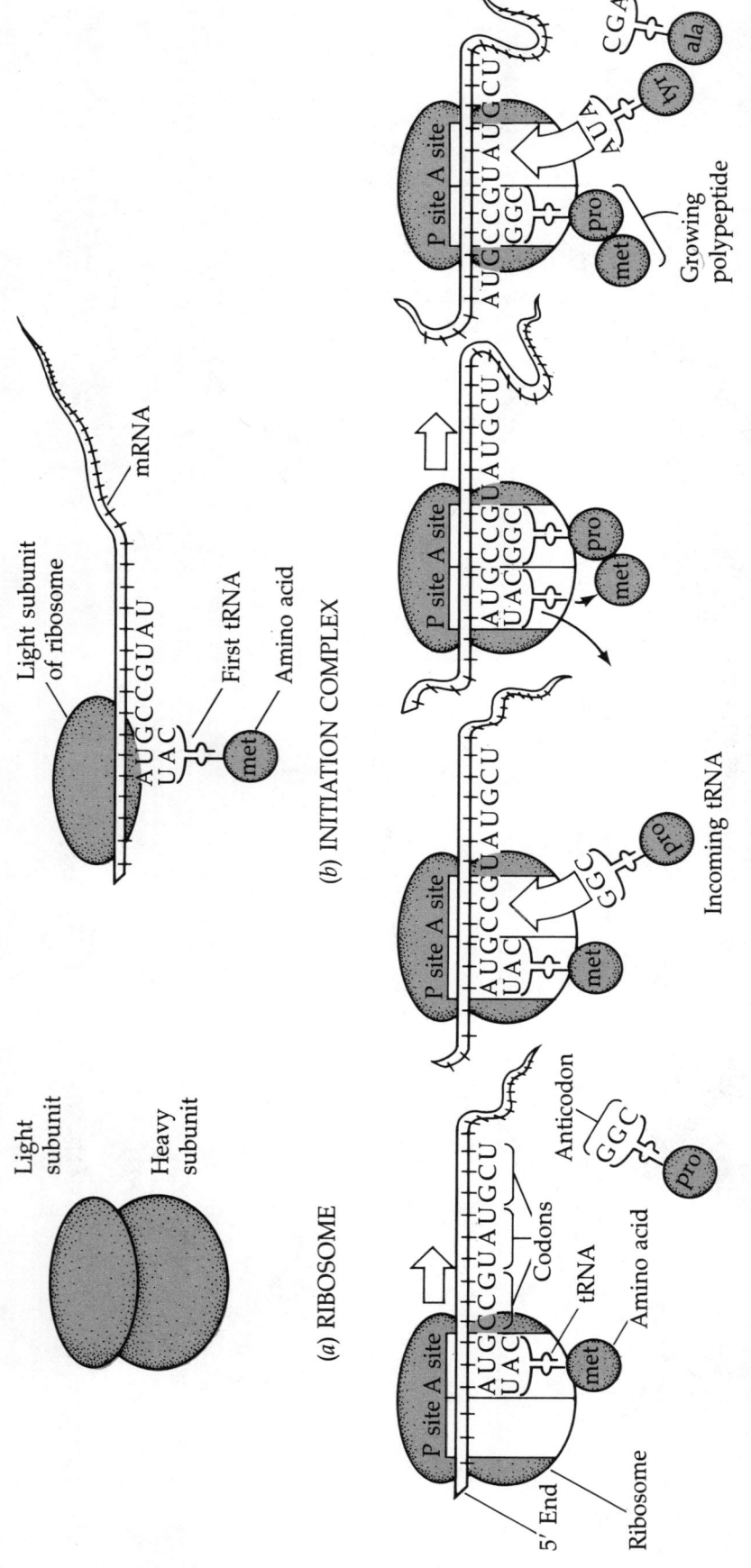

(a) RIBOSOME

Light subunit

Heavy subunit

(b) INITIATION COMPLEX

Light subunit of ribosome

mRNA

First tRNA

Amino acid

(c) TRANSLATION

5' End

Ribosome

Codons

tRNA

Amino acid

Anticodon

Incoming tRNA

Growing polypeptide

FIGURE 11.19   TRANSLATION OF GENETIC INFORMATION   (text page 291)

From *Life: The Science of Biology.* Copyright © 1983 by William K. Purves and Gordon H. Orians.

FIRST LETTER	SECOND LETTER				THIRD LETTER
	U	C	A	G	
U	Phenylalanine	Serine	Tyrosine	Cysteine	U
	Phenylalanine	Serine	Tyrosine	Cysteine	C
	Leucine	Serine	(End chain)	(End chain)	A
	Leucine	Serine	(End chain)	Tryptophan	G
C	Leucine	Proline	Histidine	Arginine	U
	Leucine	Proline	Histidine	Arginine	C
	Leucine	Proline	Glutamine	Arginine	A
	Leucine	Proline	Glutamine	Arginine	G
A	Isoleucine	Threonine	Asparagine	Serine	U
	Isoleucine	Threonine	Asparagine	Serine	C
	Isoleucine	Threonine	Lysine	Arginine	A
	Methionine	Threonine	Lysine	Arginine	G
G	Valine	Alanine	Aspartic acid	Glycine	U
	Valine	Alanine	Aspartic acid	Glycine	C
	Valine	Alanine	Glutamic acid	Glycine	A
	Valine	Alanine	Glutamic acid	Glycine	G

FIGURE 11.22   THE UNIVERSAL GENETIC CODE   (text page 296)

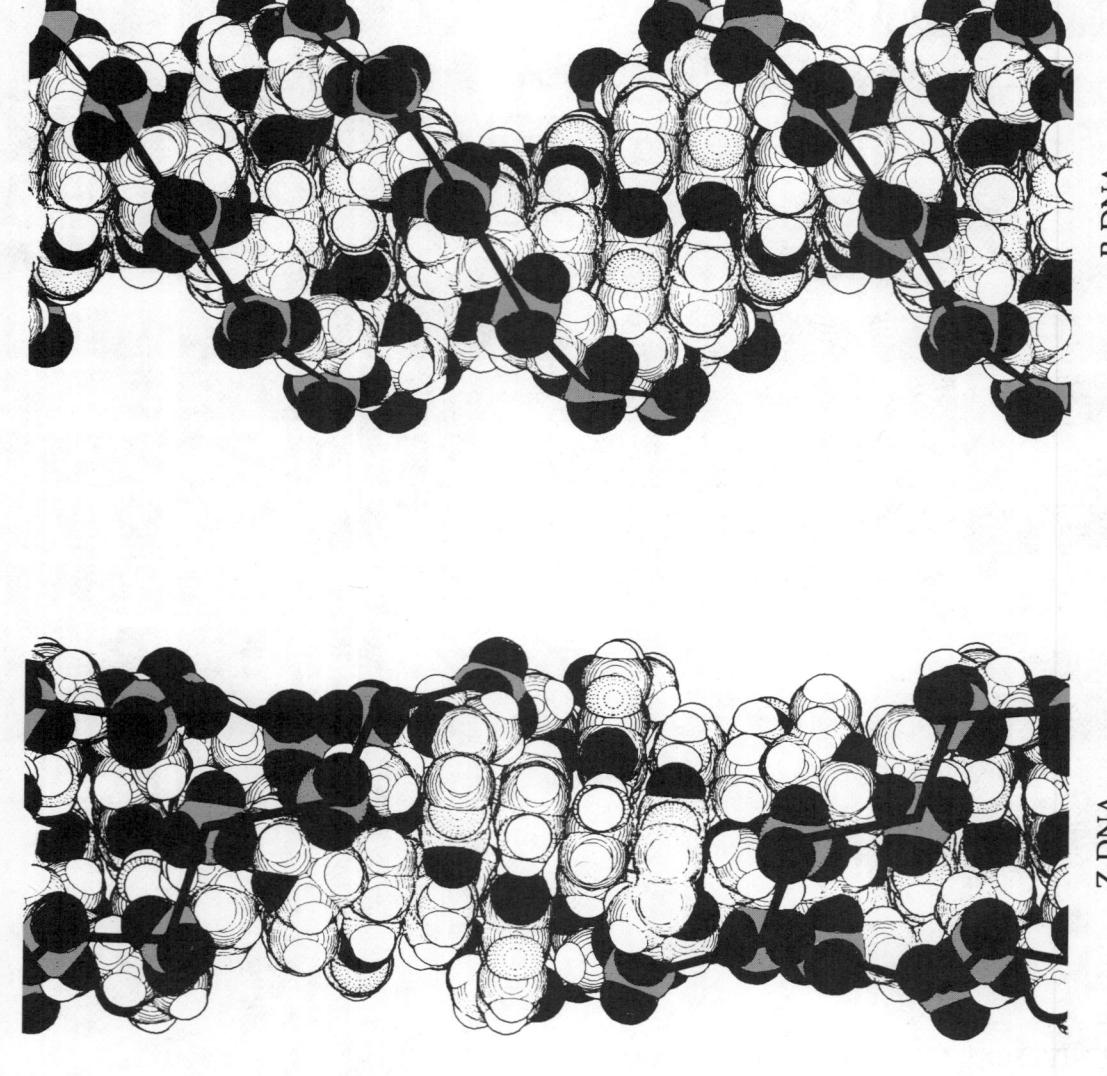

Z-DNA   B-DNA

FIGURE 11.27   THE Z-FORM OF DNA   (text page 301)

From Life: The Science of Biology, Copyright © 1983 by William K. Purves and Gordon H. Orians.

(c)

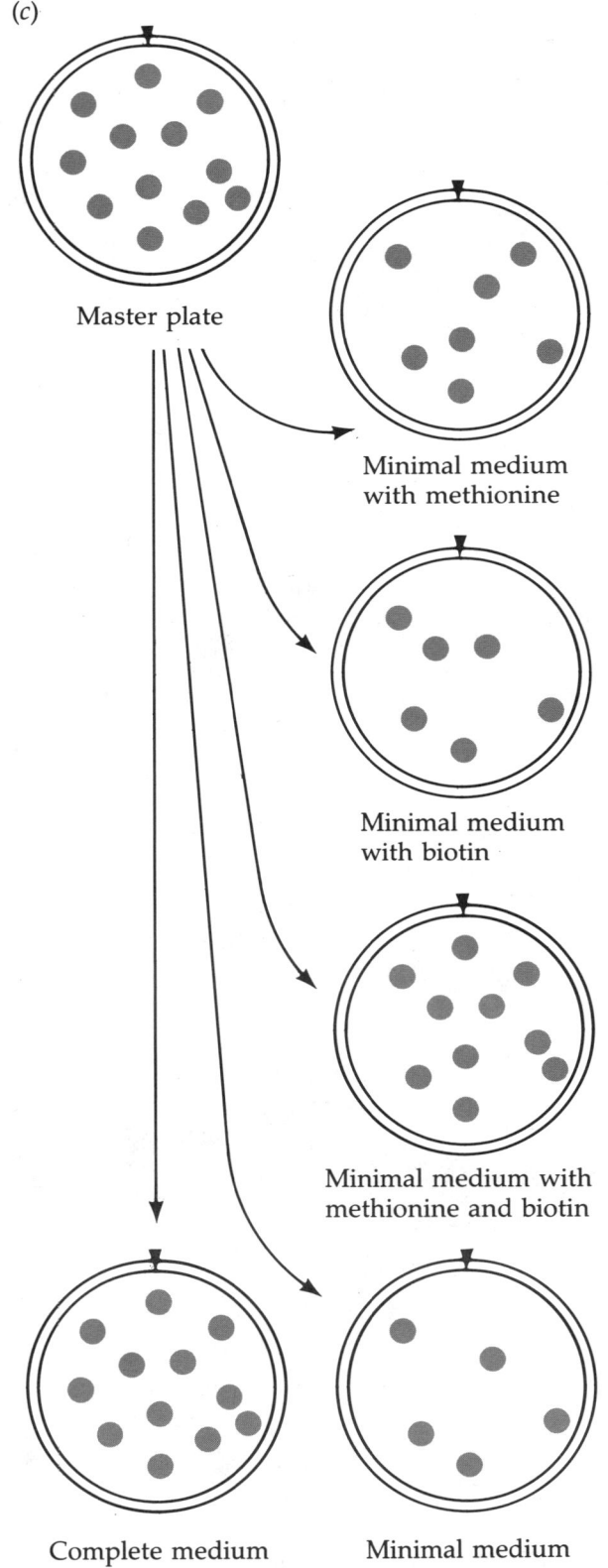

Master plate

Minimal medium
with methionine

Minimal medium
with biotin

Minimal medium with
methionine and biotin

Complete medium   Minimal medium

FIGURE 12.5c   ISOLATING AND IDENTIFYING AUXOTROPHIC MUTANTS   (text page 312)

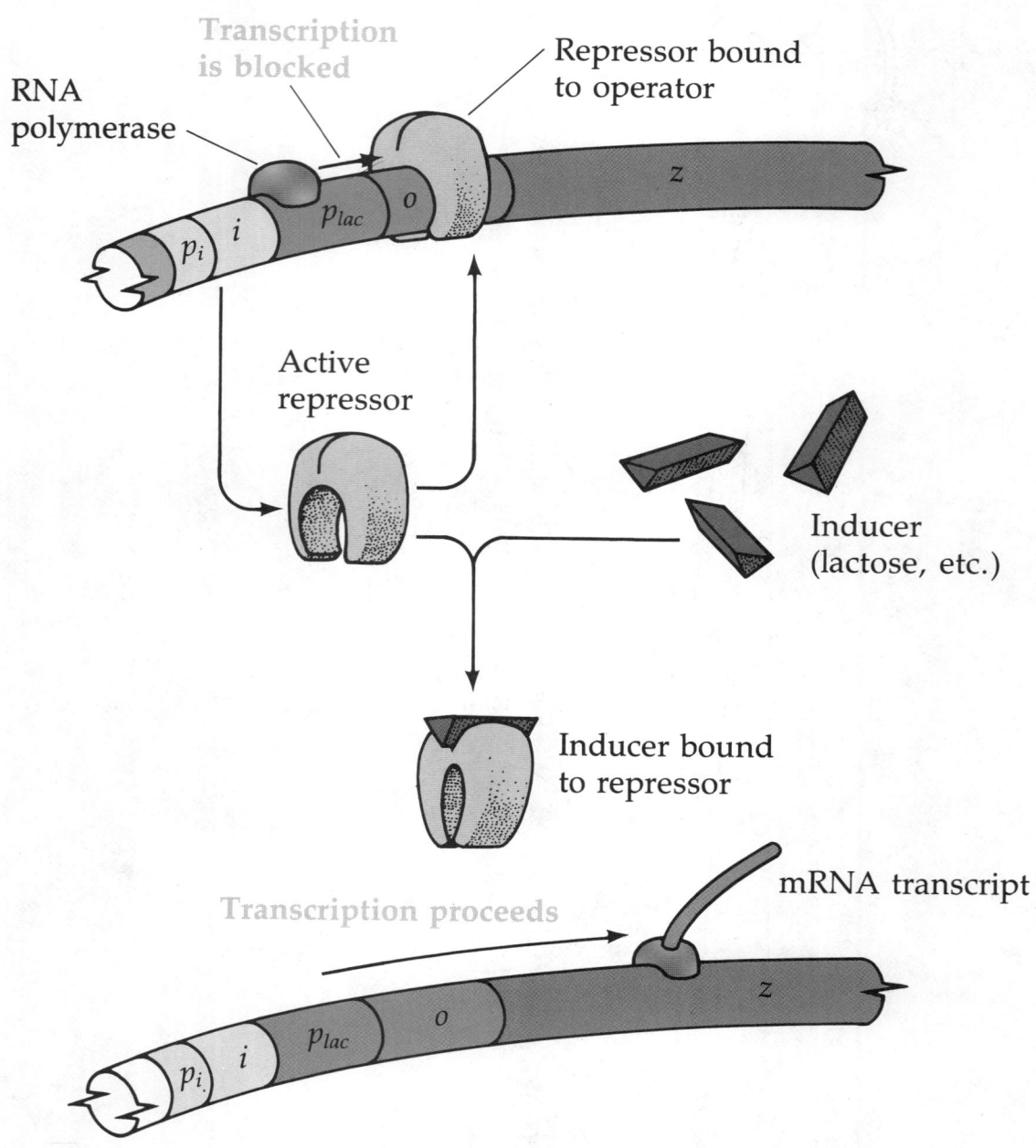

FIGURE 13.4   INDUCTION OF THE *LAC* OPERON   (text page 332)

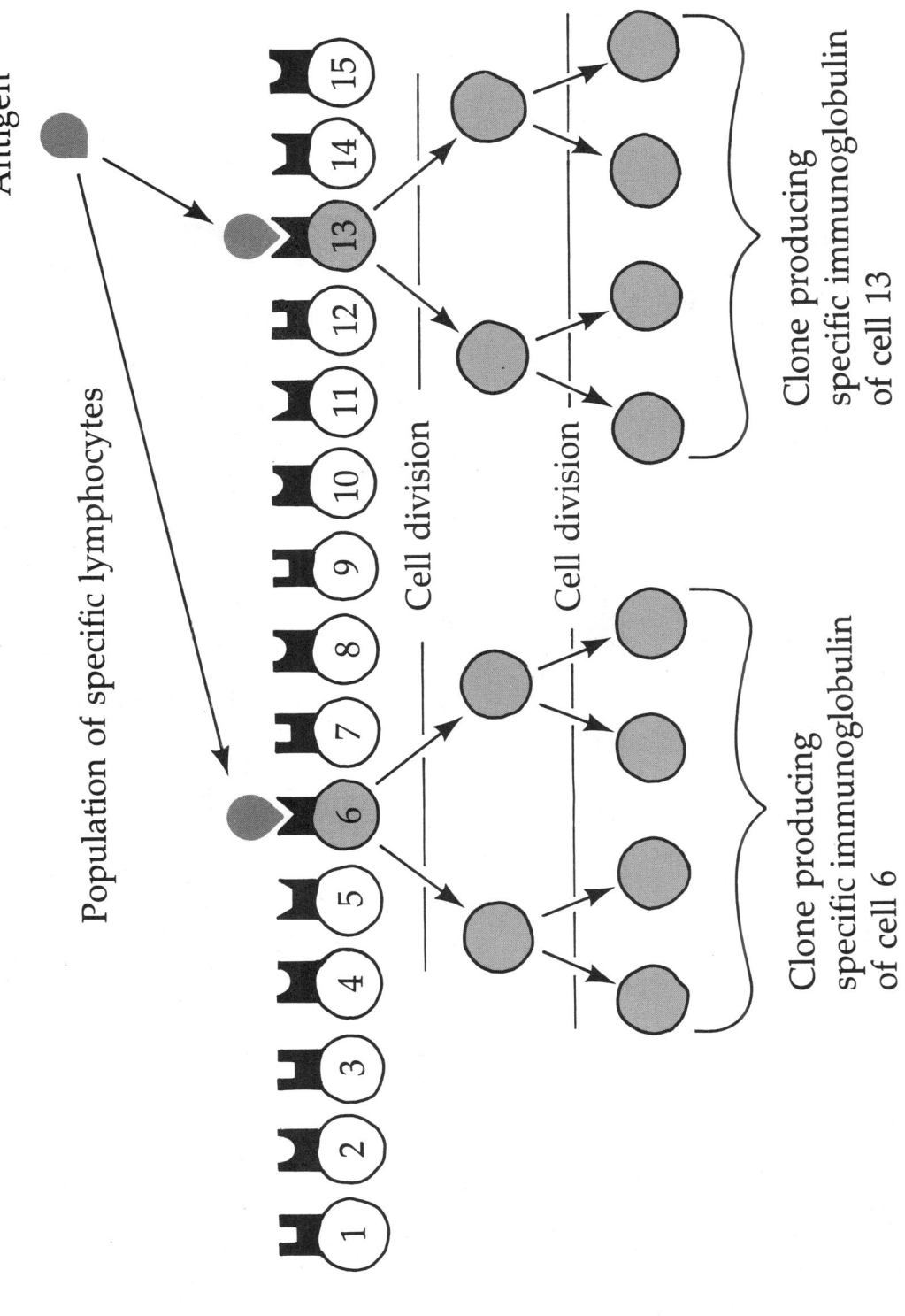

Antigen

Population of specific lymphocytes

Cell division

Cell division

Clone producing
specific immunoglobulin
of cell 6

Clone producing
specific immunoglobulin
of cell 13

FIGURE 14.6   CLONAL SELECTION   (text page 361)

From *Life: The Science of Biology*, Copyright © 1983 by William K. Purves and Gordon H. Orians.

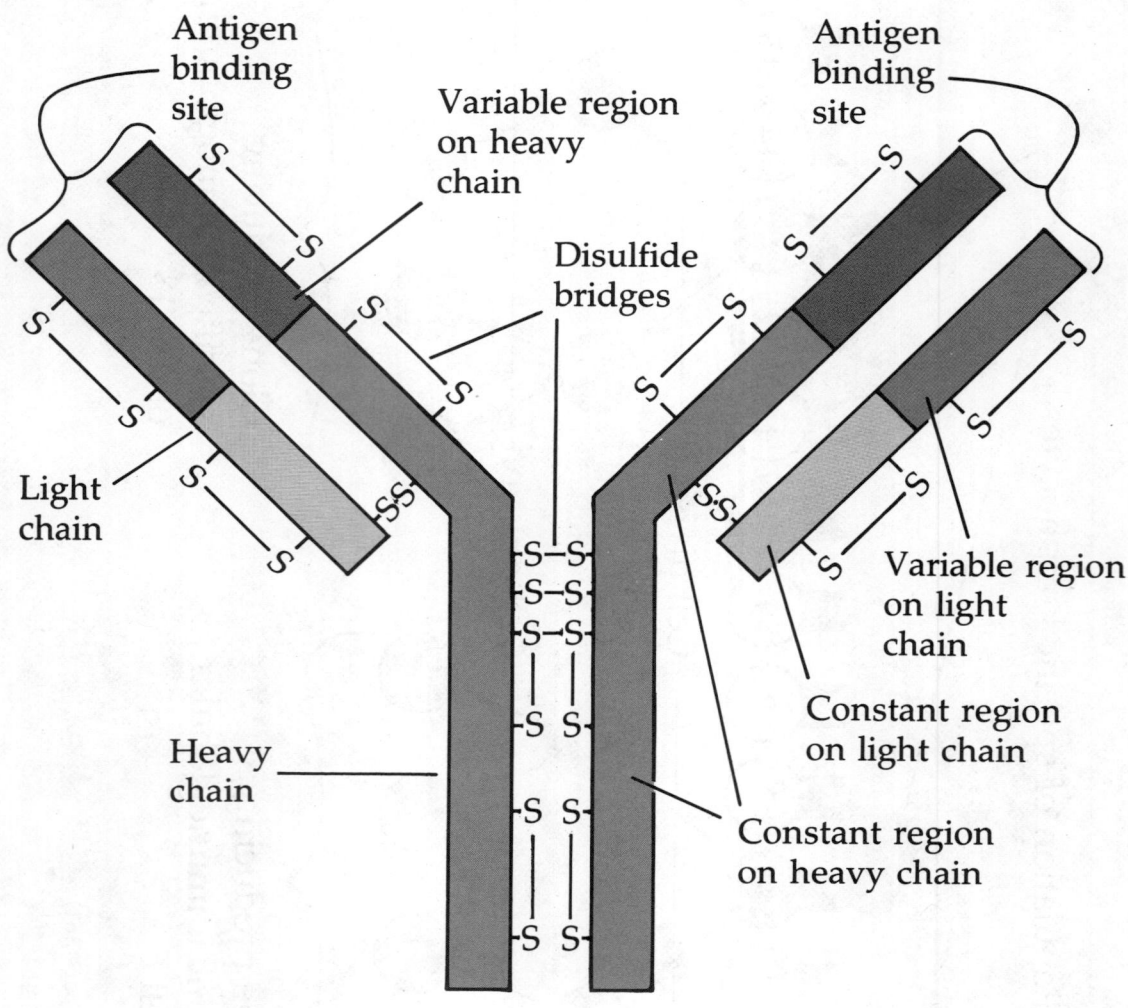

Antigen binding site

Antigen binding site

Variable region on heavy chain

Disulfide bridges

Light chain

Heavy chain

Variable region on light chain

Constant region on light chain

Constant region on heavy chain

FIGURE 14.9a   STRUCTURE OF IMMUNOGLOBULINS   (text page 364)

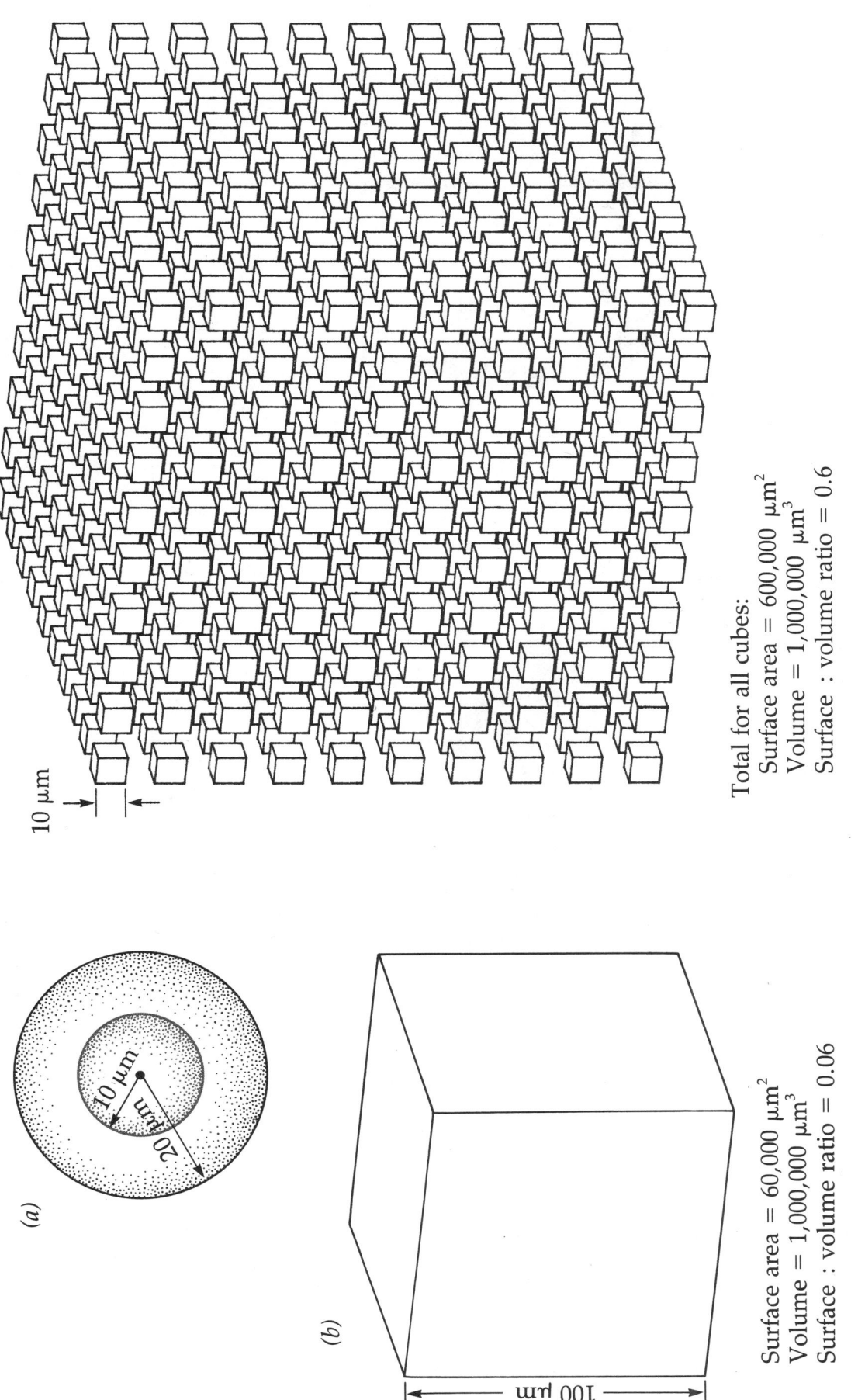

10 μm

Total for all cubes:
Surface area = 600,000 μm²
Volume = 1,000,000 μm³
Surface : volume ratio = 0.6

(a)

10 μm
20 μm

(b)

100 μm

Surface area = 60,000 μm²
Volume = 1,000,000 μm³
Surface : volume ratio = 0.06

FIGURE 15.3   SURFACE-TO-VOLUME RATIOS   (text page 384)

From *Life: The Science of Biology*, Copyright © 1983 by William K. Purves and Gordon H. Orians.

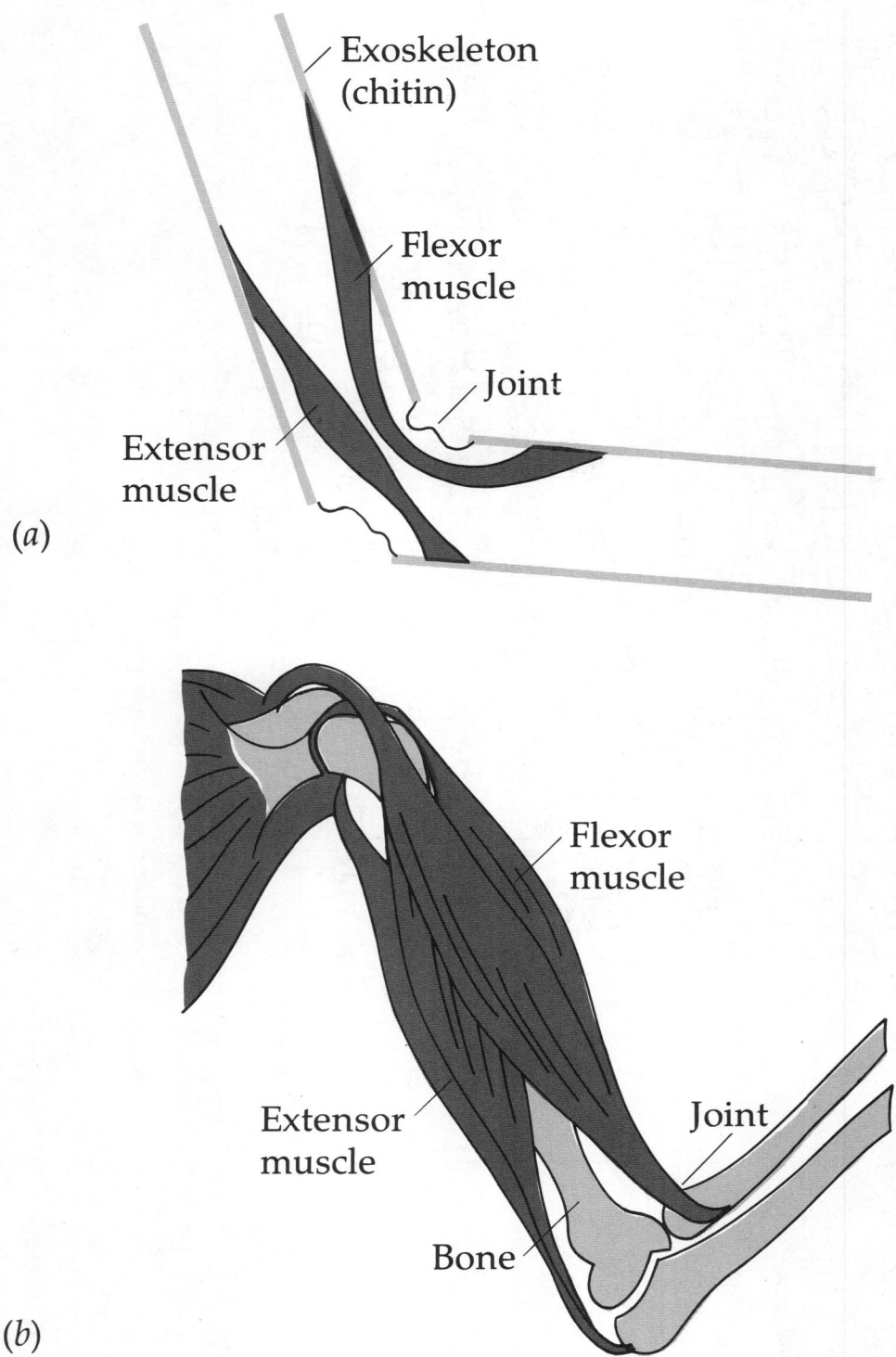

Exoskeleton
(chitin)

Flexor
muscle

Joint

Extensor
muscle

(a)

Flexor
muscle

Extensor
muscle

Joint

Bone

(b)

FIGURE 15.19   ALTERNATIVE WAYS TO FLEX APPENDAGES   (text page 396)

From *Life: The Science of Biology*, Copyright © 1983 by William K. Purves and Gordon H. Orians.

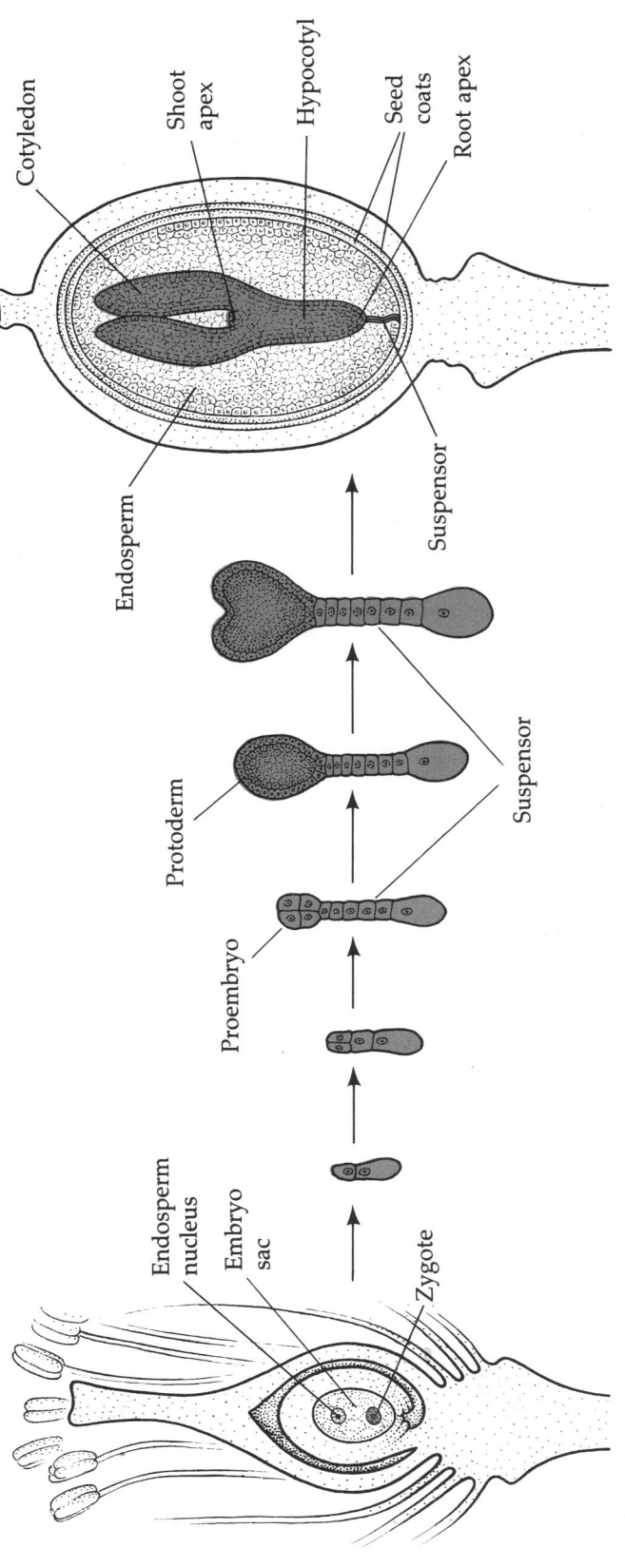

FIGURE 16.1   EARLY DEVELOPMENT OF ANGIOSPERMS   (text page 401)

From *Life: The Science of Biology*, Copyright © 1983 by William K. Purves and Gordon H. Orians.

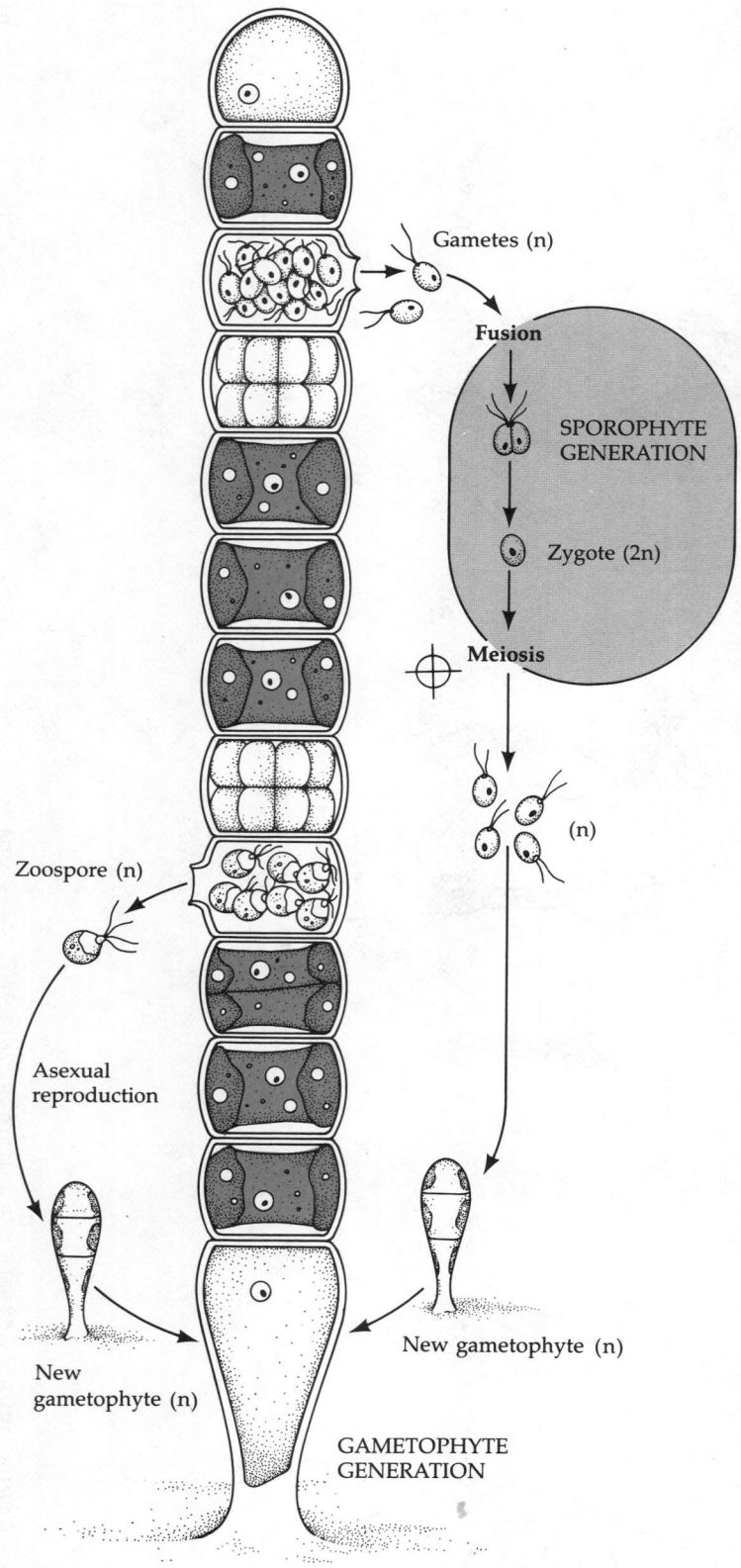

Gametes (n)

Fusion

SPOROPHYTE
GENERATION

Zygote (2n)

Meiosis

(n)

Zoospore (n)

Asexual
reproduction

New gametophyte (n)

New
gametophyte (n)

GAMETOPHYTE
GENERATION

FIGURE 17.6   A HAPLONTIC LIFE CYCLE   (text page 428)

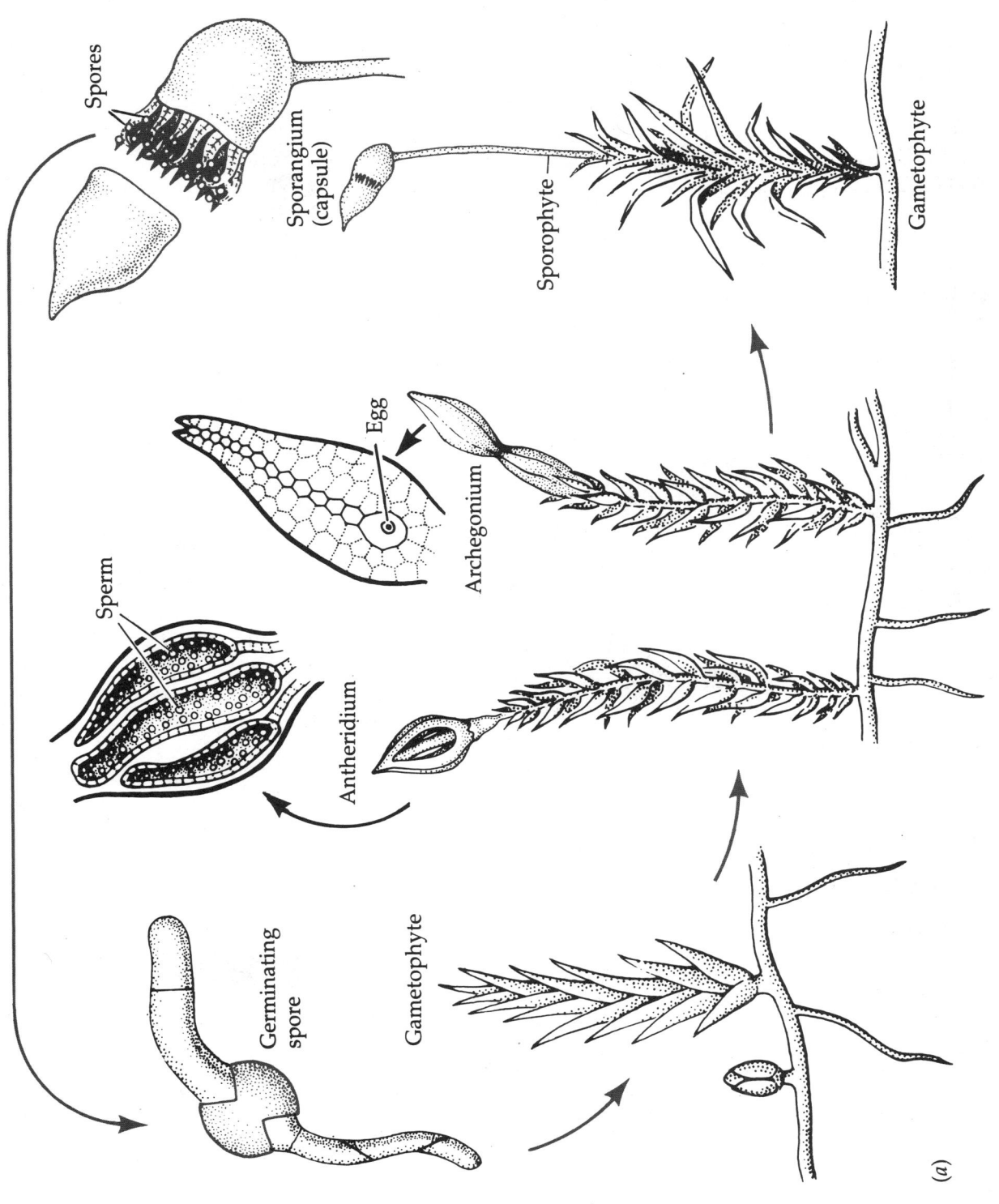

Spores

Sporangium (capsule)

Sporophyte

Gametophyte

Egg

Archegonium

Sperm

Antheridium

Germinating spore

Gametophyte

FIGURE 17.9a LIFE CYCLE OF A MOSS   (text page 432)

From Life: The Science of Biology, Copyright © 1983 by William K. Purves and Gordon H. Orians.

(a)

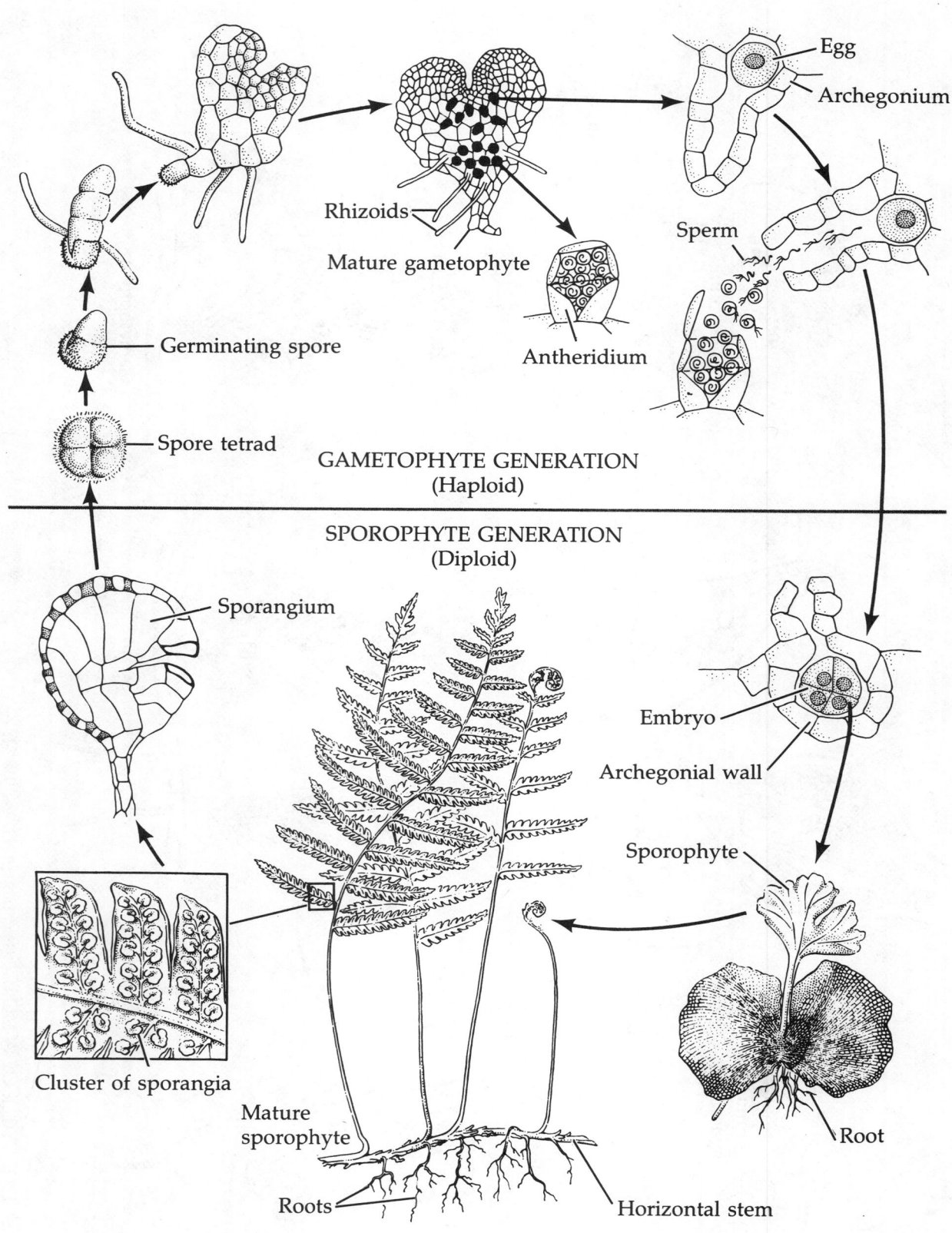

Egg

Archegonium

Rhizoids

Mature gametophyte

Sperm

Germinating spore

Antheridium

Spore tetrad

GAMETOPHYTE GENERATION
(Haploid)

SPOROPHYTE GENERATION
(Diploid)

Sporangium

Embryo

Archegonial wall

Sporophyte

Cluster of sporangia

Mature
sporophyte

Root

Roots

Horizontal stem

FIGURE 17.11a   LIFE CYCLE OF A FERN   (text page 433)

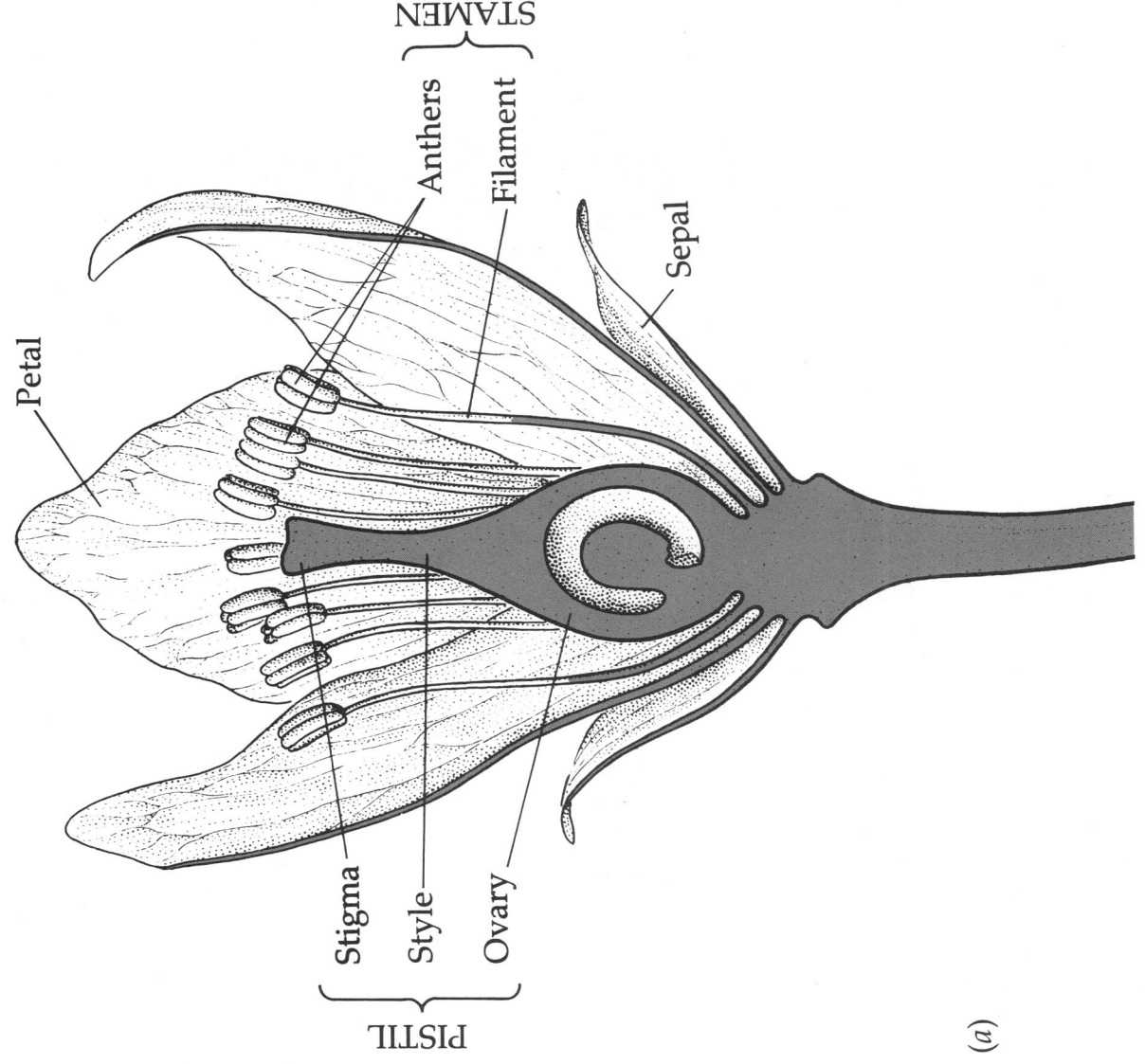

STAMEN

Anthers

Filament

Sepal

Petal

Stigma

Style

Ovary

PISTIL

(a)

FIGURE 17.13a STRUCTURES OF A FLOWER (text page 435)

From *Life: The Science of Biology,* Copyright © 1983 by William K. Purves and Gordon H. Orians.

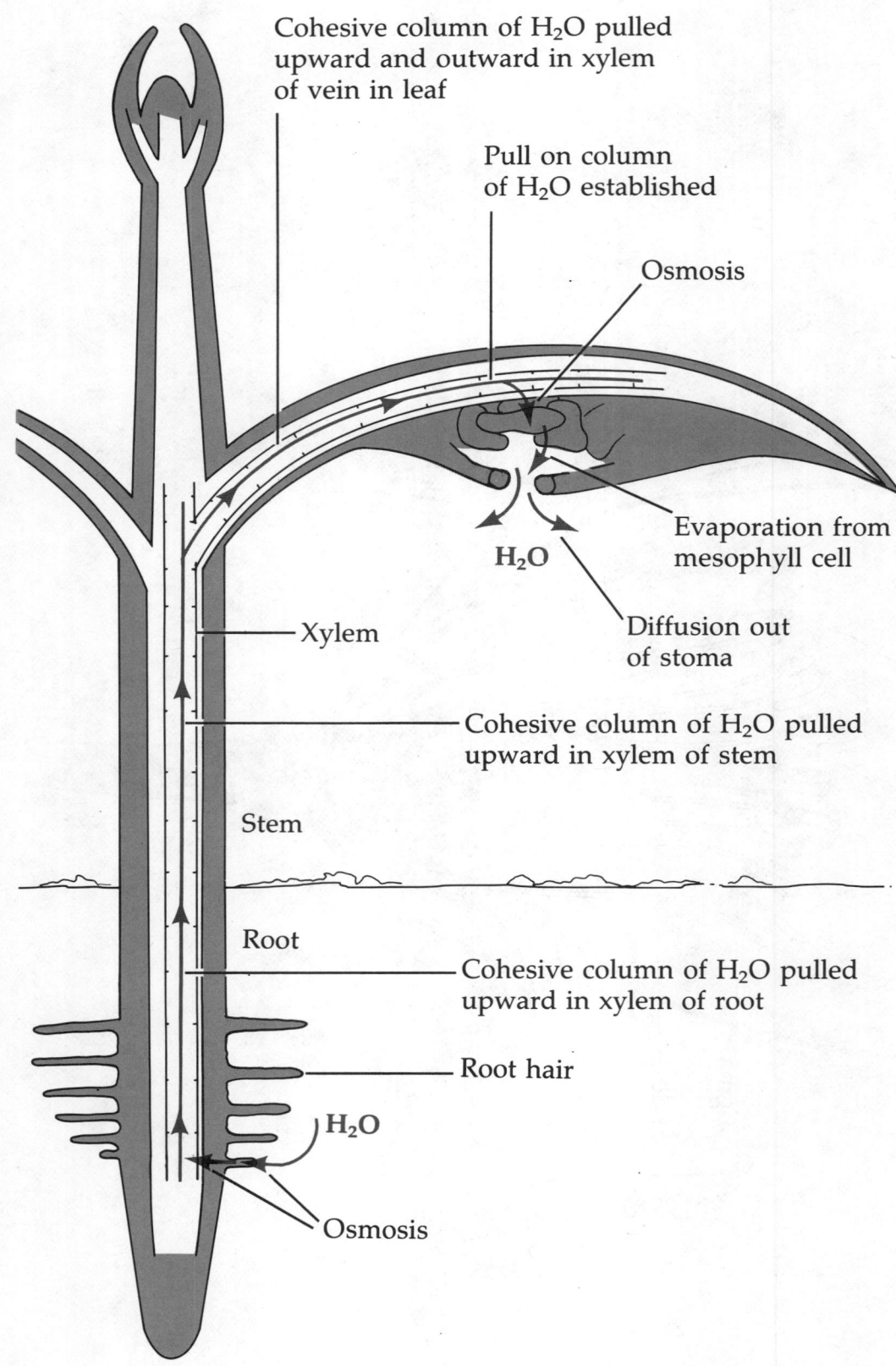

Cohesive column of H₂O pulled
upward and outward in xylem
of vein in leaf

Pull on column
of H₂O established

Osmosis

Evaporation from
mesophyll cell

H_2O

Diffusion out
of stoma

Xylem

Cohesive column of H₂O pulled
upward in xylem of stem

Stem

Root

Cohesive column of H₂O pulled
upward in xylem of root

Root hair

H_2O

Osmosis

FIGURE 19.21   WATER TRANSPORT IN PLANTS   (text page 482)

(a)

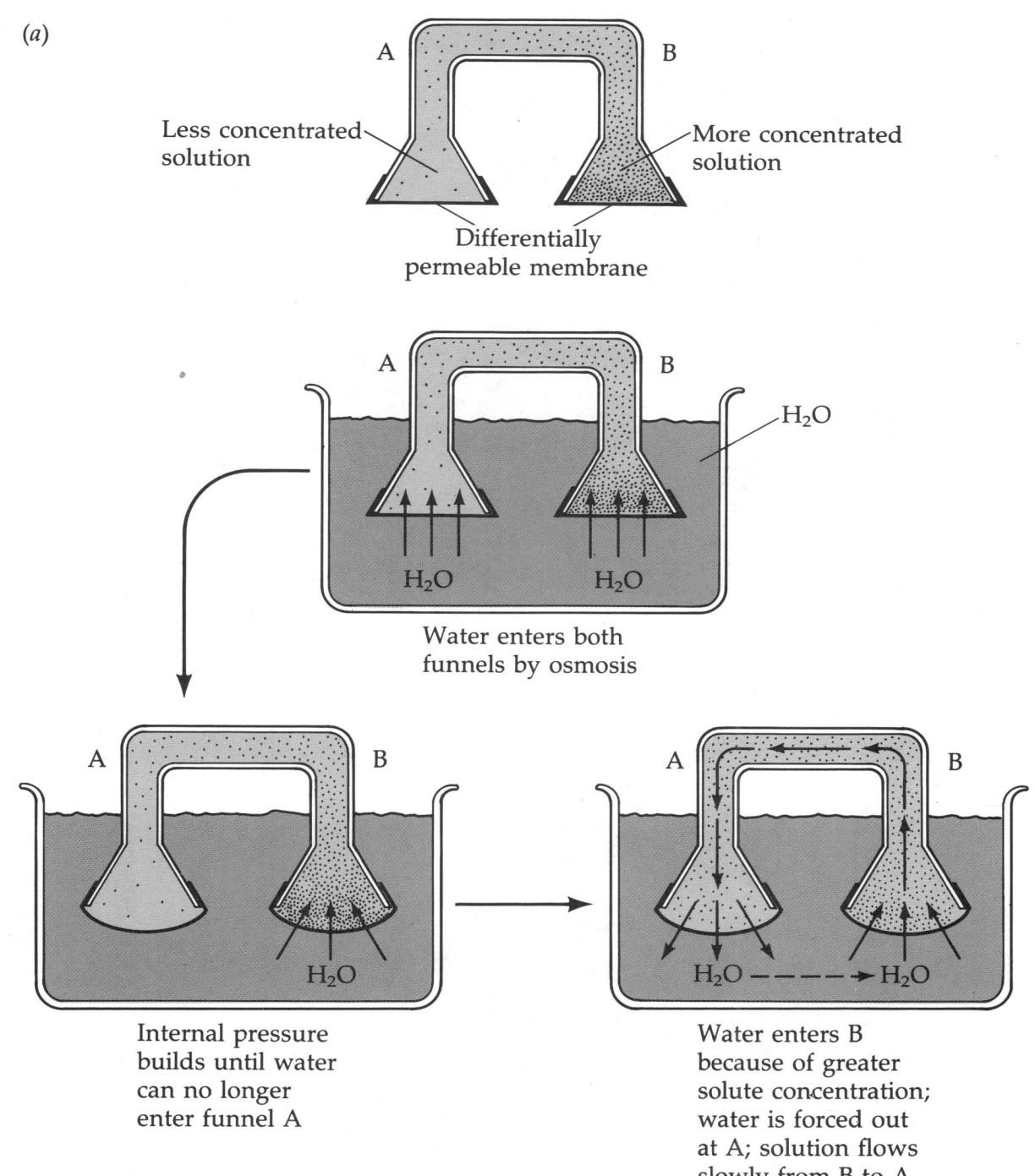

Less concentrated solution

More concentrated solution

Differentially permeable membrane

H_2O

H_2O H_2O

Water enters both funnels by osmosis

H_2O

Internal pressure builds until water can no longer enter funnel A

H_2O — — — → H_2O

Water enters B because of greater solute concentration; water is forced out at A; solution flows slowly from B to A

FIGURE 19.22a   TRANSPORT IN PHLOEM   (text page 483)

From *Life: The Science of Biology*, Copyright © 1983 by William K. Purves and Gordon H. Orians.

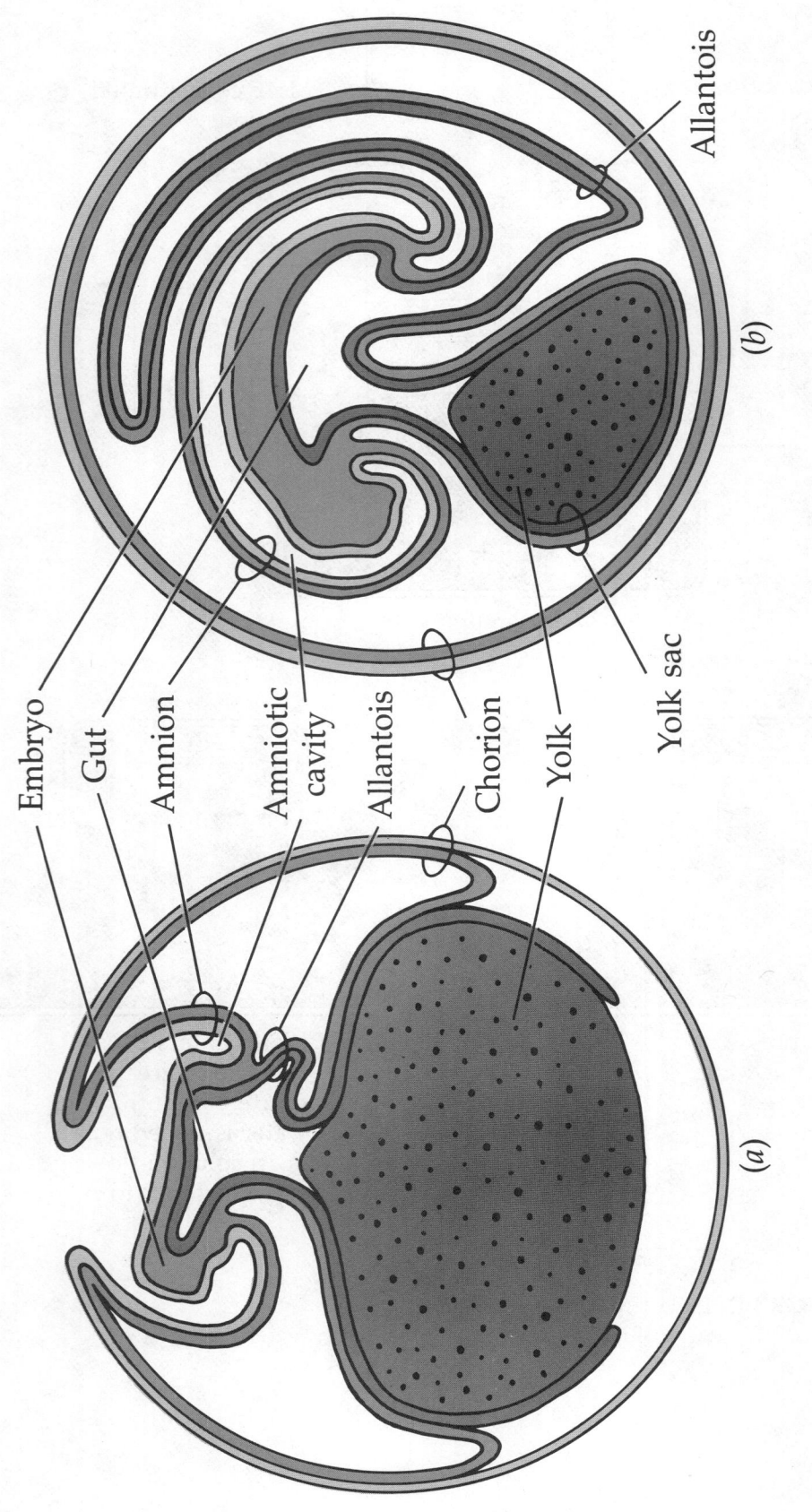

FIGURE 20.19   EMBRYONIC MEMBRANES OF THE AMNIOTIC EGG   (text page 504)

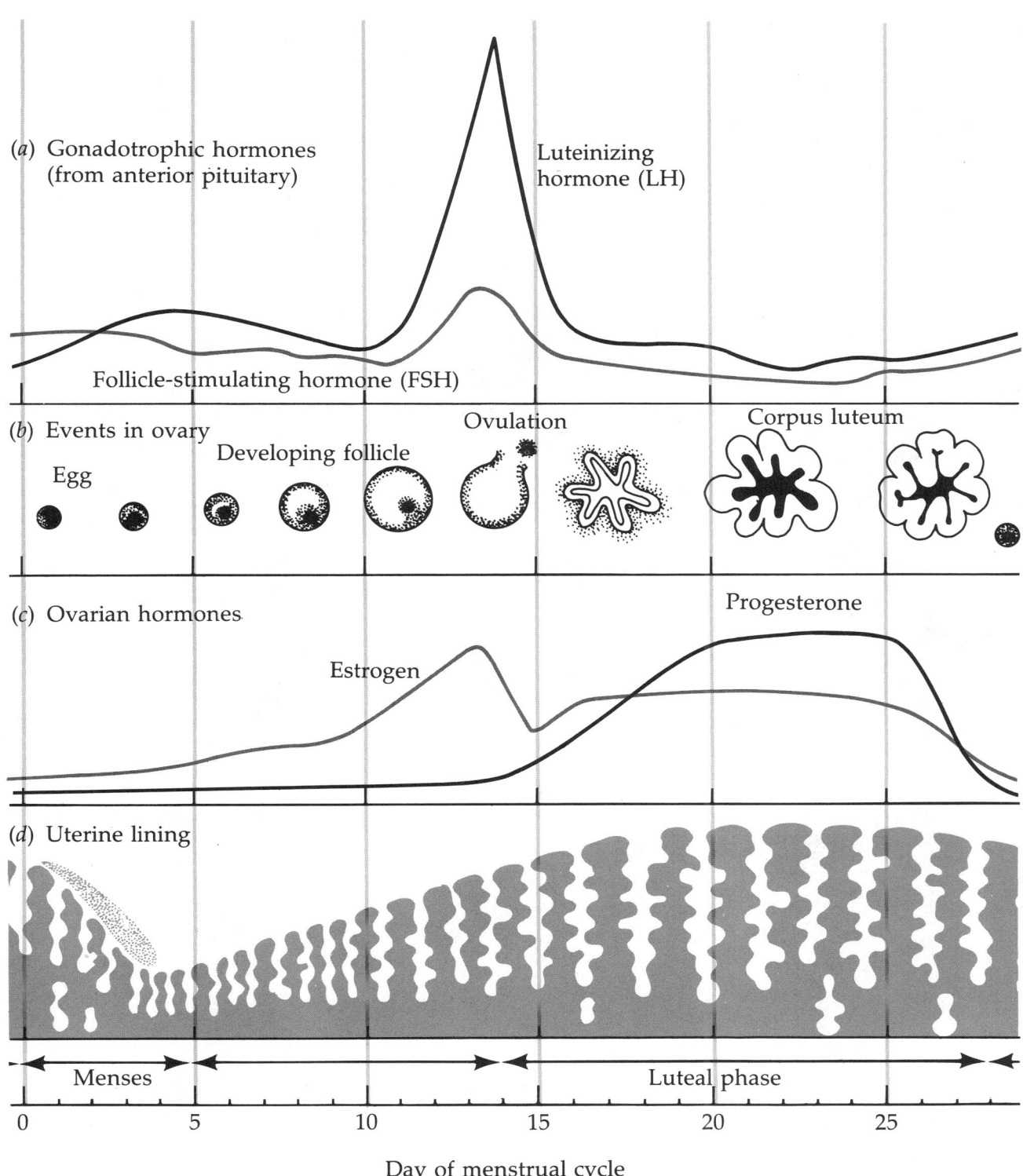

(a) Gonadotrophic hormones (from anterior pituitary)

Luteinizing hormone (LH)

Follicle-stimulating hormone (FSH)

(b) Events in ovary

Egg

Developing follicle

Ovulation

Corpus luteum

(c) Ovarian hormones

Estrogen

Progesterone

(d) Uterine lining

Menses

Luteal phase

0   5   10   15   20   25

Day of menstrual cycle

FIGURE 20.25 THE MENSTRUAL CYCLE (text page 509)

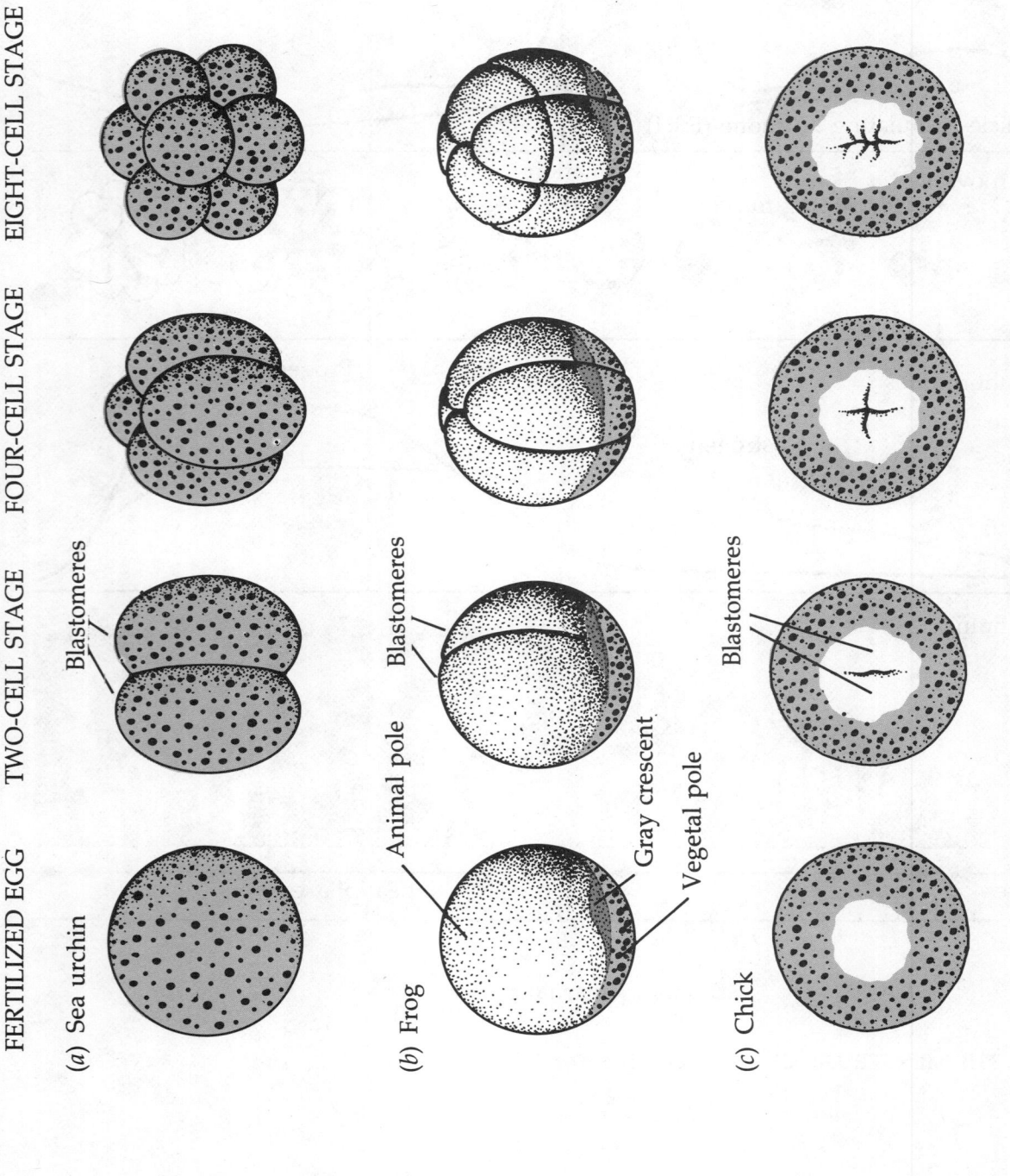

FERTILIZED EGG   TWO-CELL STAGE   FOUR-CELL STAGE   EIGHT-CELL STAGE

Blastomeres

(a) Sea urchin

Animal pole

Blastomeres

Gray crescent

Vegetal pole

(b) Frog

Blastomeres

(c) Chick

FIGURE 21.5a, b, & c PATTERNS OF CLEAVAGE (text page 522)

From *Life: The Science of Biology.* Copyright © 1983 by William K. Purves and Gordon H. Orians.

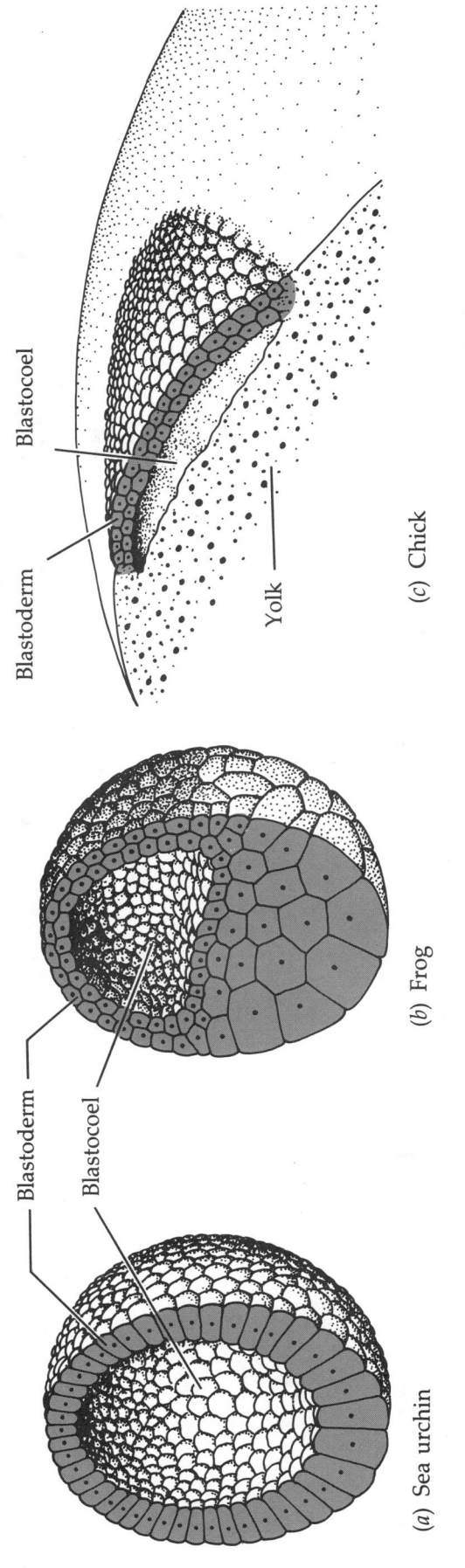

Blastocoel

Blastoderm

Yolk

(c) Chick

Blastoderm

Blastocoel

(b) Frog

Blastoderm

Blastocoel

(a) Sea urchin

FIGURE 21.7   BLASTULAS REFLECT PATTERNS OF CLEAVAGE   (text page 524)

From *Life: The Science of Biology.* Copyright © 1983 by William K. Purves and Gordon H. Orians.

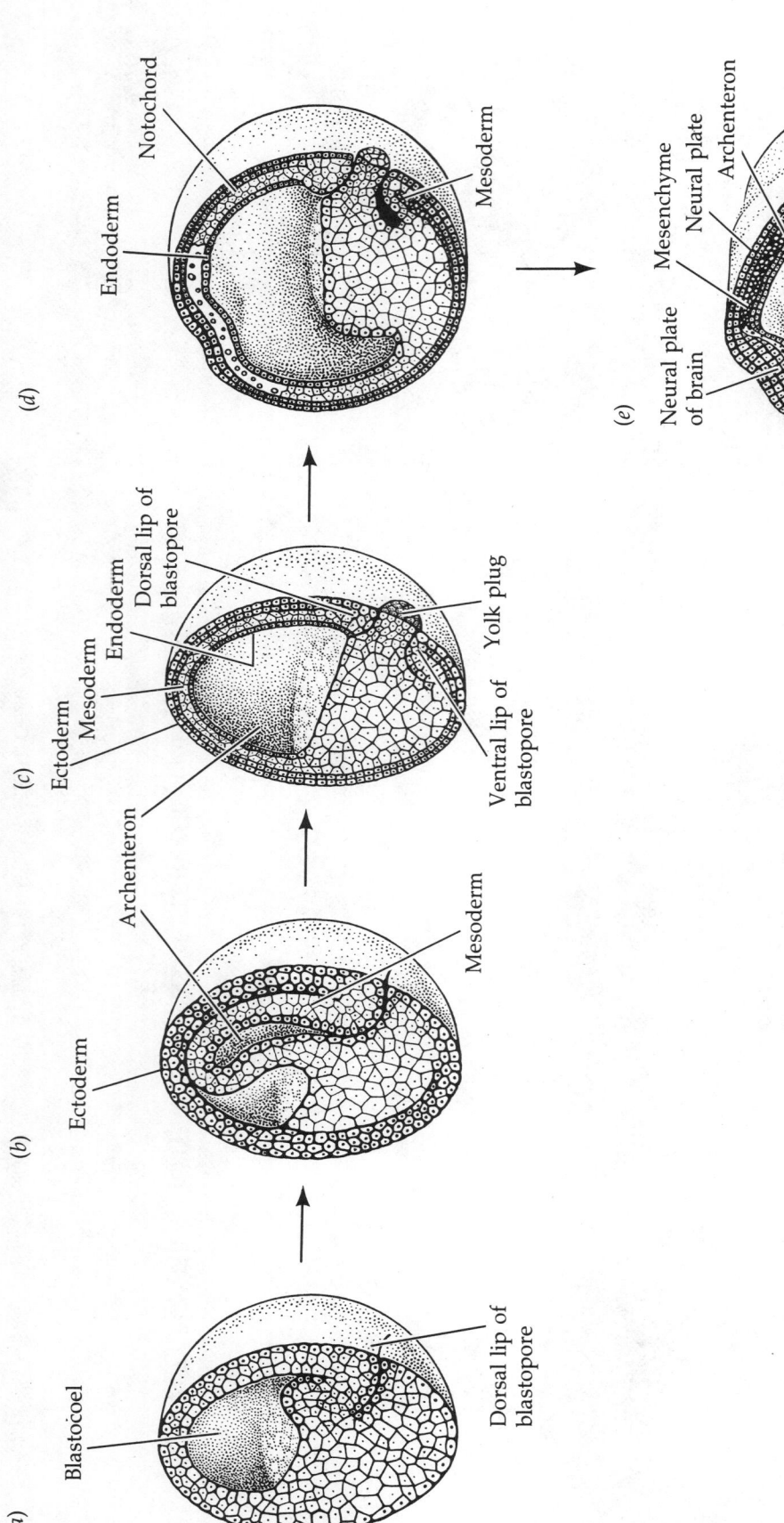

FIGURE 21.10 GASTRULATION IN THE FROG (text page 527)

From *Life: The Science of Biology*, Copyright © 1983 by William K. Purves and Gordon H. Orians.

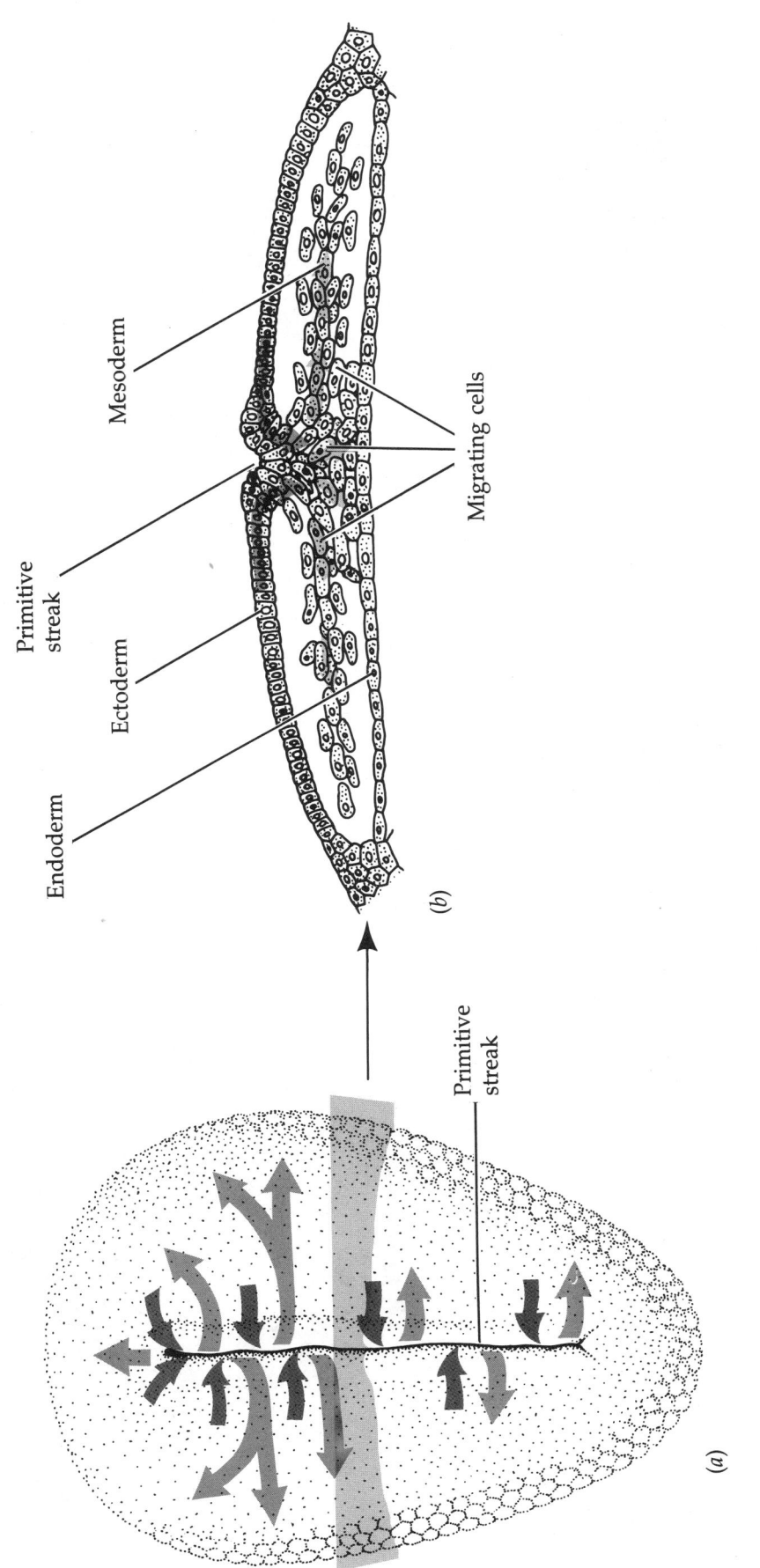

Endoderm

Primitive
streak

Ectoderm

Mesoderm

Migrating cells

(b)

Primitive
streak

(a)

FIGURE 21.11 GASTRULATION IN FISH AND BIRDS (text page 528)

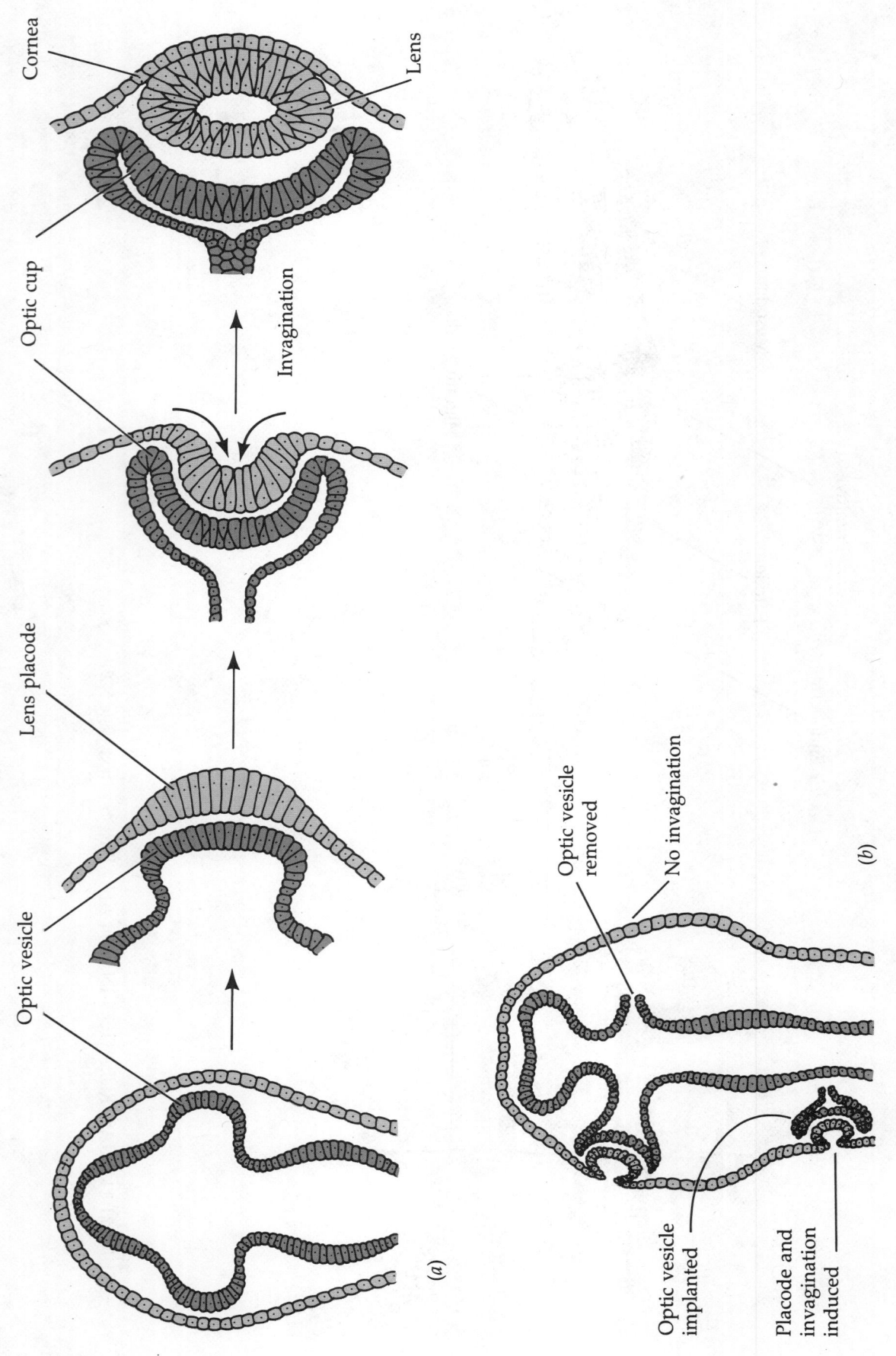

Cornea

Lens

Optic cup

Invagination

Lens placode

Optic vesicle

(a)

Optic vesicle removed

No invagination

Optic vesicle implanted

Placode and invagination induced

(b)

FIGURE 21.16 INDUCERS IN THE VERTEBRATE EYE (text page 532)

From *Life: The Science of Biology.* Copyright © 1983 by William K. Purves and Gordon H. Orians.

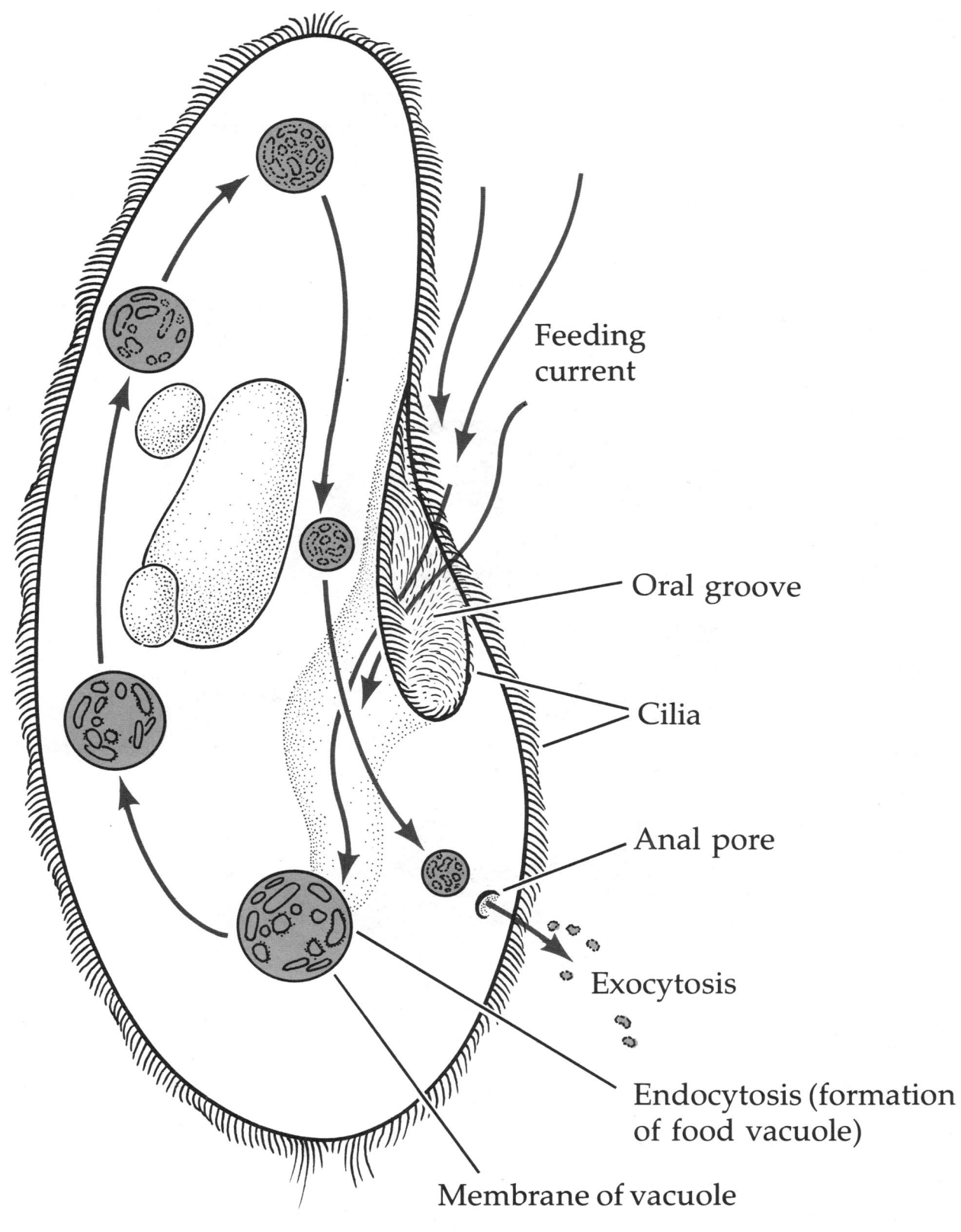

Feeding
current

Oral groove

Cilia

Anal pore

Exocytosis

Endocytosis (formation
of food vacuole)

Membrane of vacuole

FIGURE 22.8   FOOD PROCESSING IN A PROTIST   (text page 553)

From *Life: The Science of Biology*, Copyright © 1983 by William K. Purves and Gordon H. Orians.

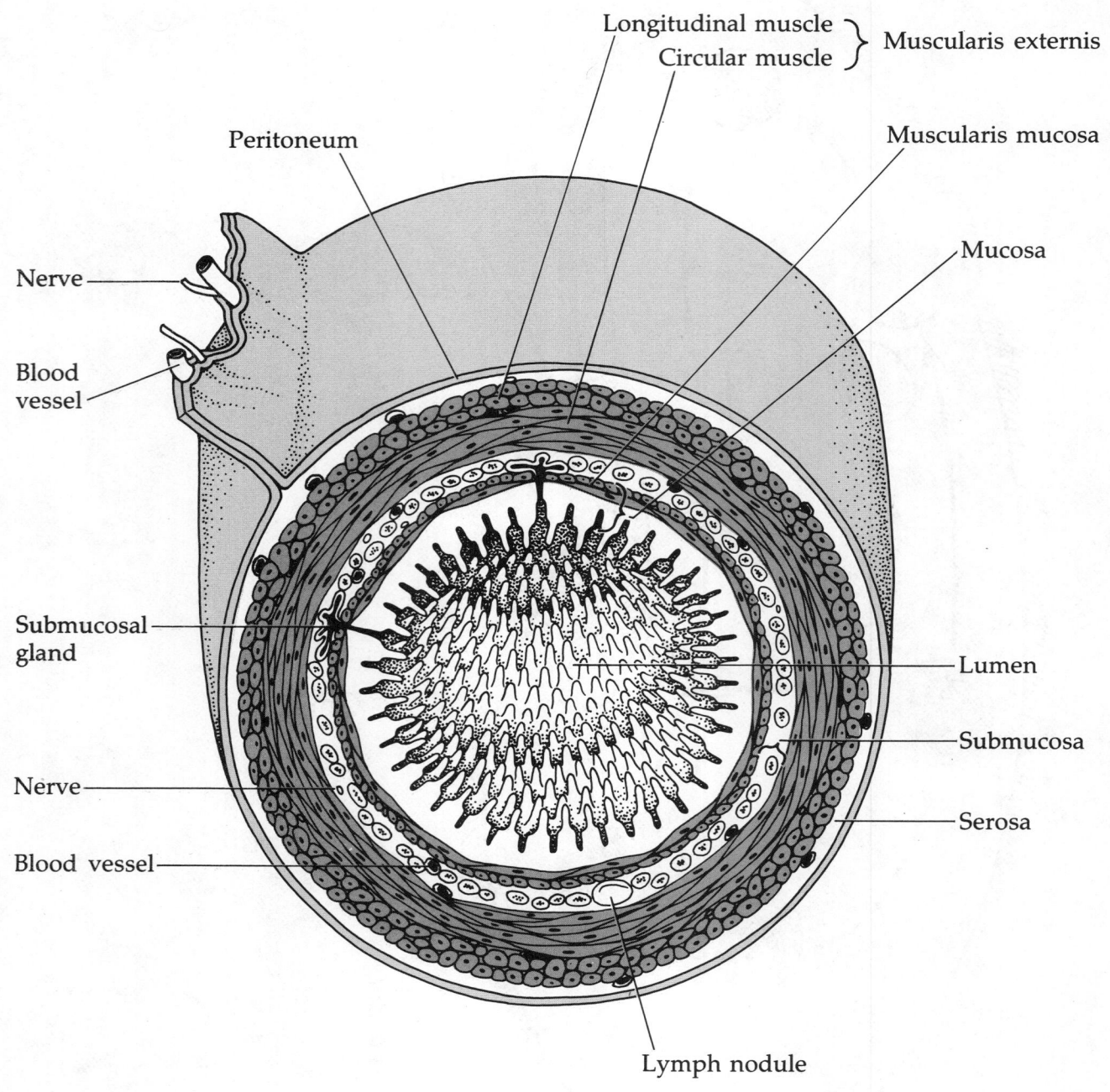

Longitudinal muscle
Circular muscle } Muscularis externis

Muscularis mucosa

Peritoneum

Mucosa

Nerve

Blood vessel

Submucosal gland

Lumen

Nerve

Submucosa

Blood vessel

Serosa

Lymph nodule

FIGURE 22.14   LAYERS OF THE HUMAN GUT   (text page 558)

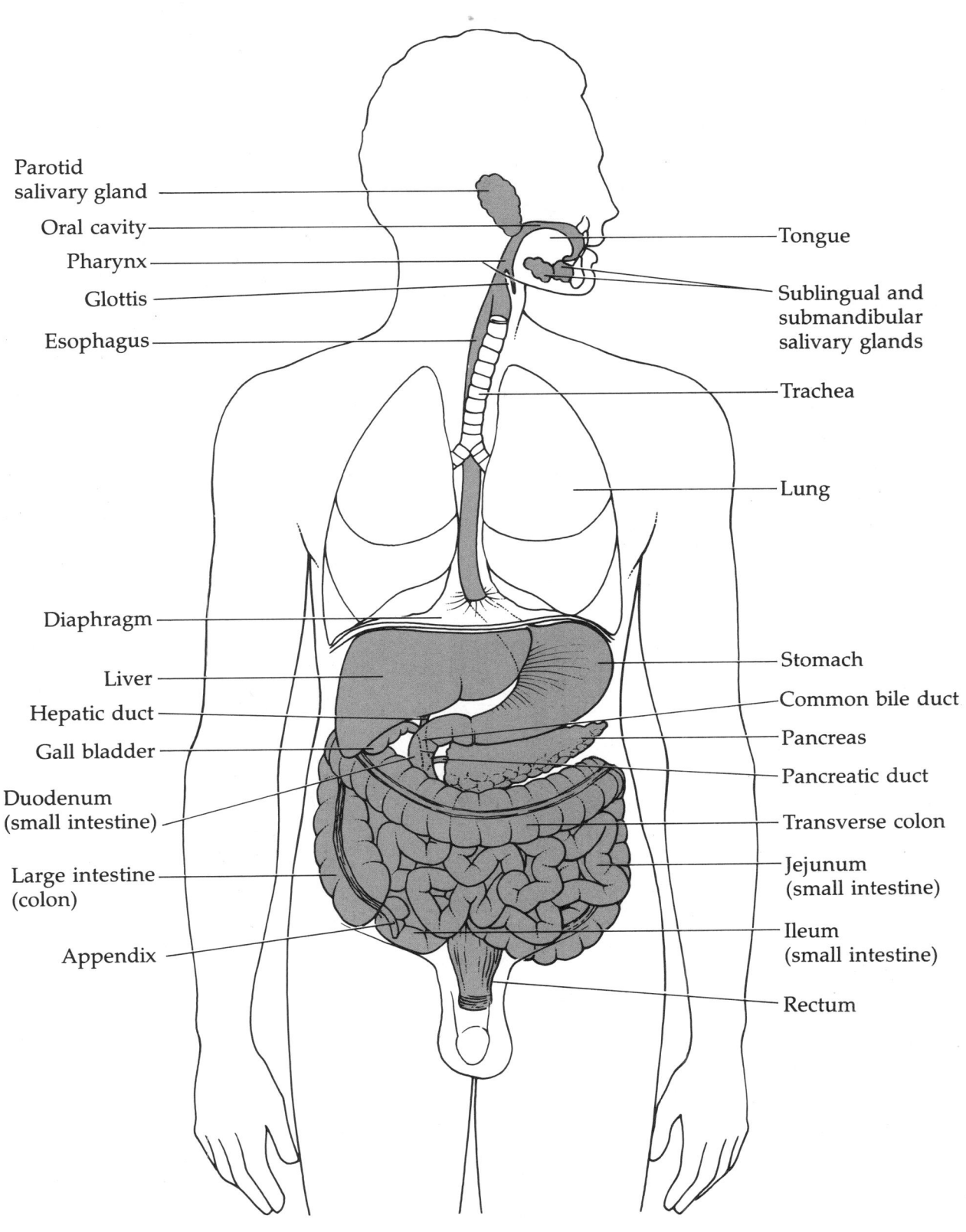

Parotid
salivary gland

Oral cavity

Pharynx

Glottis

Esophagus

Tongue

Sublingual and
submandibular
salivary glands

Trachea

Lung

Diaphragm

Liver

Hepatic duct

Gall bladder

Duodenum
(small intestine)

Large intestine
(colon)

Appendix

Stomach

Common bile duct

Pancreas

Pancreatic duct

Transverse colon

Jejunum
(small intestine)

Ileum
(small intestine)

Rectum

FIGURE 22.19   THE HUMAN DIGESTIVE SYSTEM   (text page 563)

From *Life: The Science of Biology*, Copyright © 1983 by William K. Purves and Gordon H. Orians.

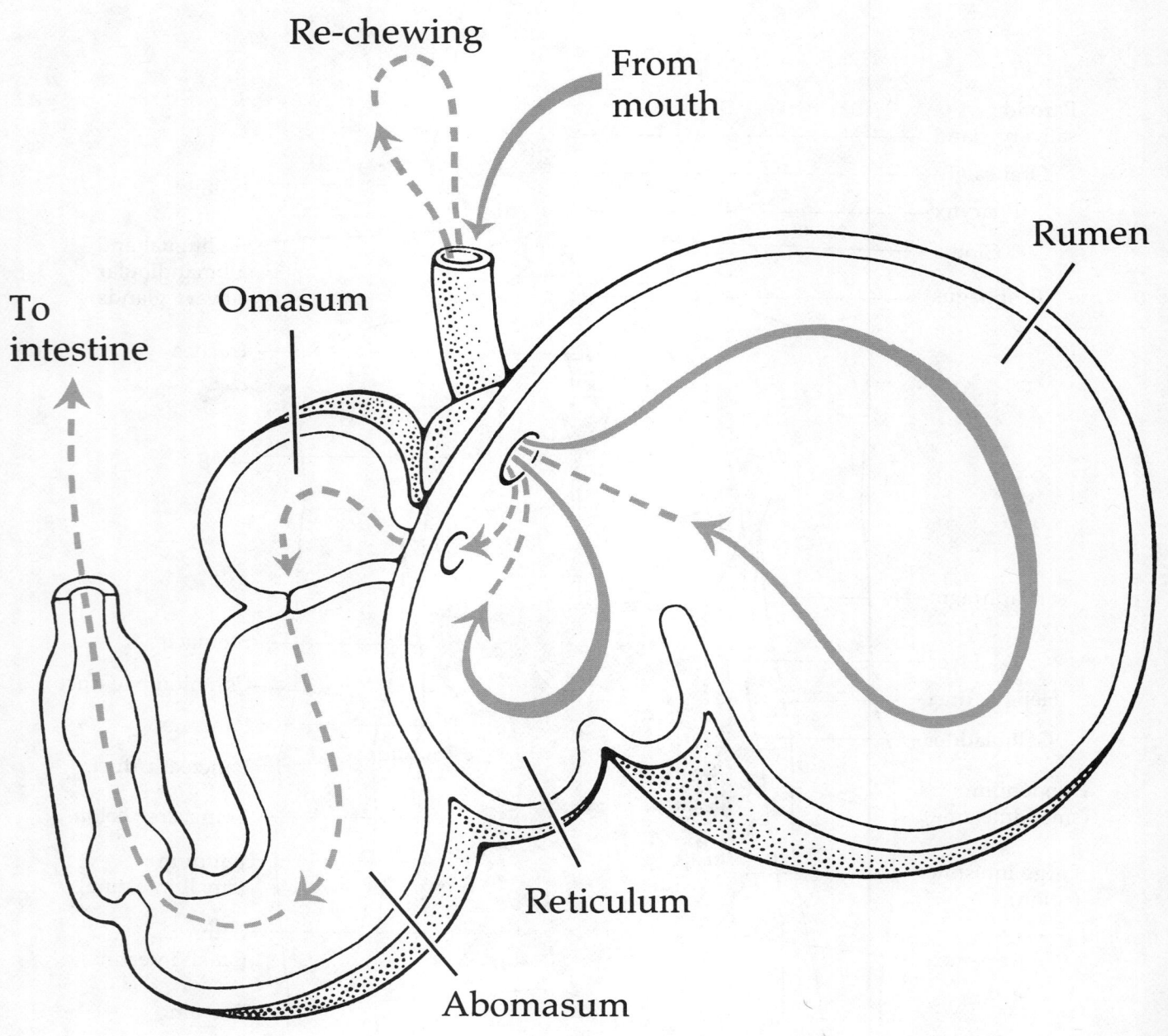

Re-chewing

From mouth

Rumen

To intestine

Omasum

Reticulum

Abomasum

FIGURE 22.20   SPECIALIZATION IN RUMINANT "STOMACHS"   (text page 566)

(a) Integument

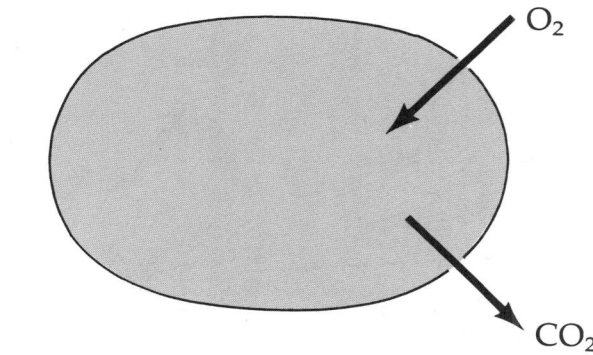

(b) Gills

(c) Lungs

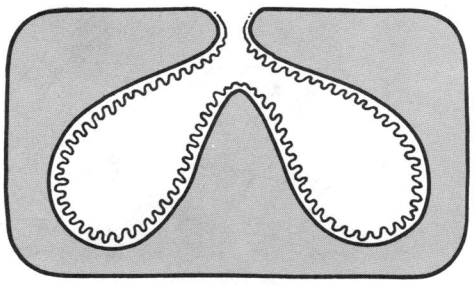

(d) Tracheae

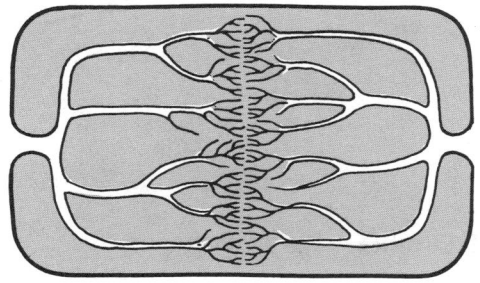

FIGURE 23.2   ADAPTATIONS FOR EXCHANGING GASES   (text page 575)

From *Life: The Science of Biology*, Copyright © 1983 by William K. Purves and Gordon H. Orians.

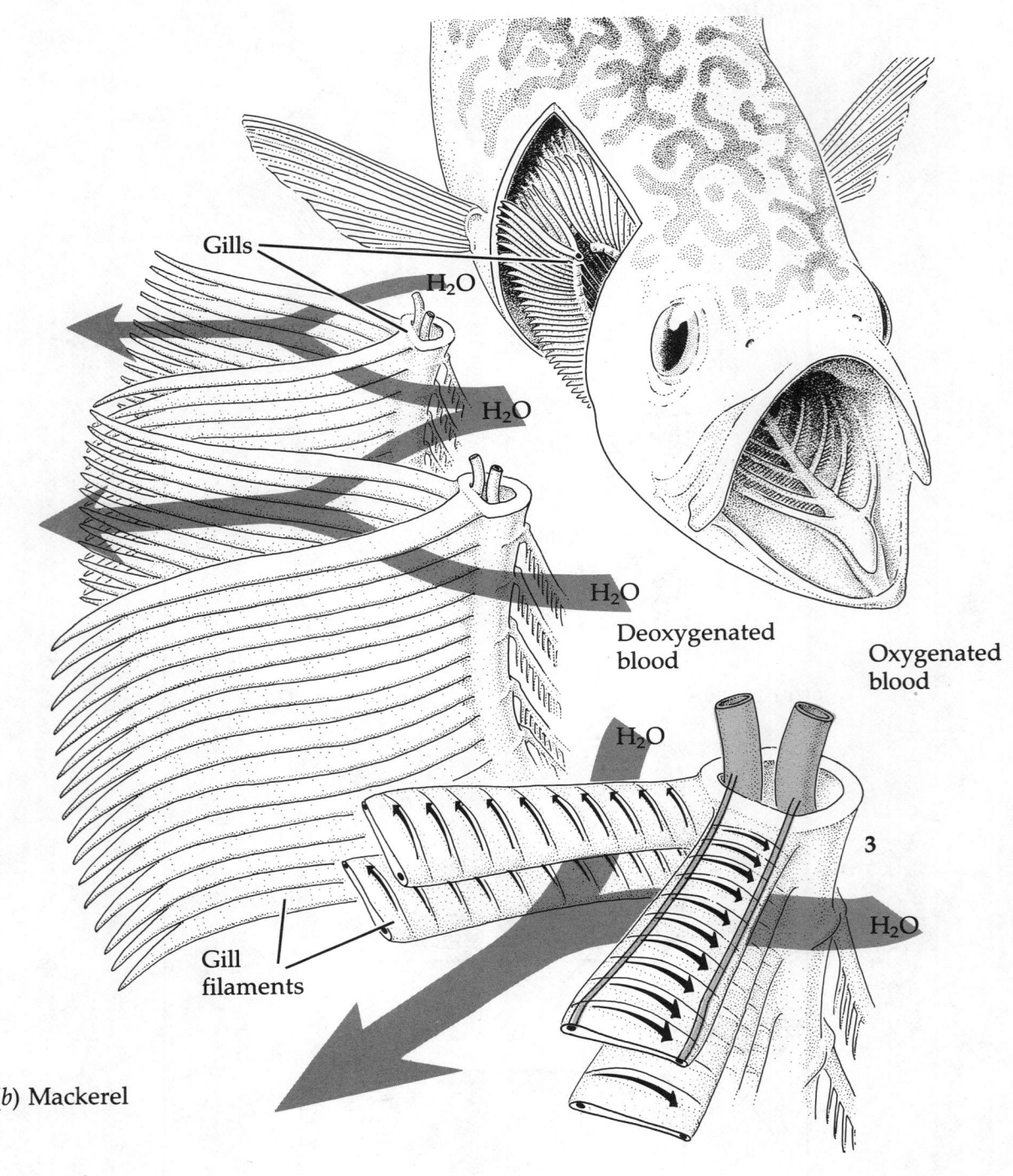

Gills

H_2O

H_2O

H_2O

Deoxygenated
blood

Oxygenated
blood

H_2O

3

H_2O

Gill
filaments

(b) Mackerel

FIGURE 23.3b   TWO KINDS OF GILLS   (text page 576)

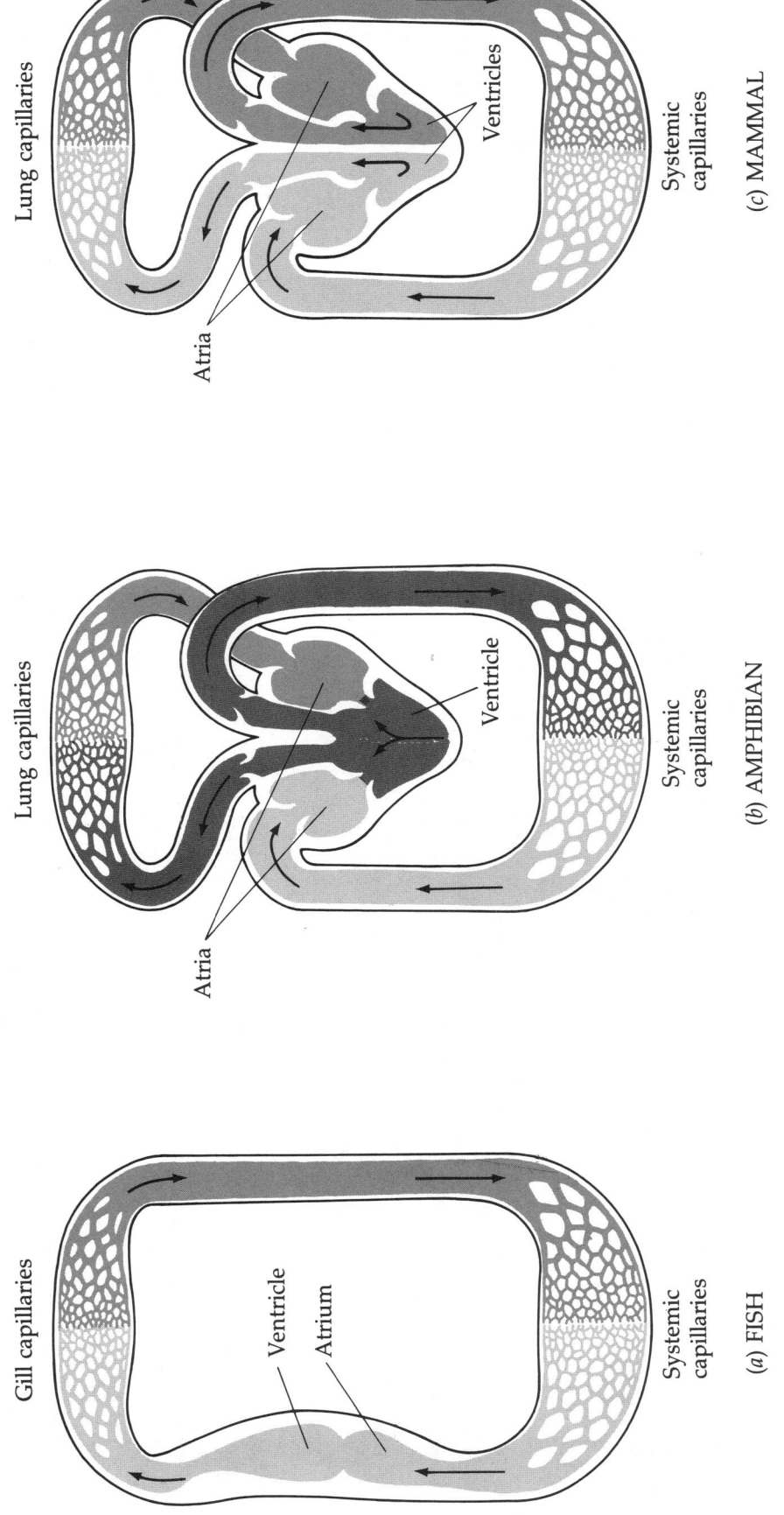

Lung capillaries

Ventricles

Atria

Systemic
capillaries

(c) MAMMAL

Lung capillaries

Ventricle

Atria

Systemic
capillaries

(b) AMPHIBIAN

Gill capillaries

Ventricle

Atrium

Systemic
capillaries

(a) FISH

FIGURE 23.11   CLOSED CIRCULATION IN VERTEBRATES   (text page 584)

From *Life: The Science of Biology*, Copyright © 1983 by William K. Purves and Gordon H. Orians.

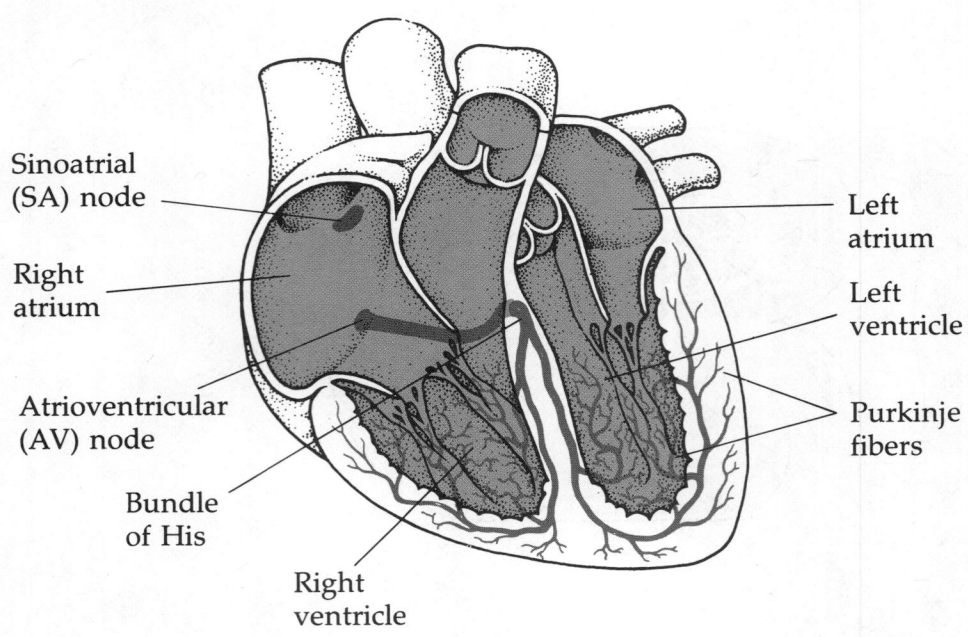

Sinoatrial
(SA) node

Right
atrium

Atrioventricular
(AV) node

Bundle
of His

Right
ventricle

Left
atrium

Left
ventricle

Purkinje
fibers

FIGURE 23.15   STIMULATION AND COORDINATION OF THE HEART   (text page 588)

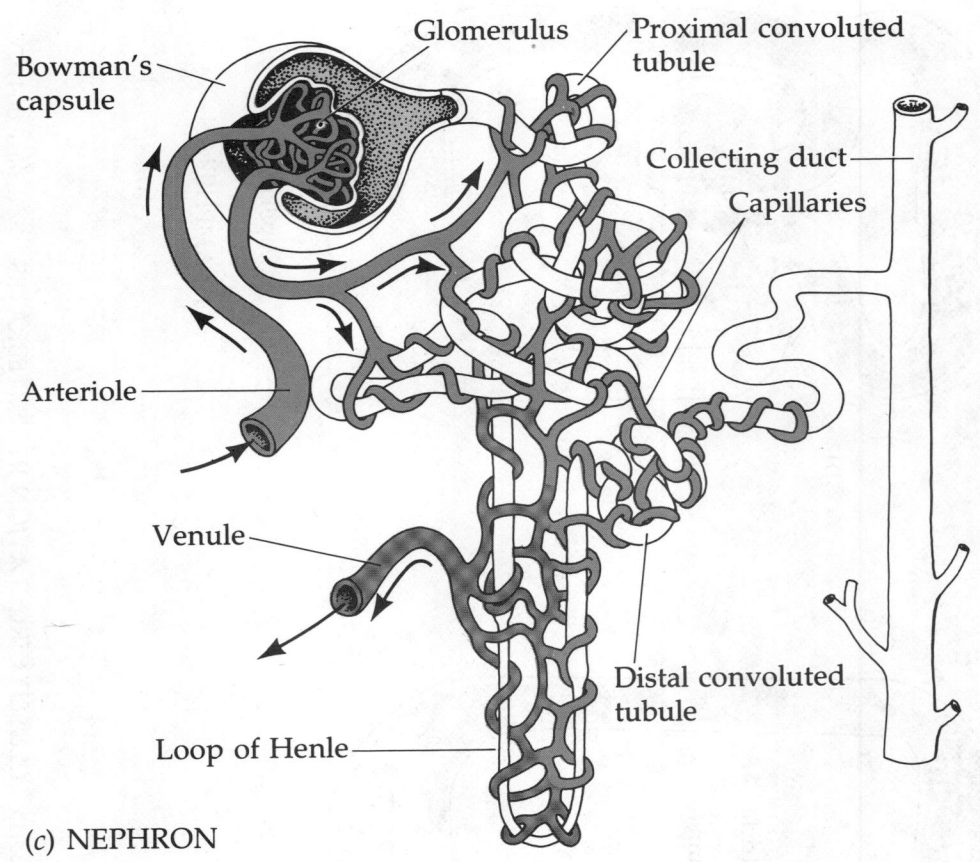

Bowman's
capsule

Glomerulus

Proximal convoluted
tubule

Collecting duct

Capillaries

Arteriole

Venule

Loop of Henle

Distal convoluted
tubule

(c) NEPHRON

FIGURE 24.12c   EXCRETORY SYSTEM OF A MAMMAL   (text page 613)

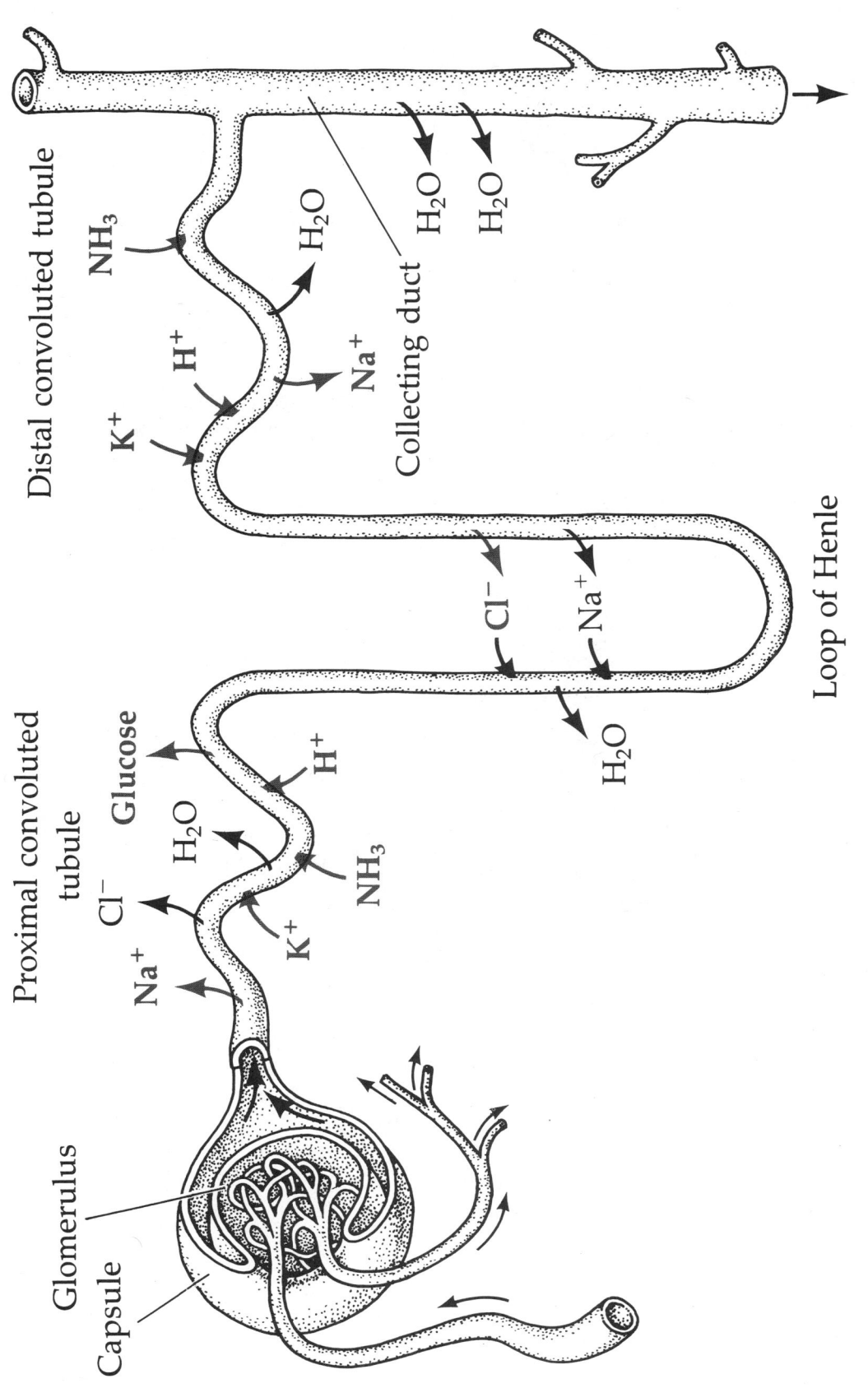

FIGURE 24.19   SUMMARY OF THE VERTEBRATE NEPHRON   (text page 620)

From *Life: The Science of Biology.* Copyright © 1983 by William K. Purves and Gordon H. Orians.

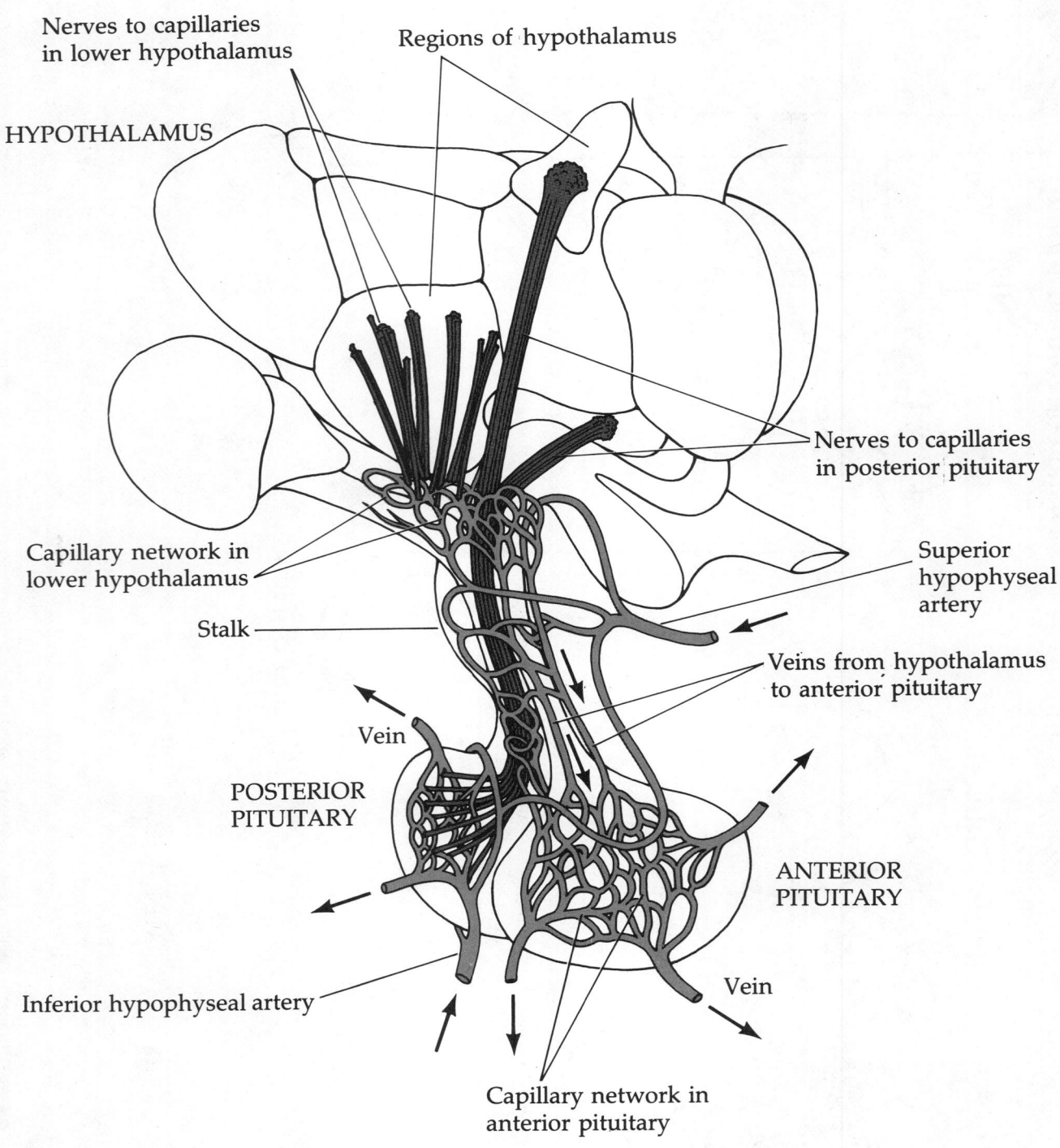

Nerves to capillaries
in lower hypothalamus

Regions of hypothalamus

HYPOTHALAMUS

Nerves to capillaries
in posterior pituitary

Capillary network in
lower hypothalamus

Superior
hypophyseal
artery

Stalk

Veins from hypothalamus
to anterior pituitary

Vein

POSTERIOR
PITUITARY

ANTERIOR
PITUITARY

Inferior hypophyseal artery

Vein

Capillary network in
anterior pituitary

FIGURE 25.4   PITUITARY AND HYPOTHALAMUS   (text page 632)
After *Hospital Practice*, 10 #4:55, April 1975.

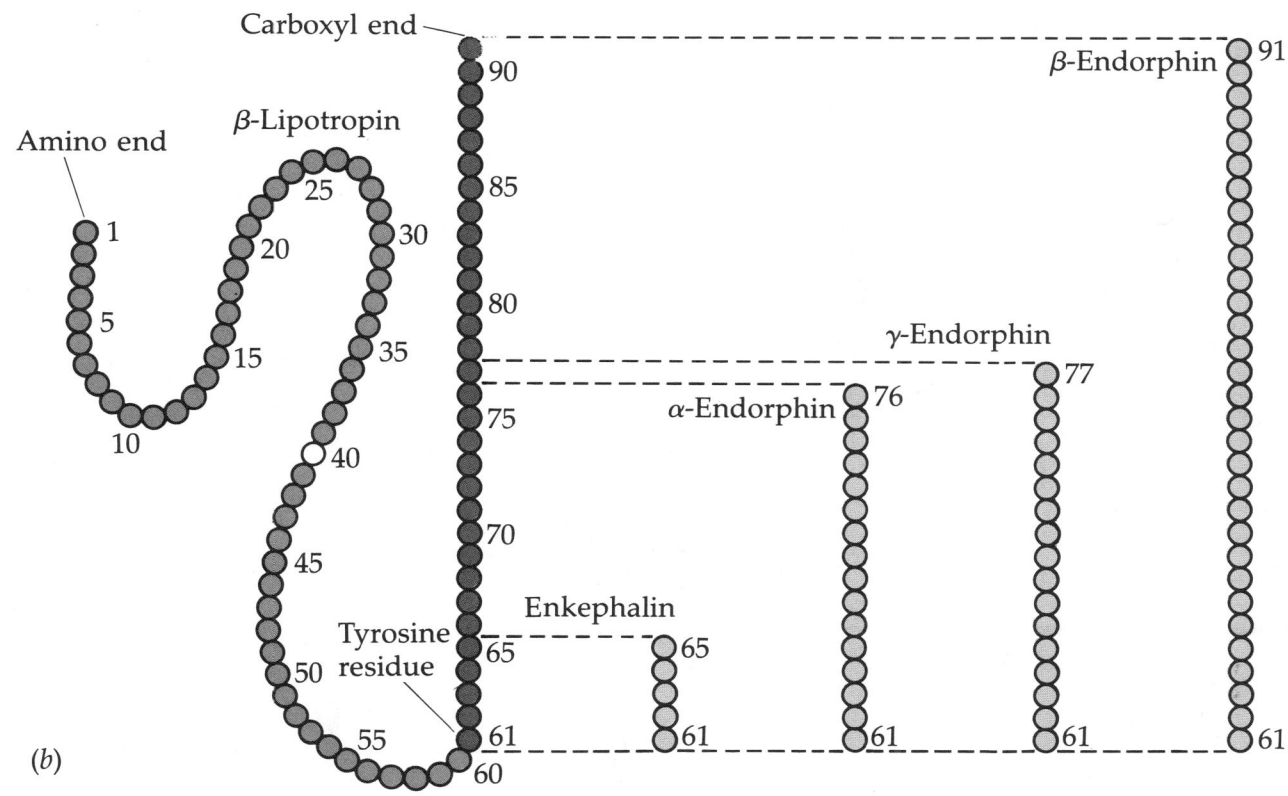

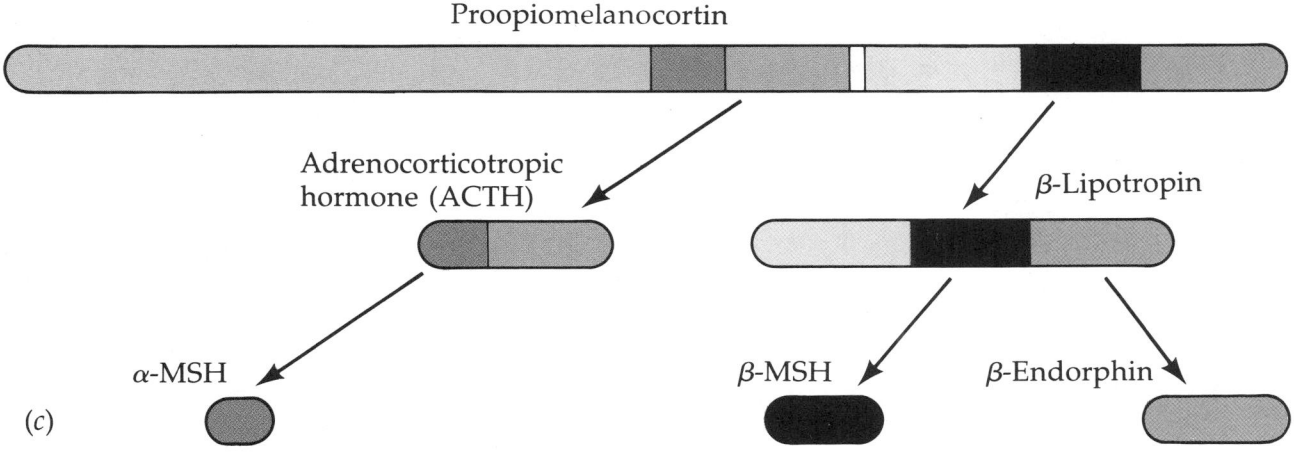

FIGURE 25.17b & c   ENDORPHINS   (text page 645)

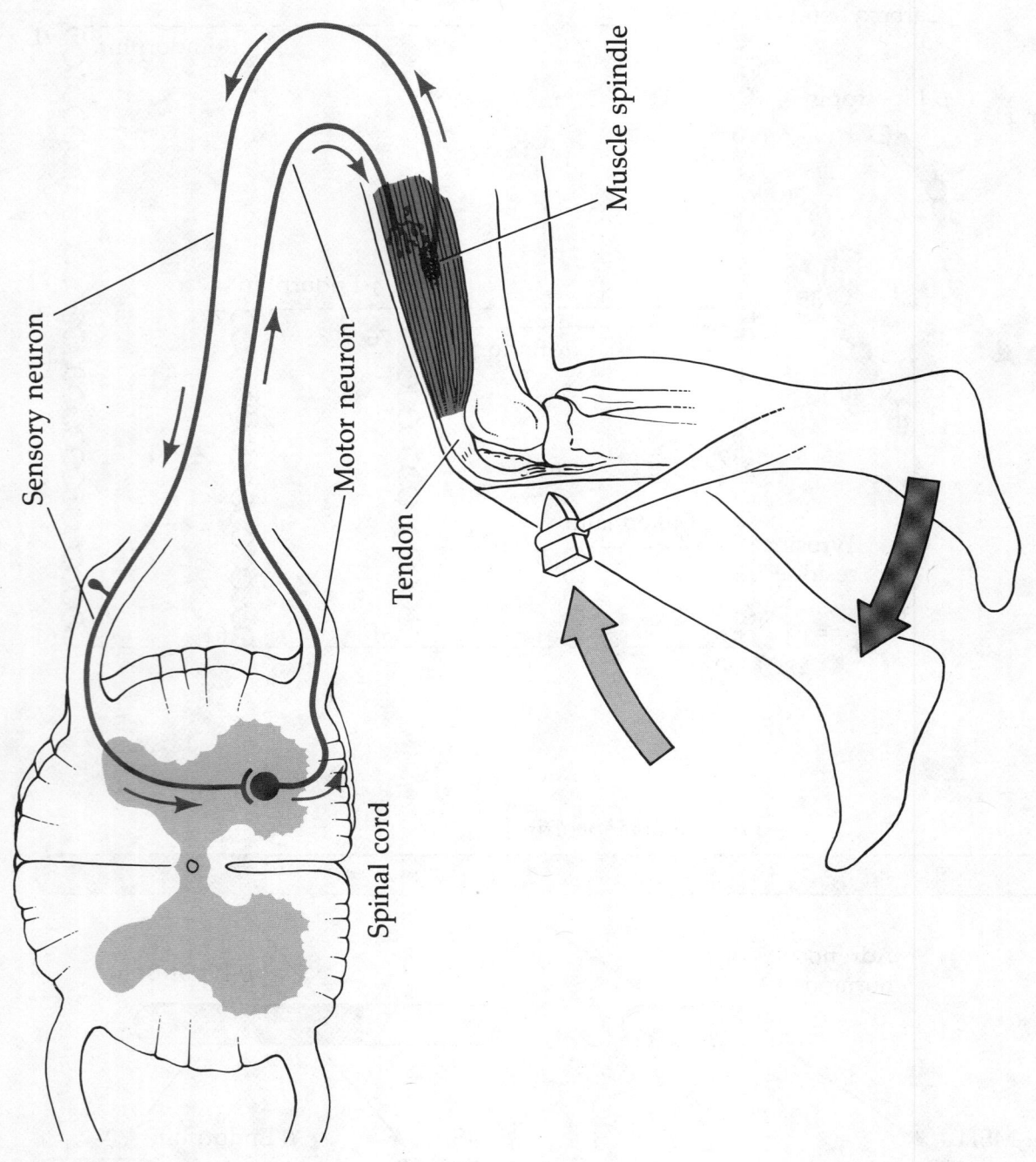

Muscle spindle

Sensory neuron

Motor neuron

Tendon

Spinal cord

FIGURE 26.5   A SIMPLE REFLEX   (text page 653)

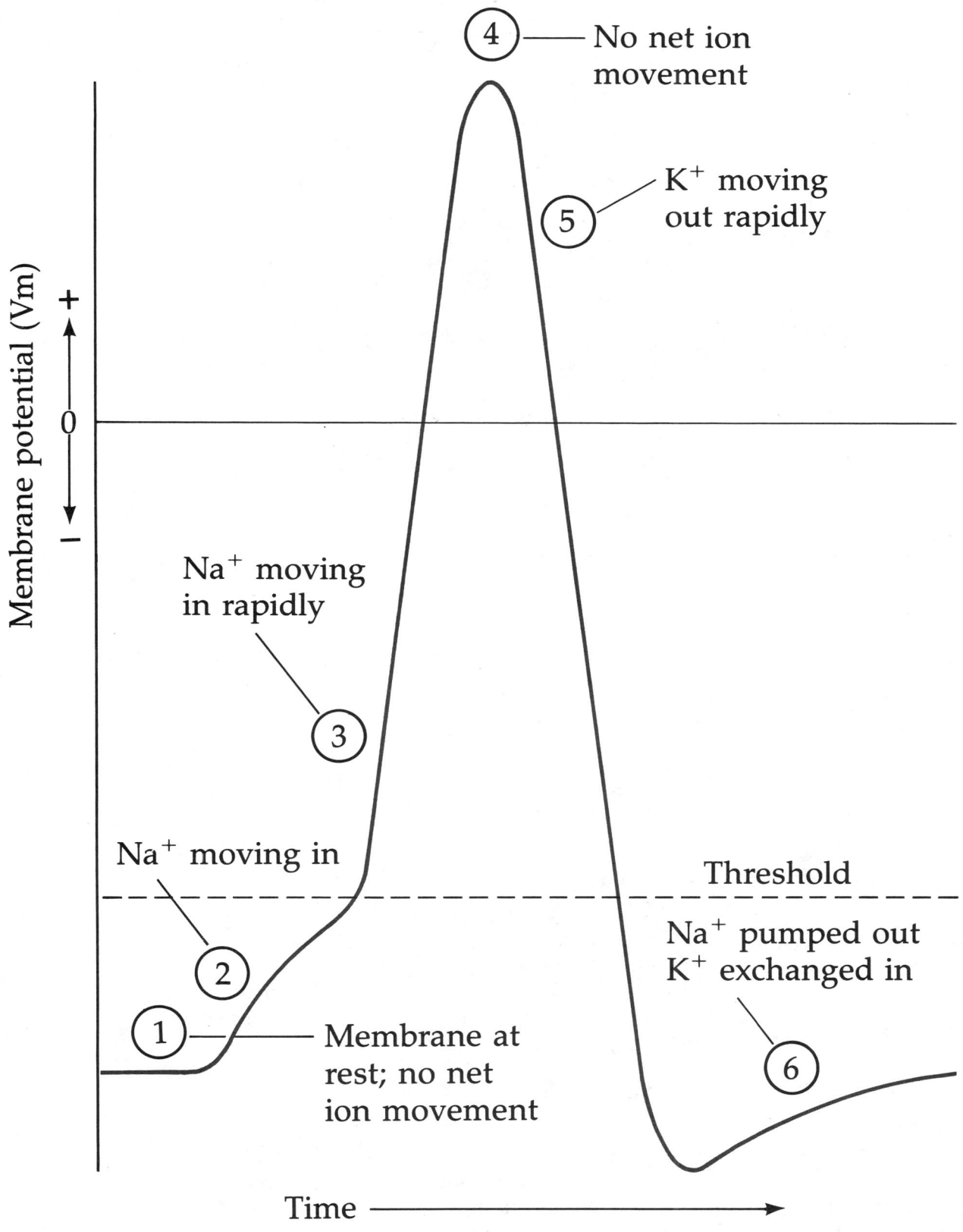

FIGURE 26.9   ION MOVEMENT AND ACTION POTENTIAL   (text page 658)

From *Life: The Science of Biology*, Copyright © 1983 by William K. Purves and Gordon H. Orians.

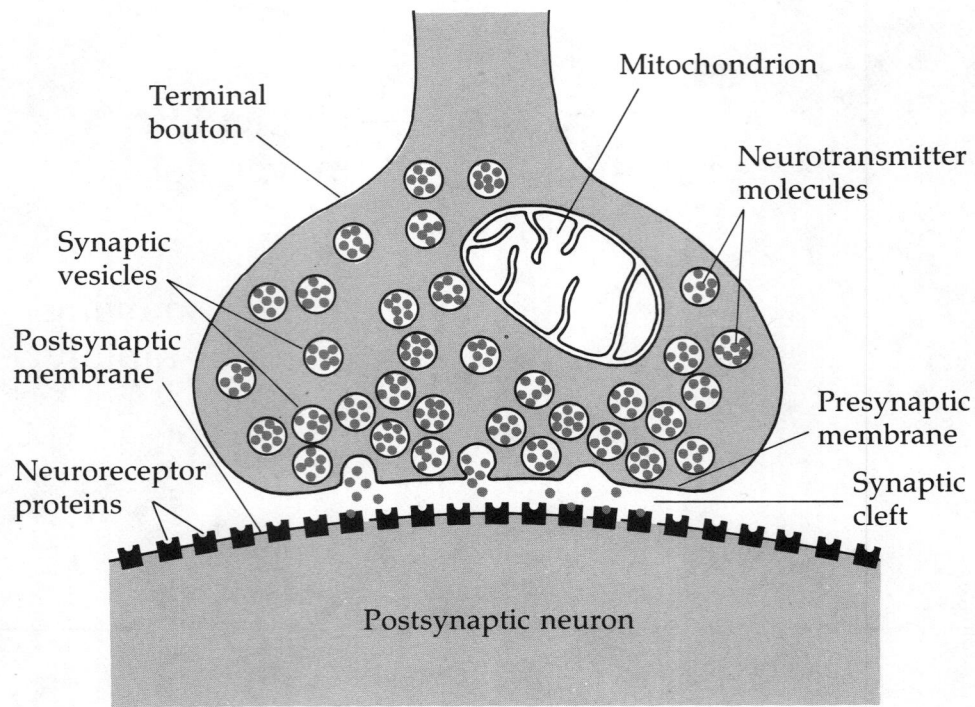

FIGURE 26.13 CHEMICAL SYNAPSES (text page 661)

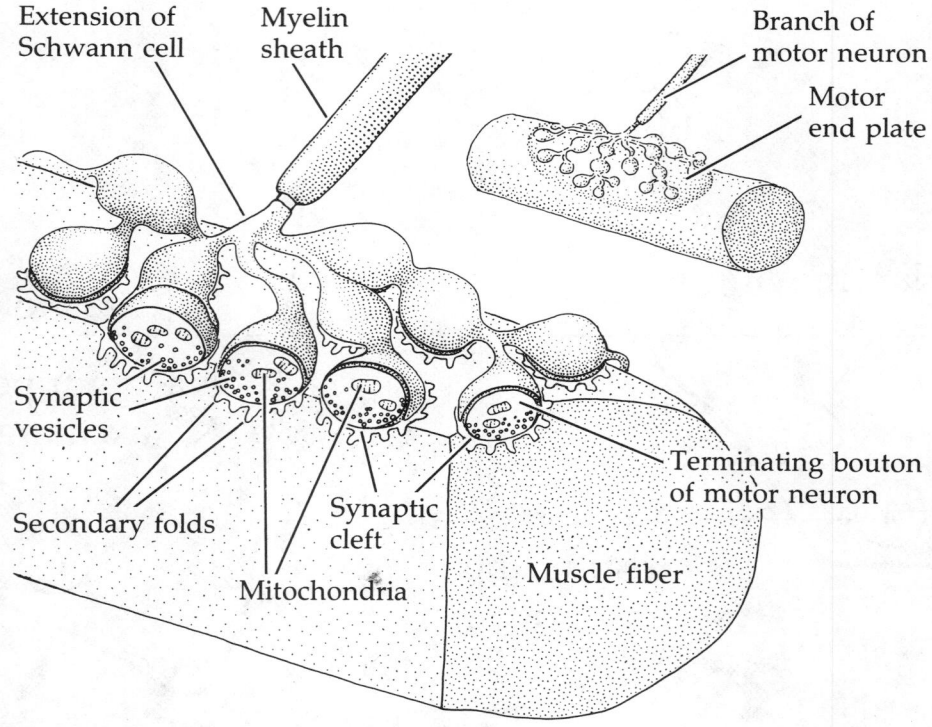

FIGURE 26.15a NEUROMUSCULAR JUNCTIONS (text page 664)

From S.W. Kuffler and J.G. Nicholls, *From Neuron to Brain*, Sinauer Associates, 1976.

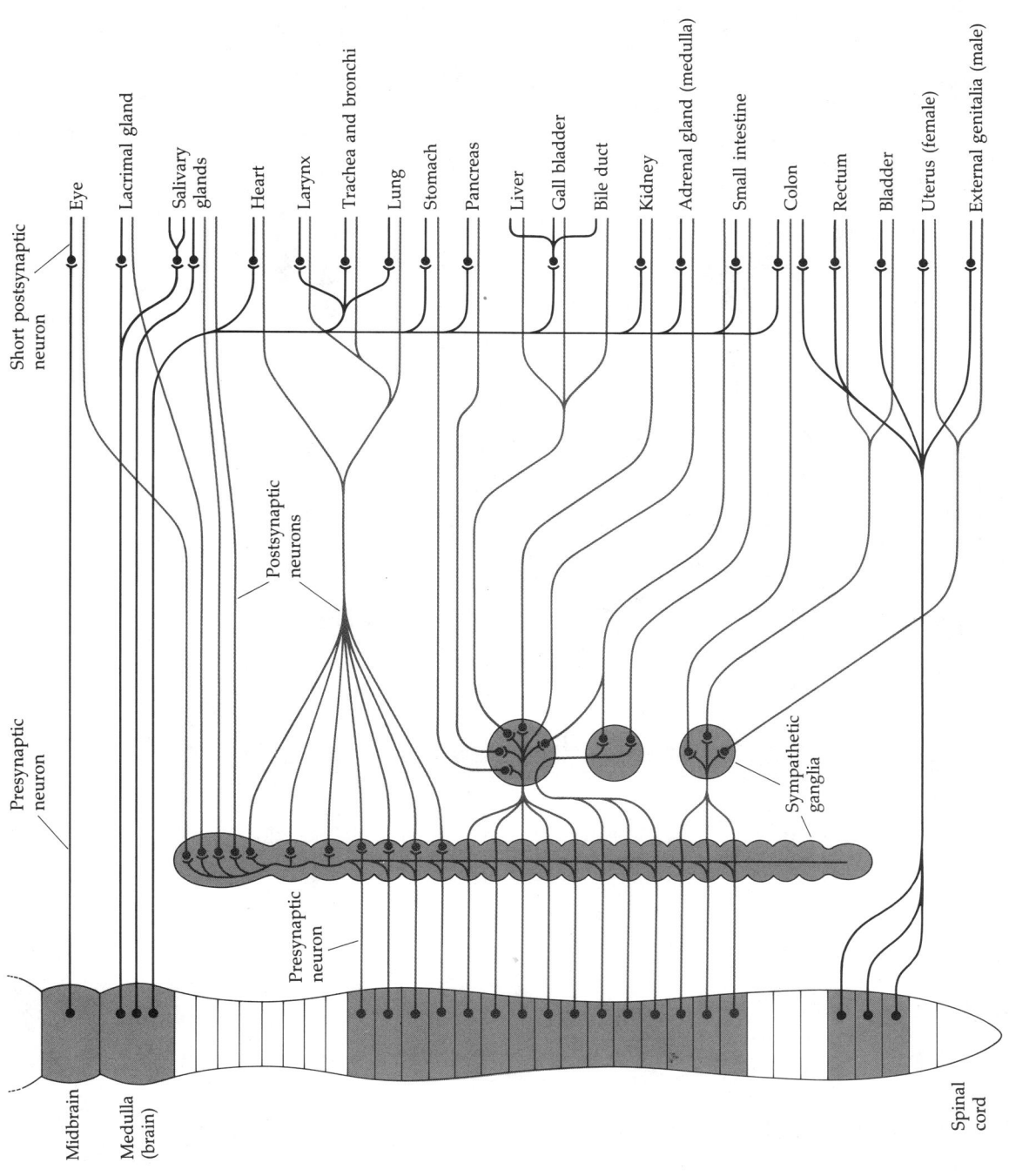

FIGURE 26.18 AUTONOMIC NERVOUS SYSTEM (text page 667)

From *Life: The Science of Biology.* Copyright © 1983 by William K. Purves and Gordon H. Orians.

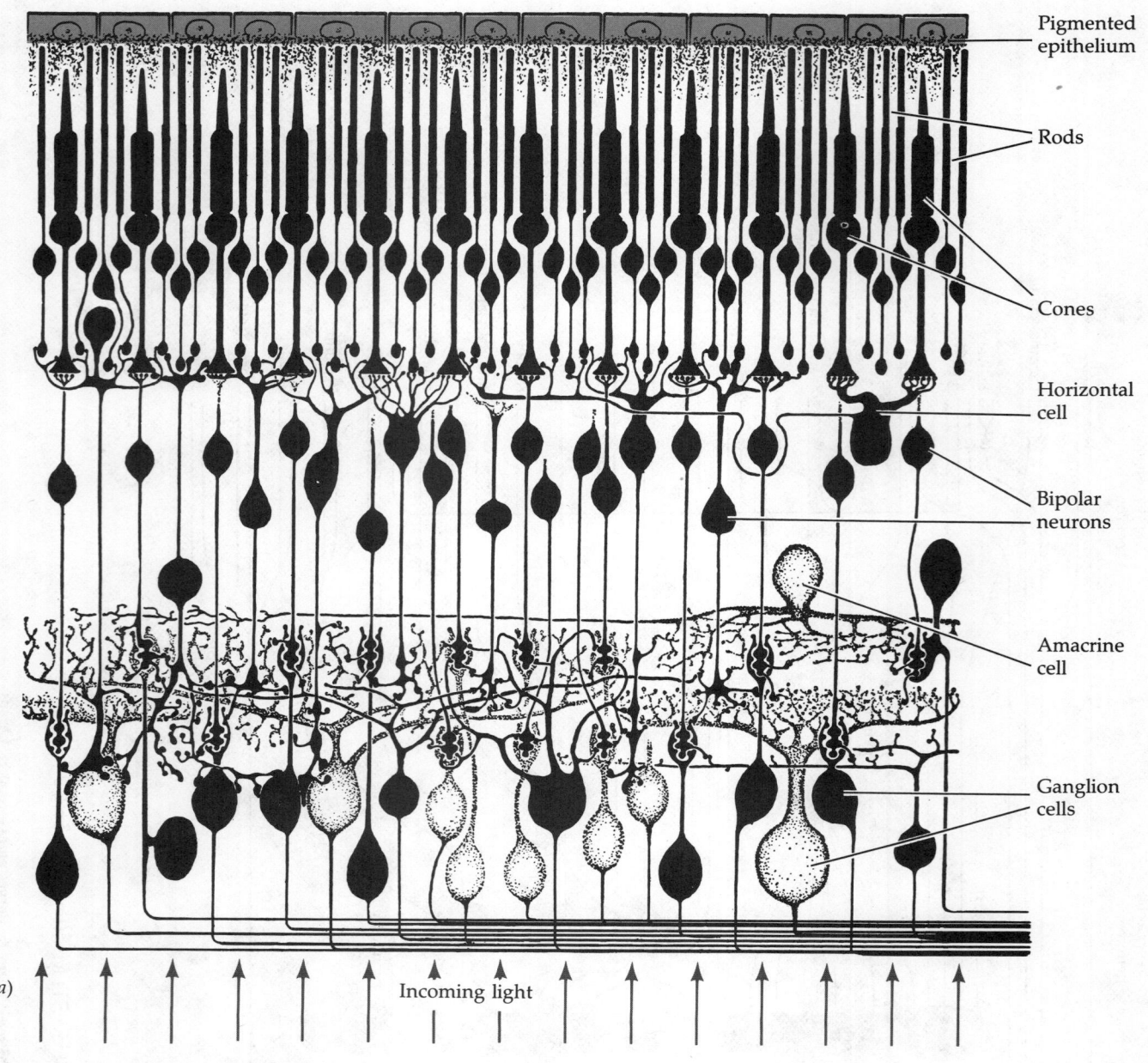

Pigmented epithelium

Rods

Cones

Horizontal cell

Bipolar neurons

Amacrine cell

Ganglion cells

(a)

Incoming light

FIGURE 27.3a   VERTEBRATE RETINA   (text page 682)

From *Life: The Science of Biology*, Copyright © 1983 by William K. Purves and Gordon H. Orians.

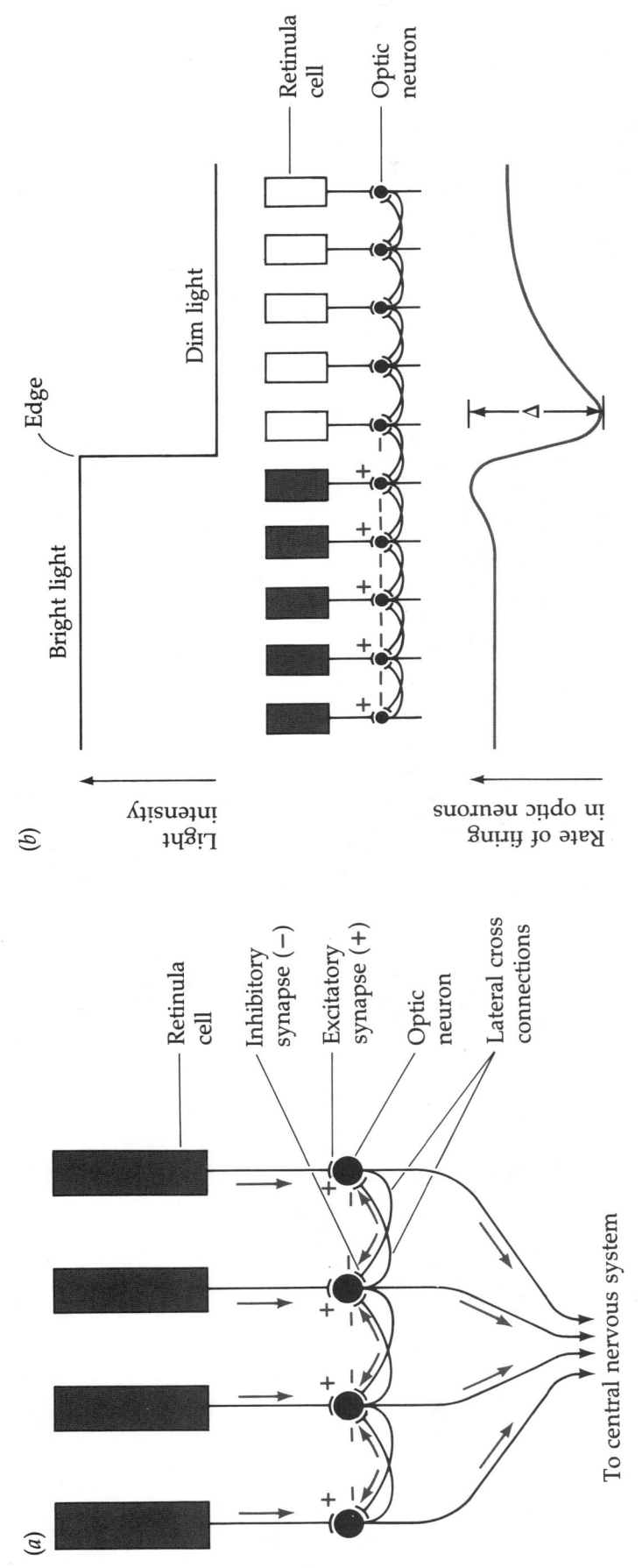

FIGURE 27.5   INFORMATION PROCESSING IN INVERTEBRATE EYES   (text page 685)

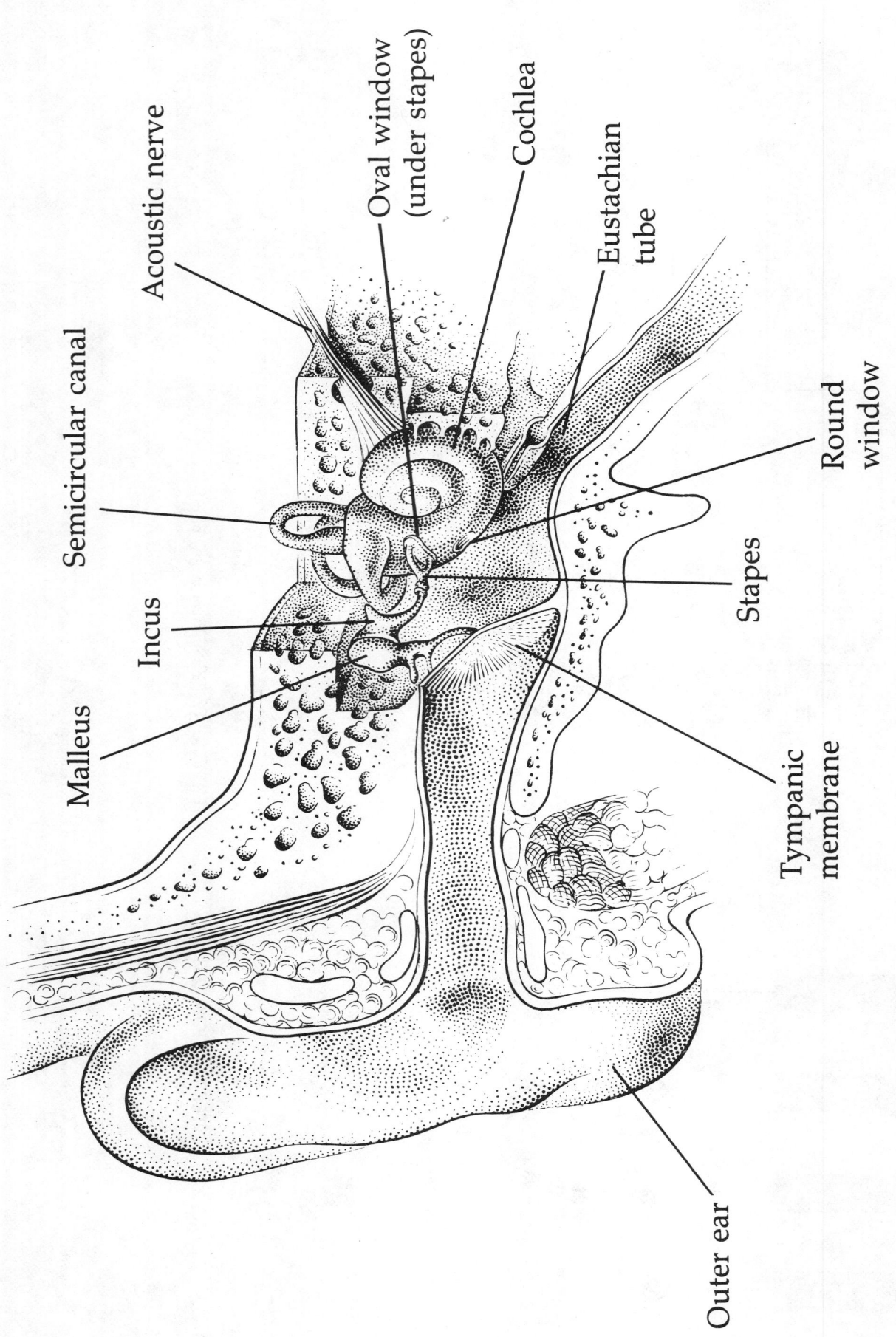

Acoustic nerve

Oval window (under stapes)

Cochlea

Eustachian tube

Semicircular canal

Round window

Incus

Malleus

Stapes

Tympanic membrane

Outer ear

FIGURE 27.9a   THE HUMAN EAR   (text page 689)

From *Life: The Science of Biology*. Copyright © 1983 by William K. Purves and Gordon H. Orians.

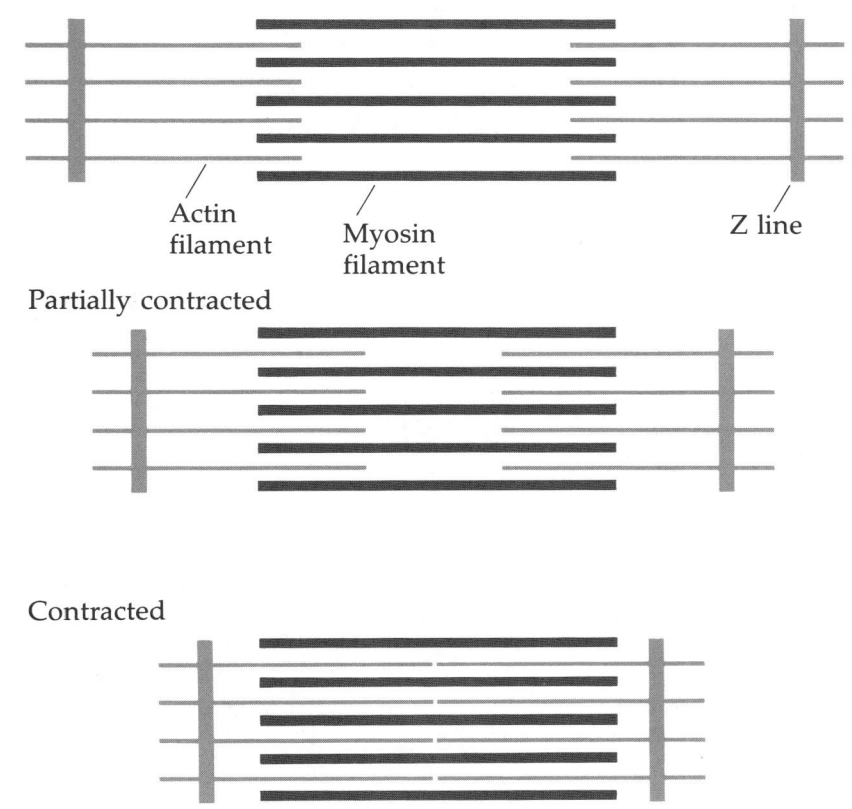

Actin filament   Myosin filament   Z line

Partially contracted

Contracted

FIGURE 28.11b   STRUCTURE AND FUNCTION OF THE SARCOMERE   (text page 709)

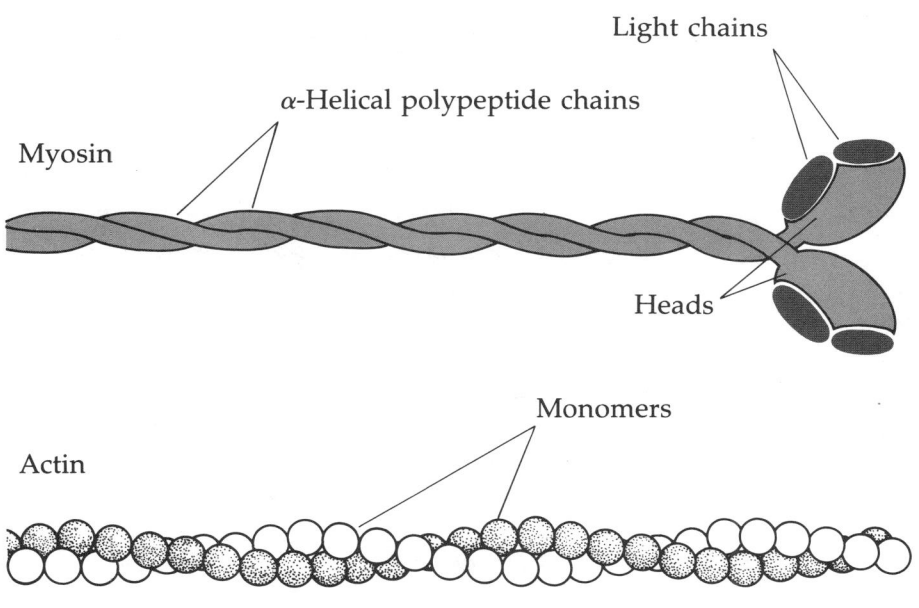

Light chains

α-Helical polypeptide chains

Myosin

Heads

Monomers

Actin

FIGURE 28.12a   ACTIN, MYOSIN, AND THE SLIDING FILAMENT MODEL   (text page 710)

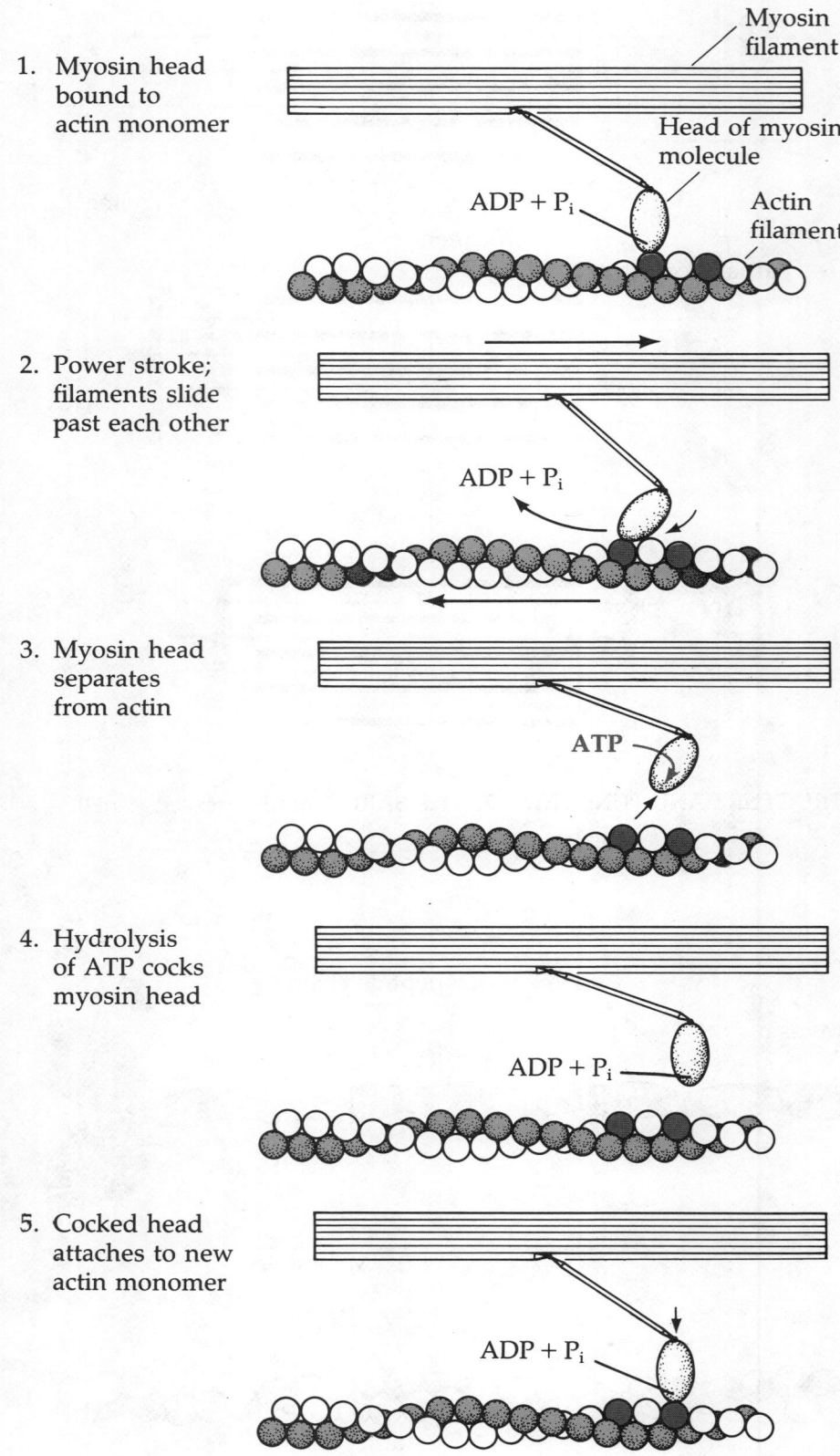

1. Myosin head bound to actin monomer

Myosin filament

Head of myosin molecule

ADP + P$_i$

Actin filament

2. Power stroke; filaments slide past each other

ADP + P$_i$

3. Myosin head separates from actin

ATP

4. Hydrolysis of ATP cocks myosin head

ADP + P$_i$

5. Cocked head attaches to new actin monomer

ADP + P$_i$

FIGURE 28.12b ACTIN, MYOSIN, AND THE SLIDING FILAMENT MODEL   (text page 710)

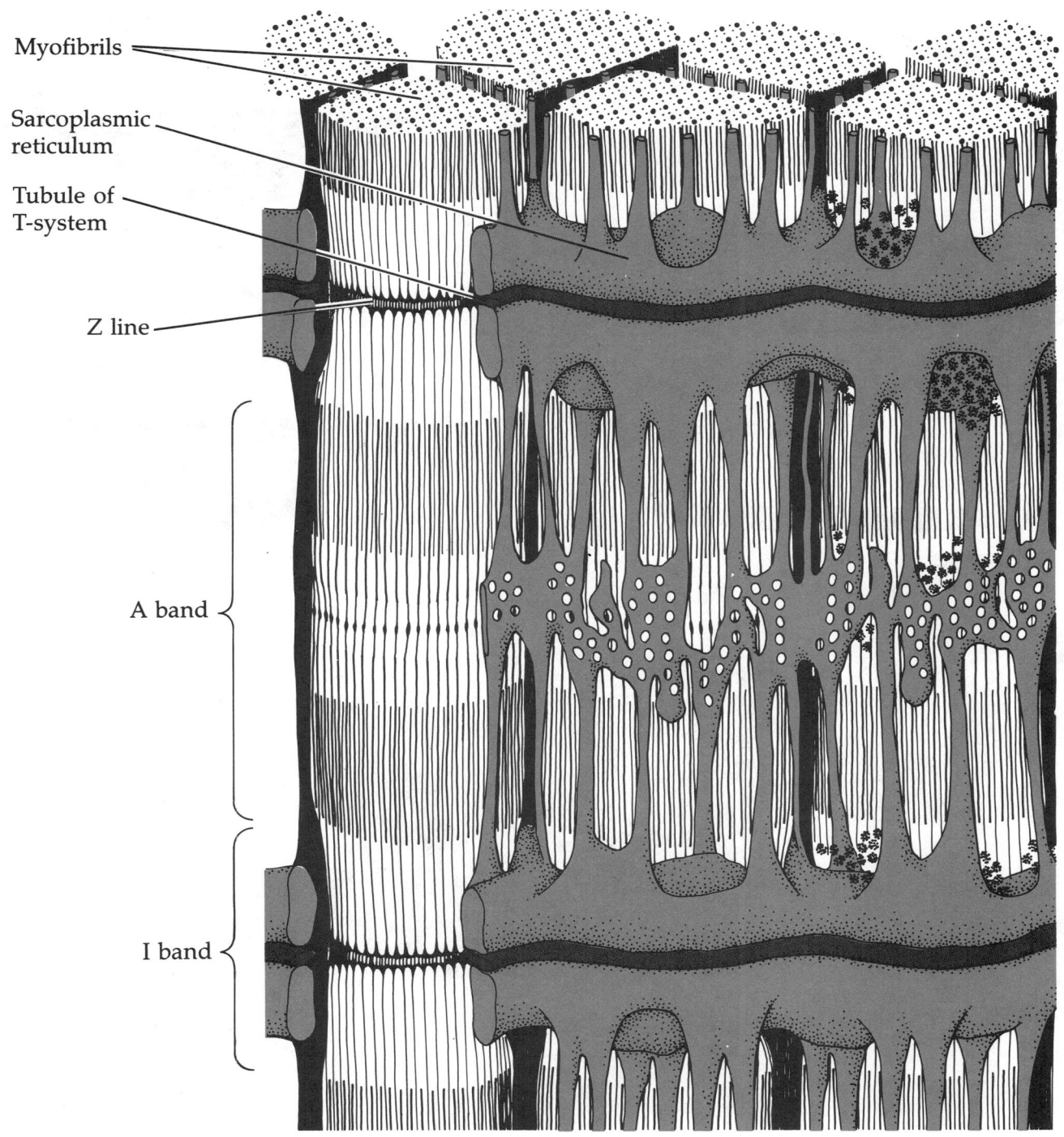

Myofibrils

Sarcoplasmic reticulum

Tubule of T-system

Z line

A band

I band

FIGURE 28.13   STRUCTURE OF MUSCLE FIBERS   (text page 711)
After L. Peachey, *J. Cell Biol.* **25**: 209, 1965.

FIGURE 29.2   SIGN STIMULI   (text page 722)

Part (b) after N. Tinbergen, *The Study of Instinct*, Oxford Univ. Press, 1951.

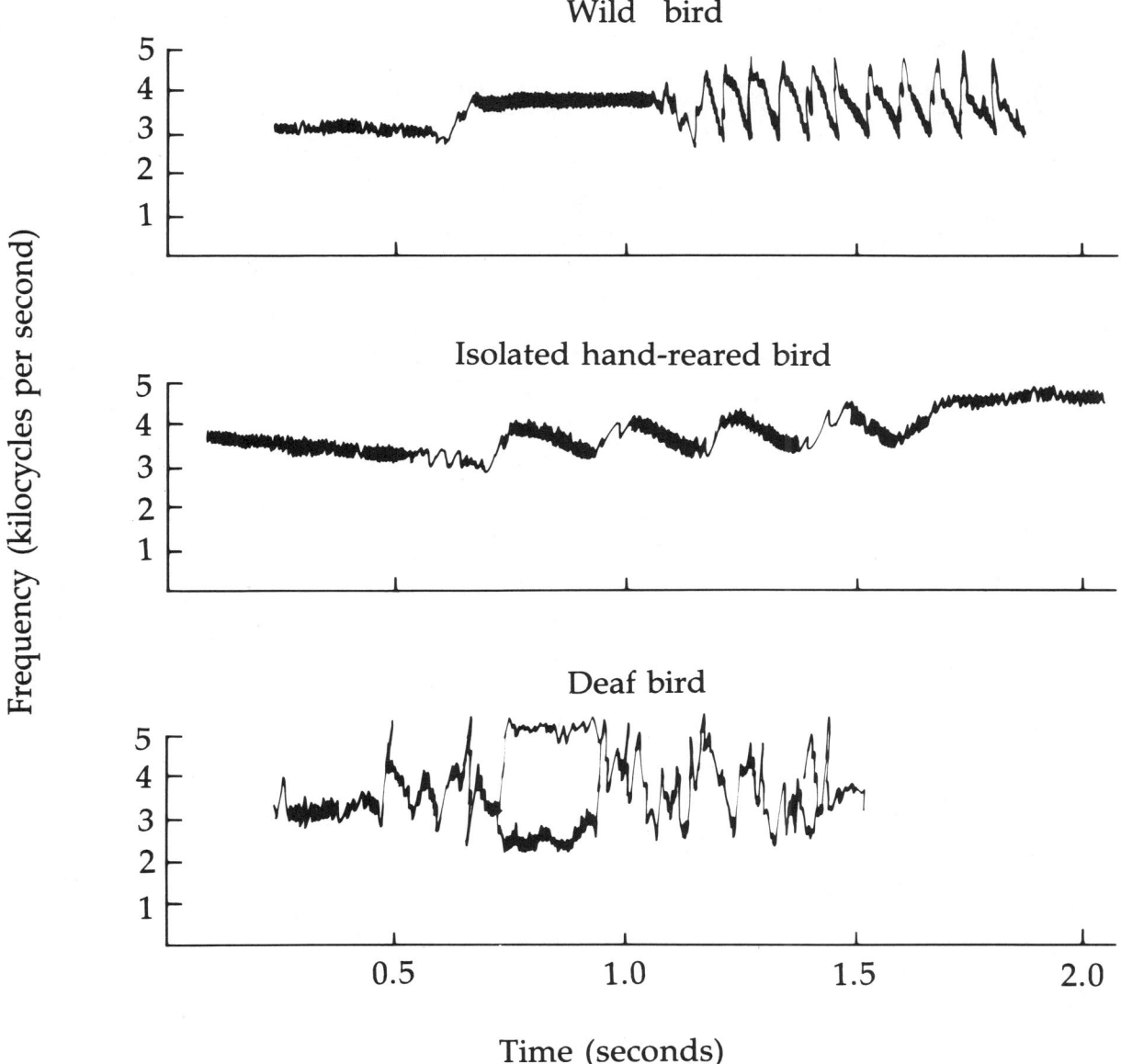

Wild bird

Isolated hand-reared bird

Deaf bird

Frequency (kilocycles per second)

Time (seconds)

FIGURE 29.9   SONG LEARNING IN THE WHITE-CROWNED SPARROW   (text page 730)

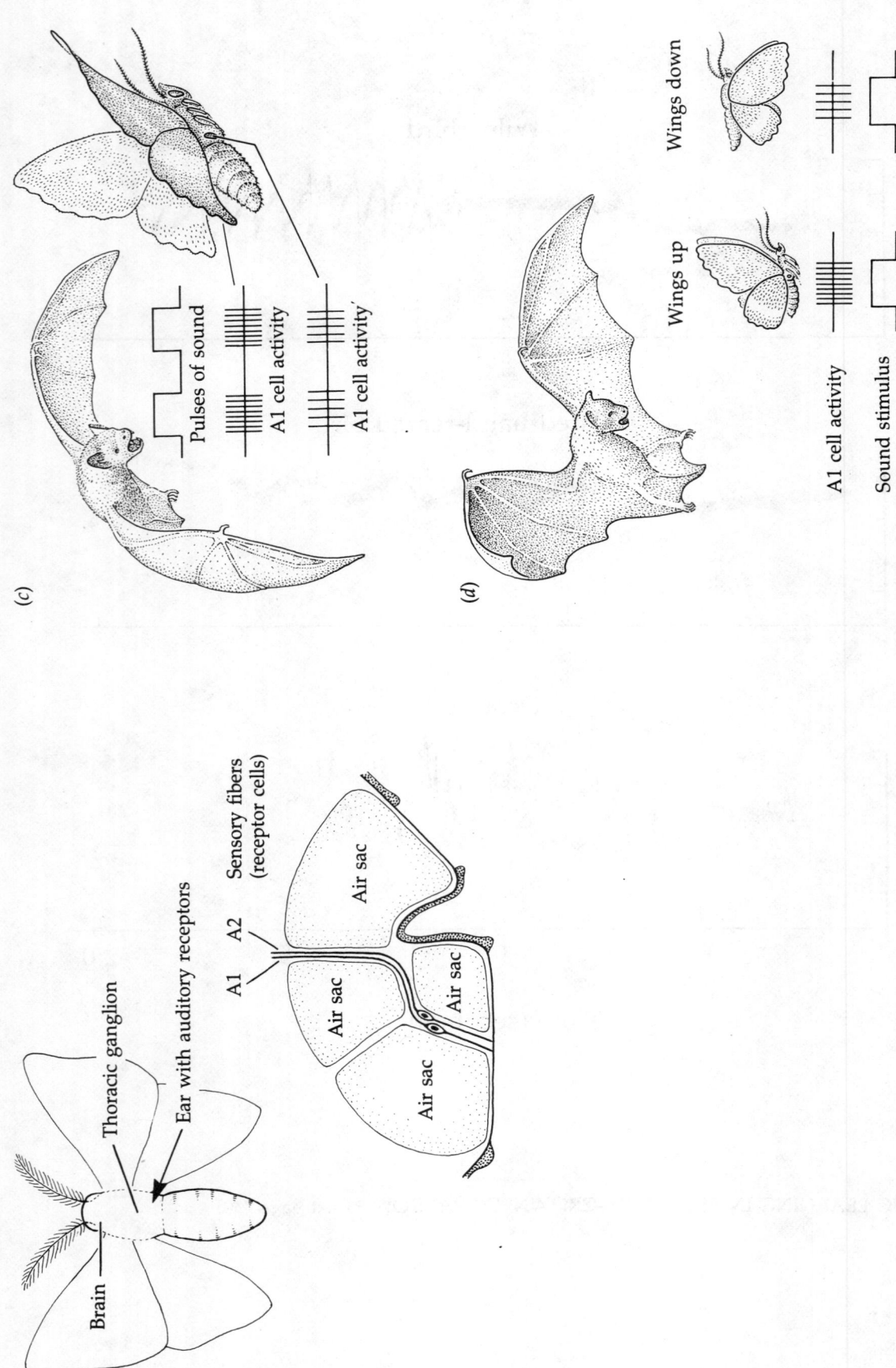

(a)

Brain

Thoracic ganglion

Ear with auditory receptors

Sensory fibers
(receptor cells)

A1 A2

Air sac

Air sac

Air sac

Air sac

(c)

Pulses of sound

A1 cell activity

A1 cell activity

(d)

Wings up

Wings down

A1 cell activity

Sound stimulus

FIGURE 29.13a, c, & d   A SIMPLE EAR THAT DETECTS PREDATORS   (text page 735)

From J. Alcock, *Animal Behaviour*, Sinauer Associates, 1979.

From *Life: The Science of Biology*, Copyright © 1983 by William K. Purves and Gordon H. Orians.

FIGURE 30.1   THE EARLIEST TRUE PEOPLE   (text page 745)
From E.O. Wilson, *Sociobiology: The New Synthesis*, Belknap Press, Harvard Univ. Press, 1975.

From *Life: The Science of Biology*. Copyright © 1983 by William K. Purves and Gordon H. Orians.

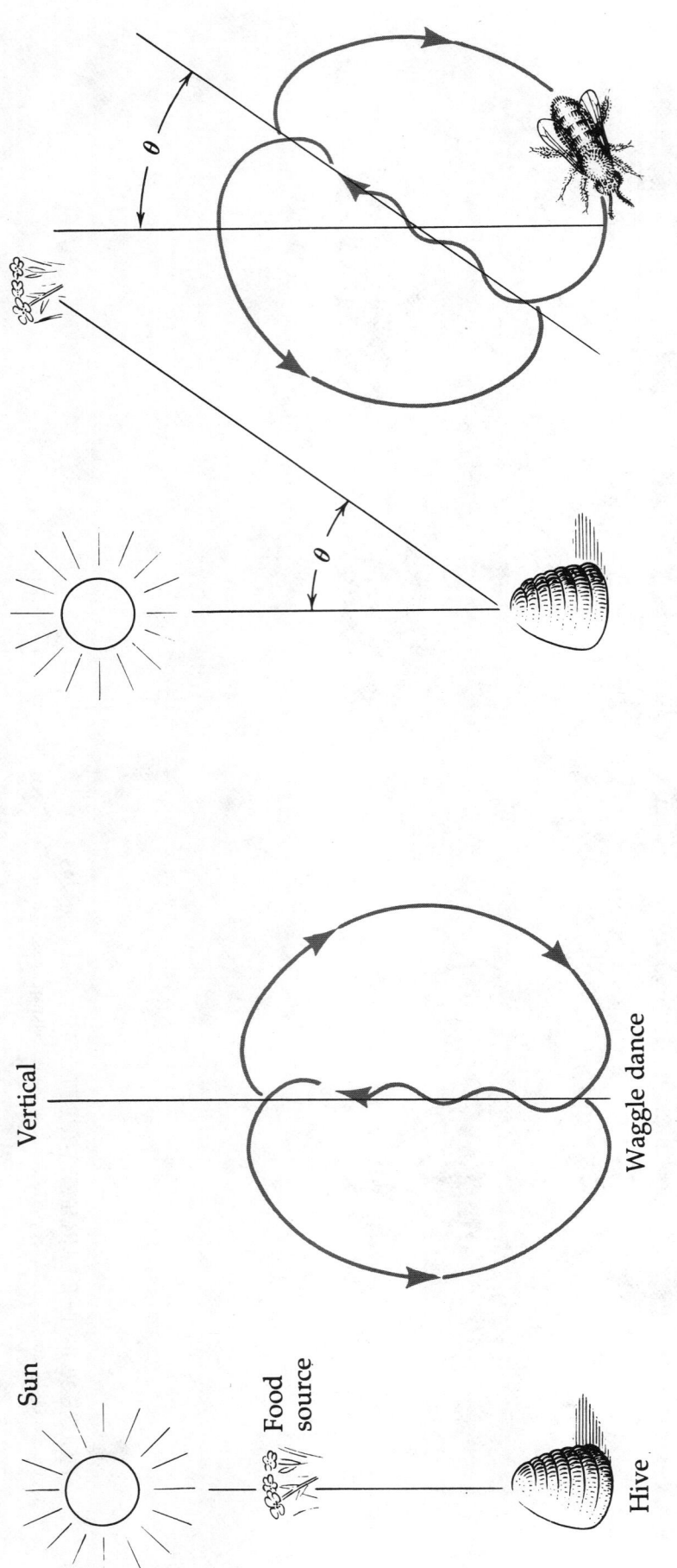

FIGURE 30.11   THE WAGGLE DANCE OF THE HONEY BEE   (text page 754)

From *Life: The Science of Biology*, Copyright © 1983 by William K. Purves and Gordon H. Orians.

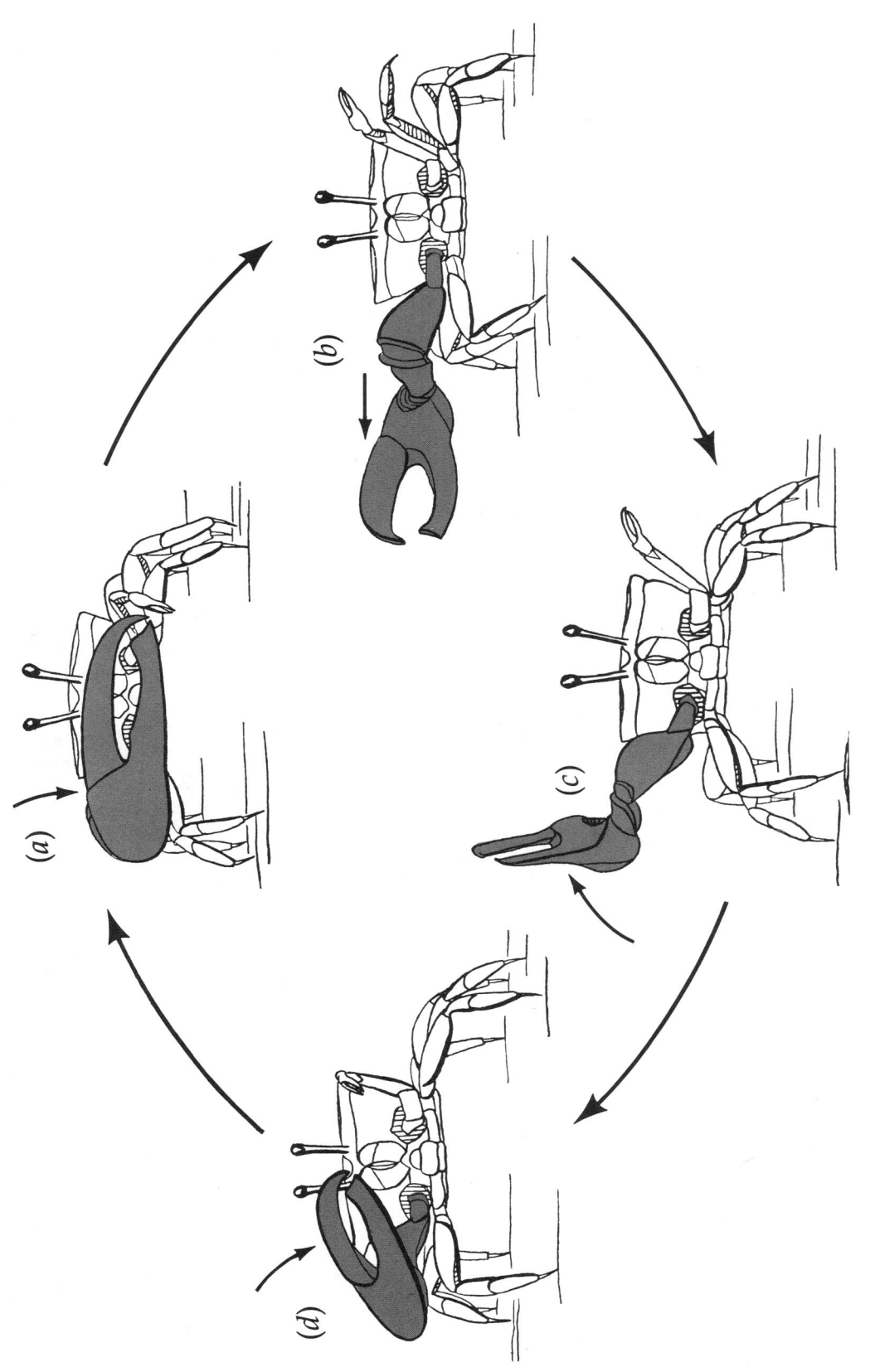

FIGURE 30.14 VISUAL DISPLAYS OF A FIDDLER CRAB (text page 756)

After J. Crane, *Zoologica* 42: 69–82, 1957.

From *Life: The Science of Biology.* Copyright © 1983 by William K. Purves and Gordon H. Orians.

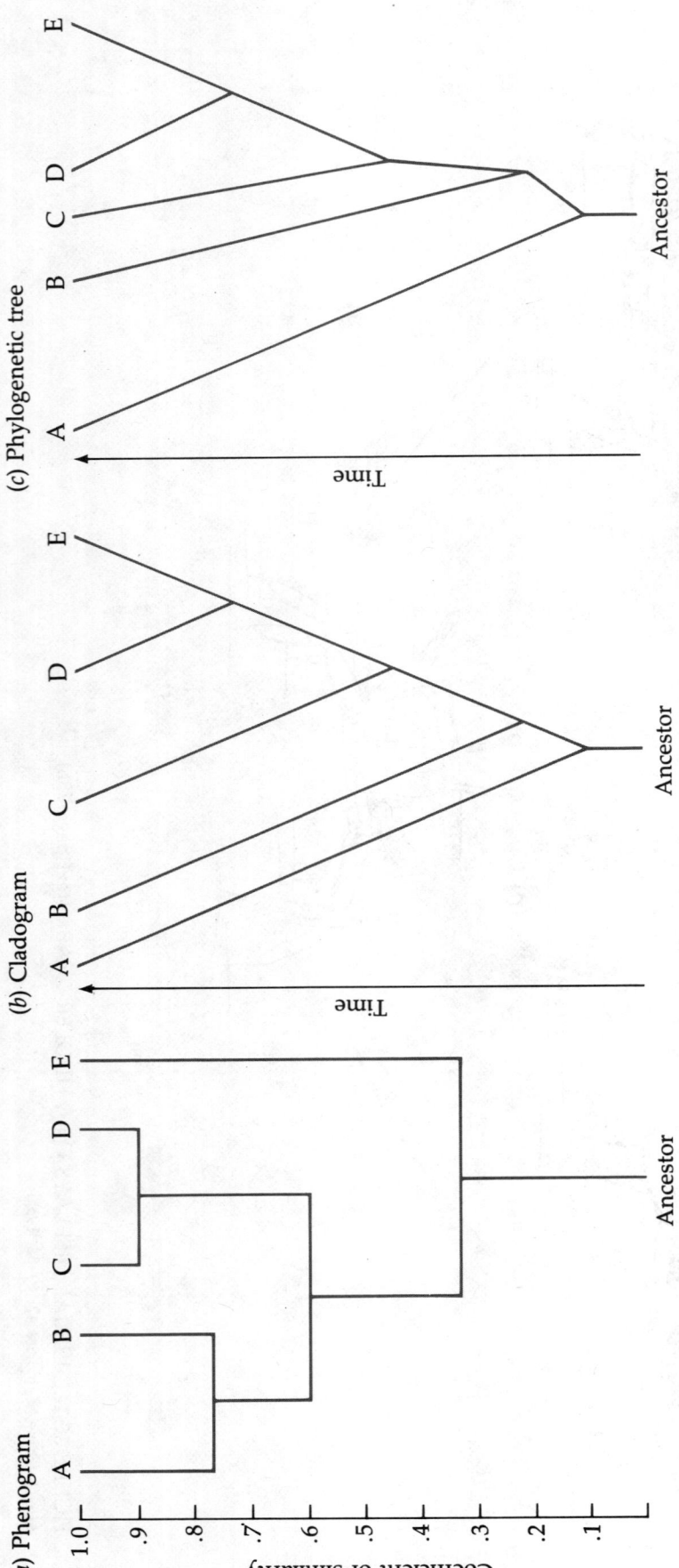

FIGURE 32.1 TYPES OF CLASSIFICATIONS (text page 796)

(*a*) Lumping 10 species into a genus

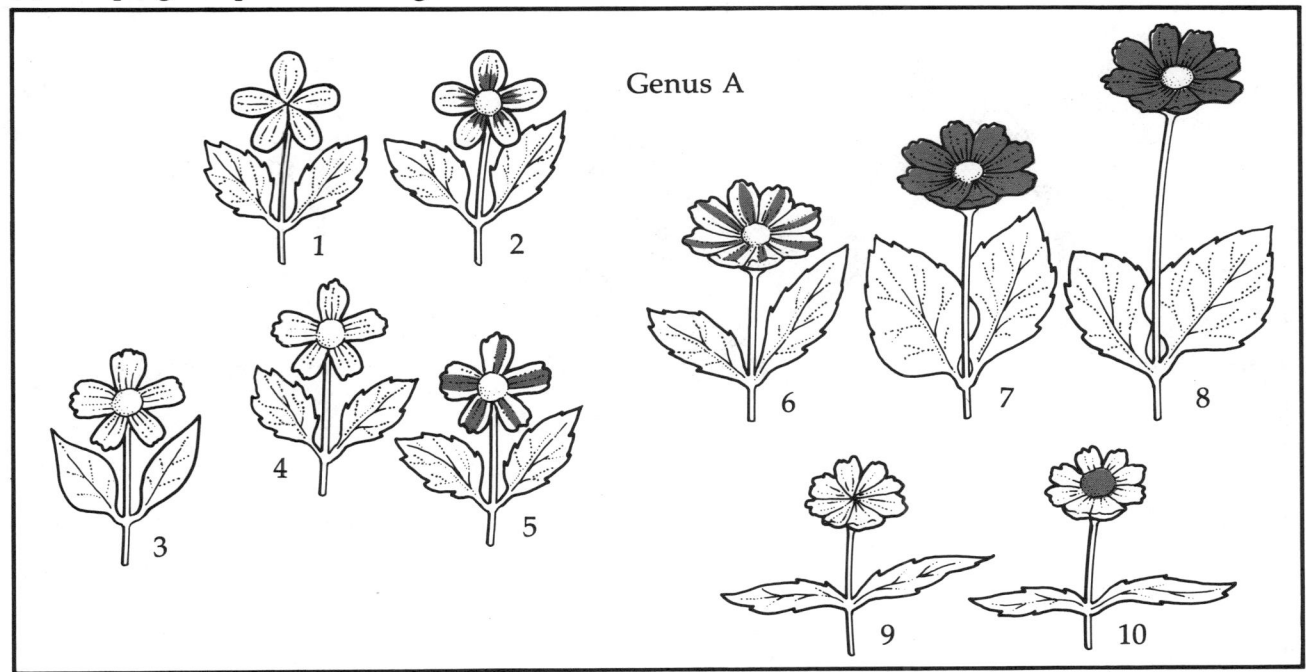

(*b*) Splitting 10 species into 3 genera

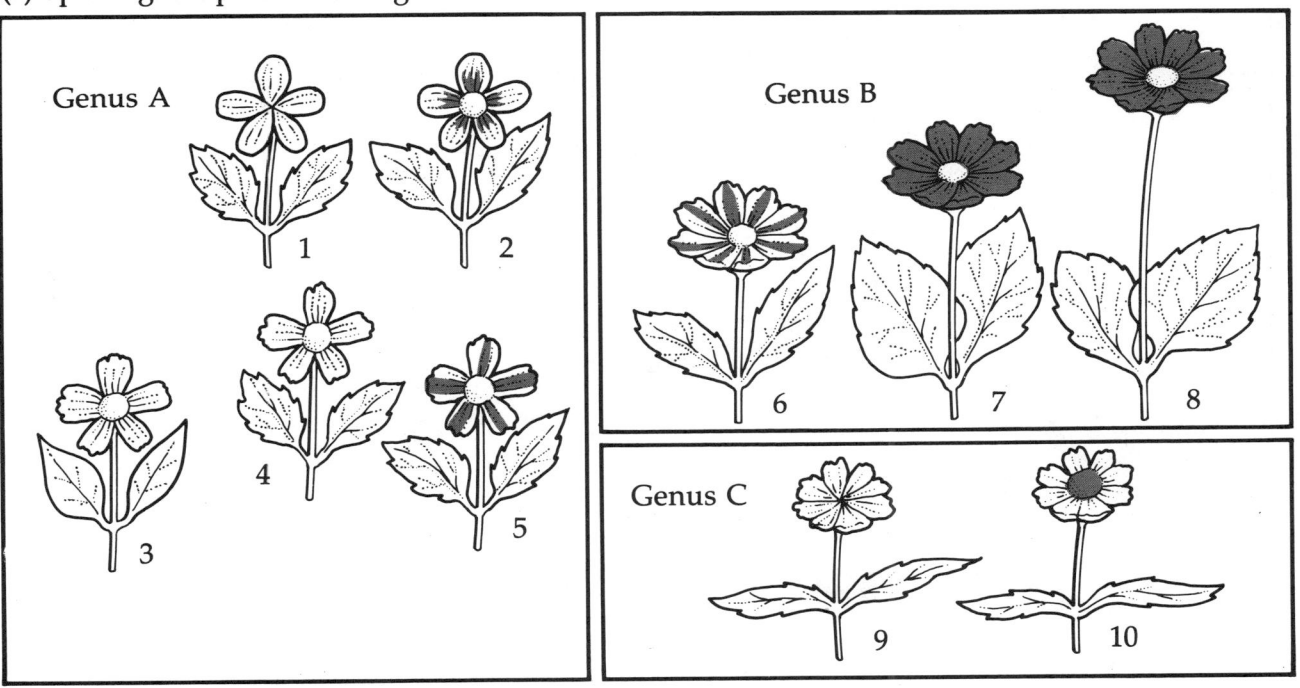

FIGURE 32.3   CLUSTERING OF IMAGINARY SPECIES INTO GENERA   (text page 798)

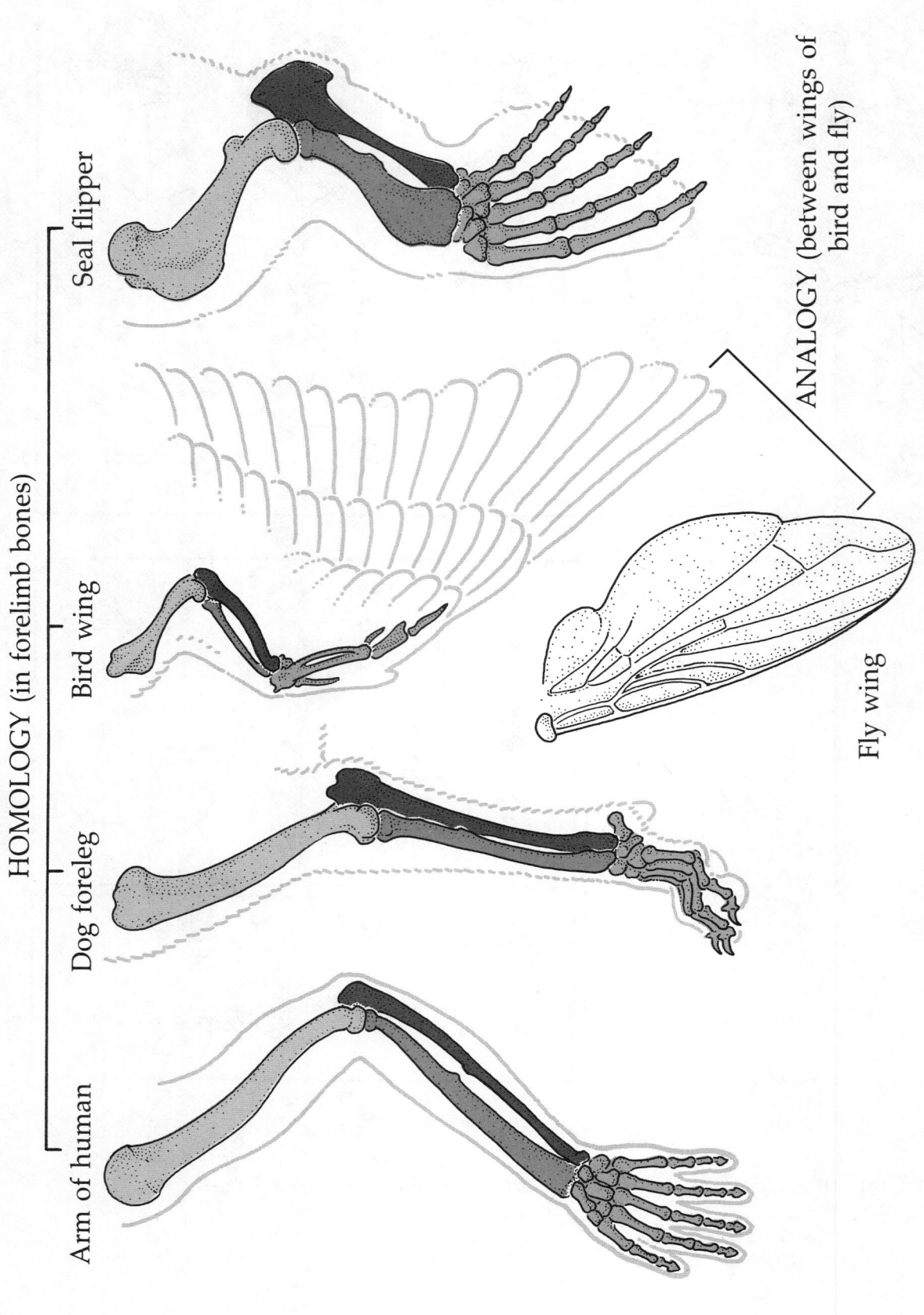

HOMOLOGY (in forelimb bones)

Arm of human   Dog foreleg   Bird wing   Seal flipper

Fly wing

ANALOGY (between wings of bird and fly)

FIGURE 32.13   HOMOLOGY AND ANALOGY   (text page 806)

From *Life: The Science of Biology.* Copyright © 1983 by William K. Purves and Gordon H. Orians.

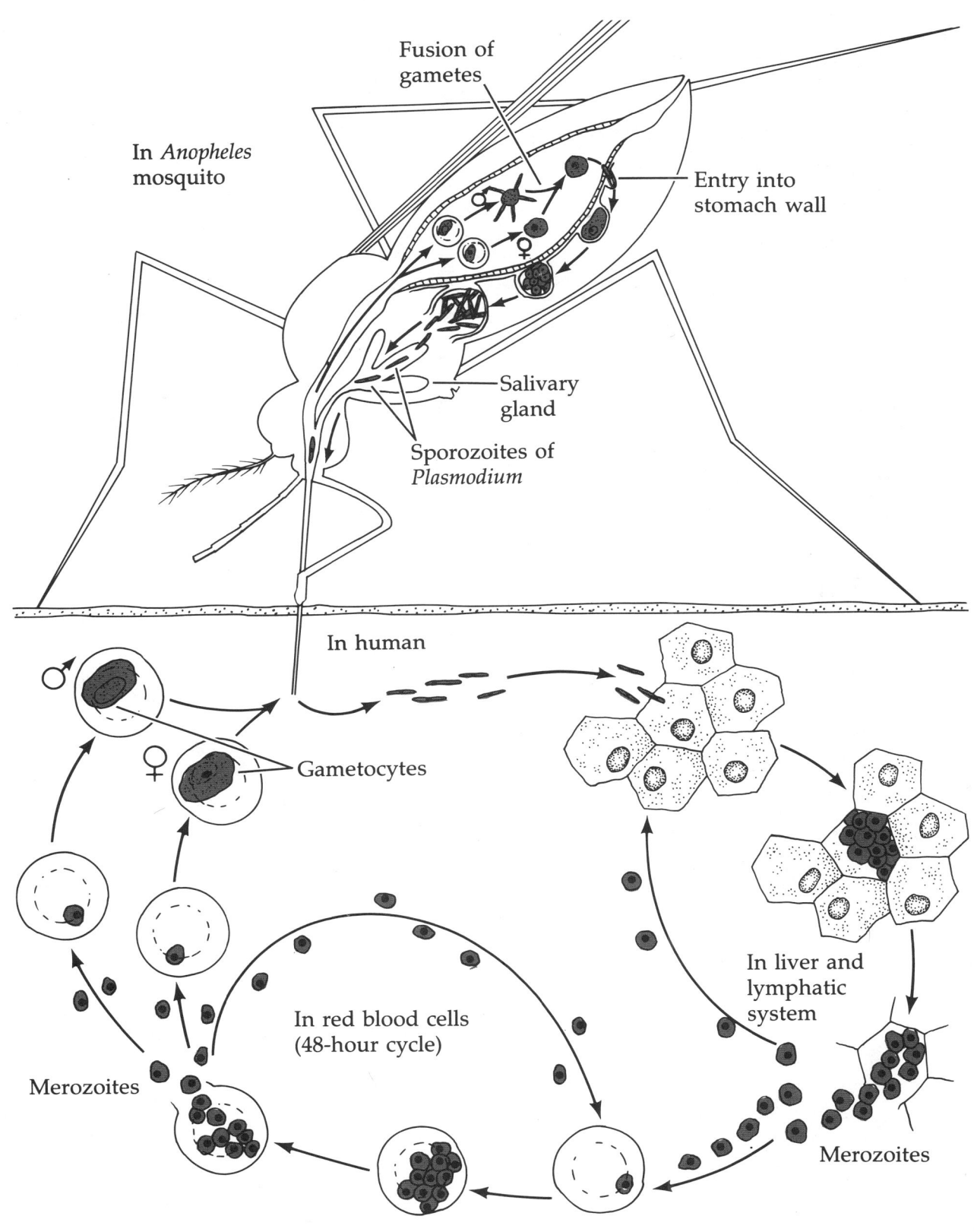

In *Anopheles* mosquito

Fusion of gametes

Entry into stomach wall

Salivary gland

Sporozoites of *Plasmodium*

In human

Gametocytes

In liver and lymphatic system

Merozoites

In red blood cells (48-hour cycle)

Merozoites

FIGURE 33.27   LIFE CYCLE OF *PLASMODIUM*   (text page 836)

From *Life: The Science of Biology*, Copyright © 1983 by William K. Purves and Gordon H. Orians.

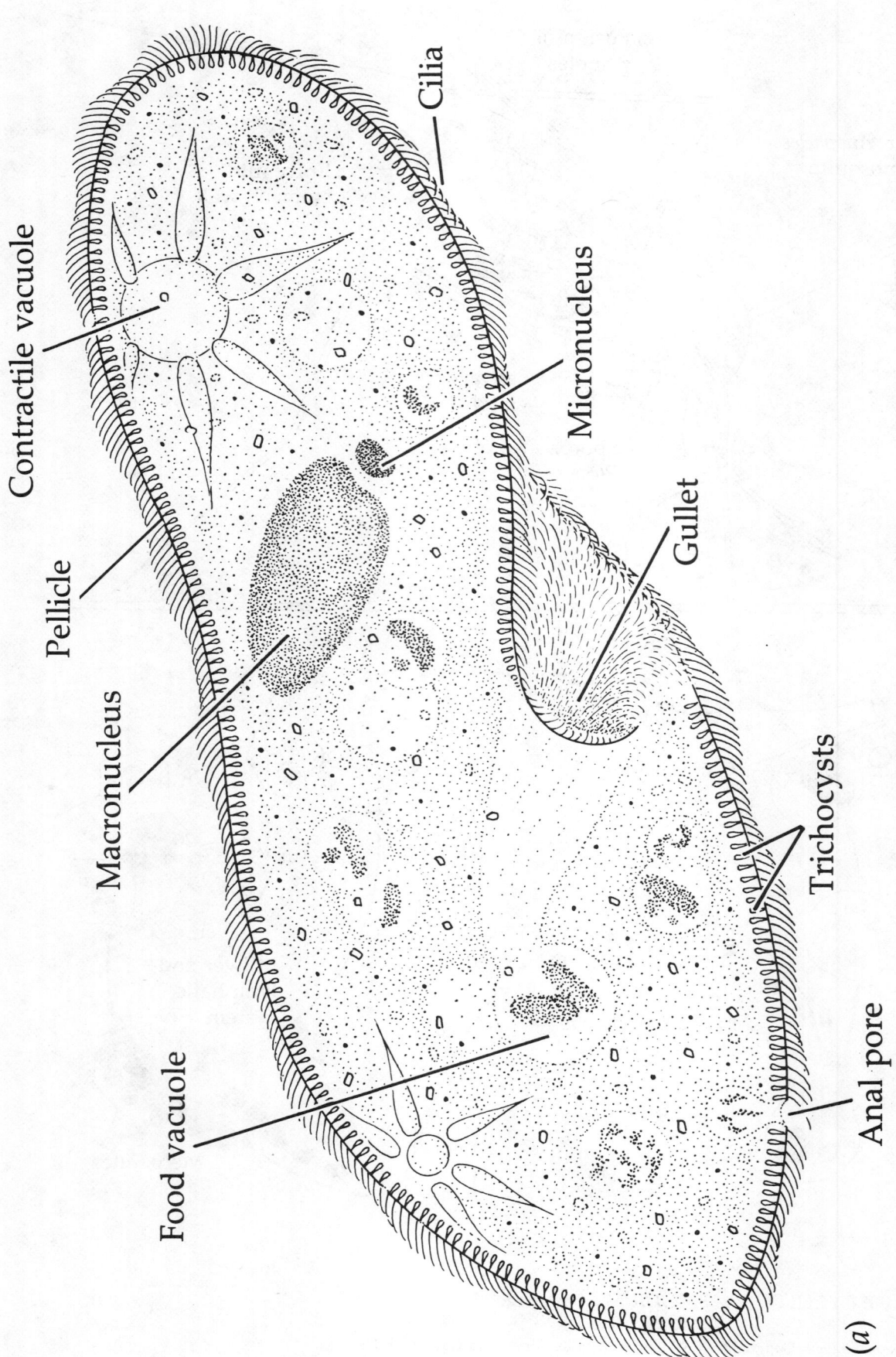

Contractile vacuole

Pellicle

Macronucleus

Food vacuole

Cilia

Micronucleus

Gullet

Trichocysts

Anal pore

(a)

FIGURE 33.29a ANATOMY OF *PARAMECIUM* (text page 838)

From *Life: The Science of Biology,* Copyright © 1983 by William K. Purves and Gordon H. Orians.

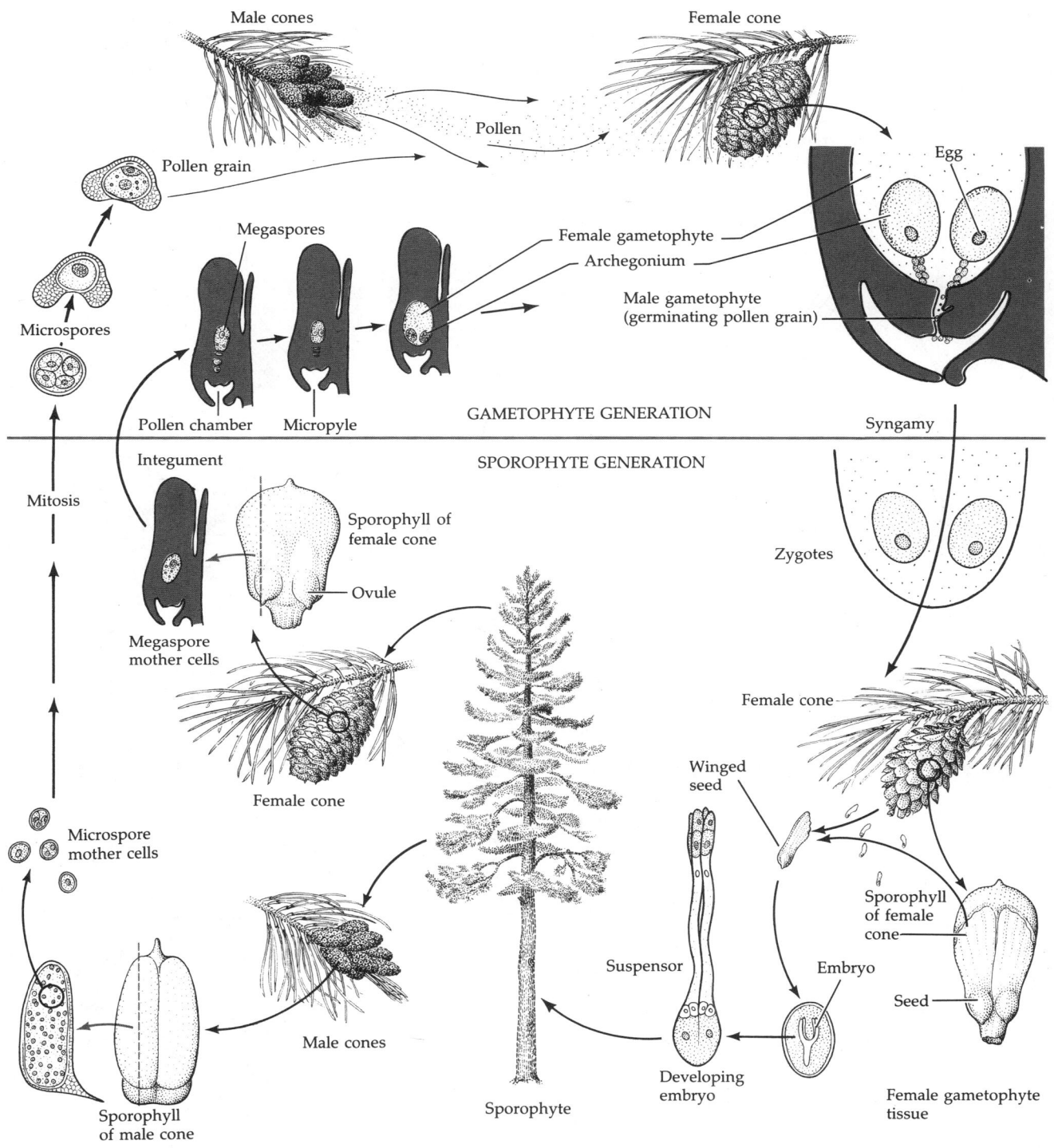

Male cones

Female cone

Pollen

Pollen grain

Egg

Microspores

Megaspores

Female gametophyte

Archegonium

Male gametophyte
(germinating pollen grain)

Pollen chamber

Micropyle

GAMETOPHYTE GENERATION

Syngamy

Mitosis

Integument

SPOROPHYTE GENERATION

Sporophyll of
female cone

Ovule

Megaspore
mother cells

Zygotes

Female cone

Female cone

Winged
seed

Microspore
mother cells

Sporophyll
of female
cone

Suspensor

Embryo

Seed

Male cones

Developing
embryo

Sporophyll
of male cone

Sporophyte

Female gametophyte
tissue

FIGURE 34.31   LIFE CYCLE OF A GYMNOSPERM   (text page 870)

From *Life: The Science of Biology*, Copyright © 1983 by William K. Purves and Gordon H. Orians.

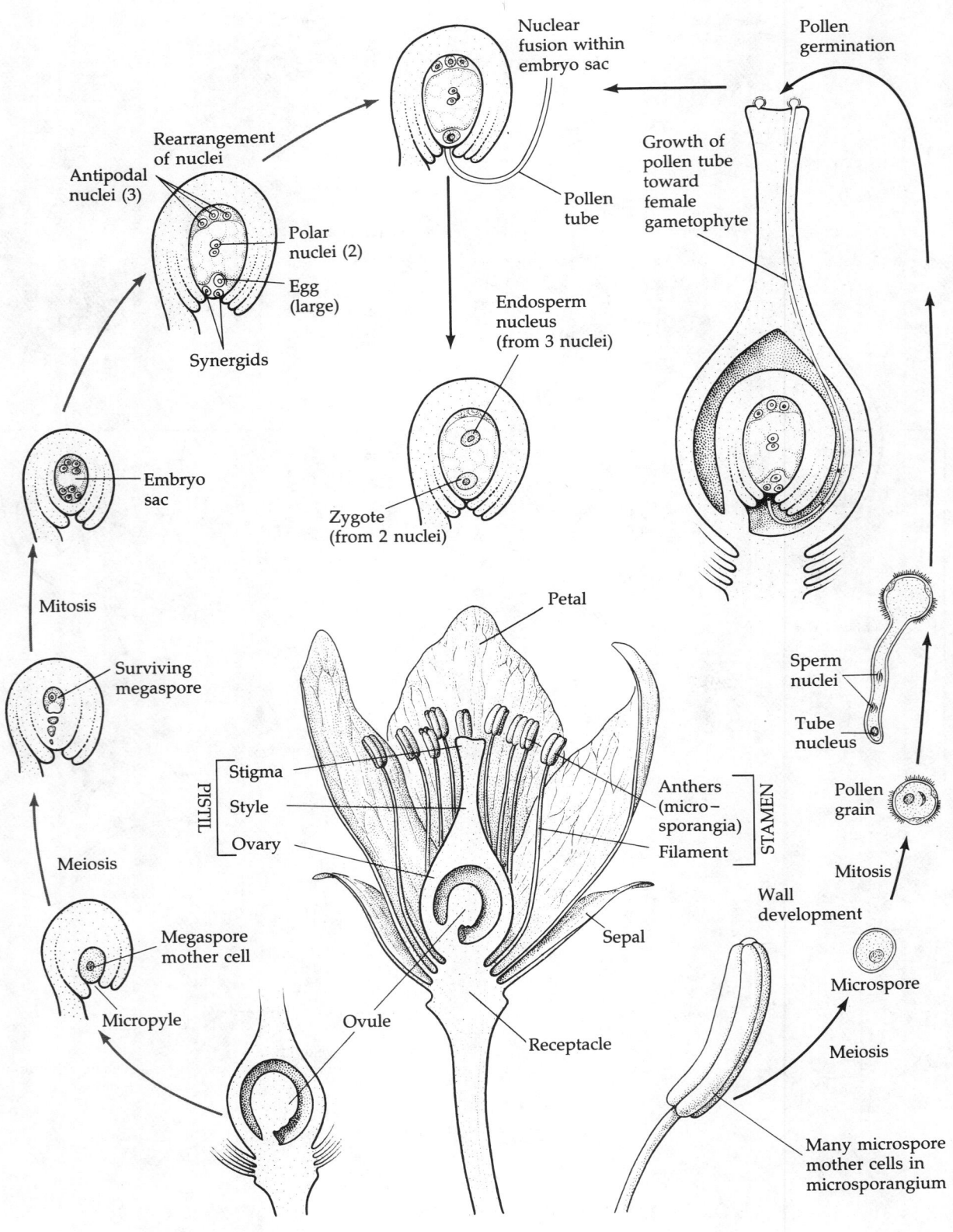

FIGURE 34.32a   GAMETOPHYTES IN AN IDEAL FLOWER   (text page 872)

From *Life: The Science of Biology*, Copyright © 1983 by William K. Purves and Gordon H. Orians.

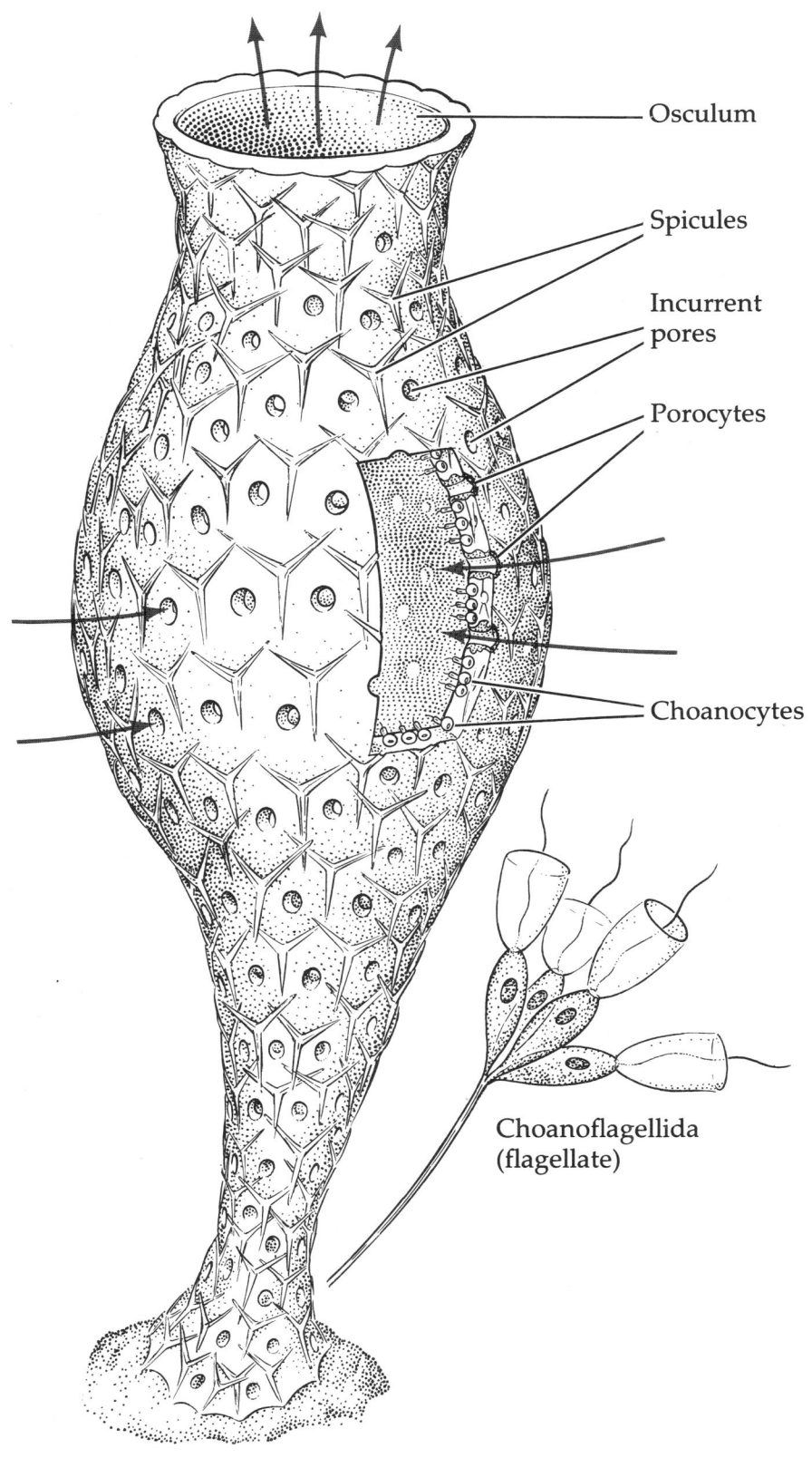

Osculum

Spicules

Incurrent
pores

Porocytes

Choanocytes

Choanoflagellida
(flagellate)

FIGURE 35.7a   SPONGES   (text page 887)

From *Life: The Science of Biology*, Copyright © 1983 by William K. Purves and Gordon H. Orians.

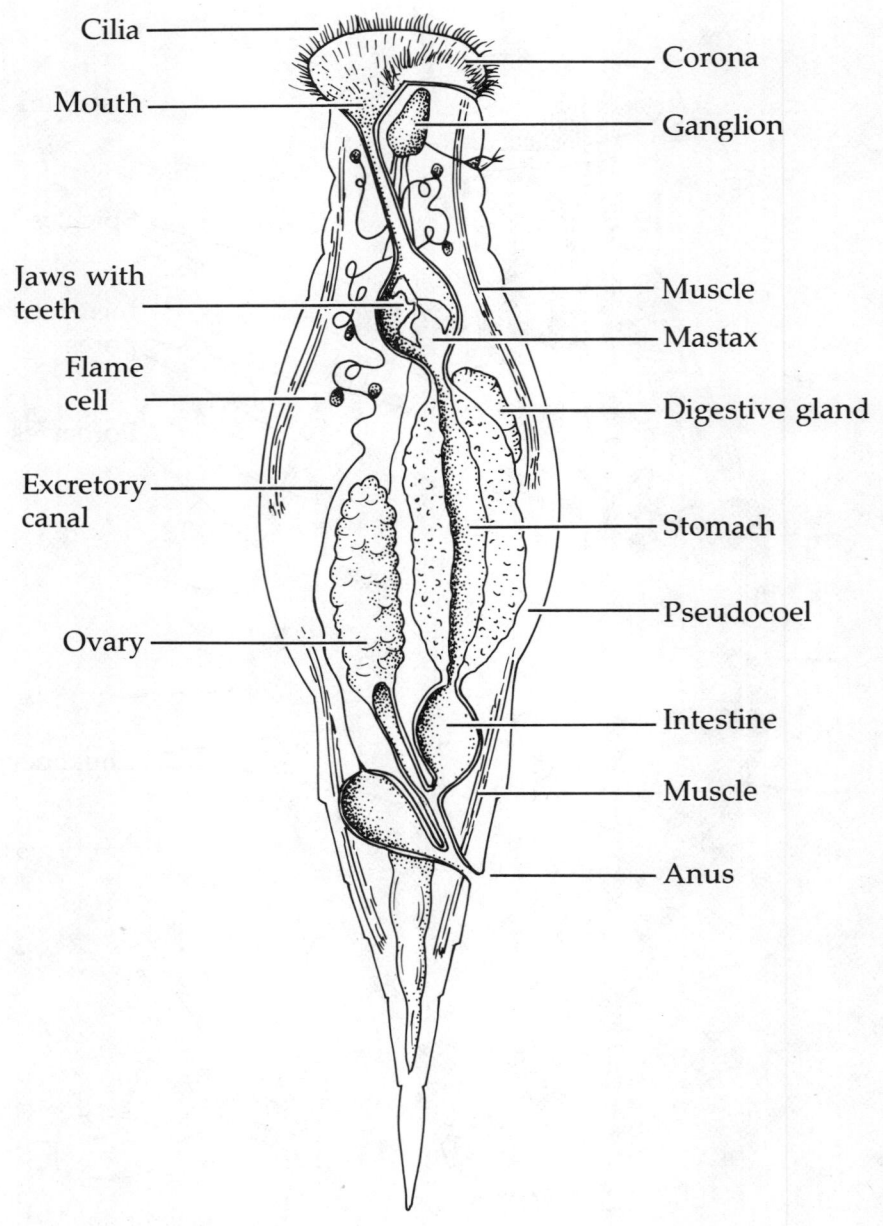

Cilia

Mouth

Jaws with
teeth

Flame
cell

Excretory
canal

Ovary

Corona

Ganglion

Muscle

Mastax

Digestive gland

Stomach

Pseudocoel

Intestine

Muscle

Anus

FIGURE 35.13a ROTIFERS (text page 893)

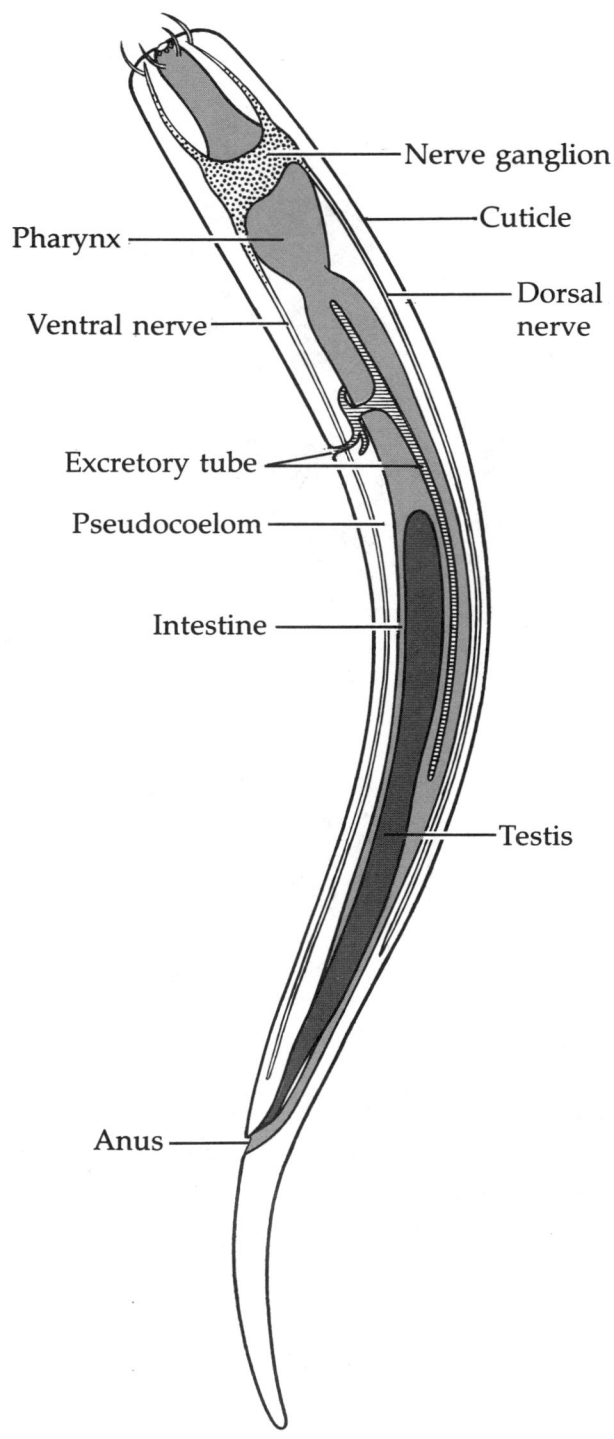

Nerve ganglion

Cuticle

Pharynx

Dorsal nerve

Ventral nerve

Excretory tube

Pseudocoelom

Intestine

Testis

Anus

FIGURE 35.14a   A REPRESENTATIVE NEMATODE   (text page 894)

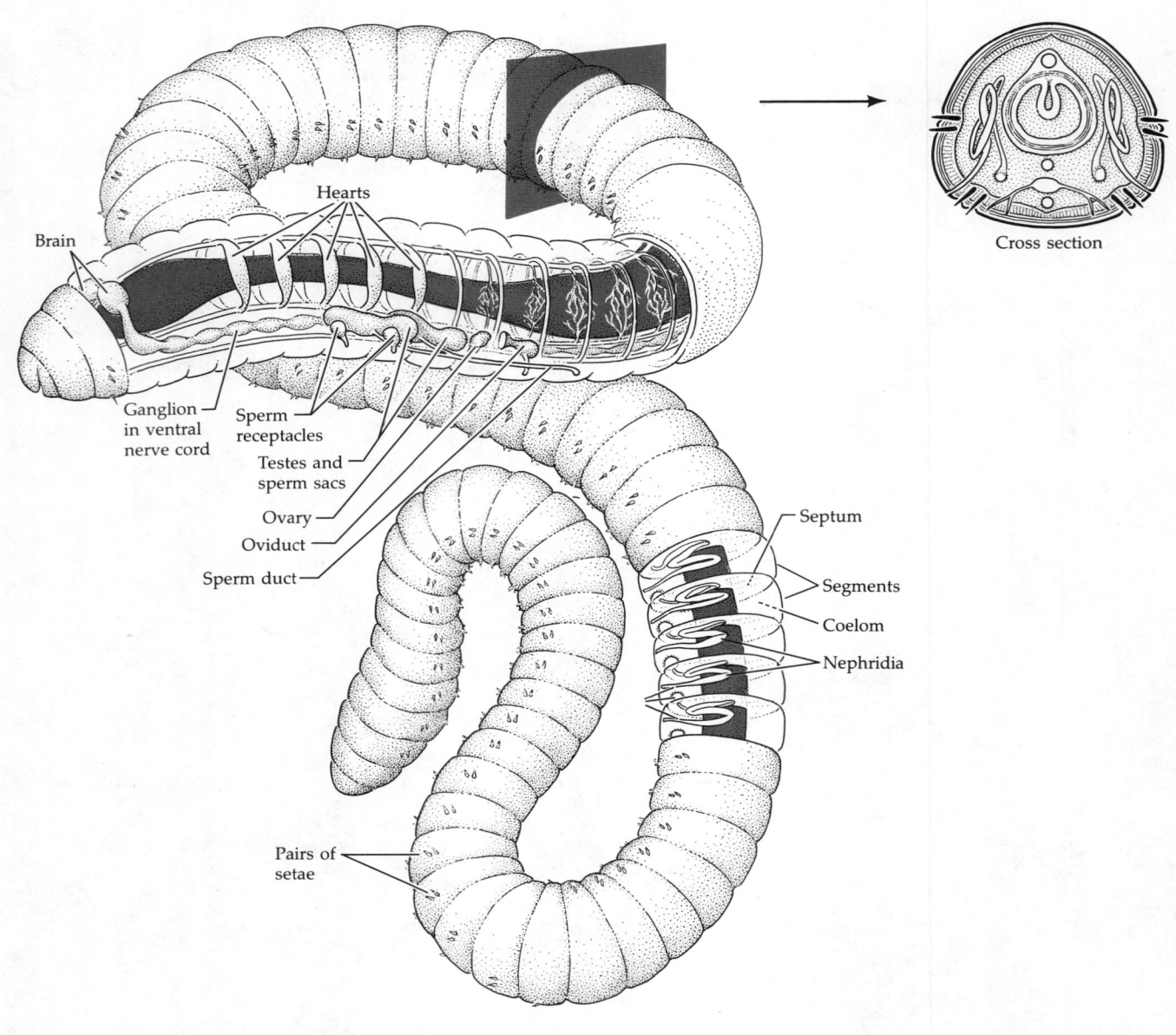

Brain

Hearts

Cross section

Ganglion
in ventral
nerve cord

Sperm
receptacles

Testes and
sperm sacs

Ovary

Oviduct

Sperm duct

Septum

Segments

Coelom

Nephridia

Pairs of
setae

FIGURE 35.16a   ANNELIDS   (text page 896)

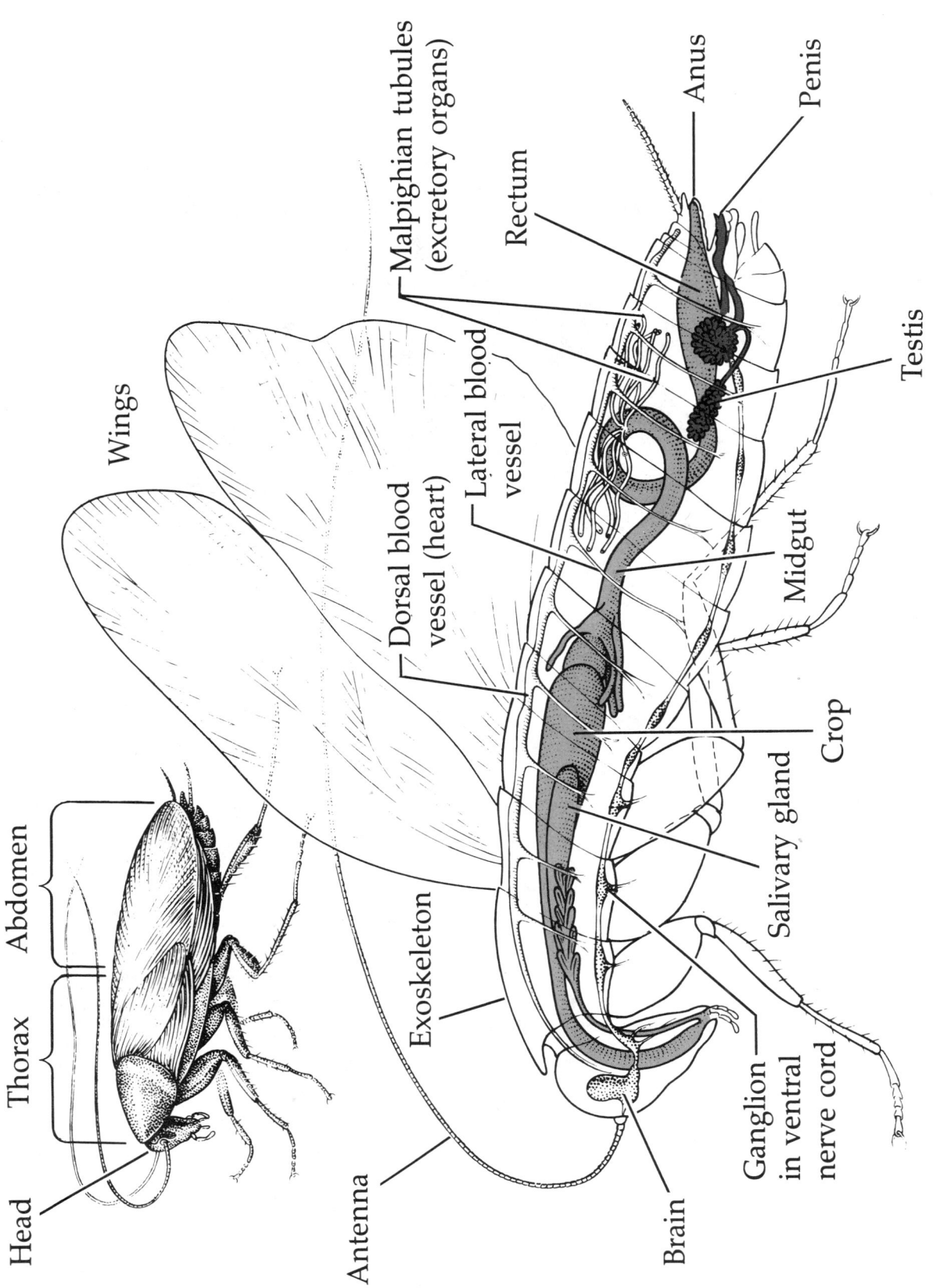

Head

Thorax

Abdomen

Wings

Antenna

Exoskeleton

Malpighian tubules (excretory organs)

Rectum

Anus

Penis

Dorsal blood vessel (heart)

Lateral blood vessel

Testis

Midgut

Crop

Salivary gland

Ganglion in ventral nerve cord

Brain

FIGURE 35.19   ARTHROPOD ANATOMY   (text page 900)

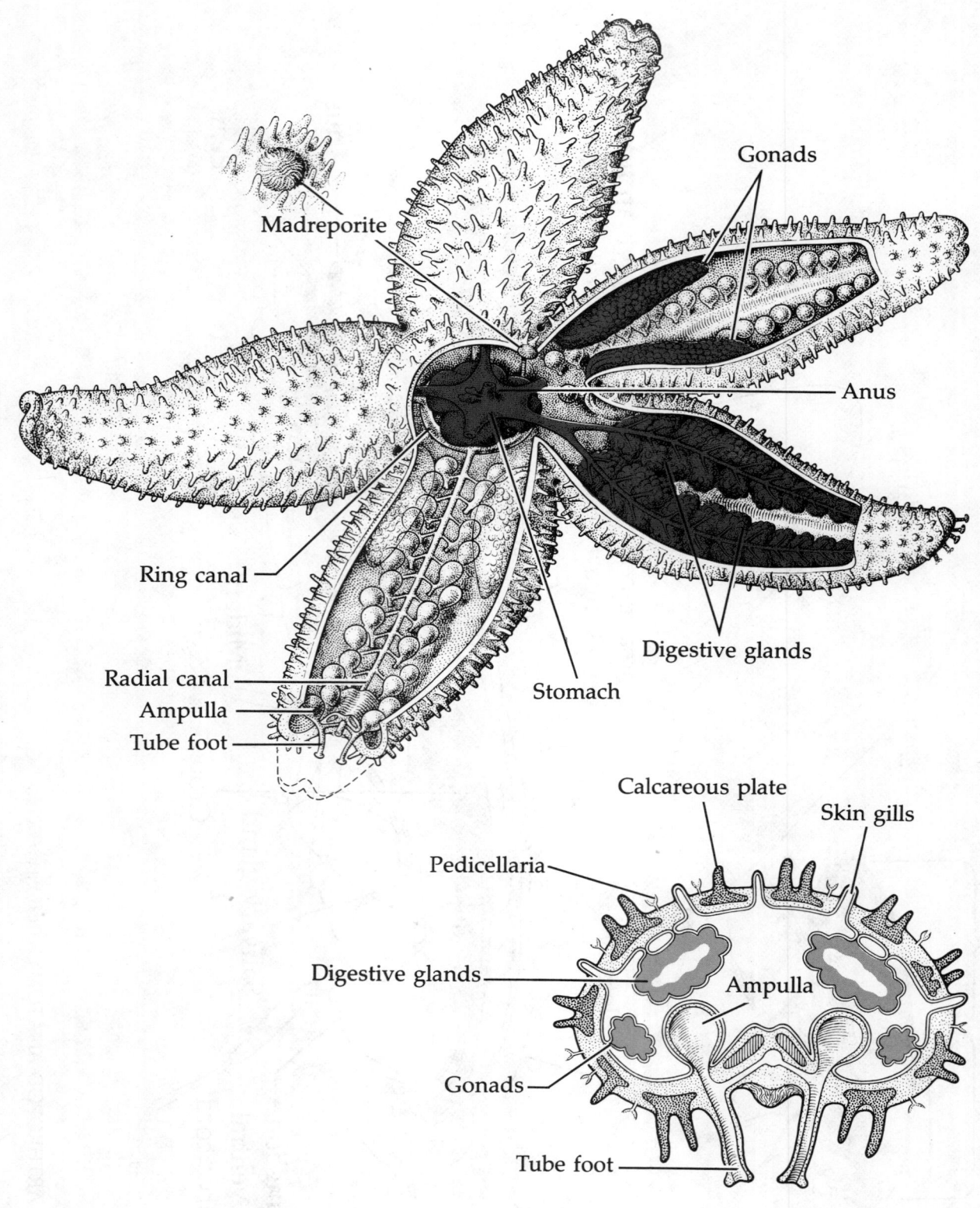

Gonads

Madreporite

Anus

Ring canal

Digestive glands

Radial canal

Ampulla

Stomach

Tube foot

Calcareous plate

Skin gills

Pedicellaria

Digestive glands

Ampulla

Gonads

Tube foot

FIGURE 35.27   STRUCTURE OF A SEASTAR   (text page 908)

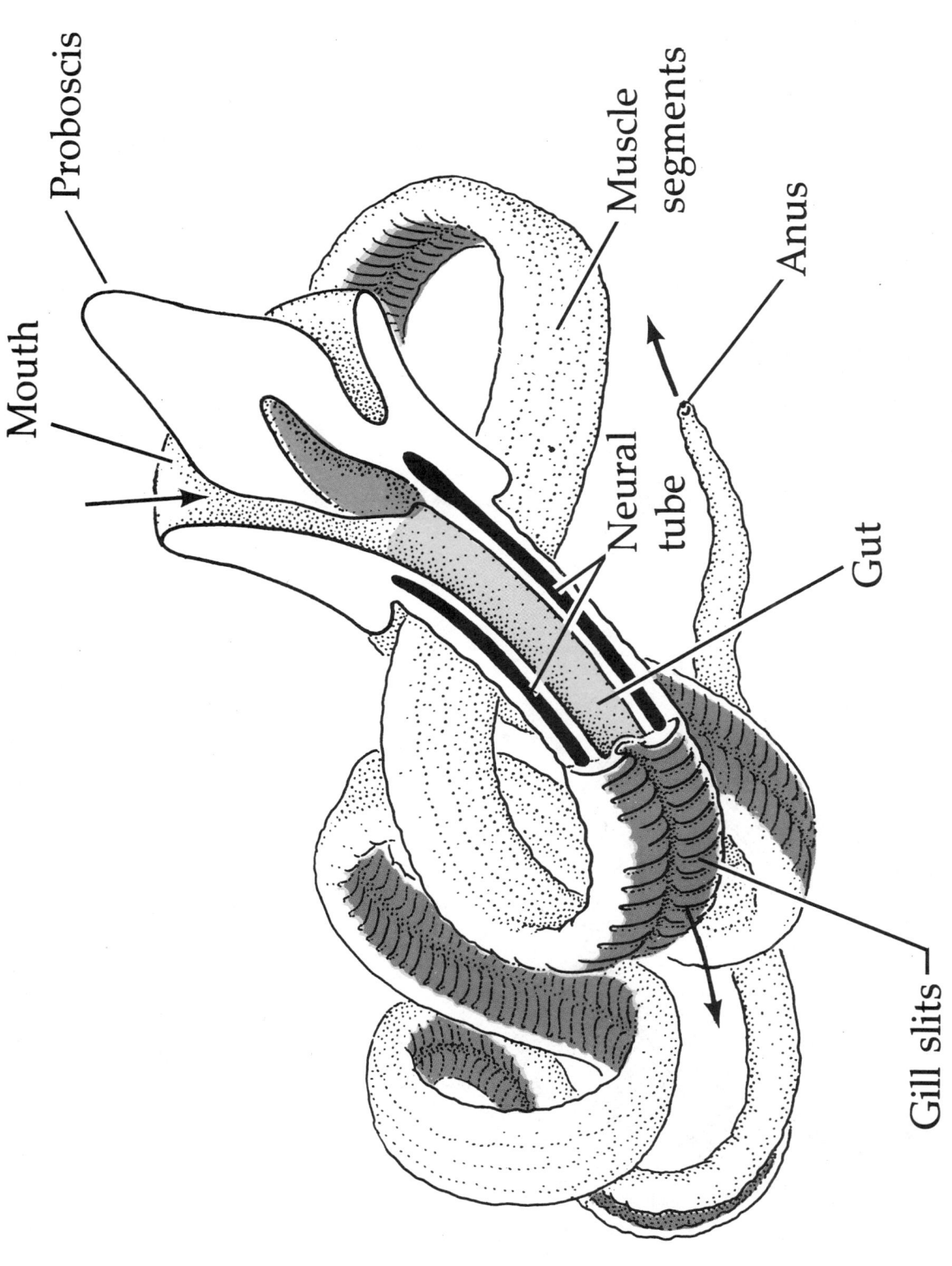

Mouth

Proboscis

Muscle segments

Anus

Neural tube

Gut

Gill slits

FIGURE 35.28   AN ACORN WORM   (text page 909)

From *Life: The Science of Biology*. Copyright © 1983 by William K. Purves and Gordon H. Orians.

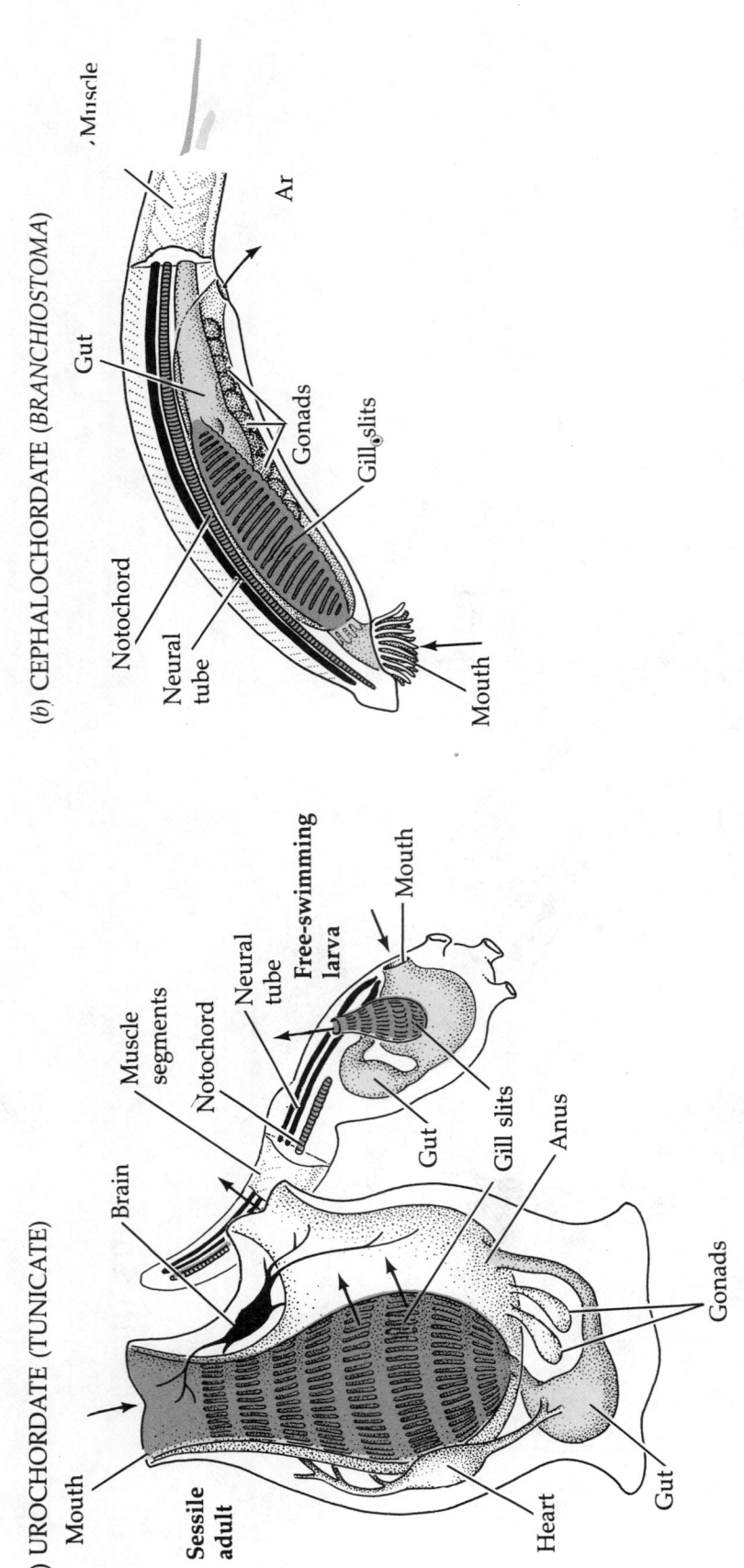

(b) CEPHALOCHORDATE (*BRANCHIOSTOMA*)

Muscle

Ar

Gut

Gonads

Gill slits

Notochord

Neural tube

Mouth

(a) UROCHORDATE (TUNICATE)

Muscle segments

Neural tube

Notochord

Free-swimming larva

Mouth

Brain

Gut

Gill slits

Anus

Gonads

Sessile adult

Mouth

Heart

Gut

FIGURE 35.29a & b   INVERTEBRATE CHORDATES   (text page 910)

Jawless fishes (agnaths)

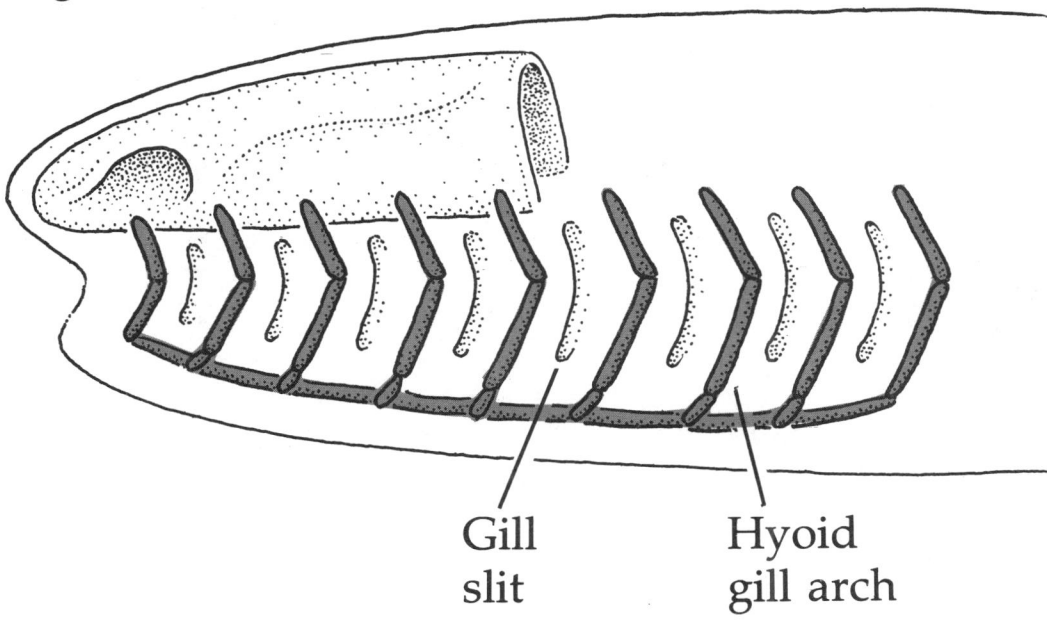

Gill
slit

Hyoid
gill arch

Primitive jawed fishes (placoderms)

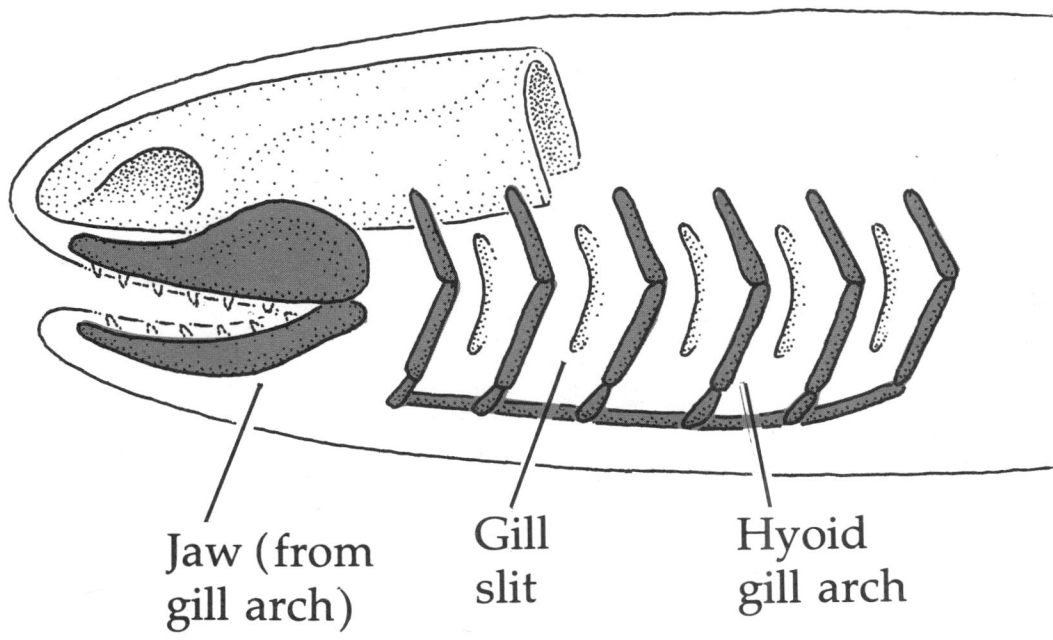

Jaw (from
gill arch)

Gill
slit

Hyoid
gill arch

FIGURE 35.32   JAWS FROM HYOID GILL ARCHES   (text page 912)

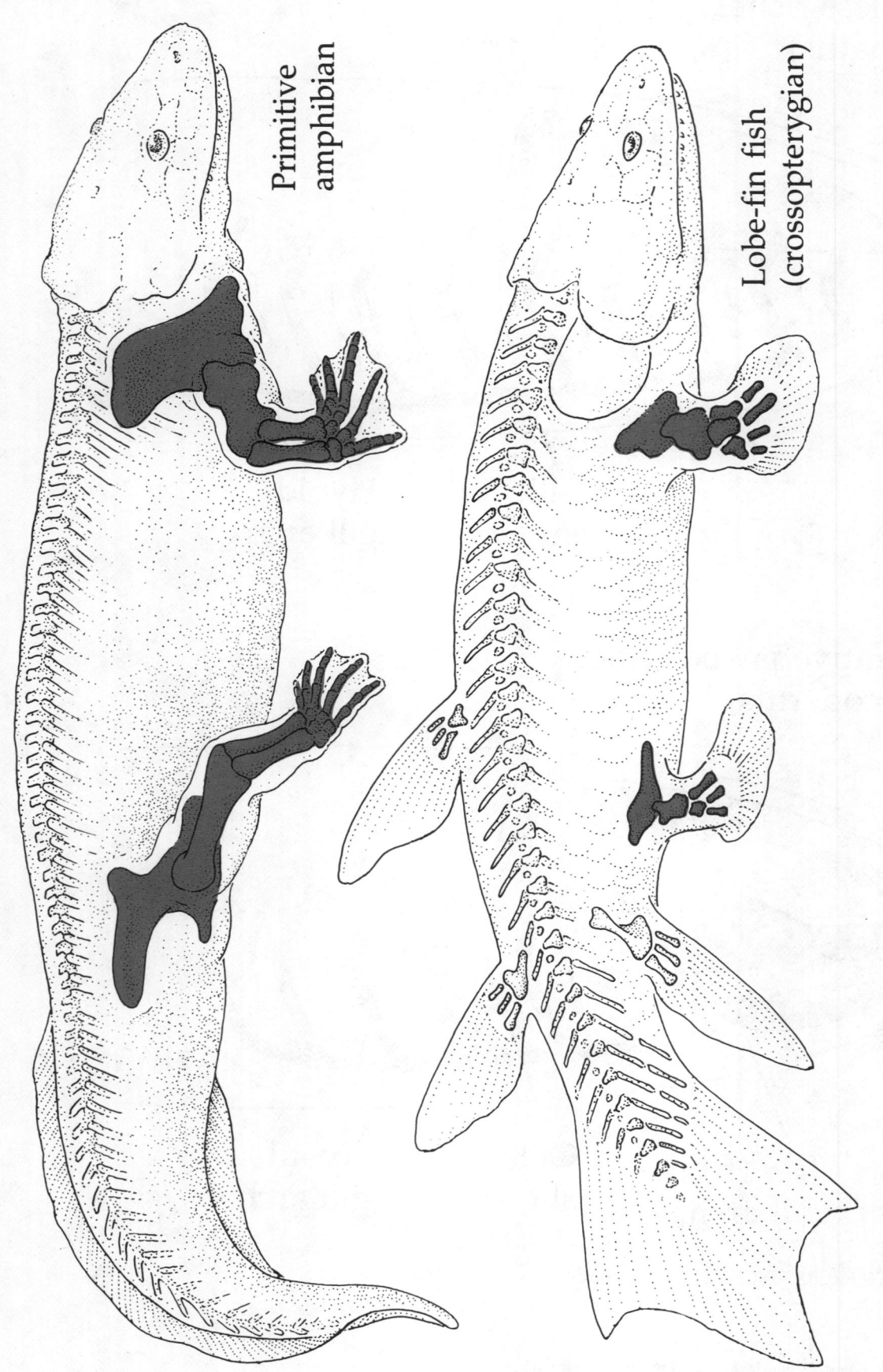

Primitive amphibian

Lobe-fin fish (crossopterygian)

FIGURE 35.35b   LEGS FROM FINS   (text page 915)

From *Life: The Science of Biology.* Copyright © 1983 by William K. Purves and Gordon H. Orians.

Ornithischian

Pterosaur

Saurischian

Ichthyosaur

FIGURE 35.38   MESOZOIC REPTILES   (text page 918)

From *Life: The Science of Biology*, Copyright © 1983 by William K. Purves and Gordon H. Orians.

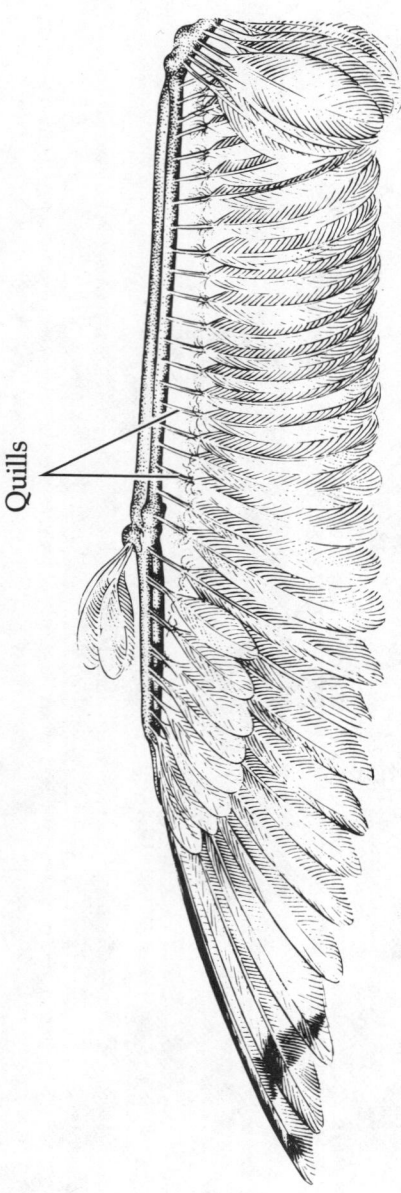

Quills

(a) Arrangement of feathers in wing

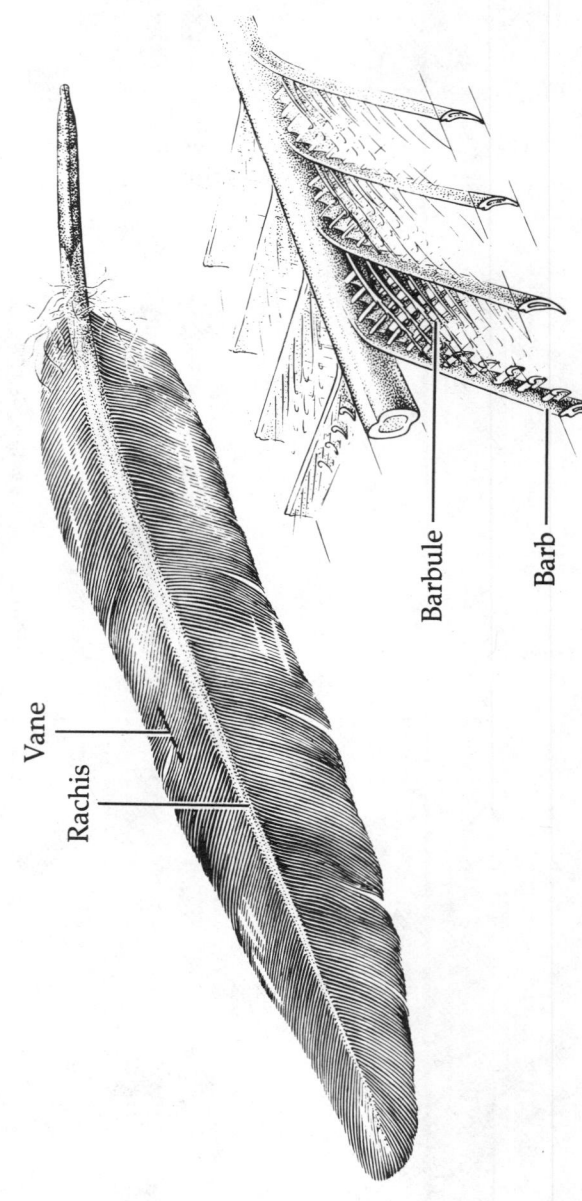

Vane

Rachis

Barbule

Barb

(b) Anatomy of a feather

FIGURE 35.40   FEATHERS AND WINGS   (text page 920)

From *Life: The Science of Biology.* Copyright © 1983 by William K. Purves and Gordon H. Orians.

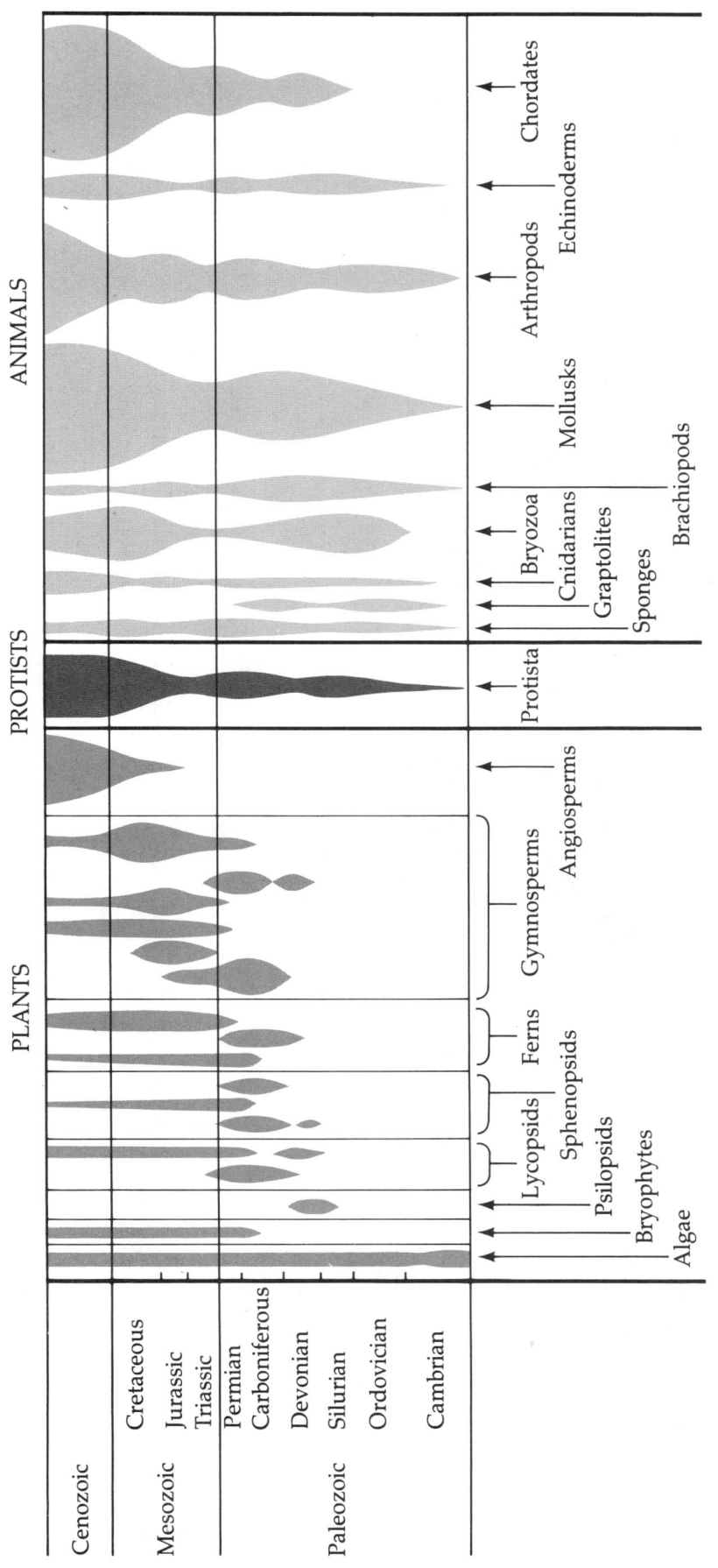

FIGURE 38.4   BROAD FEATURES OF THE FOSSIL RECORD   (text page 974)

From *Life: The Science of Biology*, Copyright © 1983 by William K. Purves and Gordon H. Orians.

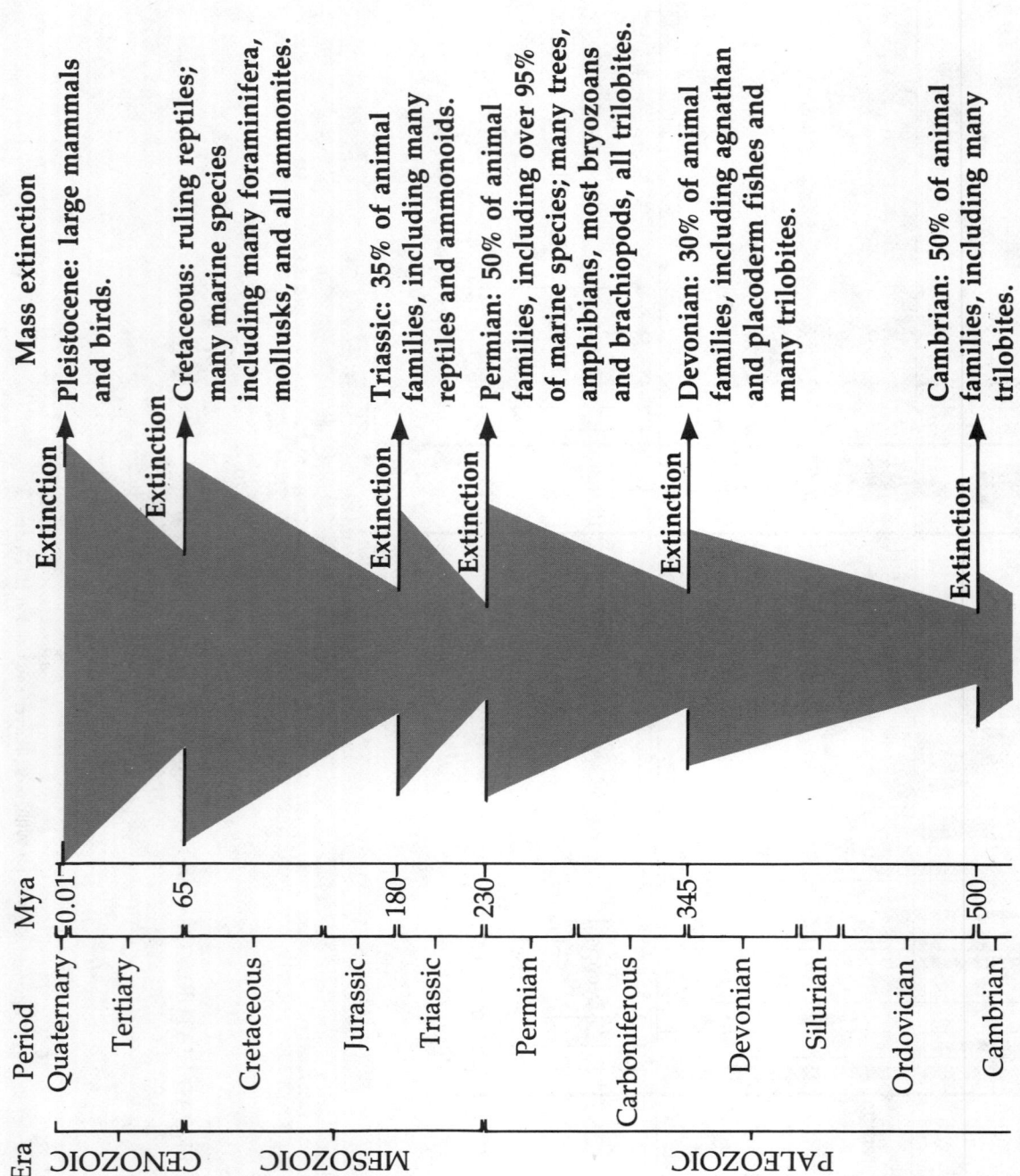

FIGURE 38.12   THE SIX MASS EXTINCTIONS   (text page 982)

From Life: The Science of Biology. Copyright © 1983 by William K. Purves and Gordon H. Orians.

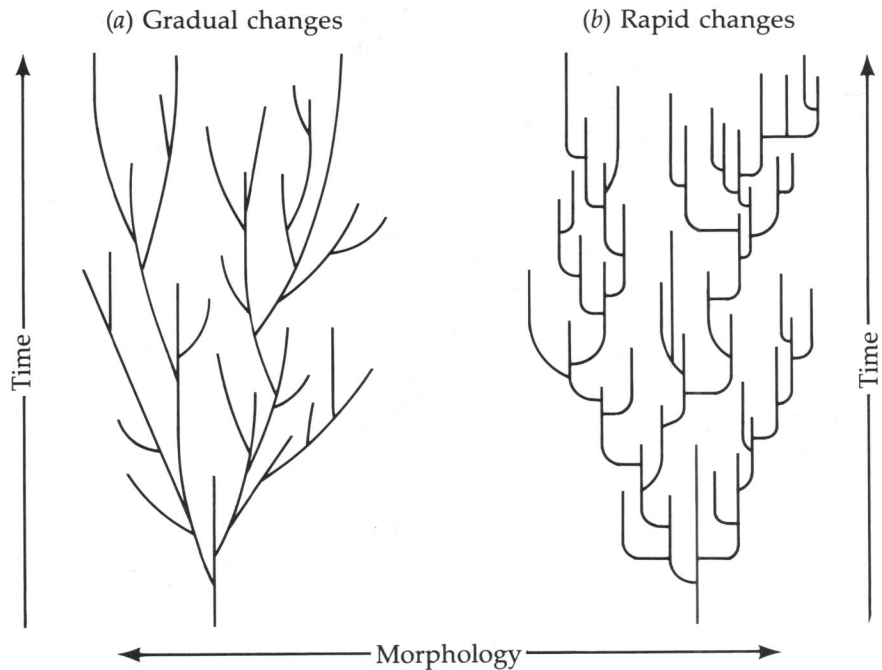

(a) Gradual changes   (b) Rapid changes

Time

Time

Morphology

FIGURE 38.19   TWO MODELS OF ADAPTIVE RADIATION   (text page 989)

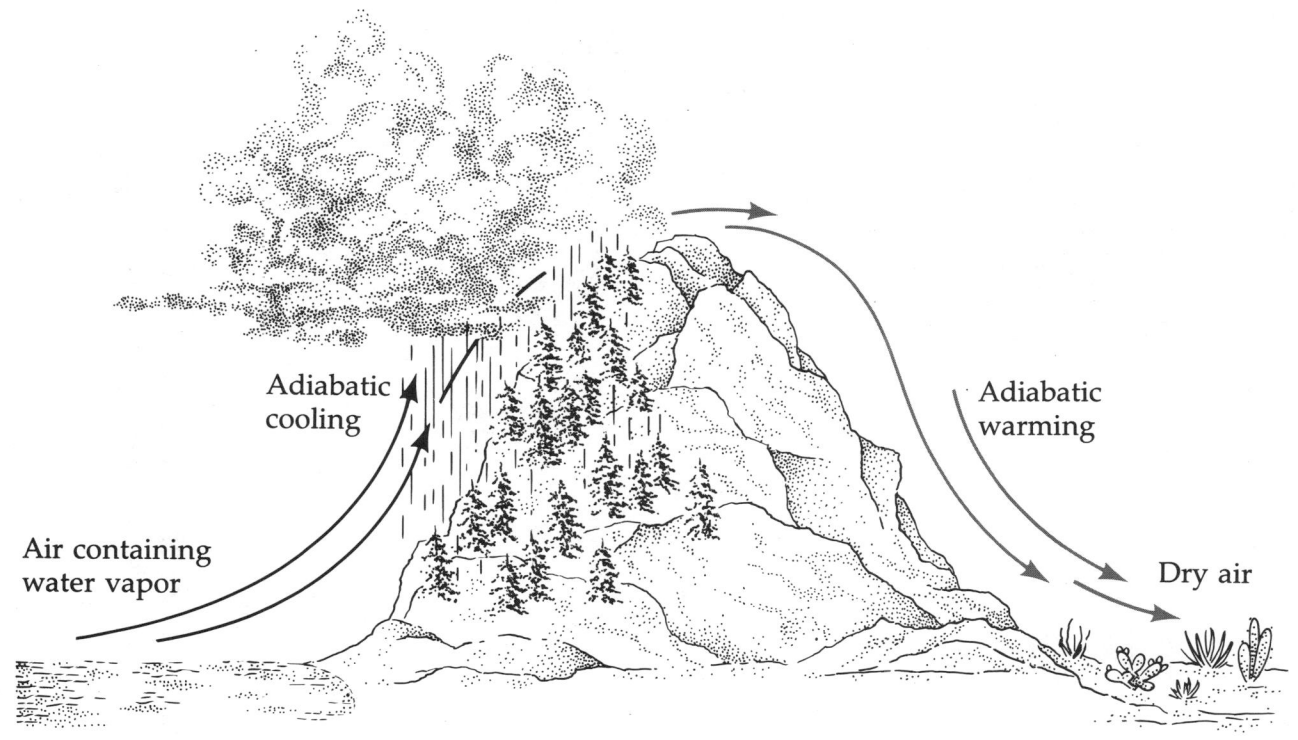

Adiabatic
cooling

Adiabatic
warming

Air containing
water vapor

Dry air

FIGURE 39.2   ADIABATIC CHANGES IN AIR TEMPERATURE   (text page 996)

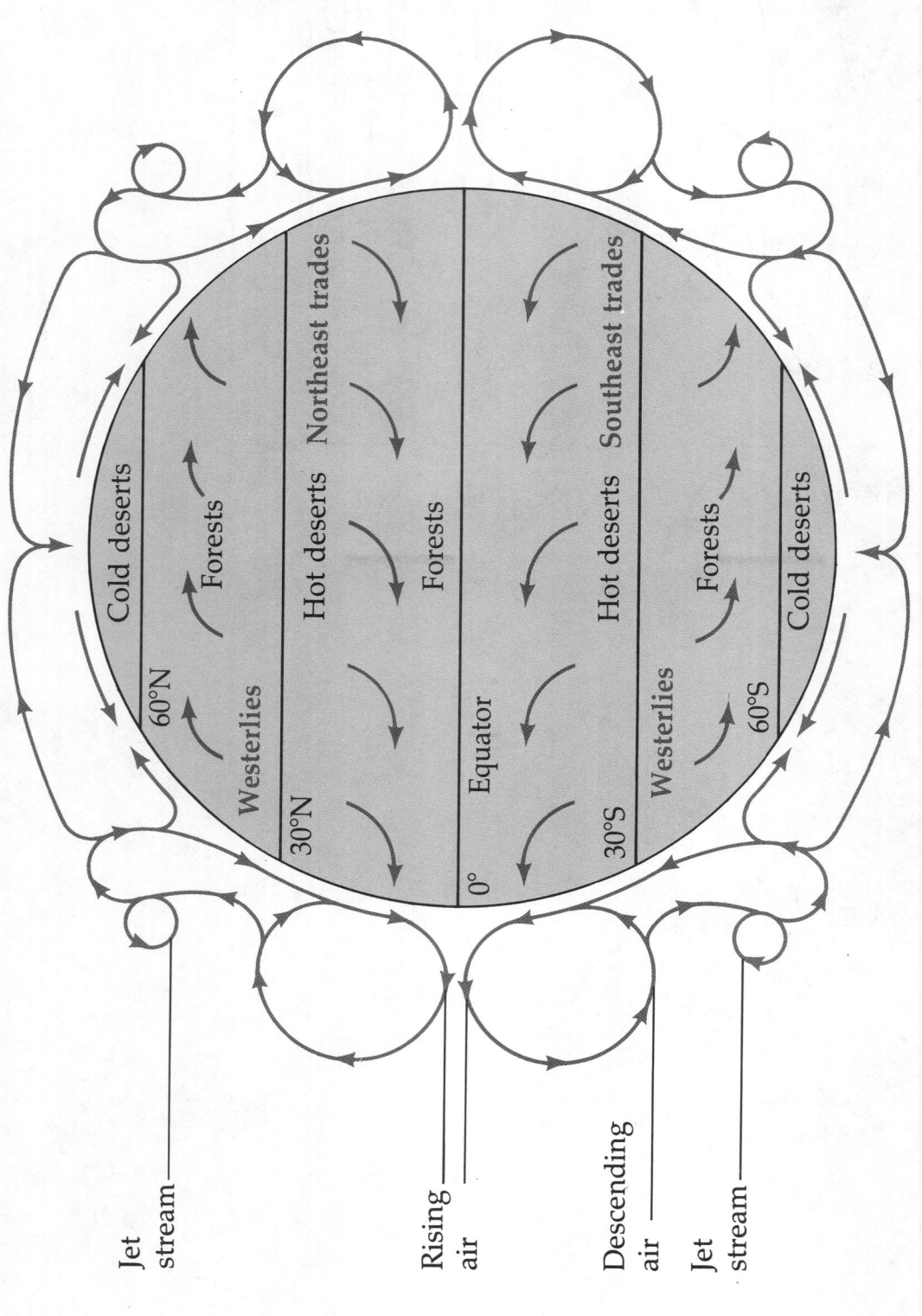

FIGURE 39.3   CIRCULATION OF EARTH'S ATMOSPHERE   (text page 997)

After R. MacArthur, *Geographical Ecology,* Harper and Row, 1972.

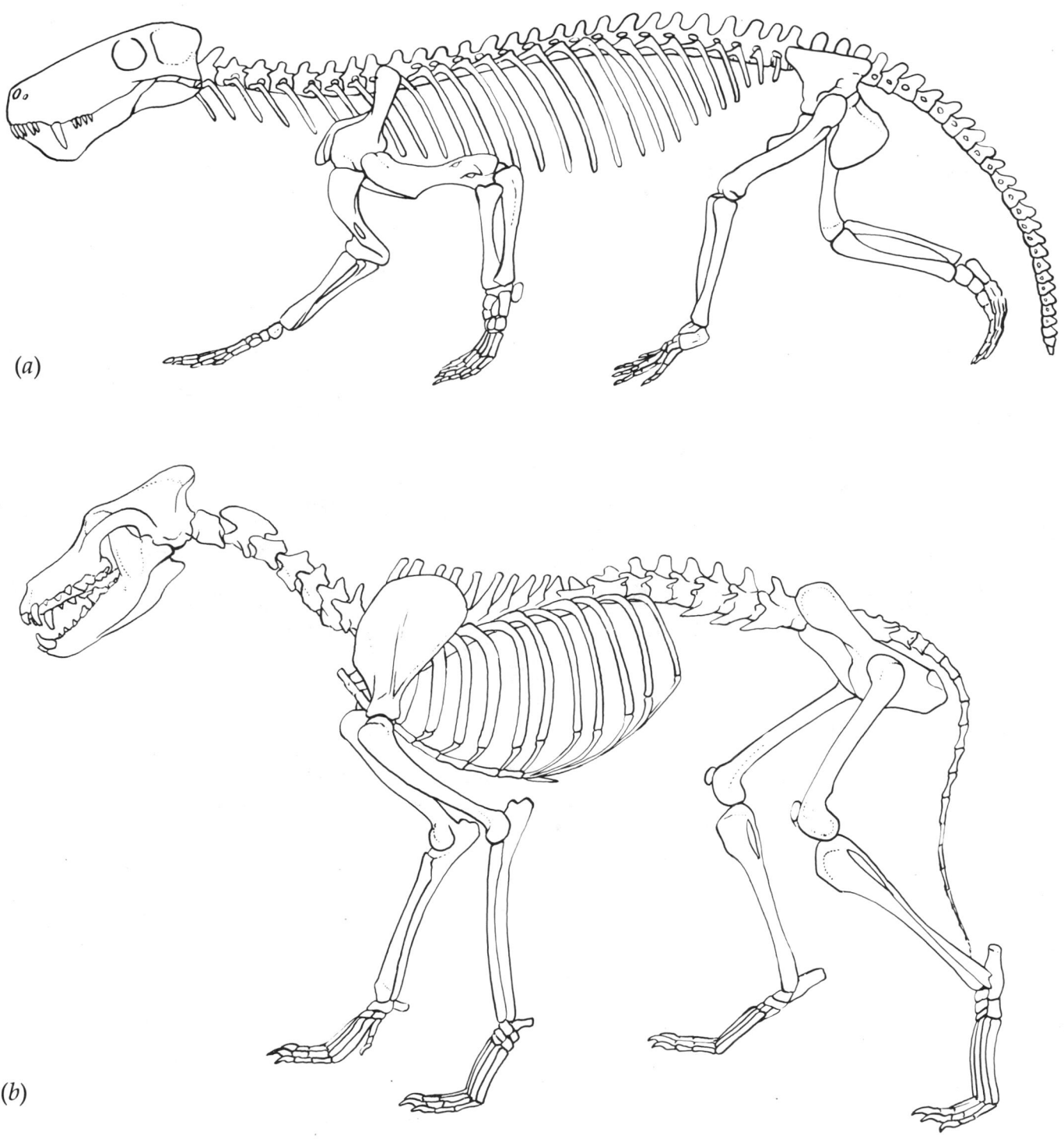

(a)

(b)

FIGURE 35.41   THERAPSID AND MAMMAL SKELETONS   (text page 921)
From T. Vaugn, *Mammalogy*, W.B. Saunders, 1973.

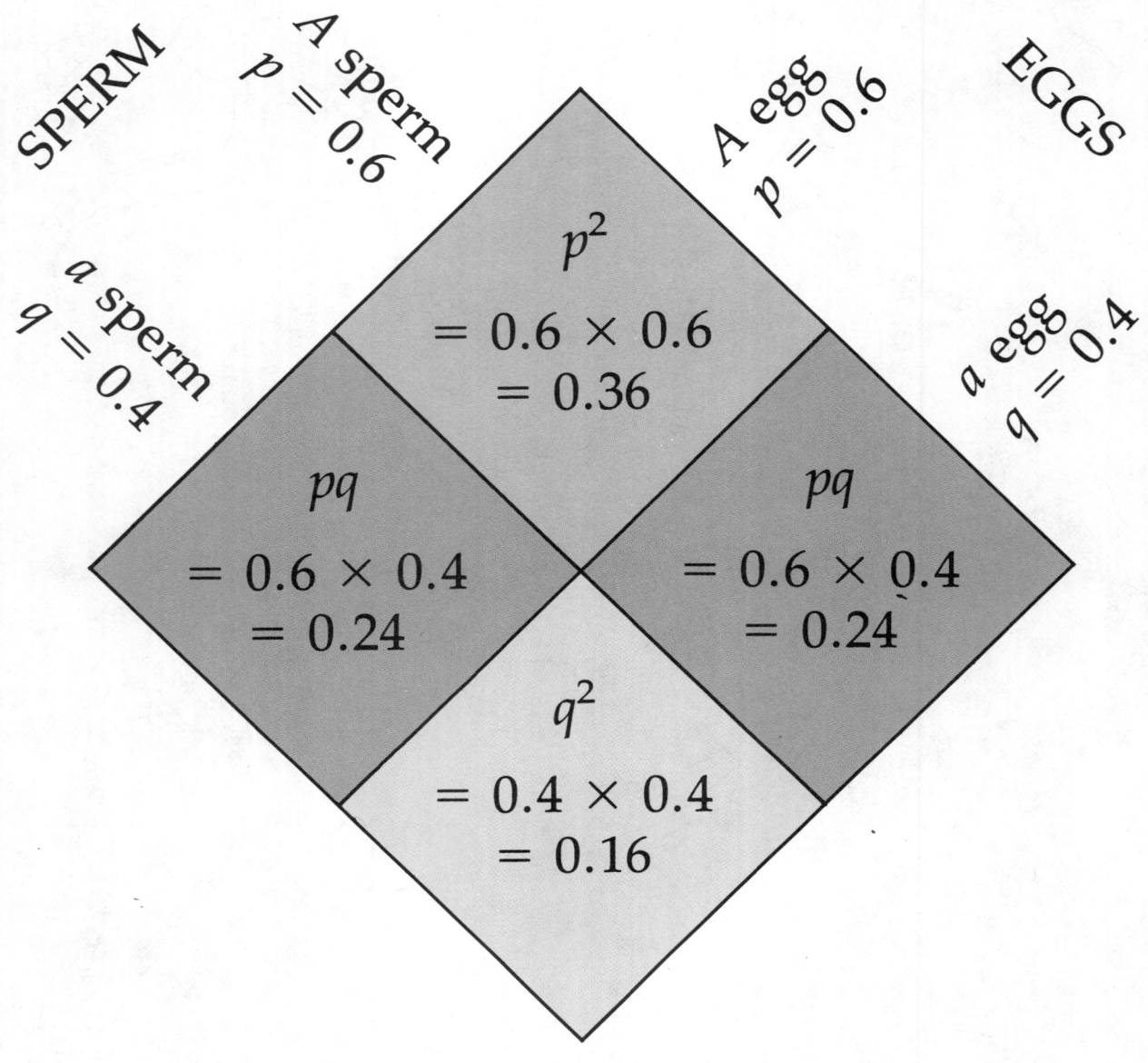

SPERM

A sperm
$p = 0.6$

a sperm
$q = 0.4$

EGGS

A egg
$p = 0.6$

a egg
$q = 0.4$

p^2
$= 0.6 \times 0.6$
$= 0.36$

pq
$= 0.6 \times 0.4$
$= 0.24$

pq
$= 0.6 \times 0.4$
$= 0.24$

q^2
$= 0.4 \times 0.4$
$= 0.16$

FIGURE 36.2   THE HARDY-WEINBERG LAW   (text page 933)

(a) Stabilizing selection

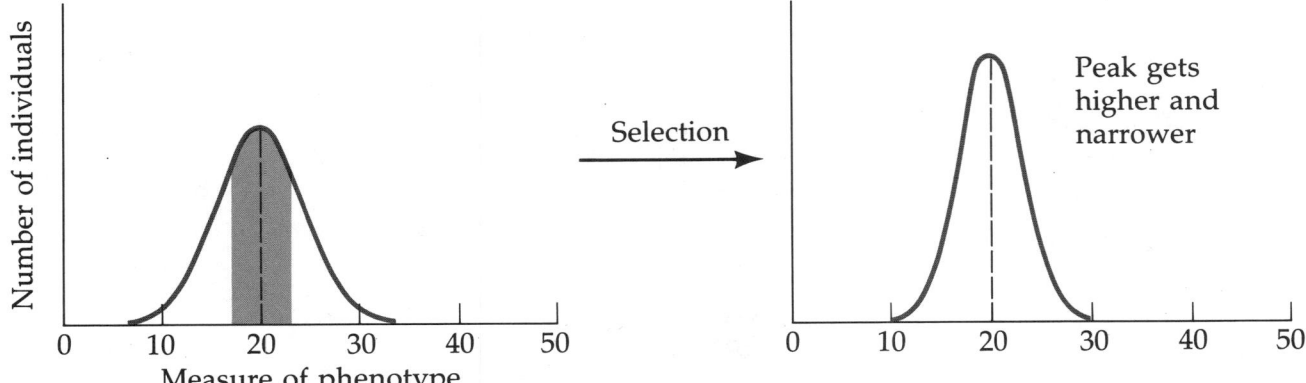

Selection

Peak gets
higher and
narrower

Number of individuals

Measure of phenotype

(b) Directional selection

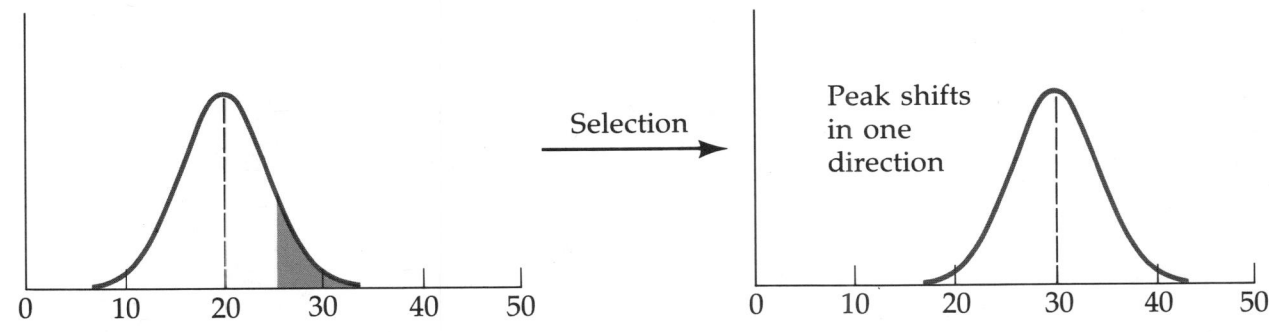

Selection

Peak shifts
in one
direction

(c) Disruptive selection

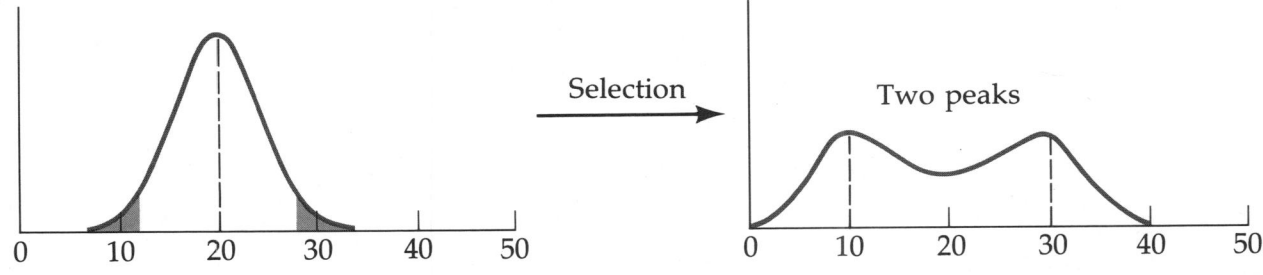

Selection

Two peaks

FIGURE 36.8   NATURAL SELECTION AND POPULATION VARIABILITY   (text page 940)

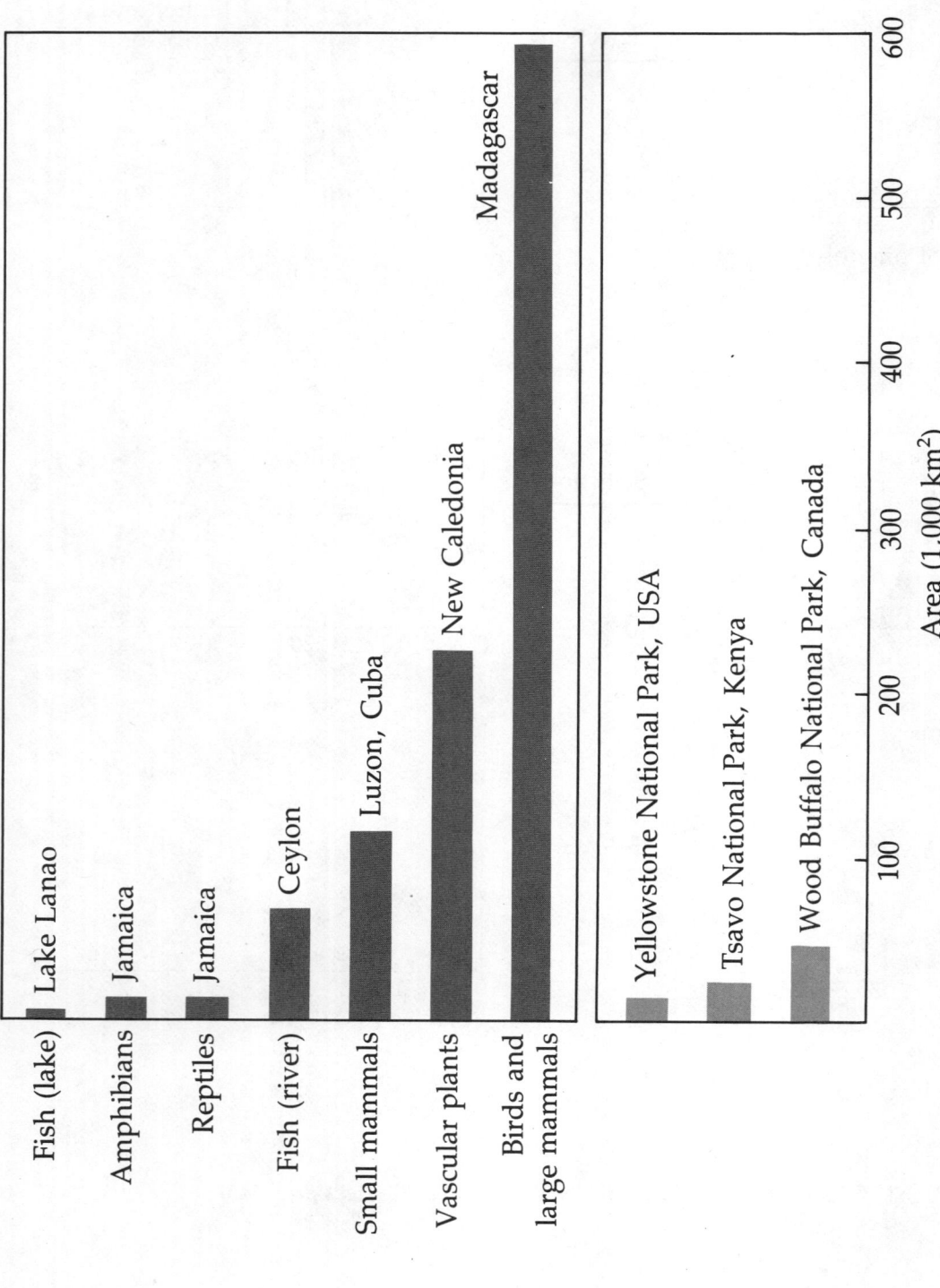

Fish (lake) — Lake Lanao

Amphibians — Jamaica

Reptiles — Jamaica

Fish (river) — Ceylon

Small mammals — Luzon, Cuba

Vascular plants — New Caledonia

Birds and large mammals — Madagascar

Yellowstone National Park, USA

Tsavo National Park, Kenya

Wood Buffalo National Park, Canada

Area (1,000 km²)

FIGURE 37.4 SPECIATION REQUIRES LARGE AREAS (text page 957)
From M. Soulé, in *Conservation Biology*, M. Soulé and B. Wilcox, eds., Sinauer Associates, 1980.

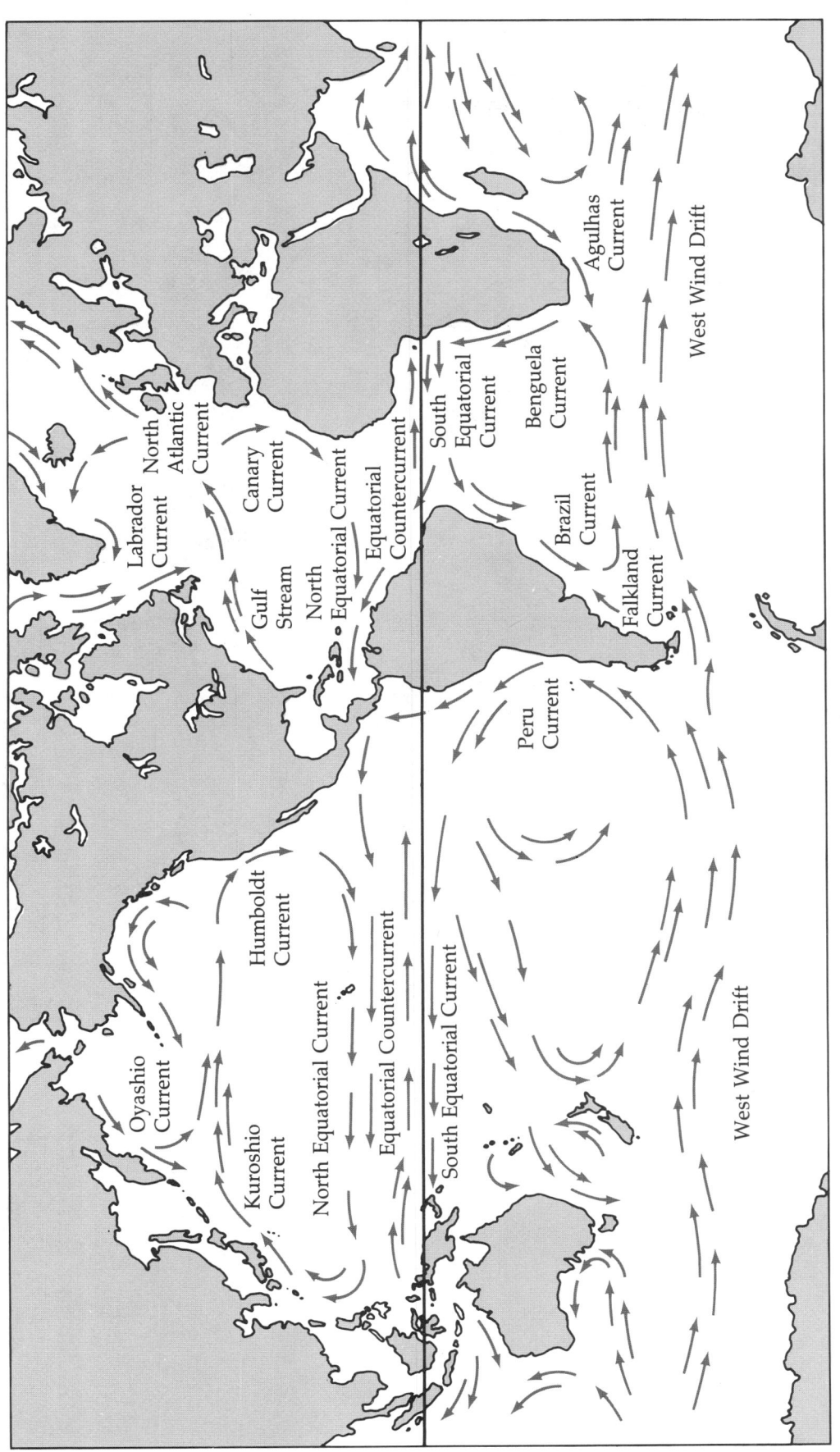

FIGURE 39.4   GLOBAL OCEAN CIRCULATION   (text page 998)

From *Life: The Science of Biology*, Copyright © 1983 by William K. Purves and Gordon H. Orians.

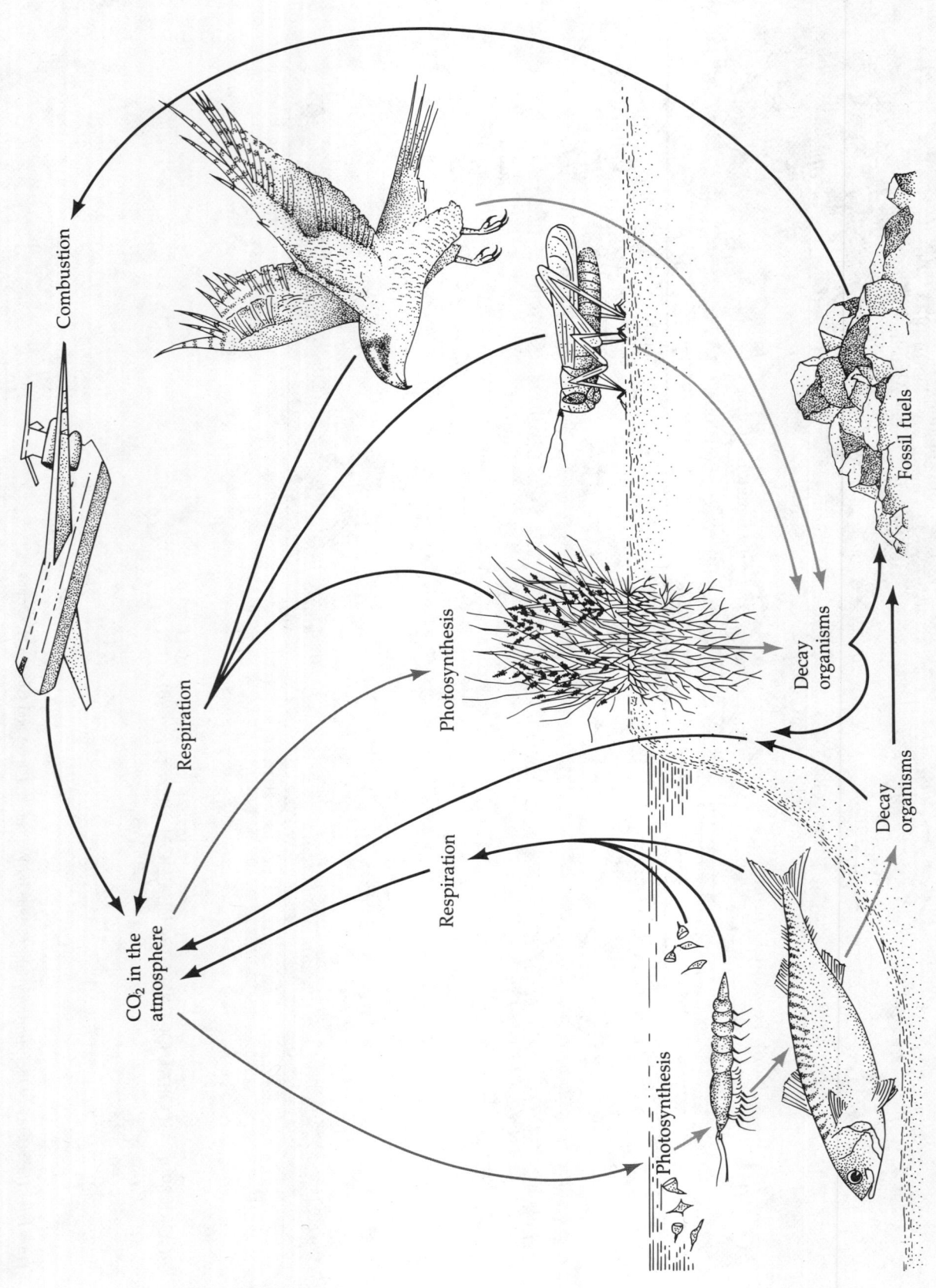

Combustion

Respiration

Photosynthesis

CO$_2$ in the atmosphere

Respiration

Photosynthesis

Fossil fuels

Decay organisms

Decay organisms

FIGURE 39.11   GLOBAL CARBON CYCLE   (text page 1006)

From *Life: The Science of Biology.* Copyright © 1983 by William K. Purves and Gordon H. Orians.

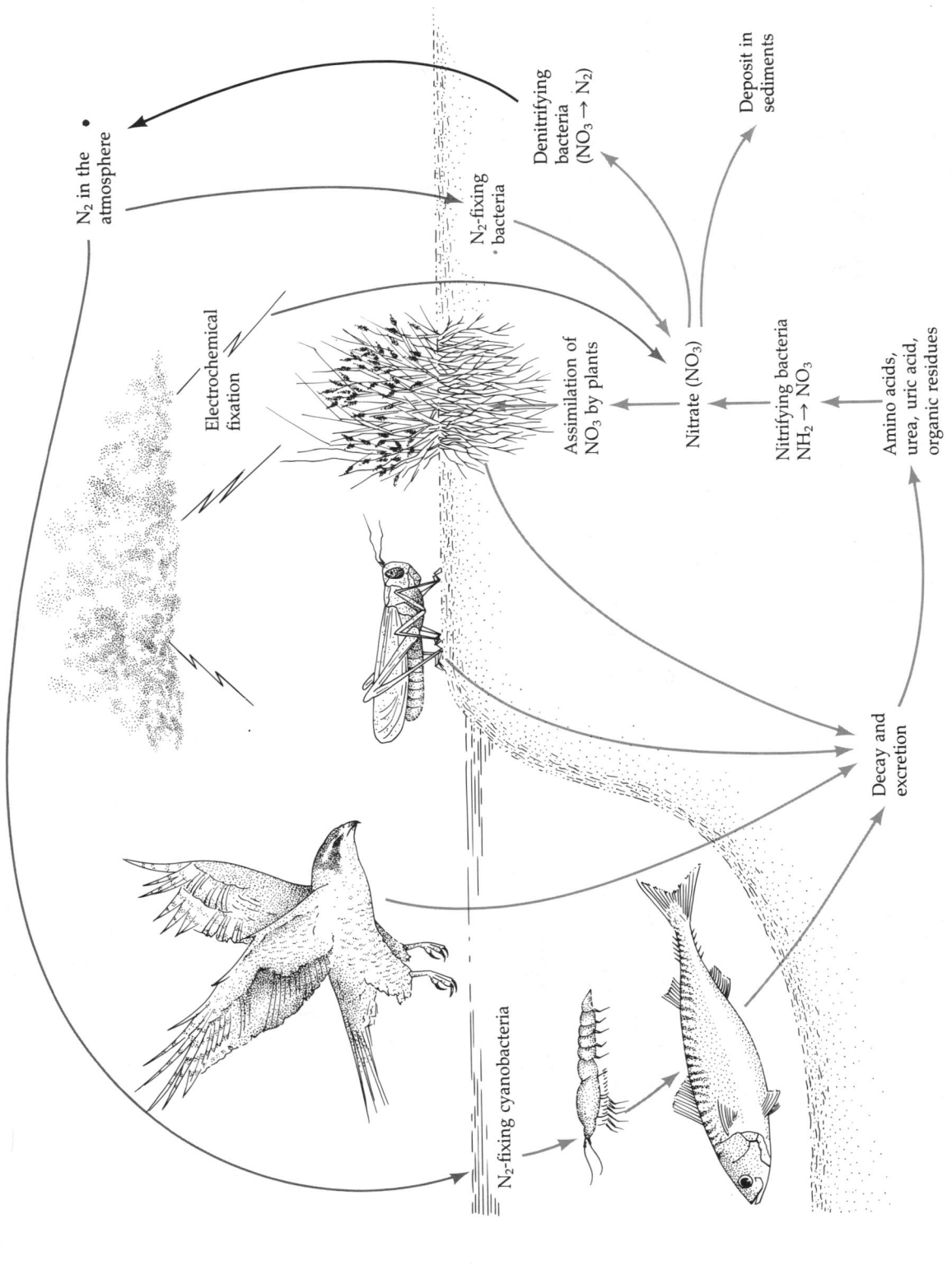

FIGURE 39.12 GLOBAL NITROGEN CYCLE   (text page 1007)

From Life: The Science of Biology, Copyright © 1983 by William K. Purves and Gordon H. Orians.

Labels within figure:

N₂ in the atmosphere

N₂-fixing bacteria

Denitrifying bacteria (NO₃ → N₂)

Deposit in sediments

Electrochemical fixation

Assimilation of NO₃ by plants

Nitrate (NO₃)

Nitrifying bacteria NH₂ → NO₃

Amino acids, urea, uric acid, organic residues

Decay and excretion

N₂-fixing cyanobacteria

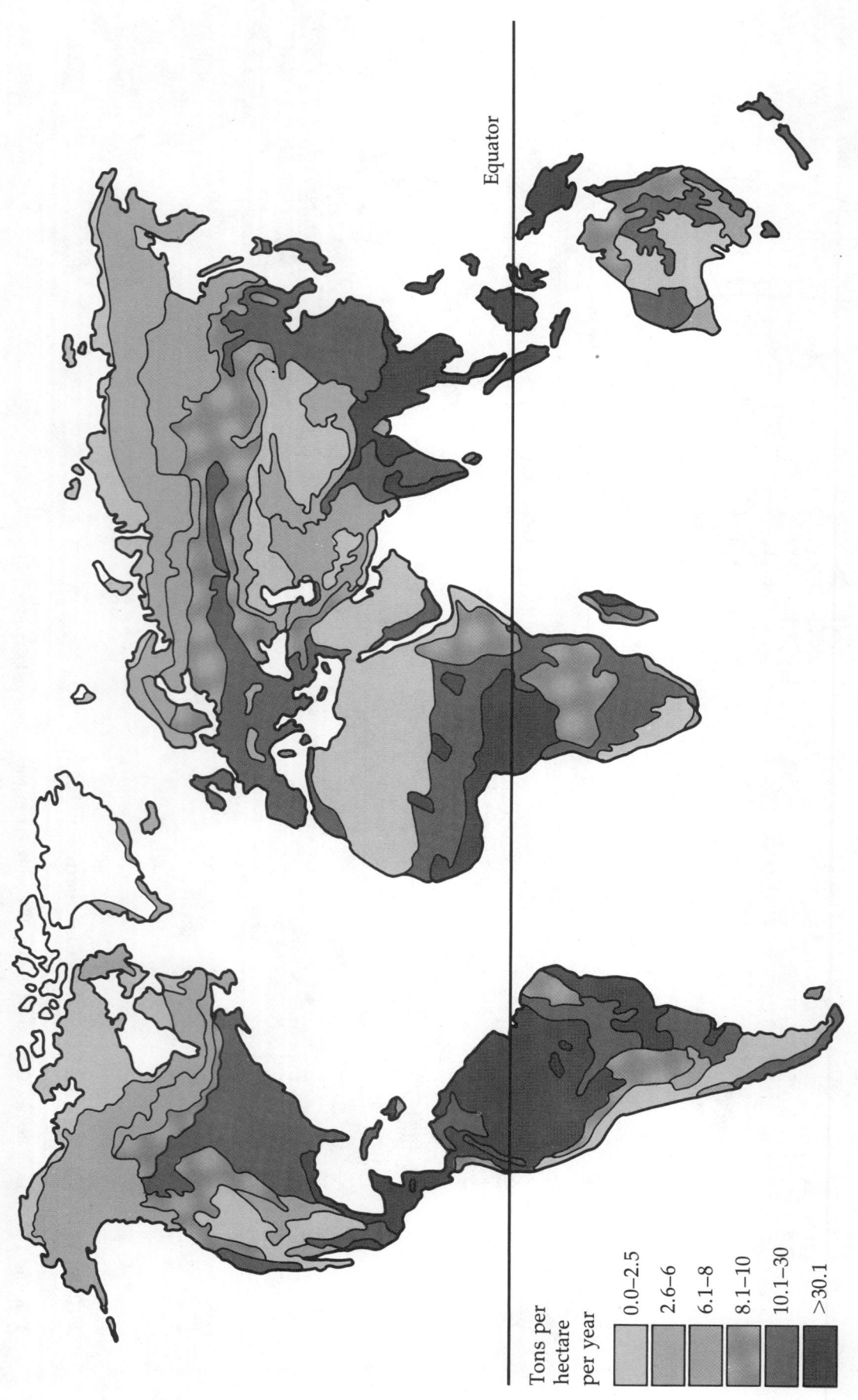

Tons per
hectare
per year

0.0–2.5

2.6–6

6.1–8

8.1–10

10.1–30

>30.1

Equator

FIGURE 39.15 PRODUCTIVITY OF WORLD ECOSYSTEMS (text page 1011)

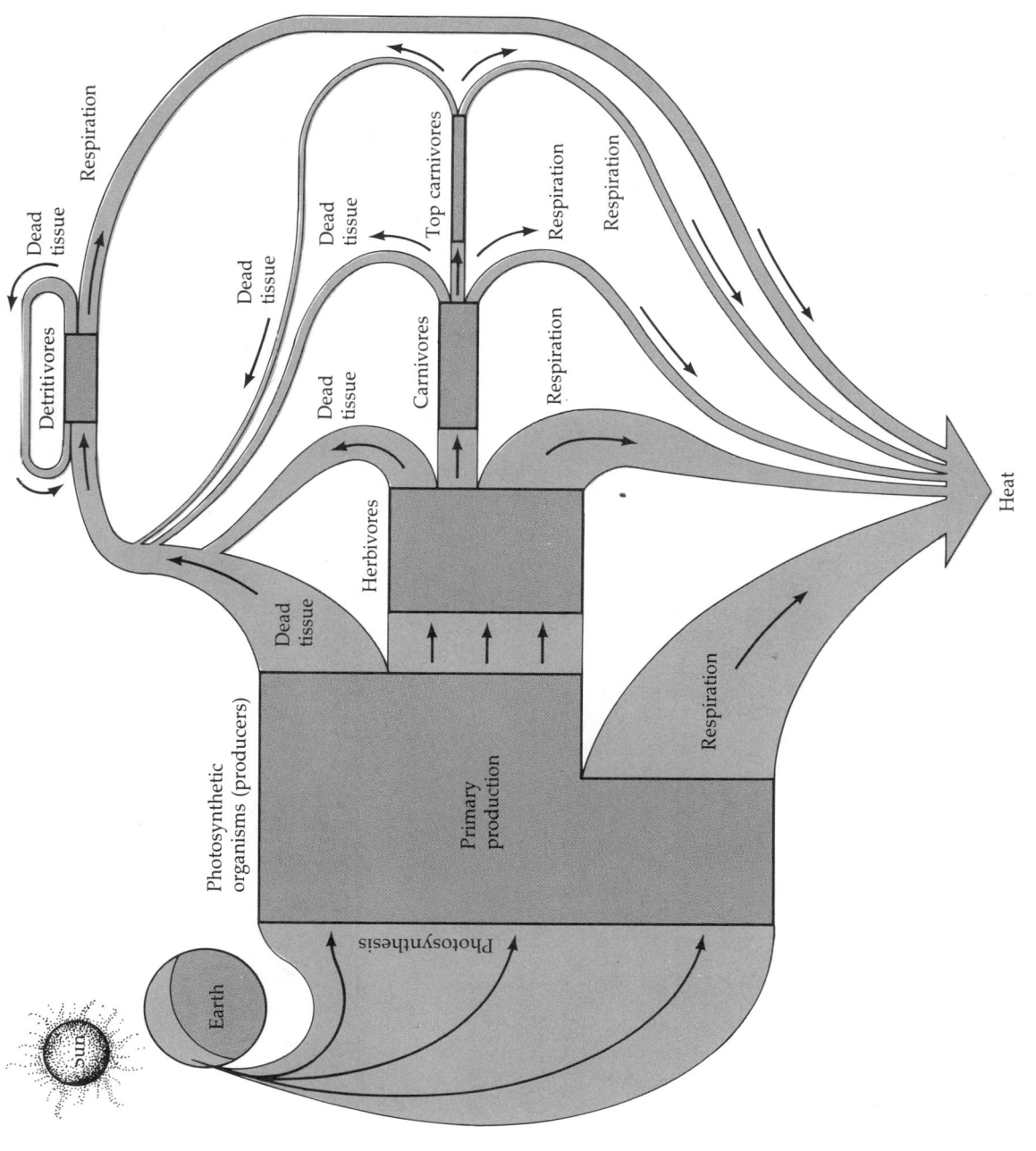

FIGURE 39.18   ENERGY FLOW THROUGH AN ECOSYSTEM   (text page 1015)

From Life: The Science of Biology, Copyright © 1983 by William K. Purves and Gordon H. Orians.

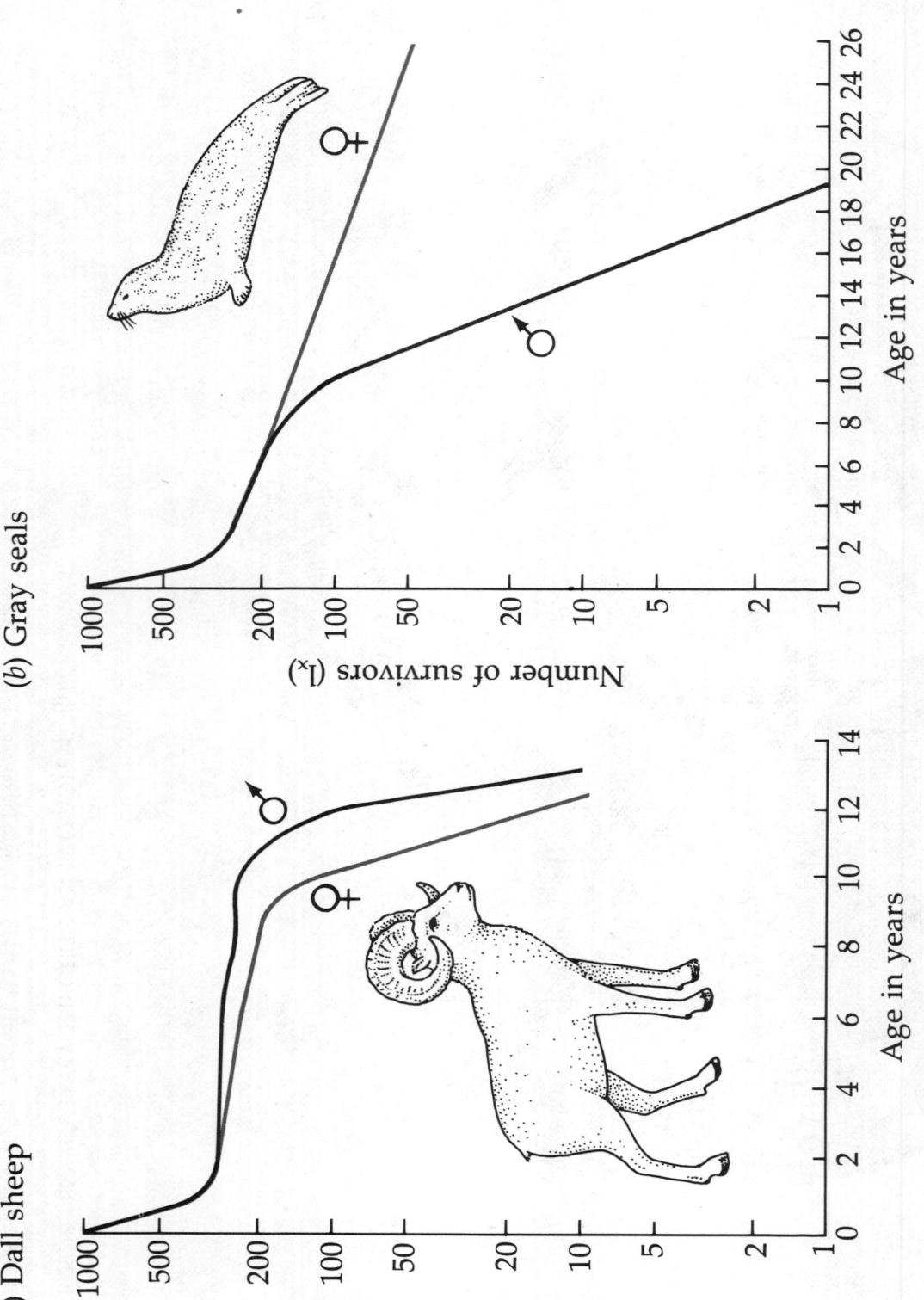

(a) Dall sheep

(b) Gray seals

Number of survivors (l_x)

Age in years

FIGURE 40.3   MAMMALIAN SURVIVORSHIP CURVES   (text page 1023)
After G.E. Hutchinson, *An Introduction to Population Ecology.* Yale Univ. Press, 1978.

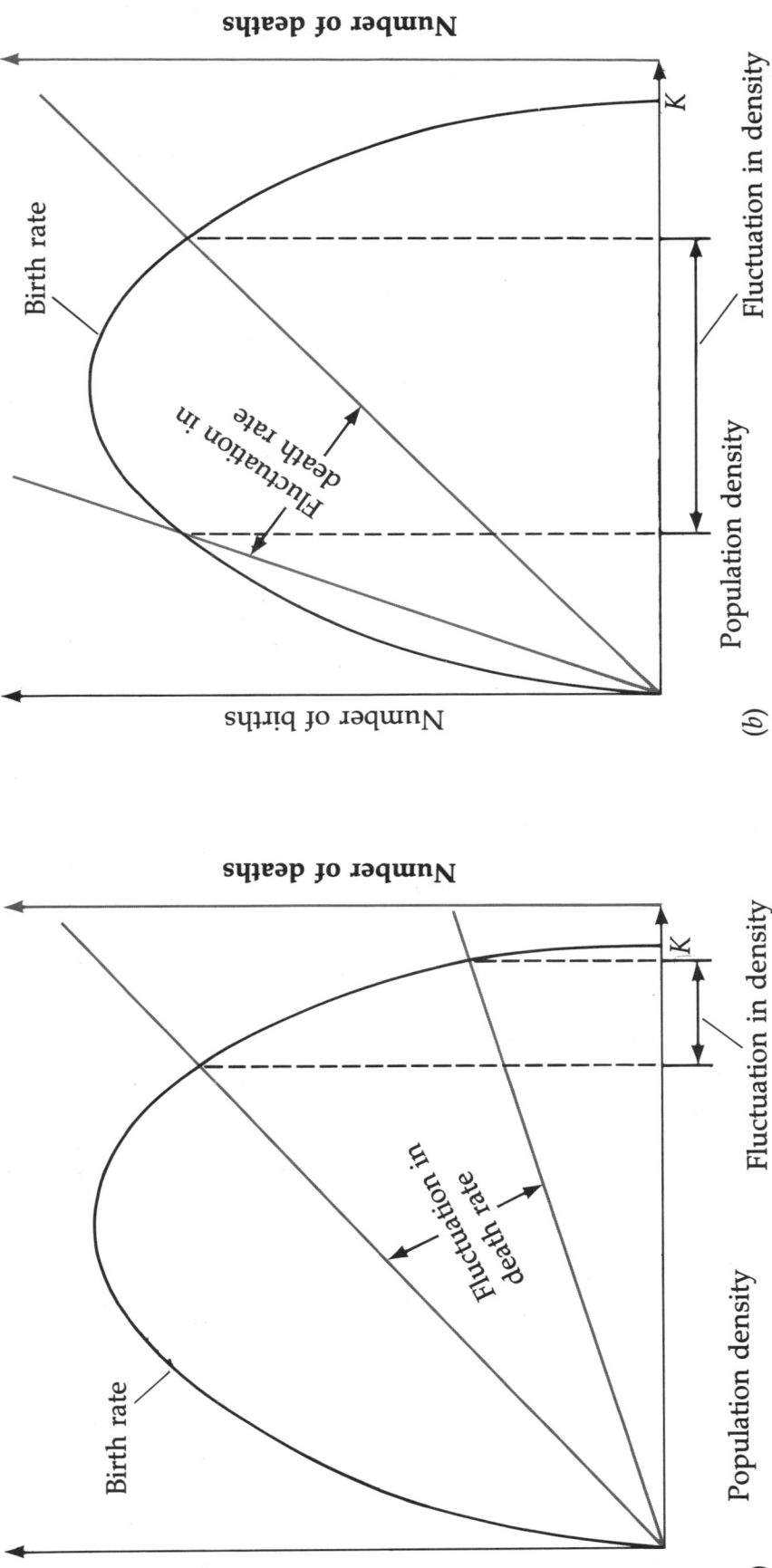

FIGURE 40.16   REGULATION OF POPULATION DENSITY   (text page 1038)

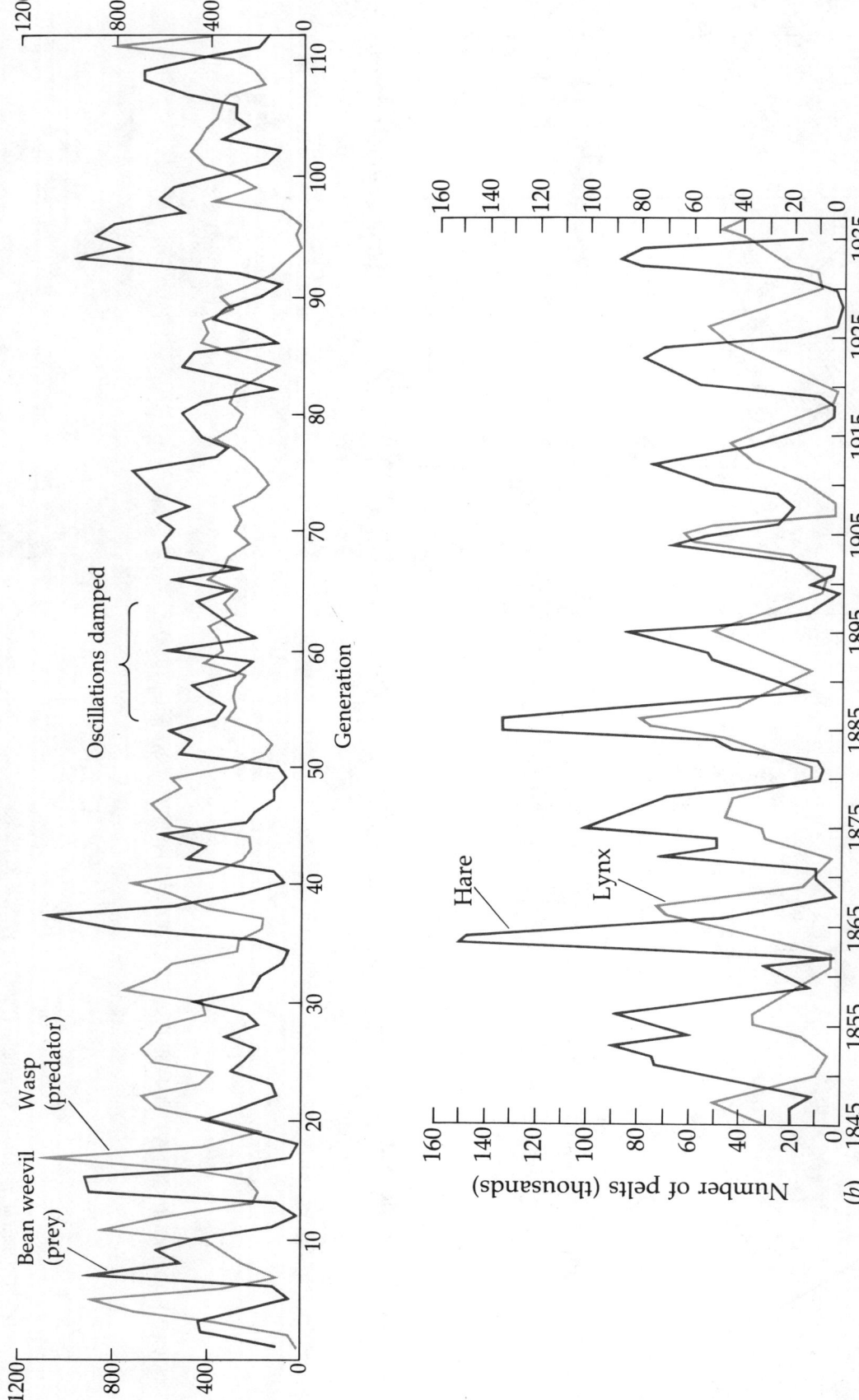

FIGURE 41.5 PREDATOR-PREY OSCILLATIONS (text page 1049)
Part (a) after G.E. Hutchinson, *An Introduction to Population Ecology*, Yale Univ. Press, 1978; part (b) after MacLulich, 1937.

From *Life: The Science of Biology*, Copyright © 1983 by William K. Purves and Gordon H. Orians.

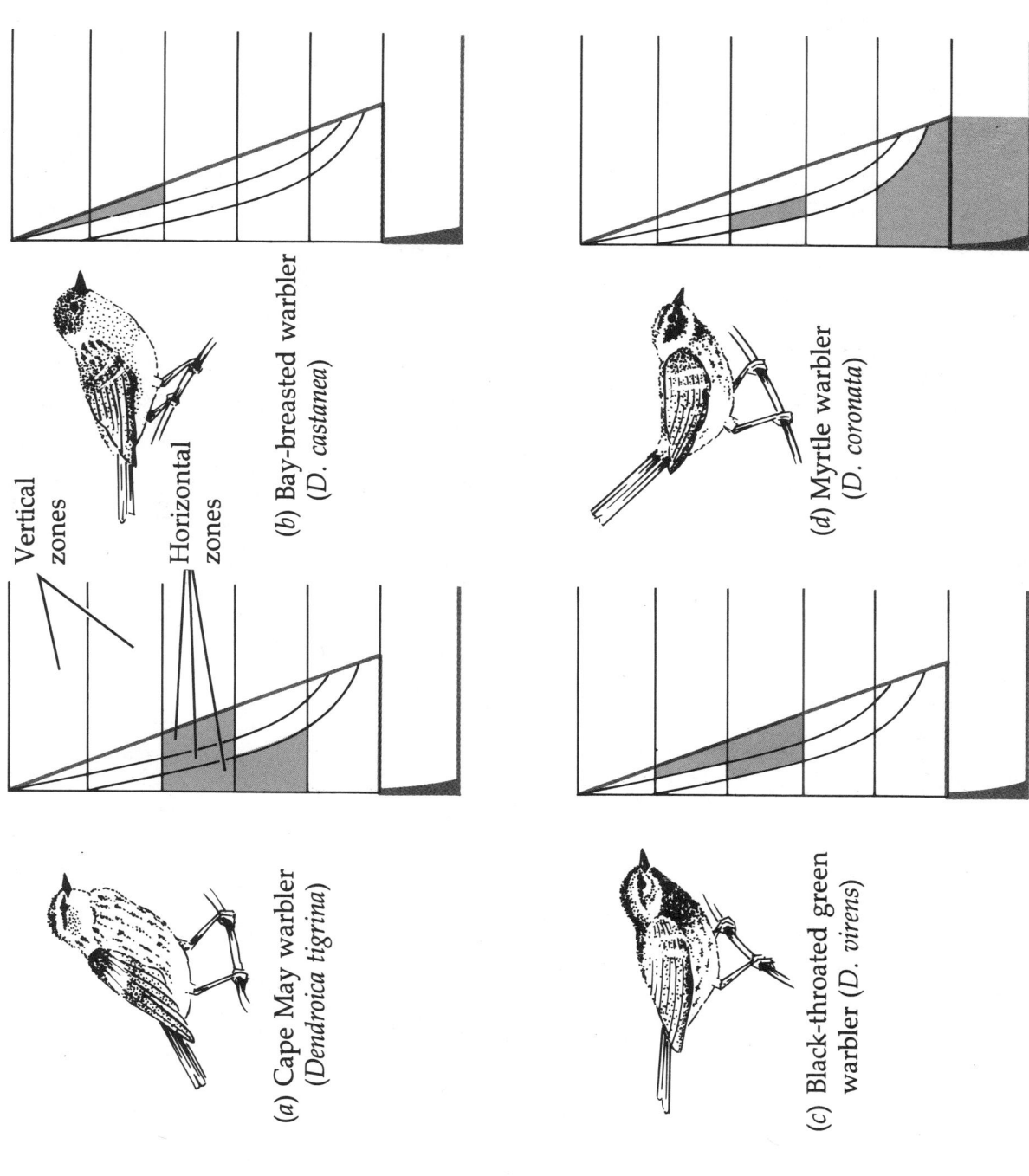

Vertical
zones

Horizontal
zones

(a) Cape May warbler
(Dendroica tigrina)

(b) Bay-breasted warbler
(D. castanea)

(c) Black-throated green
warbler (D. virens)

(d) Myrtle warbler
(D. coronata)

FIGURE 41.21   FORAGING NICHES OF FOUR WARBLERS   (text page 1063)

From *Life: The Science of Biology*, Copyright © 1983 by William K. Purves and Gordon H. Orians.

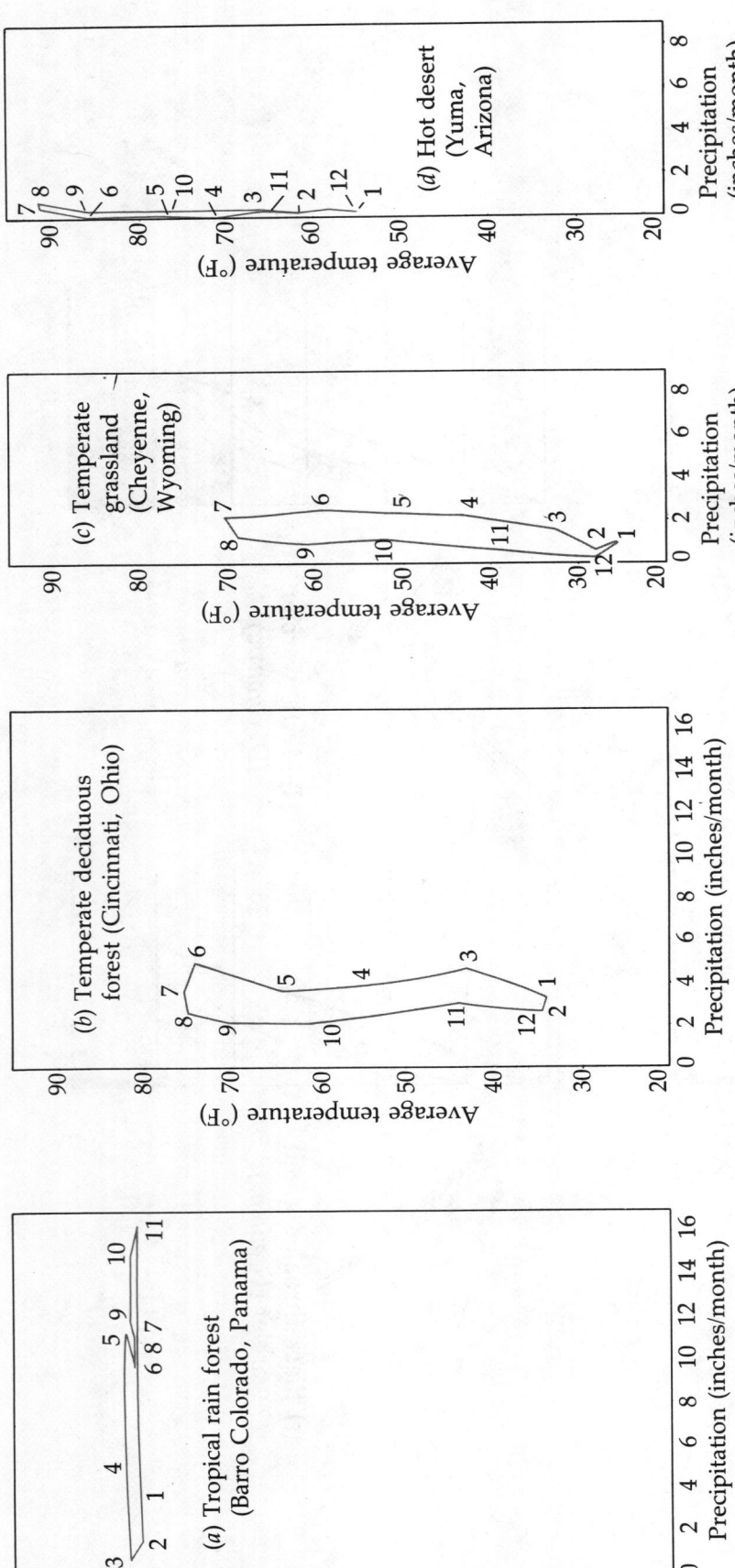

FIGURE 42.14   CLIMOGRAPHS OF SOME MAJOR BIOMES   (text page 1081)

From *Life: The Science of Biology.* Copyright © 1983 by William K. Purves and Gordon H. Orians.

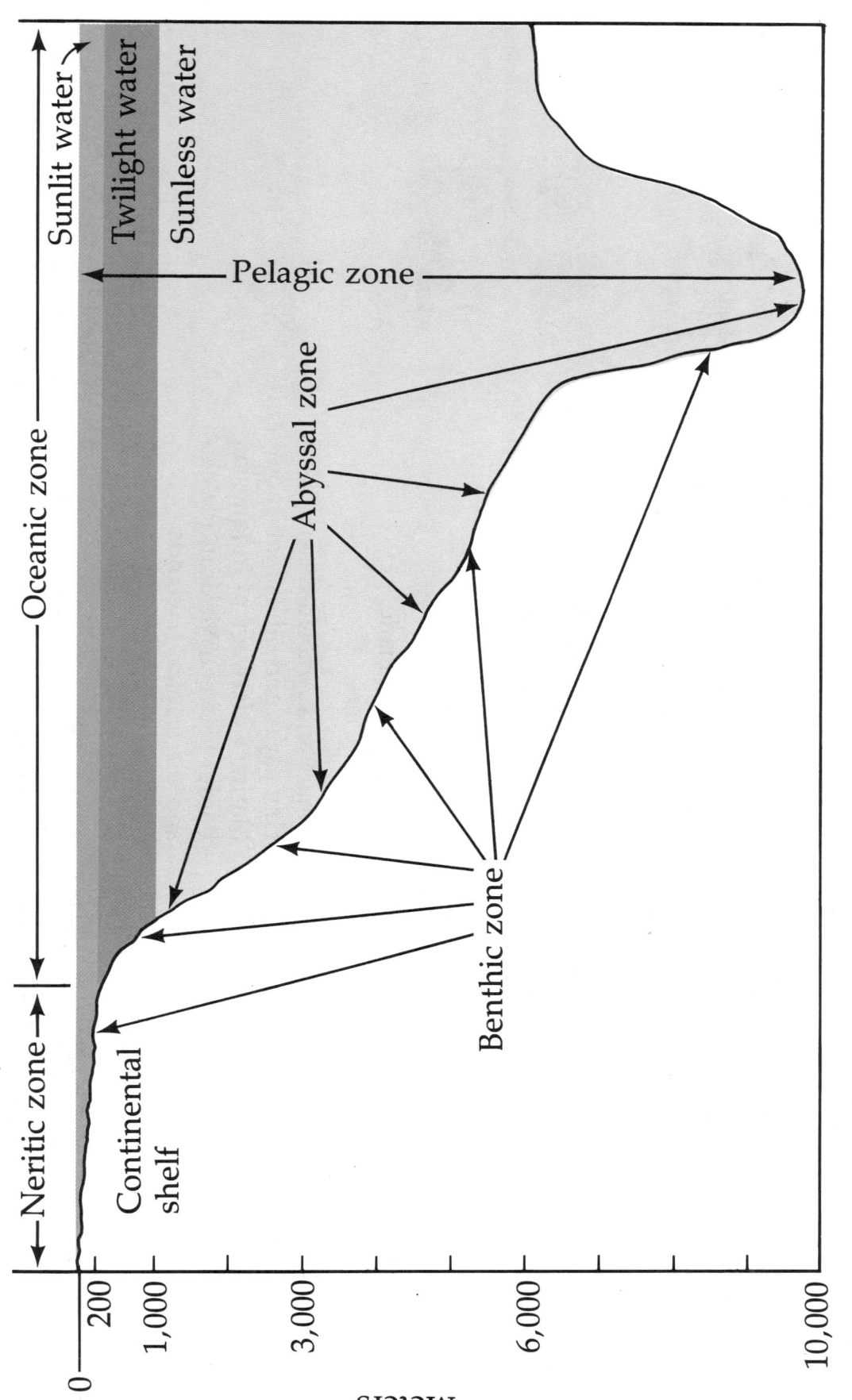

FIGURE 42.28   ZONES OF THE OCEANS   (text page 1091)

From *Life: The Science of Biology*, Copyright © 1983 by William K. Purves and Gordon H. Orians.

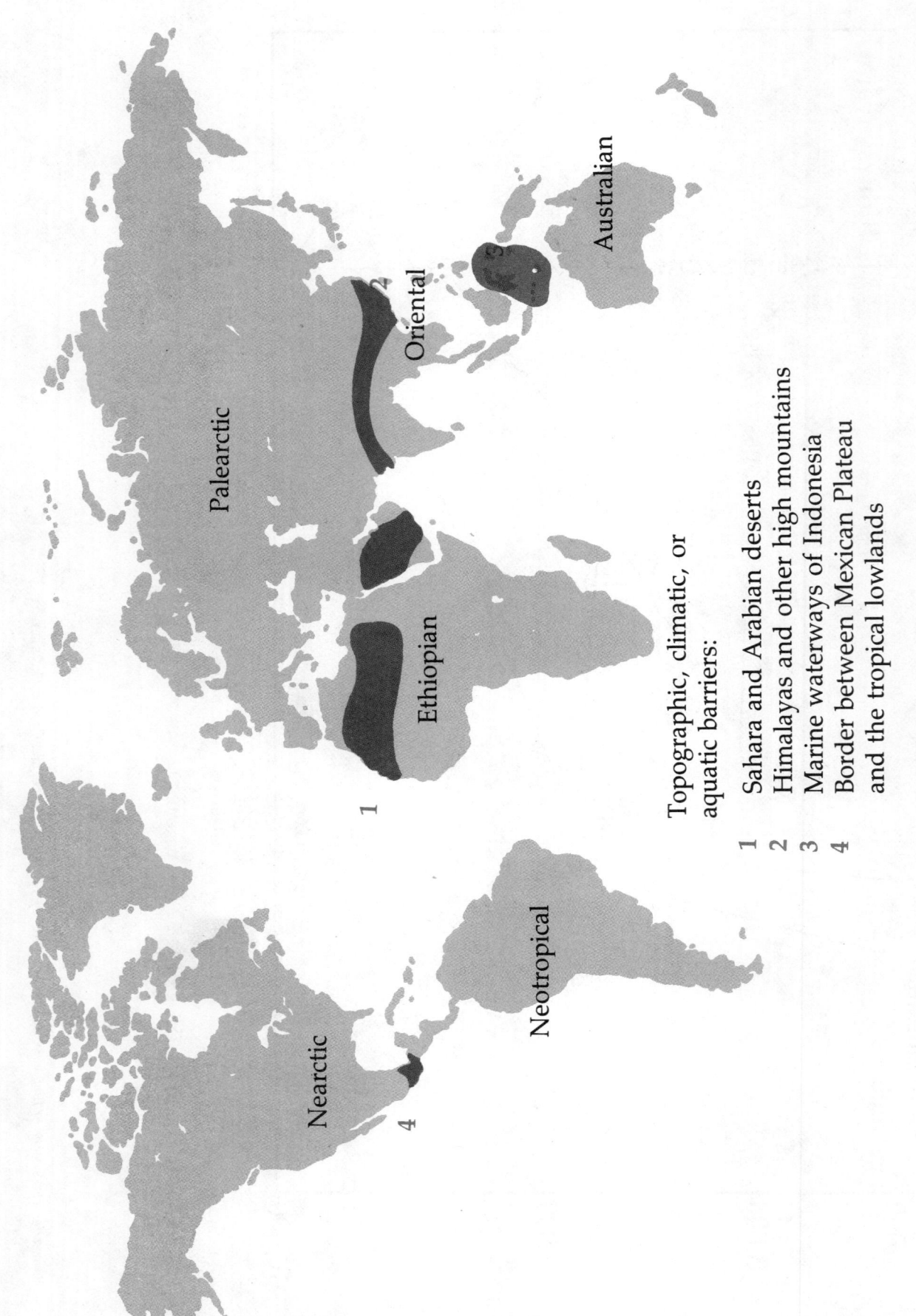

FIGURE 43.3   MAJOR BIOGEOGRAPHIC REGIONS   (text page 1100)

From *Life: The Science of Biology*, Copyright © 1983 by William K. Purves and Gordon H. Orians.

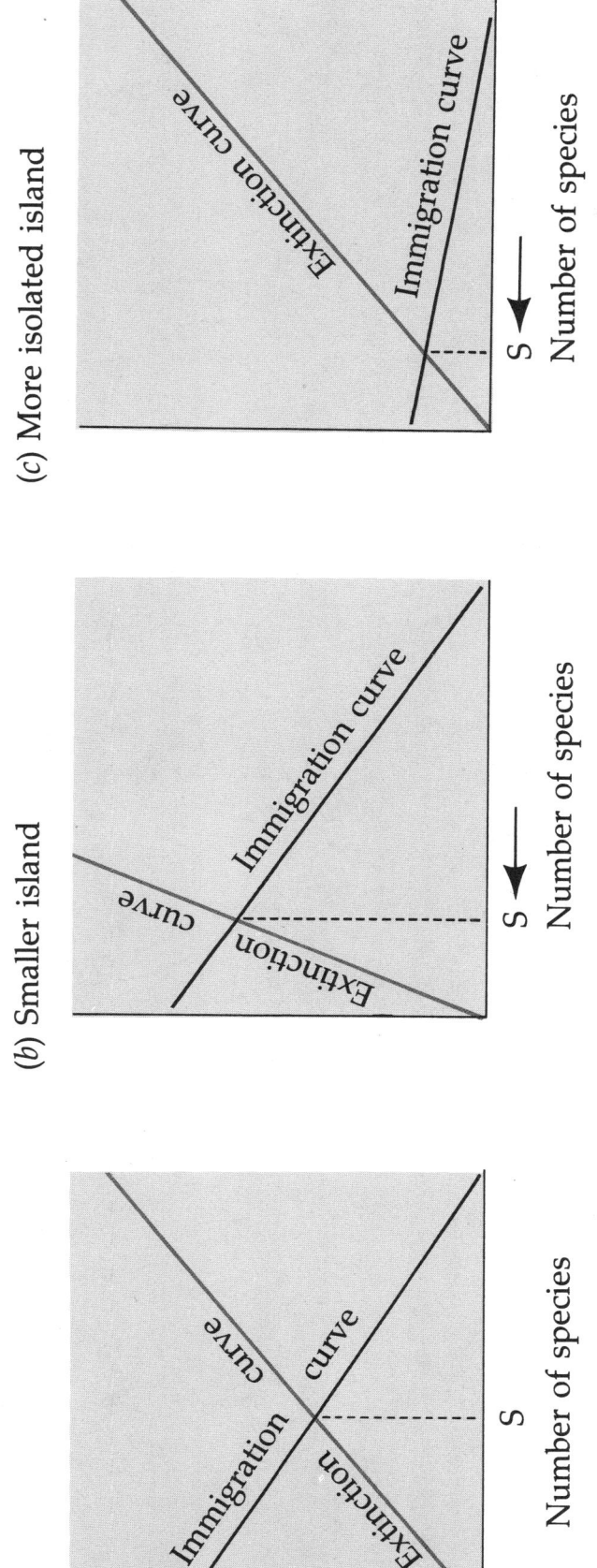

(a)

(b) Smaller island

(c) More isolated island

FIGURE 43.12 SPECIES EQUILIBRIUM THEORY (text page 1106)

From Life: The Science of Biology, Copyright © 1983 by William K. Purves and Gordon H. Orians.